Springer Series in
SOLID-STATE SCIENCES 126

Springer
Berlin
Heidelberg
New York
Barcelona
Hong Kong
London
Milan
Paris
Singapore
Tokyo

Springer Series in
SOLID-STATE SCIENCES

126 **Physical Properties of Quasicrystals**
Editor: Z.M. Stadnik

Volumes 1–125 are listed at the end of the book.

Zbigniew M. Stadnik (Ed.)

Physical Properties of Quasicrystals

With 221 Figures

Professor Dr. Zbigniew M. Stadnik
University of Ottawa, Department of Physics
150 Louis Pasteur, Ottawa
Ontario K1N 6N5, Canada
e-mail: stadnik@joule.physics.uottawa.ca

ISSN 0171-1873

ISBN 3-540-65188-8 Springer-Verlag Berlin Heidelberg New York

Library of Congress Cataloging-in-Publication Data applied for

Die Deutsche Bibliothek – CIP-Einheitsaufnahme
Physical properties of quasicrystals / Zbigniew M. Stadnik (ed.). Berlin ; Heidelberg ; New York ; Barcelona ; Hong Kong ; London ; Milan ; Paris ; Singapore ; Tokyo : Springer 1999
(Springer series in solid-state sciences ; 126)
ISBN 3-540-65188-8

Typesetting: Data conversion by Satztechnik Katharina Steingraeber, Heidelberg,
using the Springer TEX macro package "clmult01"
Cover concept: eStudio Calamar Steinen
Cover production: *design & production* GmbH, Heidelberg

SPIN: 10653237   57/3144 - 5 4 3 2 1 0 – Printed on acid-free paper

Preface

The discovery in 1984 of crystals with "forbidden" symmetry, materials that were dubbed quasicrystals, opened a new branch of crystallography and solid-state physics. These structures possess long-range aperiodic order and crystallographically forbidden rotational symmetries, and are thus fundamentally different from the two other known types of solid structure, crystalline and amorphous. The literature devoted to the physical properties of quasicrystals has been growing so rapidly that it is very difficult for a scientist to remain well informed on more than a narrow speciality within this area. Some definite and intriguing new physical phenomena have already been discovered, and other surprising results will probably follow. Our knowledge of the physical properties of quasicrystals has now reached such a maturity that it seems necessary to provide a comprehensive review. I think this is an opportune time to produce a research-level monograph on the subject. As yet, no comprehensive treatise exists at this level. It is very unlikely that a single author could produce a monograph at the level and depth that is needed to do justice to the physical properties of quasicrystals. I hope that this book comes close to being such a monograph.

This book is intended for researchers in the field of the physics of quasicrystals, solid-state physicists, materials scientists, crystallographers, as well as for graduate students working in the area of new materials. Written by active researchers in their respective fields, it summarizes in a critical fashion much of our present knowledge and understanding of the complex, yet exciting physical properties of quasicrystals. I hope that it conveys some of the fascination I have found during my work in this field. We are still far from a complete understanding of these materials; this book will have served its purpose if it stimulates new research and helps to guide that research in fruitful directions.

I wish to thank the authors for their thorough and competent work. I thank my departmental colleague, Dr. Brian A. Logan, for his assistance with the English of the book. I am grateful to Dr. Claus Ascheron from Springer-Verlag for his advice and friendly patience concerning some unexpected delays. Finally, I express my thanks to Dr. Gerhard Stroink from Dalhousie University who introduced me in 1986 to the fascinating research area of quasicrystals.

Ottawa, September 1998 *Zbigniew M. Stadnik*

Contents

List of Contributors

Martin Bartsch
Max-Planck-Institut für
Mikrostrukturphysik, Weinberg 2
D-06120 Halle/S.
Germany

Michel Boudard
Laboratoire de Thermodynamique et
de Physico-Chimie Métallurgiques
UMR 5614 CNRS, INPG, UJF, BP
75, 38402 Saint Martin d'Hères Cedex
France

Marc de Boissieu
Laboratoire de Thermodynamique et
de Physico-Chimie Métallurgiques
UMR 5614 CNRS, INPG, UJF, BP
75, 38402 Saint Martin d'Hères Cedex
France

Michael Feuerbacher
Institute für Festkörperforschung
Forschungszentrum Jülich GmbH
D-52425 Jülich
Germany

Takeo Fujiwara
Department of Applied Physics
University of Tokyo
Hongo 7-3-1, Bunkyo-ku
Tokyo 113
Japan

Kazuaki Fukamichi
Department of Materials Science
Faculty of Engineering
Tohoku University
Sendai 980-77
Japan

Patrick C. Gibbons
Department of Physics
Washington University
St. Louis, Missouri 63130
USA

Alan I. Goldman
Department of Physics and
Astronomy, and Ames Laboratory
Iowa State University
Ames, IA 50011
USA

Jürgen Hafner
Institute für Theoretische Physik
and Center for
Computational Materials Science
Technische Universität Wien
A-1040 Wien
Austria

Torsten Haibach
Laboratory of Crystallography
ETH-Zürich, 8092-Zürich
Switzerland

Cynthia Jenks
Department of Chemistry
and Ames Laboratory
Iowa State University
Ames, IA 50011
USA

Kenneth F. Kelton
Department of Physics
Washington University
St. Louis, Missouri 63130
USA

Marian Krajčí
Institute of Physics
Slovak Academy of Sciences
SK-84228 Bratislava
Slovak Republic

Ulrich Messerschmidt
Max-Planck-Institut für
Mikrostrukturphysik, Weinberg 2
D-06120 Halle/S.
Germany

Östen Rapp
Physics Department
Kungliga Tekniska Högskolan
SE 100 44 Stockholm
Sweden

Zbigniew M. Stadnik
Department of Physics
University of Ottawa
Ottawa, Ontario K1N 6N5
Canada

Walter Steurer
Laboratory of Crystallography
ETH-Zürich, 8092-Zürich
Switzerland

Patricia A. Thiel
Department of Chemistry
and Ames Laboratory
Iowa State University
Ames, IA 50011
USA

An Pang Tsai
National Research Institute
for Metals
Tsukuba 305
Japan

Knut Urban
Institute für Festkörperforschung
Forschungszentrum Jülich GmbH
D-52425 Jülich
Germany

Markus Wollgarten
Centre d'Etudes
de Chimie Métallurgique
15, rue Georges Urbain
F-94407 Vitry-sur-Seine Cedex
France

1. Introduction

Zbigniew M. Stadnik

One of the main developments in solid-state physics in the last two decades has been the discovery of quasicrystals (QCs) by Shechtman et al. (1984). The discovery of new compounds often leads to the emergence of new fields in solid-state physics. This has been the case with heavy fermions, oxide superconductors, and fullerenes. It is clearly also the case with QCs.

These compounds are a new form of matter which differs from the other two known forms, crystalline and amorphous, by possessing a new type of long-range translational order, *quasiperiodicity*, and a noncrystallographic orientational order associated with the classically forbidden fivefold (icosahedral), eightfold (octagonal), tenfold (decagonal), and 12-fold (dodecagonal) symmetry axes. Although the first icosahedral QC in the binary alloy Al-Mn was discovered by Shechtman on April 8, 1982 (Hargittai 1997), there was a delay of two years before the result was published (Shechtman et al. 1984). Only after it was realized that Bragg diffraction is possible from aperiodic structures and that the periodicity paradigm of crystallography has to be revised, did it become widely accepted that QCs are indeed a new form of matter. This testifies to the revolution that was brought about by Shechtman's discovery in our way of thinking about the structure of solids.

Since their discovery, an enormous amount of effort has gone into determining the atomic structure of QCs. In spite of significant progress, the determination of the structure of QCs still remains an outstanding problem (Jeong and Steinhardt 1997, Cockayne and Widom 1998) in solid-state physics.

A central problem in solid-state physics is to determine whether quasiperiodicity leads to physical properties which are significantly different from those of crystalline and amorphous materials. The first few years of studies of QCs revealed that their physical properties are disappointingly similar to those of the corresponding crystalline or amorphous counterparts. It was only realized later that the first QCs, which were thermodynamically *metastable*, possessed significant structural disorder, as manifested in the broadening of x-ray and/or electron diffraction lines. In addition, they contained non-negligible amounts of second phases. These poor-quality samples impeded the detection of those properties which could be intrinsic to quasiperiodicity. A significant development in the studies of both structural and physical properties of QCs occurred at the end of the 1980s when the thermodynamically *stable* QCs were discovered. These new QCs possess a high degree of structural perfection comparable to that found in the best periodic alloys. Intensive investigations of them led to the discovery of many unusual physical properties. The most salient feature of many QCs, one which would not be expected for alloys consisting of normal metallic elements, is the very high value of the electrical resistivity (the very small value of the electrical conductivity). For some QCs, the electrical conductivity is smaller than Mott's minimum metallic conductivity for the metal–insulator transition. Second, the temperature coefficient of the electrical resistivity is generally negative and often of large magnitude. Third, the electrical resistivity and other electronic transport parameters are extremely sensitive to small changes in the sample composition and to its thermal treatment. Fourth, the electrical resistivity of many QCs increases as their structural quality improves, which is in contrast to the behavior of typical metallic alloys. Fifth, a large and unusual temperature dependence of magnetoresistance, Hall coefficient, and thermoelectric power, and low thermal conductivity are observed for many QCs. Sixth, diamagnetism is observed in QCs which have a large concentration of transition-metal atoms. Seventh, anisotropy in various physical parameters is observed in decagonal alloys in the periodic directions and the quasiperiodic planes. Eighth, a very low density of states at the Fermi level is found in many QCs. The interpretation of these properties poses a great challenge since quasiperiodicity precludes the use of well-established concepts based on the translational invariance of periodic matter. No equivalent theory has yet been developed for quasiperiodic matter.

There are a number of books on QCs (Jarić 1988, 1989, Jarić and Gratias 1989, Fujiwara and Ogawa 1990, Hargittai 1990, DiVincenzo and Steinhardt 1991, Hippert and Gratias 1994, Janot 1994, Axel and Gratias 1995, Senechal 1995) and reviews (Goldman and Kelton 1993, Kelton 1993, Steurer 1996, Yamamoto 1996, Ranganathan et al. 1997). However, these are mainly concerned with the structural and/or theoretical aspects of these materials. There are only a few reviews (Poon 1992, Takeuchi 1993, 1994, Mizutani 1994, Stadnik 1996, Goldman et al. 1997, Quilichni and Janssen 1997) dealing with selected aspects of the physical properties of QCs. These properties have been studied extensively as shown by the numerous papers published in conference proceedings (Chapuis and Paciorek 1995, Janot and Mosseri 1995, de Boissieu et al. 1998, Takeuchi and Fujiwara 1998). This book represents an attempt to assemble in one volume an account of the large variety of unusual physical properties of QCs, with an emphasis on the experimental results, and also to discuss our, still incomplete, understanding of these properties.

The second chapter by A.-P. Tsai discusses the metallurgy of QCs. Chapter 3 by W. Steurer and T. Haibach provides a background for the crystallographic description and analysis of QCs. In the fourth chapter M. Boudard and M. de Boissieu present the experimental results on the determination of the structure of QCs. Chapter 5 by Ö. Rapp discusses the unusual electronic transport properties of QCs. The theory of the electronic structure of QCs is presented in Chap. 6 by T. Fujiwara and in Chap. 7 by J. Hafner and M. Krajčí. The latter chapter also discusses the dynamical properties of QCs. Spectroscopic results on the electronic structure of QCs are presented in Chap. 8 by Z.M. Stadnik. Magnetic properties of QCs are reviewed by K. Fukamichi in Chap. 9. P.A. Thiel, A.I. Goldman, and C.J. Jenks discuss the surface science of QCs in Chap. 10. Mechanical characteristics of QCs are reviewed in Chap. 11 by K. Urban, M. Feuerbacher, M. Wollgarten, M. Bartsch, and U. Messerschmidt. Finally, in Chap. 12, the potential industrial applications of QCs are disussed by P.C. Gibbons and K.F. Kelton.

References

Axel, F., Gratias, D. (eds.) (1995): Beyond Quasicrystals. Les Editions de Physique, Les Ulis

Chapuis, G., Paciorek, W. (eds.) (1995): Aperiodic'94, Proceedings of the International Conference on Aperiodic Crystals. World Scientific, Singapore

Cockayne, E., Widom, M. (1998): Phys. Rev. Lett. **81**, 598

de Boissieu, M., Verger-Gaugry, J.L., Currat, R. (eds.) (1998): Aperiodic'97, Proceedings of the International Conference on Aperiodic Crystals. World Scientific, Singapore

DiVincenzo, D.P., Steinhardt, P.J. (eds.) (1991): Quasicrystals, The State of the Art. World Scientific, Singapore

Fujiwara, T., Ogawa, T. (eds.) (1990): Quasicrystals. Springer-Verlag, Berlin

Goldman, A.I., Kelton, K.F. (1993): Rev. Mod. Phys. **65**, 213

Goldman, A.I., Sordelet, D.J., Thiel, P.A., Dubois, J.M. (eds.) (1997): New Horizons in Quasicrystals. World Scientific, Singapore

Hargittai, I. (ed) (1990): Quasicrystals, Networks, and Molecules of Fivefold Symmetry. VCH, Weinheim

Hargittai, I. (1997): Chem. Intelligencer **3**, 25

Hippert, F., Gratias, D. (eds.) (1994): Lectures on Quasicrystals. Les Editions de Physique, Les Ulis

Janot, C. (1994): Quasicrystals, A Primer, 2nd ed. Oxford University Press, New York

Janot, C., Mosseri, R. (eds.) (1995): Proceedings of the 5th International Conference on Quasicrystals. World Scientific, Singapore

Jarić, M.V. (ed) (1988): Introduction to Quasicrystals. Academic, Boston

Jarić, M.V. (ed) (1989): Introduction to the Mathematics of Quasicrystals. Academic, Boston

Jarić, M.V., Gratias, D. (eds.) (1989): Extended Icosahedral Structures. Academic, Boston

Jeong, H.-C., Steinhardt, P.J. (1997): Phys. Rev. B **55**, 3520

Kelton, K.F. (1993): Int. Mater. Rev. **38**, 105

Mizutani, U. (1994): in Materials Science and Technology, Vol. 3B, Cahn, R.W., Haasen, P., Kramer, E.J. (eds.). VCH, Weinheim, p 97

Poon, S.J. (1992): Adv. Phys. **41**, 303

Quilichini, M., Janssen, T. (1997): Rev. Mod. Phys. **69**, 277

Ranganathan, S., Chattopadhyay, K., Singh, A., Kelton, K.F. (1997): Prog. Mater. Sci. **41**, 195
Senechal, M. (1995): Quasicrystals and Geometry. Cambridge University Press, New York
Shechtman, D., Blech, I., Gratias, D., Cahn, J.W. (1984): Phys. Rev. Lett. **53**, 1951
Stadnik, Z.M. (1996): in Mössbauer Spectroscopy Applied to Magnetism and Materials Science, Vol. 1, Long, G.J., Grandjean, F. (eds.). Plenum, New York, p 125
Steurer, W. (1996): in Physical Metallurgy, Vol. 1, Cahn, R.W., Haasen, P. (eds.). Elsevier Science BV, Amsterdam, p 371
Takeuchi, S. (1993): in Current Topics in Amorphous Metals, Physics and Technology, Sakurai, Y., Hamakawa, Y., Masumoto, T., Shirae, K., Suzuki, K. (eds.). Elsevier Science BV, Amsterdam, p 65
Takeuchi, S. (1994): Mater. Sci. Forum **150-151**, 35
Takeuchi, S., Fujiwara, T. (eds.) (1998): Proceedings of the 6th International Conference on Quasicrystals. World Scientific, Singapore
Yamamoto, A. (1996): Acta Crystallogr. A **52**, 509

2. Metallurgy of Quasicrystals

An Pang Tsai

2.1 Introduction

Discovered in 1984 (Shechtman et al. 1984), icosahedral (i) quasicrystals (QCs) exhibit a forbidden fivefold diffraction pattern with sharp diffraction peaks indicating the presence of a long range-order similar to the one of crystals. This discovery has stimulated interactions between scientists from many different fields. Intense efforts over the last decade by scientists dedicated to QCs have firmly established this new form of solid in solid state physics. Since QCs are found in metallic alloys, metallurgy plays an important role in this field. Great progress has been made by the discovery of new and highly perfect QCs. We now have a better understanding of QCs which leads us to address some very basic issues concerning QCs. QCs with structural quality comparable to single-grain silicon have been discovered. Like crystalline solids, QCs have several kinds of lattices and like intermetallic compounds, QCs are synthesized from different groups of elements with different chemistry. As a result, the name "quasicrystal" has been included as a new form of solid for the first time in a text book of solid state physics (Kittel 1995). In this article, I shall describe results obtained in the last decade and some aspects of QCs that have been of particular importance to the metallurgical community.

QCs can be formed through various processes and by various techniques. The structure and the intrinsic disorder of QCs are very sensitive to the process of formation. I will discuss how the structure of QCs is affected by the processing and how it is related to the growth mechanism.

An open question since the discovery of QCs has concerned their stability with respect to the crystalline phases. A great number of experimental and theoretical investigations have been performed to resolve this question. A discovery of stable QCs allows us to address this question more effectively. I shall discuss this on the basis of the formation tendency from the metallurgical viewpoints.

QCs have now been shown to exist in the equilibrium phase diagram. Studies to revise the phase diagrams to include QCs will be described. This will also concern the formation of QCs from the melt through a nucleation and growth process. I shall also describe the phase transformation kinetics between an amorphous (a) phase and the QC. Then I shall discuss the phase equilibria and the task of producing single-grain QCs.

2.2 Preparation of Quasicrystals

QCs have been produced by various processes involving different routes. Figure 2.1 shows a schematic diagram exhibiting different kinetics routes for obtaining QCs.

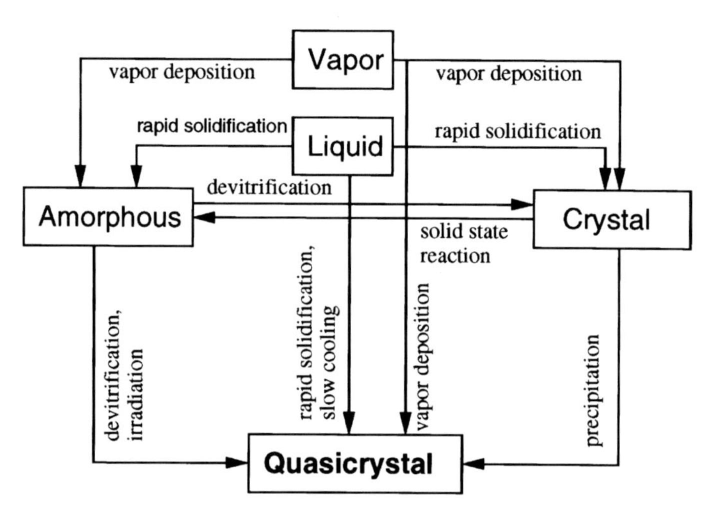

Fig. 2.1. Schematic illustration for the formation of QCs via various kinetic routes.

2.2.1 Rapid Solidification

Rapidly solidified alloys of Al with transition metals (TMs) are of interest as high specific strength materials. It was the study of such alloys that led to the discovery of the *i* phase in the alloys of Al with Mn. It was expected that this phase will also occur in the Al-TM alloys with other TMs, such as Cr, Fe, or Co (Bancel and Heiney 1986). Initial studies were conducted on the melt-quenched $Al_{86}Mn_{14}$ alloys which show a mixture of fcc Al and the *i* phase. The technique generally used was melt-spinning. Alloys of Al with Mn 21.5–22 at.% showing no excess Al were taken to be the stoichiometric (Kimura et al. 1985). However, as the Mn content was increased towards stoichiometry of the *i* phase, a competition with another metastable Al-Mn phase developed. This metastable phase was called the "T phase" and later was identified to be a decagonal (*d*) phase, a 2D QC (Bendersky 1985, Chattopadhyay et al. 1985). The *d* phase forms at slightly lower quenching rates than the *i* phase and can be suppressed by an appropriate change of quenching parameters. In another study the stoichiometry of the *i* phase was determined to be around 15.4 at.% Cr in alloys of Al with Cr (Inoue et al. 1987). Apart from these two alloys, the *i* phase was also found in other Al-TM alloys. The *i* phase of binary alloys was poor in structural quality and less in volume fraction (as compared to ternary alloys where almost 100% quasicrystalline phase can be formed).

The study was then extended to ternary alloys. Substitution by Si of a few at.% of Al was found to produce an *i* phase (Chen and Chen 1986) that was better in structural quality and more stable with respect to crystalline phases. Positive effects of the third element were confirmed with a number of elements in several Al-TM alloys. This approach finally led to the discovery of stable QCs.

2.2.2 Vapor Condensation

The unusual properties of QCs have been studied mainly with bulk samples. Since QCs have the potential for application as a functional material, it is crucial to develop a thin film technology for these materials. The technique for vapor condensation can be vapor deposition, laser ablation, electron-beam evaporation, or sputtering. Although the techniques used for preparing thin films are different, the routes to form QCs are almost the same. QCs prepared by the vapor condensation have succeeded in several Al-based alloys. In order to control the composition and the geometry of the QC, various approaches have been tried.

2.2.2.1 Multilayers Deposited from Each Element and Subsequent Annealing. Since the vapor pressure and the vaporization rate are different among the constituent elements, the films are deposited sequentially in layers of each element from separate sources to control the total composition. Follstaedt and Knapp (1986) used vapor deposition to produce thin films of Al-Mn alloys. Layers of Al and Mn were sequentially deposited by evaporation from separate sources and the quasicrystalline state was formed by heating the specimens at temperatures in the range 523–698 K. The formation of QCs was induced by interdiffusion between the Al (~8 nm) and Mn (~2 nm) layers. As this QC is metastable, further annealing led to the transformation to a crystalline phase. Blanpain and Mayer (1990) studied the formation of the *d* phase of Al_3Pd_2 by reacting thin films of the constituent elements formed by the same technique. The initial film was a sandwich multilayer with deposition of subsequent layers of 42.5 nm-thick Al, 25 nm-thick Pd, and 42.5 nm-thick Al on a sodium chloride substrate. A crystalline phase Al_3Pd_2 formed first as the film was heated to 473 K and then the *d* phase formed at a temperature of ~653 K. These experiments demonstrate that the *i* and *d* phases can nucleate from solid state. Recently, Klein and Symko (1994) formed 300 nm-thick quasicrystalline Al-Cu-Fe thin films by sputtering from three separate rf magnetron guns, one for each element, and a post-annealing at 873 K for 2 h.

2.2.2.2 Direct Formation of Quasicrystals by Vapor Evaporation. Saito et al. (1986) reported the formation of Al-Mn alloy particles with spherical shape prepared by evaporation of an $Al_{90}Mn_{10}$ alloy in a He atmosphere.

The evaporation was carried out in a tungsten boat at a temperature in the range 1873–2173 K. The metal vapor flowed outward from its source by diffusion through the He gas. Since the heating of the source brought about a convective flow of the gas, the vapor was exposed to this gas and cooled rapidly. A supersaturation sufficient for nucleation was attained, and nuclei of the particles were formed. The experiment demonstrated that QCs can be formed by direct condensation from vapor.

A thin film of the Al-Pd-Mn *i* phase has been prepared by KrF excimer laser ablation. Thin films were obtained only when the substrate temperature was lowered to liquid nitrogen temperature during the ablation (Ichikawa et al. 1994). The films were 500 nm-thick and the composition was within a few at.% of that of the target.

2.2.2.3 Crystallization from Amorphous Phase. Crystallization from the *a* to a metastable quasicrystalline phase as a result of ion beam irradiation or thermal ion beam assisted annealing for a number of sputtered or evaporated thin films has been reported in the Al-Mn, Al-Cr, and Al-Si-Mn alloys (Lilienfeld et al. 1985, 1986). However, the grain size of QCs in these alloys is too small to be characterized and could only be observed by electron diffraction. Large grains are not expected since QCs in these alloys are metastable.

An $Al_{65}Cu_{20}Fe_{15}$ alloy prepared by magnetron sputtering and subsequent annealing has been observed in three distinct states: *a*, metastable crystalline, and quasicrystalline (Chien and Lu 1992). Thin films prepared by sputter deposition from an $Al_{60}Cu_{20}Fe_{15}$ alloy were amorphous. This *a* state is not attainable by solidification from the melt. Upon annealing at 723 K, the *a* phase transformed to a cubic crystalline phase with a CsCl-type structure. Upon further annealing at 873 K, the formation of the quasicrystalline phase followed. Yoshioka et al. (1995) produced thin films of *a* alloy of composition $Al_{63}Cu_{25}Fe_{12}$ with thickness of about 500–1000 Å by an electron beam sputtering technique. The film surface was coated with alumina to prevent compositional changes due to selective evaporation. The *a* phase transformed to an *i* QC by annealing at 973 K for 1 h (Yoshioka et al. 1995). Since an *a* phase can be easily obtained by vapor evaporation process, this route to the quasicrystalline phases is available for most QC-forming alloys. However, the disadvantage of this process is that controlling the composition of the films is very critical.

2.2.3 Mechanical Alloying

QCs have been prepared by mechanical alloying of Al-Cu-Mn powders. The quasicrystalline phase forms directly by milling via a solid-state reaction from the crystalline elemental powders. A quasicrystalline single phase was obtained for 15–25 at.% Cu and 10–20 at.% Mn (Eckert et al. 1991).

The formation of QCs and a number of related crystalline phases in the Al-Mg-Zn system by mechanical alloying have been systematically investigated (Mizutani et al. 1993). In this system, the QC is formed over a wide composition range and the related crystalline phases are formed by a subsequent annealing of the QC. It has also been found that during the mechanical alloying the C14 Laves phase Zn_2Mg frequently grows prior to the formation of the QC. It was suggested that the growth of the Zn_2Mg plays a key role as the precursor for the formation of the quasicrystalline phase.

QCs have been formed in the Al-Cu-Fe and Al-Cu-Co systems by annealing of a CsCl-type supersaturated cubic phase synthesized by ball-milling a mixture of the equilibrium cubic phase and fcc-Al powders (Tsai et al. 1993a). In both the alloy systems, first a peritectoid reaction occured between the residual Al and the preliminary milled supersaturated cubic phase to form a tetragonal $Al_7Cu_2(Fe,Co)_1$ phase around 623 K. Subsequently, a similar reaction between the tetragonal phase and the supersaturated phase occurred to form QCs.

2.2.4 Crystallization of Melt-Quenched Amorphous Ribbons

In the classical *a* alloys the nucleation and growth of the crystalline phases are suppressed in the solidification process. These *a* phases are most easily formed at compositions close to deep eutectics, where the liquid state is relatively stable. On the other hand, even though most of QCs are also produced by rapid solidification, they are generally formed at the compositions close to those of intermetallic compounds. This is clearly in contrast to the case of the *a* alloys. Nevertheless, an *i* phase can be formed by crystallization of an *a* phase, as in the Pd-U-Si (Poon et al. 1985) and Al-Si-Mn alloys (Chen et al. 1987b, Dunlap and Dini 1986). In contrast to QCs prepared by vapor deposition, the grain size of the *i* phase obtained from the *a* phase is as large as ~1 μm and the transformation is most likely complete. However, these alloys cannot yield the QC as a single phase directly from the melt, and therefore formation via the *a* phase is necessary.

Figure 2.2 shows a schematic of the time-temperature-transformation diagram for an *a* $Al_{75}Cu_{15}V_{10}$ alloy. This diagram is based on experimental results. Lines a and b represent the solidification rates achieved by melt-spinning. Line c represents the cooling rates achieved in conventional solidification. The cooling rates represented by these lines are in the order a>b>c. Lines d and e represent the isothermal heating and continuous heating from the *a* state, respectively. It is to be noted that a single QC can only be generated from an *a* state.

2.2.5 Conventional Solidification

Unlike the processes described previously, the conventional solidification can only be used to produce stable QCs. It is now clear that some of QCs are

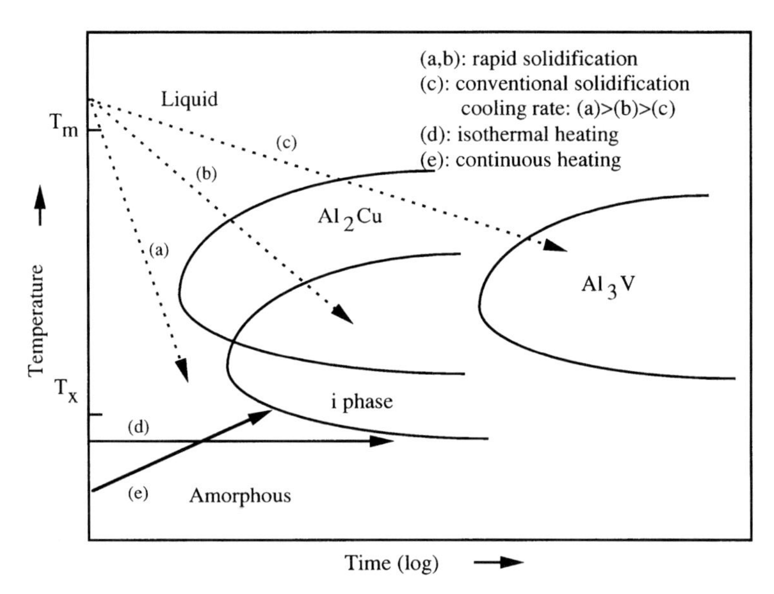

Fig. 2.2. Schematic of the time-temperature-transformation diagram for the $Al_{75}Cu_{15}V_{10}$ alloy.

stable and they are in an equilibrium state in the specific temperature and composition ranges. For example, the quasicrystalline phases such as Al-Li-Cu or Al-Cu-Fe can be obtained using conventional solidification. At present, more than ten alloys are known to form QCs in this way. This allows us to produce large single-grain samples by conventional crystal growth methods. QCs prepared by conventional solidification involve segregation and inhomogeneity resulting from the solidification process. They therefore display broad diffraction peaks or diffuse scattering due to structural disorder and imperfections. Generally, these defects or disorder are removed upon annealing.

2.3 Structural Classification from Diffraction Patterns

The first QCs had a small grain size and were discovered by using the electron diffraction technique. Although the QCs can now be obtained in large grain sizes, electron diffraction is still the most powerful technique to study them. QCs found to date can be classified by their diffraction features and symmetries, which can be easily inspected by the electron diffraction technique. Three-dimensional (3D), two-dimensional (2D), and one-dimensional (1D) QCs have been found in several alloys using this technique.

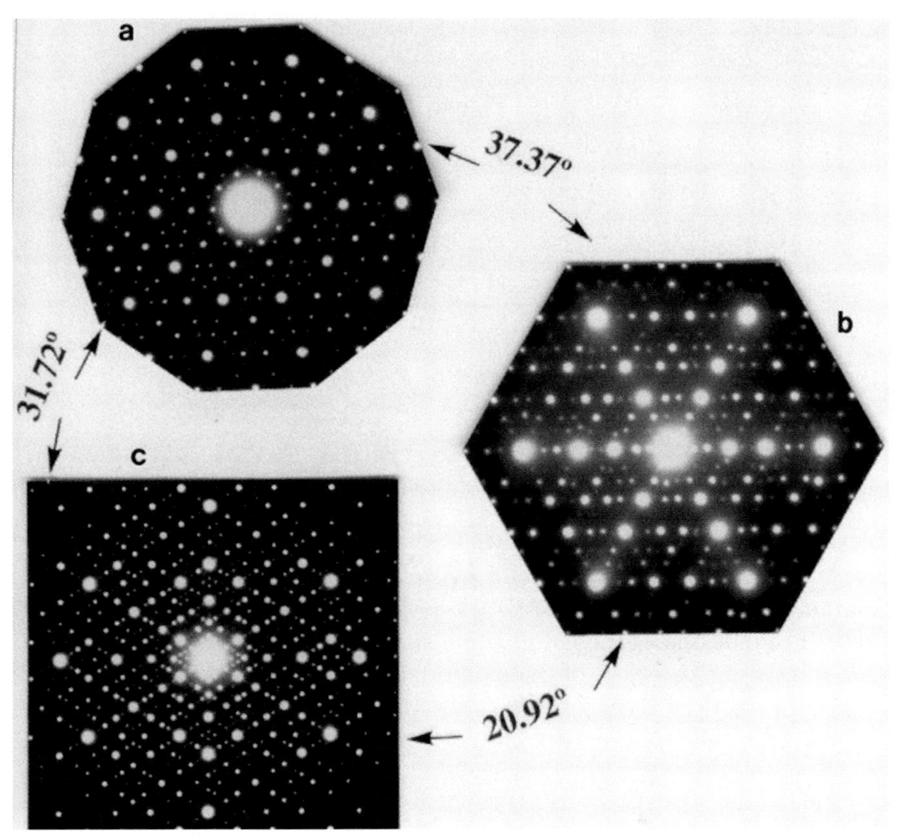

Fig. 2.3. Electron diffraction patterns of a typical *i* QC taken along the high symmetry axes: (a) fivefold, (b) threefold, and (c) twofold. The angles at which these axes occur are marked.

2.3.1 Three-Dimensional Quasicrystals

3D QCs studied by electron diffraction show reflections arrayed quasiperiodically in three dimensions. With regard to the symmetry, the 3D QCs are generally icosahedral. Figure 2.3 shows typical electron diffraction patterns taken along five-, three-, and twofold symmetry axes. Clearly, the *i* phase possesses a very high symmetry and has the point group $m\overline{35}$. It was shown that there can exist three types of Bravais lattices in three dimensions consistent with the *i* point symmetry (Rokhsar et al. 1987). These three lattices correspond to the primitive, body-centered and face-centered hypercube in 6D hyperspace (Cahn et al. 1986). The primitive and face-centered *i* QCs have been discovered. These two types of *i* QCs can be distinguished by their twofold patterns. Figure 2.4 shows the twofold patterns of (a) the primitive and (b) face-centered icosahedral phases (hereafter, these are abbreviated to P-type and F-type, respectively) indexed along a fivefold direction. Along

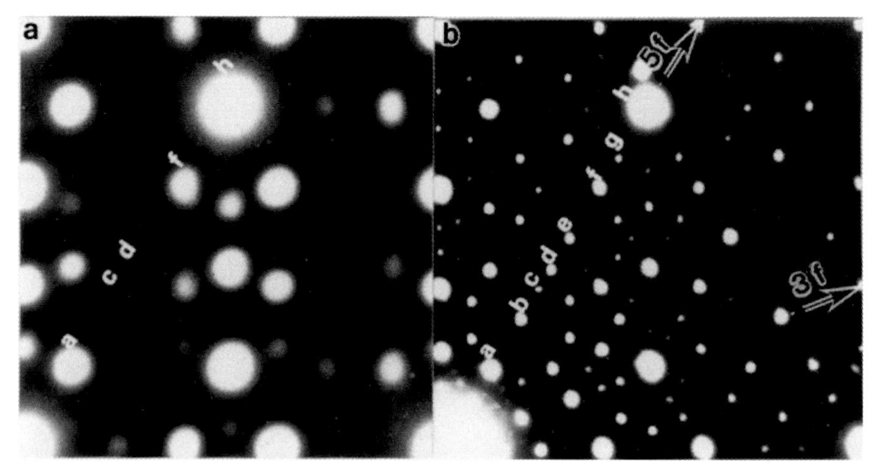

Fig. 2.4. Twofold diffraction patterns of (a) a P-type and (b) a F-type i QC.

this fivefold direction, the distance between the diffraction spots shows an inflation by τ^3 ($\tau = \frac{1+\sqrt{5}}{2}$) for the P-type i phase and by τ for the F-type i phase. The indexing of these diffraction spots is shown in Table 2.1. The 6D indices (n_1,n_2,n_3,n_4,n_5,n_6) of these diffraction spots have all combinations of integers n_i for the P-type i phase, while the superlattice spots (at τ ratios) in the F-type i phase have only odd integers.

All earlier discovered i phases, including the stable phases in Al-Li-Cu and Zn-Mg-Ga alloys, displayed a P-type lattice. The first experimental evidence of the F-type i phase was obtained in a low-temperature annealed Al-Mn i phase in which diffuse diffraction spots arranged in a τ sequence developed (Mukhopadhyay et al. 1987, Henley 1988). The F-type i phase with a highly ordered structure was first observed in a stable $Al_{65}Cu_{20}Fe_{15}$ QC (Tsai et al. 1987a, Ishimasa et al. 1988, Ebalard and Spaepen 1989). Subsequently, the F-type i phase was observed as stable QCs in several alloys in Al-Pd-TM and Zn-Mg-Rare Earth (RE) systems. These stable i alloys have face-centered lattice and all of them concurrently reveal a high structural quality without any visible phason disorder (Guryan et al. 1989). On the contrary, all the P-type i QCs which have been discovered are accompanied by a large amount of intrinsic disorder, which is present even in the stable Al-Li-Cu i phase (Heiney et al. 1987). This tendency raises an open question: does the P-type i phase really exist? The available experimental evidence seems to indicate that the P-type phase is a disordered structure of the F-type i phase in the real quasicrystalline materials.

Table 2.1. Six-dimensional indexing of the diffraction peaks for the P-type and F-type *i* QCs.

spot	Indices		
	Elser	Cahn	Ebalard
a	(100000)	(2/0 0/2 0/0)	(422222)
b	$\frac{1}{2}$(111111)	(0/2 2/2 0/0)	(733333)
c	(200000)	(4/0 0/4 0/0)	(844444)
d	(011111)	($\bar{2}$/4 4/2 0/0)	(10 44444)
e	$\frac{1}{2}$(311111)	(2/2 2/4 0/0)	(11 55555)
f	(111111)	(0/4 4/4 0/0)	(14 66666)
g	$\frac{1}{2}$(511111)	(4/2 2/6 0/0)	(15 77777)
h	(211111)	(2/4 4/6 0/0)	(18 88888)
i	(311111)	(4/4 4/8 0/0)	(22 10 10 10 10 10)

2.3.2 Two-Dimensional Quasicrystals

2.3.2.1 Tenfold Symmetry. A 2D QC, as its name implies, is a QC which is quasiperiodic in two-dimensions and periodic in another. Shortly after the discovery of the *i* QC, the first 2D QC, a *d* QC which has a point group symmetry 10/mmm, was discovered (Bendersky 1985). Figure 2.5 shows the typical diffraction patterns of the *d* QC along a tenfold axis and two characteristic twofold axes. The tenfold axis is the axis along which the diffraction spots are arranged periodically. Perpendicular to this axis is the quasiperiodic plane in which the spots are arranged in a quasiperiodic array with a tenfold symmetry, as observed in Fig. 2.5a. Along the periodic tenfold axis, several different periodicities have been obtained in different alloys or in the same alloy treated at different annealing conditions.

Apart from different periodicities along the tenfold axis, superlattice-type structures in the quasiperiodic plane itself have also been observed. Figure 2.6 shows the tenfold electron diffraction patterns for four such structures: primitive, S1, type I, and type II. The convention used here are the same as that used by Ritsch et al. (1995) and Ritsch (1996) and these structures are labelled respectively as P, S1, T1, and T2. The primitive structure reveals high structural quality characterized by a high density of sharp spots in the diffraction pattern (Fig. 2.6a). The S1 structure can be indexed to be a partially ordered superstructure without S2 reflections, as indicated by Edagawa et al. (1992). The characteristic features in the pattern of the S1 structure are very clear as compared to those of the primitive structure. First, intense reflections are surrounded by twenty small weak spots, which are partial S1

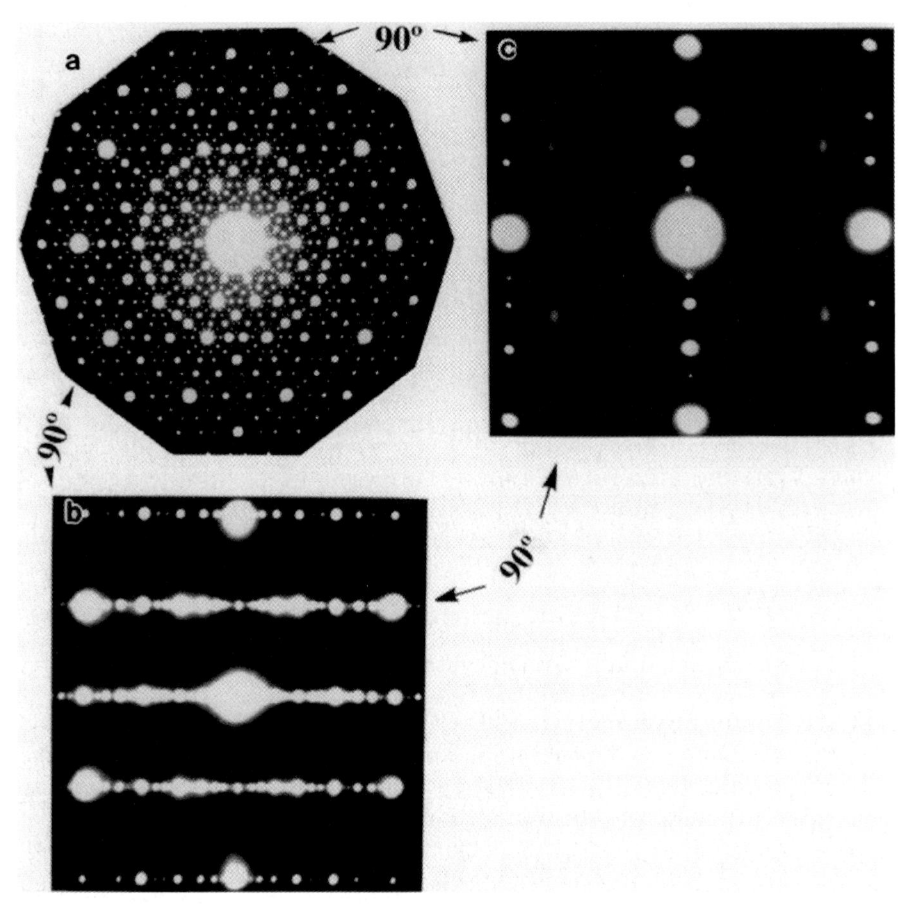

Fig. 2.5. Electron diffraction patterns of a *d* QC taken along (a) the tenfold axis and (b,c) the two twofold symmetry axes. Each of these two twofold patterns are repeated at 36° intervals about the tenfold axis.

reflections. Second, the extra spots are centered in the pentagons formed by five intensive spots (Fig. 2.6b). The T1 structure is closely related to the S1 structure. It is characterized by a fully ordered superstructure in which, in addition to the S1 reflections, the S2 reflections are also visible. For example, one of the S2 reflections indicated with an arrowhead centered in the smaller (inner) pentagons is observed (Fig. 2.6c). The T2 structure is characterized by an extra reflection located in halfway between two strong peaks, which is indicated with a small arrowhead (Fig. 2.6d).

2.3.2.2 Fivefold Symmetry. A 2D QC with a $\overline{10}/2$ point group symmetry has been observed in several alloys. This point group is equivalent to 5/mm2. This type of QC usually consists of small anti-phase domains (smaller than 100 nm) and therefore the normal electron diffraction gives only an average

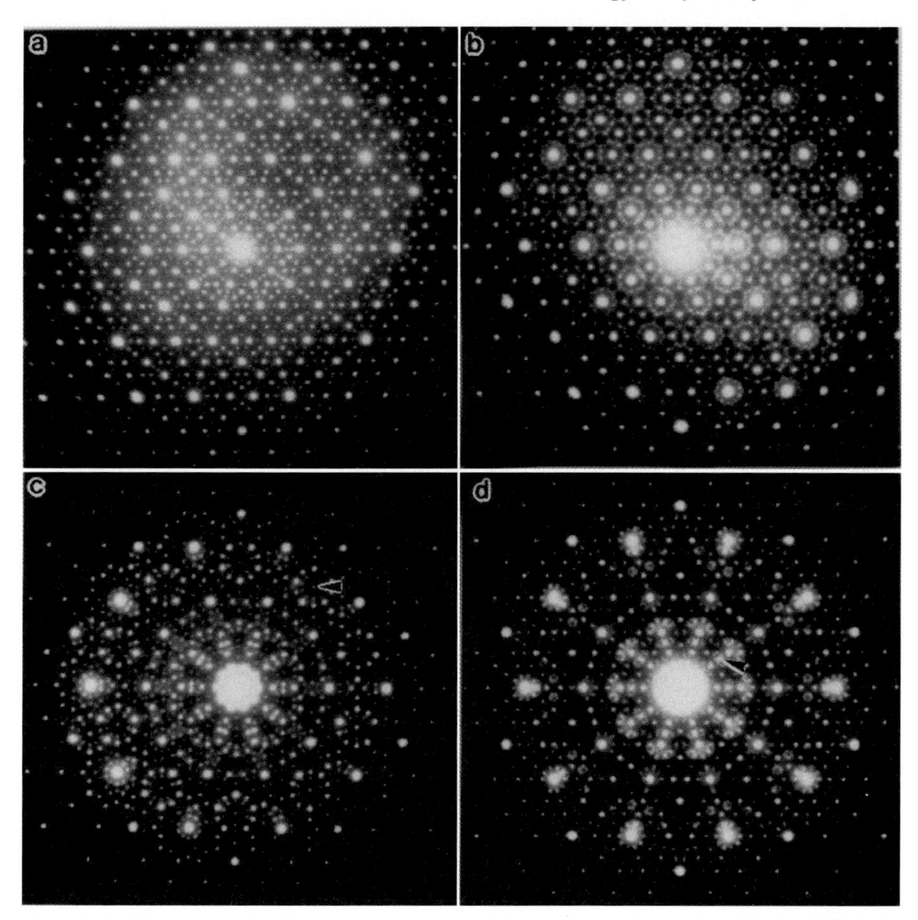

Fig. 2.6. The tenfold diffraction patterns of four types of *d* QCs: (a) P-, (b) S1-, (c) T1-, and (d) T2-type.

symmetry, which is tenfold. The difference between this QC and the *d* QC can be distinguished by means of convergent electron beam diffraction (CBED) technique, as shown in Fig. 2.7 (Saito et al. 1992, Tsuda et al. 1993). A tenfold and two twofold patterns [(a), (b) and (c)] satisfy the symmetry 10/mmm, while a fivefold and two pseudo twofold patterns [(d), (e) and (f)] clearly give the symmetry $\overline{10}/2$.

2.3.2.3 Other Two-Dimensional Quasicrystals. 2D QCs with eight- and 12-fold symmetries, which are called respectively octagonal and dodecagonal QCs, have been observed in some alloys. Their diffraction patterns are shown in Fig. 2.8.

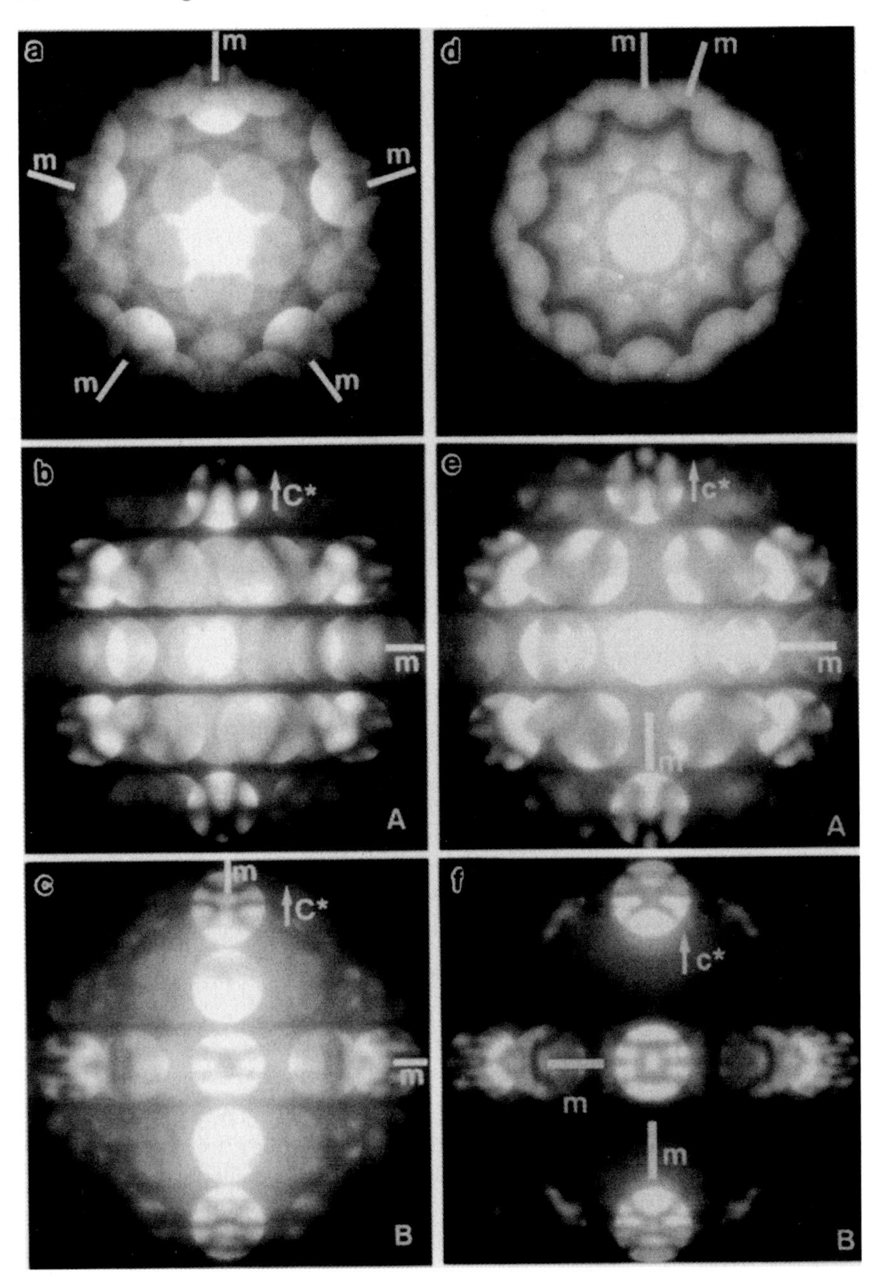

Fig. 2.7. (a) A fivefold CBED pattern and (b,c) two pseudo twofold patterns from a QC with a $\overline{10}/2$ symmetry. (d) A tenfold and (e,f) two twofold CBED patterns from a d QC with a 10/mmm symmetry. After Saito et al. (1992) and Tsuda et al. (1993).

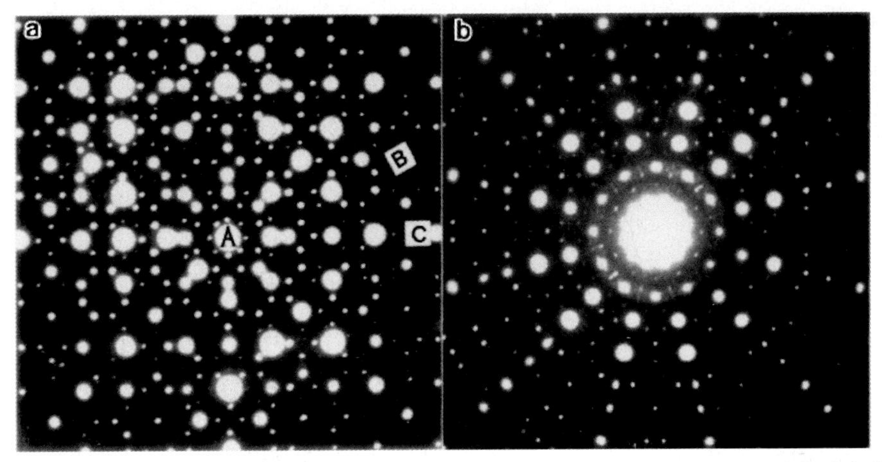

Fig. 2.8. (a) An eightfold electron difffraction pattern of an octagonal QC. (b) A 12-fold electron diffraction pattern of a dodecagonal QC. After Wang et al. (1987) and Krumeich et al. (1994).

2.3.3 One-Dimensional Quasicrystals

A 1D QC, as observed by electron diffraction, possesses a quasiperiodic arrangements of diffraction spots along one axis (He et al. 1988a). Perpendicular to this axis is a periodic plane in which spots are arranged in a periodic array. Since the 1D QC can be derived from a *d* QC using a linear phason theory (Zhang and Kuo 1990), the diffraction patterns are closely related to those of the *d* QC. In this sense, a 1D QC can be considered as an intermediate state between the *d* QC and its approximant crystals. The existence of 1D QCs has been demonstrated in several alloys (Tsai et al. 1992a, 1992b).

2.4 Quasicrystalline Alloy Systems and the Formation of Quasicrystals

QCs are known to form as metastable or stable phases in more than a hundred alloy systems. However, the number of *i* and *d* quasicrystalline samples of high structural quality for studying structure and physical properties is limited. Table 2.2 lists some important *i* and *d* QCs.

2.4.1 Metastable Quasicrystals

2.4.1.1 Icosahedral Quasicrystals. Quasicrystalline alloys were first discovered in Al-rich TM alloys in the composition range of 10–14 at.% Mn. Later, *i* phases were found in binary alloys of Al with Fe and Cr (Shechtman

Table 2.2. Alloy systems and alloys forming *i* and *d* QCs.

Stability	Al-TM Class	Mg-Al-Zn Class
Metastable	**Icosahedral Phase**	
	Al-TM (TM=V,Cr,Mn,Ru,Re,...) Al-(MN,Cr,Fe)-(Si,Ge) Al-(Cu,Pd)-TM (TM=Cr,Mn, Fe,Mo,Ru,Re,Os) Ga-Pd-Mn	Mg-Al-(Zn,Cu,An,Pd) Al-Mg-(Cu,Ag) Ga-Mg-Zn Al-Li-(Cu,Au,Zn,Mg) Zn-Mg-RE (RE=Y,Nd,Sm, Dy,Gd,Er,Ho,Tb)
	Crystallized from amorphous phase $Pd_{60}U_{20}Si_{20}$, $Al_{75}Cu_{15}V_{10}$ $Al_{55}Mn_{25}Si_{20}$ $Zr_{69.5}Cu_{12}Ni_{11}Al_{7.5}$	$Ti_{45}Zr_{38}Ni_{17}$
	Decagonal Phase	
	Al-TM (TM=Mn,Co,Fe,Pd) Al-(Cu,Ni,Pd)-TM (TM=Fe,Ru Re,Co,Rh,Ir)	$Fe_{60}Nb_{40}$ Zn-Mg-RE (RE=Y,Dy,Ho, Lu,Tb,Gd)
Stable	**Icosahedral Phase**	
	$Al_{63}Cu_{25}TM_{12}$ (TM=Fe,Ru,Os) $Al_{70}Pd_{20}TM_{10}$ (TM=Mn,Re) $Al_{70}Pd_{20}V_5Co_5$ $Al_{70}Pd_{20}Cr_5Fe_5$ $Al_{70}Pd_{20}Mo_5Ru_5$ $Al_{70}Pd_{20}W_5Os_5$	$Al_{56}Li_{33}Cu_{11}$ $Ga_{20}Mg_{37}Zn_{43}$ $Ti_{45}Zr_{38}Ni_{17}$ $Mg_{47}Al_{38}Pd_{15}$ $Zn_{60}Mg_{30}RE_{10}$ (RE=Y,Dy, Gd,Ho,Tb,Er)
	Decagonal Phase	
	$Al_{70}Ni_xCo_{30-x}$ (x=10–20) $Al_{65}Cu_{15}Co_{20}$ $Al_{75}Pd_{15}TM_{10}$ (TM=Fe,Ru,Os) $Al_{70}Pd_{13}Mn_{17}$	$Zn_{60}Mg_{38}RE_2$ (RE=Y,Dy, Ho,Lu,Tb,Gd)

et al. 1984), and Ru, V, W, and Mo (Bancel and Heiney 1986). These alloys were prepared by melt-spinning, with typical quenching rates of $\sim 10^6$ K/s. Quenching from the melt yields micron-size grains of quasicrystalline phases in an Al matrix. The *i* phase was found to display a higher structural quality and better stability with respect to decomposition when the Al atoms were substituted by a few at.% of Si in the Al-Mn system (Chen and Chen 1986, Bendersky and Kaufman 1986). Subsequently, QCs were discovered in ternary and quaternary alloys by substitution of Al with other TM elements.

Shortly after the discovery of QCs in the Al-TM system, another class of quasicrystalline alloys was discovered in the Mg-Al-Zn alloys (Ramachandrarao and Sastry 1985). Soon several alloy systems forming quasicrystalline phases were discovered and it became apparent that these had some common features. Almost all the quasicrystalline phases could be associated with crystalline phases containing icosahedrally packed groups of atoms. This correlation has been the basis for the discovery of many new *i* quasicrystalline phases. It appears that the *i* quasicrystalline phases form mostly at compositions close to the related crystalline phases.

Many QCs have been formed by identifying and rapidly solidifying alloys that have equilibrium phases containing icosahedrally arranged groups of atoms in the Mg-Al-Zn class. As mentioned earlier, the first of these was $Mg_{32}(Al, Zn)_{49}$ (Ramachandrarao and Sastry 1985). This was followed by several others, including Mg_4Al_6Cu (Sastry et al. 1986) and $Mg_{32}(Al,Zn,Cu)_{49}$ (Mukhopadhyay et al. 1986). Quantum structural diagrams (Villars et al. 1986) have been used to predict new QCs based on a structural mapping of ternary intermetallic compounds formed by *p* elements (Al,Ga,Ge,Sn), *s* elements (Li,Na,Mg), and near-noble metals (Ni,Cu,Zn, etc.). The success of these predictions has been demonstrated by the discovery of new *i* QCs of Al_6Li_3Au, $Al_{51}Li_{32}Zn_{17}$, $Ga_{16}Mg_{32}Zn_{52}$, and $Ag_{15}Mg_{35}Al_{15}$ (Chen et al. 1987a). The metastable *i* QCs were also discovered in Ti-TM alloys (Dong et al. 1986) and in Cu-Cd alloys (Bendersky and Biancaniello 1987).

Great progress in understanding the structure of the *i* phase was made by recognizing (Elser and Henley 1985, Henley and Elser 1986) that two complex compounds known for a long time, α-$Mn_{12}(Al,Si)_{57}$ (Cooper and Robinson, 1966) and $Mg_{32}(Al,Zn)_{49}$ (Bergman et al. 1957), were indeed approximant structures of the *i* QCs in the Al-Mn class and the Mg-Al-Zn class, respectively. Each of these two structures is a bcc packing of clusters consisting of two concentric atomic shells of full *i* symmetry. The atomic arrangements of the clusters are different in the Al-TM class (54 atoms) and Mg-Al-Zn class (44 atoms), so that these fall into different structural classes. Henley and Elser (1986) suggested a convenient criterion for grouping the *i* phases into these two classes. If d denotes the average atomic diameter and a_R is the quasilattice constant, then a_R/d is 1.65 for the Al-TM class and 1.75 for the Zn-Mg-Al class. The difference in the a_R/d ratio is due to different clusters with the *i* symmetry and a different packing geometry of these clusters in the two classes of the *i* phase. This criterion is valid for almost all *i* QCs.

2.4.1.2 Decagonal Quasicrystals. As described above, a survey of approximant crystals was made intensively for the formation of metastable QCs. Soon after the discovery of the Al-Mn *i* QC, a *d* QC was found in the same alloy with a somewhat slower cooling or subsequent annealing (Chattopadhyay et al. 1985, Bendersky et al. 1985). In many instances, the *d* phase coexisted with the *i* phase and thus the *d* phase was believed to be an intermediate state between the *i* phase and the crystalline phases. Interestingly, the *i* phase was formed in the Al-early TM systems (Al-Cr, Al-V, Al-Mn, and Al-Mo), whereas the *d* phase was mainly formed in the Al-late TM systems (Al-Fe, Al-Co, Al-Ni, Al-Pd, and Al-Rh). This indicates that the quasicrystalline phase formation is closely related to the location of the elements in the periodic table. At the same time, the structure of QCs has been found to be related to the structure of their corresponding approximant crystals. The approximant crystals of *i* QCs are cubic or orthorhombic, whereas those of *d* QCs are orthorhombic or monoclinic. The structural relationship between QCs and their approximant crystals can be described by a higher dimensional approach and can be interpreted in terms of phason strains.

Almost all binary *d* phases are formed as small grains and a large amount of disorder is introduced by rapid solidification. However, the imperfections are removed remarkably when the same formation technique is extended to ternary systems, e.g., Al-Ni-Fe, Al-Ni-Co (Tsai et al. 1989a), and Al-Cu-Co (He et al. 1988b). This extension to the ternary systems led to the discovery of the stable *d* phases.

2.4.1.3 Other Metastable Quasicrystals. In addition to the *d* and *i* QCs, QCs with other symmetries have also been observed as metastable phases. 2D octagonal QCs were first found in rapidly solidified Cr-Ni-Si and V-Ni-Si alloys (Wang et al. 1987) and later in a Mn_4Si alloy (Cao et al. 1988). A dodecagonal QC was observed by lattice imaging in laser vaporized and then condensed Cr-Ni particles (Ishimasa et al. 1985). A similar phase was observed in an annealed thin film of Bi-Mn alloy obtained by vacuum vaporization (Yoshida and Taniguchi 1991).

2.4.2 Stable Quasicrystals

Metastable QCs are generally prepared by melt-quenching. A drawback of this technique is that the resulting alloys are in the form of short ribbons containing small grains of QCs which have quenched-in defects. This makes a study of the intrinsic structure and properties of QCs very difficult. Therefore, stable QCs with larger single grains were strongly desired.

2.4.2.1 Zn-Mg-Al Class. *(a) Icosahedral phase.* The first stable QC was discovered in an $Al_{5.1}Li_3Cu$ alloy (Ball and Lloyd 1985). The existence of this phase had been known for a long time, but its structure had remained undetermined (Hardy and Silcock 1955-56). It was soon evident that this phase is closely related in structure to the cubic $Al_{4.8}Li_3Cu$ phase (R phase), which is isostructural with $Mg_{32}(Al,Zn)_{49}$, containing large i groups. In this case, the QC and the approximant crystal have slightly different composition and are both stable. Subsequently, stable i phases of this class were found in Zn-Mg-Ga (Ohashi and Spaepen 1987) and Mg-Al-Pd (Koshikawa et al. 1992) systems. However, the i phases in these alloys normally possess a correlation length ζ ($\zeta=2\pi$/FWHM, where FWHM is the full width at half maximum of the diffraction peak) smaller than 50 nm.

A second generation of stable i QCs of this class was found in the Zn-Mg-RE system (Luo et al. 1993). These QCs are grouped into the Zn-Mg-Al class according to their quasilattice parameters. The new group of stable QCs in the Zn-Mg-RE (RE=Y,Gd,Tb,Dy,Ho,Er) system (Niikura et al. 1994, Tsai et al. 1994a) exhibit a face-centered icosahedral lattice that was observed in the Zn-Mg-Al class for the first time. In addition, the i phase reveals sharp diffraction peaks corresponding to the correlation length of 200 nm, which is the largest in the Zn-Mg-Al class reported to date. New ways of joining cluster centers and making linkages, or new style of decoration of the clusters, are expected in this system since they show a unique ratio of atomic sizes and atomic fractions. In this way, this new group should be important, more so because RE metals are involved and therefore interesting magnetic properties are expected.

(b) Decagonal Phase. The first Frank-Kasper-type d QC has been found in the Zn-Mg-Dy alloy (Sato et al. 1997). This novel d phase is not formed directly from the melt or by annealing at higher temperatures close to the melting point. It forms through a solid state reaction between a primary i phase (close to $Zn_{60}Mg_{30}RE_{10}$) and a monoclinic Zn_7Mg_4 phase at intermediate temperatures. It is very interesting that the concentration of RE varies in the order from the i QC to the d QC and then to the monoclinic phase. In Al-TM alloys the composition of the i and the d QC and of the related crystalline phase remains nearly the same. In a systematic investigation, the presence of a d phase was comprehensively identified in the alloy $Zn_{58}Mg_{40}RE_2$ (RE=Y,Dy,Ho,Er,Tm,Lu) annealed at temperatures around 623 K (Sato et al. 1998a). The d phase disappears on annealing at higher temperatures ($\sim$723 K) but reappears on subsequent annealing at lower temperatures ($\sim$623 K). Thus it seems that this d QC is stable in an intermediate temperature range.

2.4.2.2 Al-Transition Metals Class. *(a) Icosahedral Phase.* Following the identification of the Al-Li-Cu i phase, another stable i phase was discovered in an Al-Cu-Fe alloy. A bulk alloy of $Al_{65}Cu_{20}Fe_{15}$ prepared by conventional solidification and then fully annealed at temperatures just below the melting temperature revealed an *i* phase with a grain size of a few mm (Tsai et al. 1987b). Since Ru and Os elements are located in the same column of the periodic table as Fe, they have a similar electronic structure to that of Fe. Icosahedral QCs were formed when Fe was substituted by Ru or Os in $Al_{65}Cu_{20}TM_{15}$ alloy (Tsai et al. 1988). This indicates that the stability of the *i* QC is dominated by the electronic structure. This new series of stable *i* alloys was shown to possess very sharp diffraction peaks comparable to those of a crystalline material and exhibited no detectable peak shifts from the *i* symmetry positions. Recently, detailed studies on the phase diagram showed that the ideal composition of the *i* phase is close to $Al_{63}Cu_{25}Fe_{12}$ which is stable at elevated temperatures (Calvayrac et al. 1990).

Following the discovery of the Al-Cu-TM systems, a stable *i* phase was found in the Al-Pd-Mn and Al-Pd-Re alloys, as well as in some Al-Pd-TM quarternary alloys (Tsai et al. 1990a). The stable Al-Pd-Mn *i* QC is particularly important because of its high structural quality (Tsai et al. 1990b). A striking feature of this system is the high resistivity of the order of $\Omega\,$cm observed in the Al-Pd-Re alloy (Akiyama et al. 1993). These stable *i* phases possess a large amount of disorder introduced during solidification, which is then completely removed by post-annealing. The correlation length of the *i* phase in this class is close to 300 nm, which is comparable to that of a single crystal of excellent structural quality. Only P-type *i* QCs were known before the discovery of the stable QCs in the Al-Cu-TM and Al-Pd-Mn alloys, which exhibit the F-type lattice.

An as-cast $Ti_{45}Zr_{38}Ni_{17}$ alloy, containing initially only a C14 hexagonal Laves phase and an α solid solution phase, was found to transform to an *i* phase upon annealing at 843 K for 64 h. Based on this observation, this *i* phase was claimed to be stable (Kelton et al. 1997).

(b) Decagonal phase. Stable *d* QCs were discovered in the ternary alloys $Al_{65}Cu_{15}Co_{20}$ and $Al_{70}Ni_{15}Co_{15}$ (Tsai et al. 1989b). The *d* phase generally grows in the form of an elongated decaprism which can easily reach $\sim$2 mm in length and 0.1 mm in diameter. Unlike the stable *i* QC, the *d* QC forms in these two systems in a wide composition range, with a constant amount of Al concentration.

A stable *d* Al-Pd-Mn QC with a composition of $Al_{70.5}Mn_{16.5}Pd_{13}$ was found in an $Al_{70}Pd_{15}Mn_{15}$ alloy annealed at 1073 K for 168 h (Beeli et al. 1991, Hiraga et al. 1991b). As in the case of the Zn-Mg-RE *d* QCs, the *d* Al-Pd-Mn QC cannot be formed directly from the melt. On cooling down from the melt of an $Al_{70.5}Mn_{16.5}Pd_{13}$ alloy, a primary $Al_{11}(Pd,Mn)_4$ orthorhombic phase embedded in an *i* QC is formed. The *d* QC is formed only when solidified at a very slow cooing rate ($\sim$5 K/h) or after a subsequent annealing which improves the solid state reaction between the $Al_{11}(Pd,Mn)_4$ crystal and the *i* QC. The Al-Pd-Mn system is the only one in which both the stable *d* and the stable *i* phases are known to exist. In addition to this system, stable *d* QCs were found in the Al-Pd-TM (TM=Fe,Ru,Os) systems. The periodicities along the tenfold axis are different among these d phases. They are 0.4 nm for Al-Ni-Co and Al-Cu-Co, 1.2 nm for Al-Pd-Mn, and 1.6 nm for Al-Pd-TM (TM=Fe,Ru,Os). A classification according to the periodicity has been summarized by Steurer (1996).

(c) Dodecagonal Phase A possible stable dodecagonal QC (Krumeich et al. 1994) was found in a $Ta_{1.6}Te$ alloy. The dodecagonal QC has a translational period of 2.07 nm along the 12-fold axis. This phase is the first chalcogenide QC.

2.4.3 Quasicrystals as Hume-Rothery Phases

Hume-Rothery (1926) derived a series of rules, based mainly on the investigations of Cu and Ag alloys, indicating the factors which favor the formation of a specific structure in a certain composition range. The most important one is about the preferential formation of specific structures in a certain range of electron-per-atom ratio e/a. This quantity indicates the number of all valence electrons in the alloy (corresponding to the number of the elements in the periodic table) per number of atoms. This rule is applicable to many intermetallic compounds (Hume-Rothery 1926).

Formation of stable QCs in the Al-Cu-TM system with a strict stoichiometry indicates that the electronic structure plays a significant role in the stabilization of the i QC. The values of e/a for these QCs were calculated based on the valencies for the TM elements estimated by Raynor (1949). By using a phenomenological model for TMs developed by Pauling (Pauling 1938) and assuming that a charge transfer from the conduction band into the d-band occurs to compensate for the unpaired spin of the TM atoms, Raynor obtained the valence values of -4.66 for Cr, -3.66 for Mn, -2.66 for Fe, -1.71 for Co, and -0.62 for Ni. Remarkably, all the three stable i QCs in the Al-Cu-TM system, ($Al_{65}Cu_{20}Fe_{15}$, $Al_{65}Cu_{20}Ru_{15}$, and $Al_{65}Cu_{20}Os_{15}$) have the e/a ratio of $\sim$1.75. The strict stoichiometric composition of the i phases has been indicated to be $Al_{63}Cu_{25}TM_{12}$, which still has the e/a ratio close to 1.75. This suggests that the important criteria for the formation of the i QC could possibly be the same as that for a Hume-Rothery electronic compound. Using this criterion, alloys with e/a ratio of 1.75 in the Al-Pd-TM system were designed (Tsai et al. 1990b, Yokoyama et al. 1991a). Two ternary and four quaternary stable i QCs, together with a few tens of metastable i and d phases, were found in this system (Tsai et al. 1990b, 1990c, Yokoyama et al. 1991a). The phase formation for the Al-Cu-TM and Al-Pd-TM systems is summarized in Fig. 2.9. The formation tendency of the i phase in quaternary alloys is of interest when Mn is replaced by the TM pairs located at the two opposite sides of Mn or Re. For example, quarternary $Al_{70}Pd_{20}V_5Co_5$, which has the e/a ratio $\sim$1.75, is a stable i alloy. The formation of new stable i alloys indicates that the Hume-Rothery rule is applicable to the stable i QCs in these alloy systems.

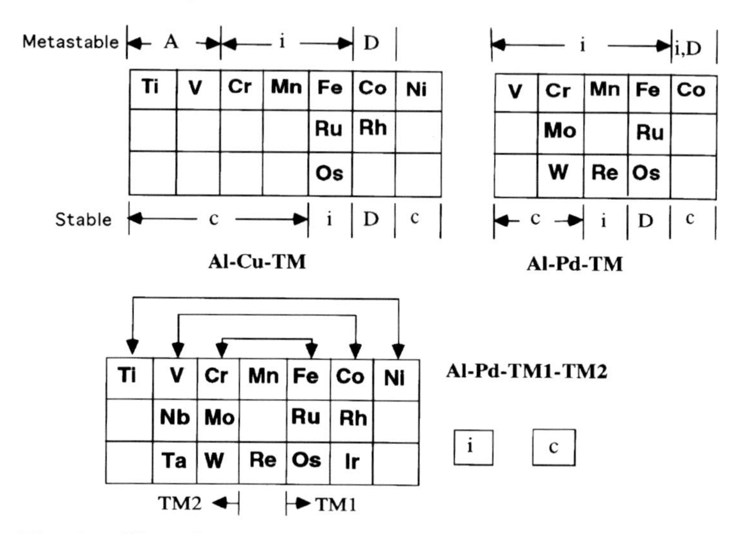

Fig. 2.9. Phase formation in ternary $Al_{65}Cu_{20}TM_{15}$ and $Al_{70}Pd_{20}TM_{10}$, and quaternary Al-Pd-TM1-TM2 systems. Here i, D, c, and A refer respectively to *i*, *d*, crystalline, and *a* phase.

The e/a ratio for the stable *i* alloys in the Zn-Mg-Al class is different from that in the Al-TM class. The composition of the new QCs is reported to be stoichiometrically close to $Zn_{60}Mg_{30}RE_{10}$ (Tsai et al. 1997). Since the valency of Zn and Mg is 2 and that of the RE elements is 3, the e/a value for the *i* phases is estimated to be ~2.1. We should note that the stable QCs of $Al_{5.1}Li_3Cu$ or $Zn_{43}Mg_{37}Ga_{20}$ also have an e/a ratio of about 2.2. Apparently, stable *i* QCs in the Zn-Mg-Al class also obey the Hume-Rothery rule. It is significant to note that this small amount (~10 at.%) of RE atoms makes a crucial difference in the formation of the QC. Since the size of these RE atoms is significantly larger than that of other atoms in the lattice, it is expected that these RE atoms occupy some special sites in the perfect *i* lattice. The importance of the atomic size of the RE elements in phase formation is brought out from the observation that the *i* phase could not be formed in alloys containing RE elements with atomic radii between 1.75 Å and 1.8 Å, but only with radii in the range 1.73–1.79 Å, even though the e/a was the same in all these alloys (Fig. 2.10). However, the e/a ratio remains the dominant factor and thus the stable *i* QCs should be described as Hume-Rothery compounds. Since the stable *d* phase forms over a wide composition range, the e/a criterion is not as rigid for the *d* phase as for the *i* phase.

Jones (1937) has given a physical interpretation of the empirical Hume-Rothery rule in terms of a rigid-band model. In a periodic crystal, energy gaps are formed at a Brillouin-zone boundary since the corresponding Bragg reflection mixes electron wave functions with $\frac{1}{2}\boldsymbol{K}_P$ and $\frac{1}{2}\boldsymbol{K}_P - \boldsymbol{G}$, where $\frac{1}{2}\boldsymbol{K}_P$ is a vector on the Brillouin-zone boundary and $\boldsymbol{G}$ is the corresponding

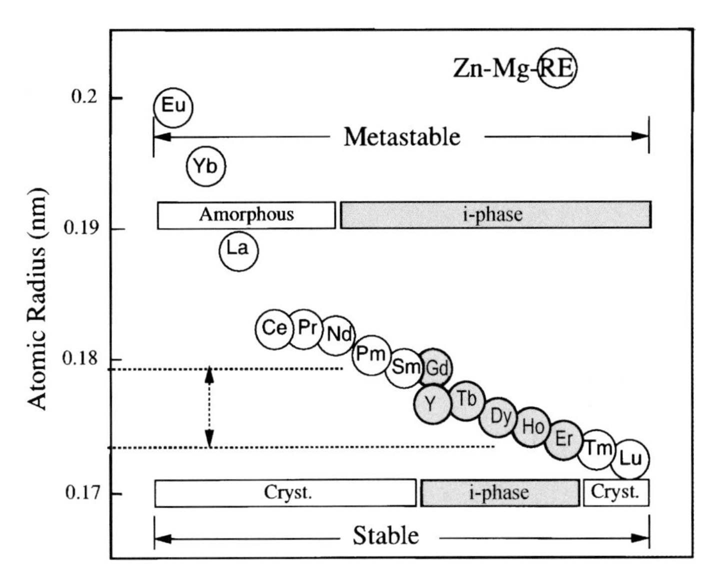

Fig. 2.10. Phase formation of Zn-Mg-RE alloys for various RE elements and their atomic radii (in nm).

Bragg reflection position. Thus, at a given electron concentration, the total energy of an electron system can be lowered with a structure that enables a large overlap of Fermi surface with the zone boundary. The system is then electronically stabilized.

Bancel and Heiney (1986) applied this mechanism to the i QCs. Even in a QC energy gaps can open on planes bisecting the line joining the origin and the intense Bragg positions, which form a pseudo-Brillouin zone. The high multiplicity of the i Bragg positions makes the pseudo-Brillouin zone nearly spherical, and thus the large overlap can be made if $|\boldsymbol{K}_P| \simeq 2|\boldsymbol{k}_\mathrm{F}|$. The i phase gives strongest reflections at a reciprocal length of 30 nm^{-1}, which can be taken as the diameter of corresponding pseudo-Brillouin zone $|\boldsymbol{K}_P|$. The value of $|\boldsymbol{k}_\mathrm{F}|$ can be estimated from the e/a ratio. The value of $\frac{|\boldsymbol{K}_P|}{2|\boldsymbol{k}_\mathrm{F}|}$ was estimated and found to be approximately 1 for all the stable i QCs in the Al-TM class. The specific heat and transport measurements for several stable i QCs have shown evidence for a gap in the density of states at the Fermi level (Wagner et al. 1988, Mizutani et al. 1990a, 1990b). This implies that the earlier empirical estimates of the $|\boldsymbol{K}_P|$ and $|\boldsymbol{k}_\mathrm{F}|$ are meaningful and confirms that the stable i phases are truly the Hume-Rothery phases dominated by the electronic structure effects.

Figure 2.11 summarizes the parameters of a_R/d and e/a for the known stable i QCs. It shows that the parameters separate into two groups. All the stable i QCs can be classified basically into two classes, as indicated by Henley and Elser (1986). The value of a_R for the whole Al-TM class is ~0.46 nm and for the Zn-Mg-Al class is ~0.52 nm. When a_R is normalized to the average atomic diameter of the constituent elements, the a_R/d ratio is ~1.65 for the former and ~1.75 for the latter. At the same time, the e/a value for the Al-TM class is 1.75 and for the Zn-Al-Mg class is 2.1. Thus, while falling into two different class of Hume-Rothery compounds, stable i QCs also have two different classes of cluster geometry.

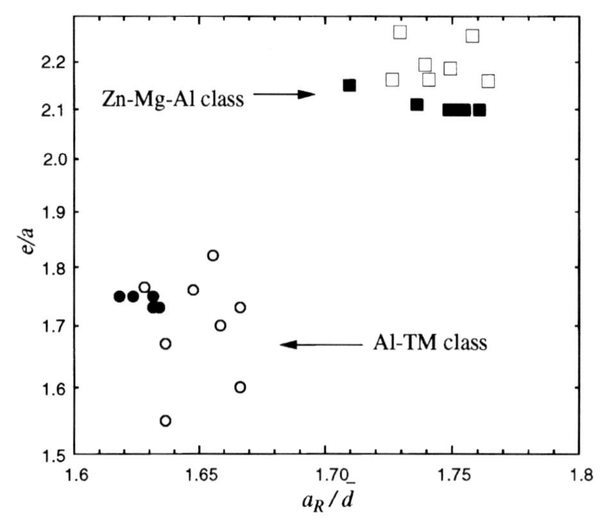

Fig. 2.11. Relationship between the ratios e/a and a_R/d.

2.5 Phase Transformation from Amorphous to Icosahedral Phase

Short-range order and i coordination of atoms in a simple undercooled liquid and metallic glasses have been discussed extensively before the discovery of QCs (Frank 1952). Later, in a molecular-dynamics simulation of the i bond-orientational order in an undercooled Lennard-Jones liquid (Steinhardt et al. 1983), a similarity in the structure factor between the metallic glasses and the i solid was observed (Nelson and Halperin 1985). Due to the suggested similarity in structure, one may expect a structural evolution from an a phase to the i phase under certain thermodynamic conditions.

Several systems are known to exhibit a transformation from an a phase to the i phase: $Pd_{60}U_{20}Si_{20}$ (Poon et al. 1985, Shen et al. 1986), $Al_{55}Si_{25}Mn_{20}$ (Inoue et al. 1988), $Al_{75}Cu_{15}V_{10}$ (Tsai et al. 1987b), $Ti_{53}Zr_{27}Ni_{20}$ (Monokanov et al. 1990), $Ti_{40}Ni_{40}Cu_{10}Si_{10}$ (Alisowa et al. 1990), and $Zr_{69.5}Cu_{12}Ni_{11}Al_{7.5}$ (Köster et al. 1996). The detailed kinetics of the transformation was studied only in a few alloy systems and was found to be very sensitive to the alloy composition.

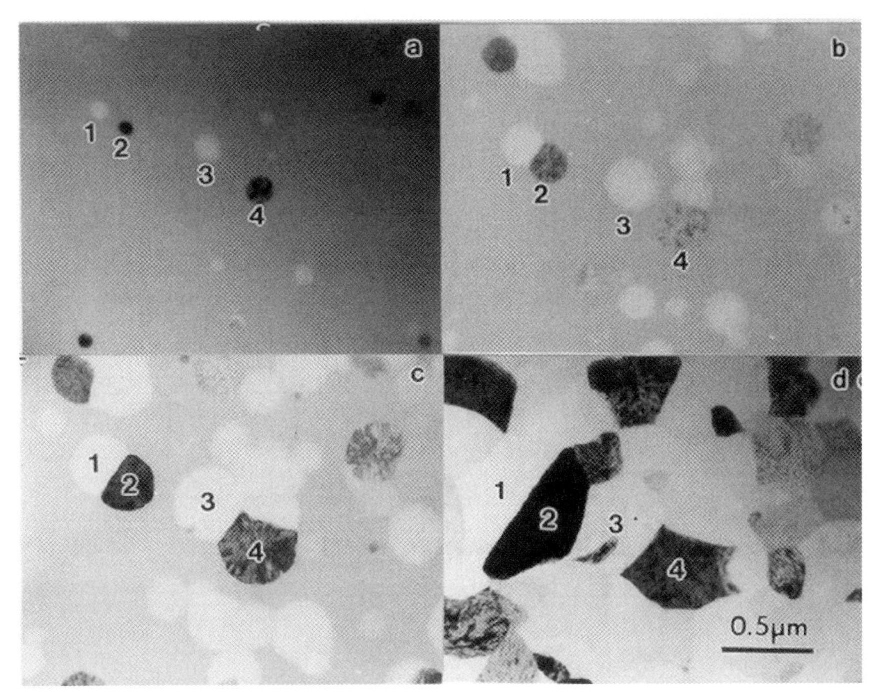

Fig. 2.12. In-situ growth of an *i* phase in an $Al_{75}Cu_{15}V_{10}$ *a* phase at 623 K for various time durations.

It has been observed in the Pd-U-Si, Al-Cu-V, and Al-Si-Mn systems that the transformation is polymorphous and proceeds via a nucleation and growth process (Shen et al. 1986, Tsai et al. 1989c). Johnson-Mehl-Avrami analysis (Avrami 1939, Johnson and Mehl 1939) gave a small Avrami exponent n in the range 1–2 for these three alloy systems just below the transformation temperatures. The small n values are inconsistent with polymorphormic transformation for which n is expected to be in the range of 3–4 for an amorphous-crystalline phase transformation. The anomalous Avrami exponent arises from the inhomogenous distribution of quenched-in nuclei in the *a* matrix. A thermodynamic analysis for the Al-Cu-V alloy shows that the free energy change for *a*-*i* phase transformation is about 0.6 times that of the amorphous-crystalline phase transformation (Tsai et al. 1989c). The interfacial energy between the *i* and *a* phases was estimated to lie in the range 0.002–0.015 J/m^2, which is much smaller than the interfacial energy of ~0.108 J/m^2 between the crystalline and liquid Al at the equilibrium melting temperature (Holzer and Kelton 1991). Such a small interfacial energy implies that the local structure of the *i* and *a* phases at the interface must be very similar. This is consistent with the existence of the short-range *i* order in the *a* state of the Pd-U-Si and Al-Cu-V alloys observed by x-ray diffraction (Kafolt et al. 1986, Matsubara et al. 1988, Chen et al. 1987a). Figure 2.12

shows an in-situ transmission electron micrograph of the growth of an i phase in the a matrix of an Al-Cu-V alloy. Initially, all the i phase grains are isolated and grow spherically before impingement of the neighboring grains occurs. Figure 2.13 shows the grain size dependence on time at various temperatures. The growth rate is estimated to be in the range 4×10^{-9}–2.5×10^{-8} cm/s at the transformation temperature of ~720 K. From the linearity of the growth, the activation energy of the grain growth of the i phase is estimated to be ~250 kJ/mol, which remains almost the same for the whole transformation process. This indicates that the role of the interface is prominent in the whole transformation process (Tsai et al. 1989c).

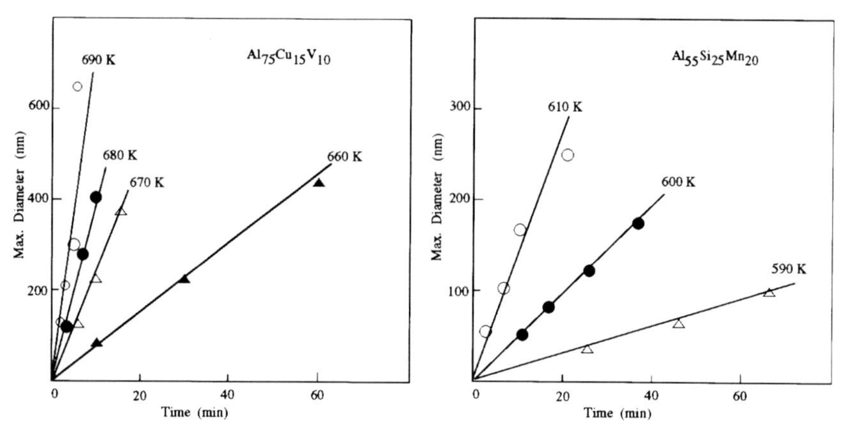

Fig. 2.13. Time dependence of the grain size of the i phase in $Al_{75}Cu_{15}V_{10}$ and $Al_{55}Mn_{20}Si_{25}$ a alloys. After Tsai et al. (1989c).

In the case of the $Zr_{69.5}Cu_{12}Ni_{11}Al_{7.5}$ alloy (Fig. 2.14), the grain size of the i phase is proportional to the square root of time. This indicates that the transformation is diffusion controlled (Köster et al. 1996). The activation energy of the diffusion, which is as large as 300 kJ/mol, is explained by the diffusion of large Zr atoms in the a matrix. The square root dependent growth in this alloy is attributed to a clear glass transition or a stable supercooled liquid in the Zr-based alloys, in which the viscosity becomes prominent and should not be ignored beyond the glass transition temperature.

The i phases formed from an a phase are all defective and show broad diffraction peaks. The high-resolution image (Tsai et al. 1994b) of such an i phase shows the lattice having a 2D tiling pattern comprised mainly of pentagons joined side to side randomly and filled with unique tear-like and crack-like defects which are expected to exist according to the i glass model (Shechtman and Blech 1985, Stephens and Goldman 1986). The i glass model is a structural model which, in contrast to the Penrose model, relies on local interactions to join clusters of atom in a somewhat random way. According to the model, all the clusters have the same orientation, but because of the

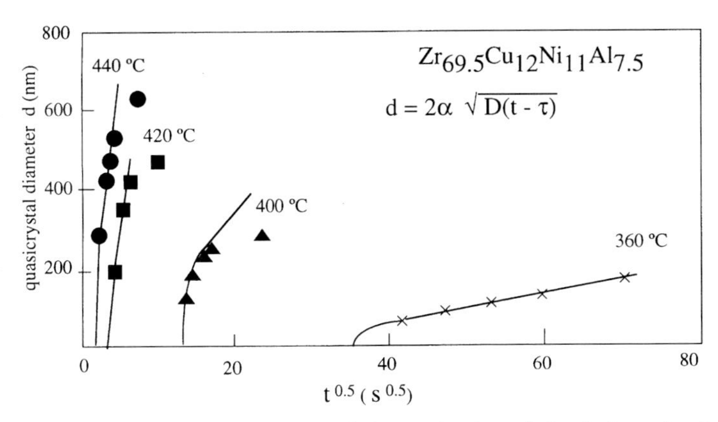

Fig. 2.14. Time dependence of the grain size of the *i* phase in the $Zr_{69.5}Cu_{12}Ni_{11}Al_{7.5}$ *a* alloy. After Köster et al. (1996).

random growth, the structure contains many defects. Although the *i* glass model cannot explain the sharpness of diffraction peaks observed in the ordered QCs, the idea that there exists a correlation in the structure of the *i* phase and the *a* phase still prevails.

2.6 Phase Diagrams

Several QCs have been found as equilibrium phases, requiring a reconsideration of the phase diagrams involving QCs. With an aim to understand the underlying growth mechanism and to facilitate the fabrication of large QCs, a few partial phase diagrams have been reconstructed.

2.6.1 Al-Li-Cu

The Al-Li-Cu system has attracted a lot of attention as it seems to have an equilibrium *i* phase. Al_6Li_3Cu was first investigated in 1955 (Hardy and Silcock 1955-56) and it was not realized that it had an *i* phase until several decades later (Dubost et al. 1986). Figure 2.15a shows a partial phase diagram of the alloy Al_xLi_3Cu (Chen et al. 1987c). It exhibits the following reactions:

$$\begin{array}{llll} liquid & \rightarrow R & \text{at } T_R \approx 900\text{ K}, & \\ liquid + R & \rightarrow i & \text{at } T_m & \text{for } x \text{ from } 6.1 \text{ to } 4.8, \\ \text{and} & & & \\ i + \text{Al} & \rightarrow liquid & \text{at } T_e \approx 810\text{ K} & \text{for } x \text{ from } 12 \text{ to } 6.1. \end{array}$$

Both the peritectic, T_m, and eutectic, T_e, temperatures are not strictly constant and show a slight decrease with increasing x. The *i* phase coexists with fcc Al in equilibrium for alloys with $x > 5.1$. This is in disagreement to the

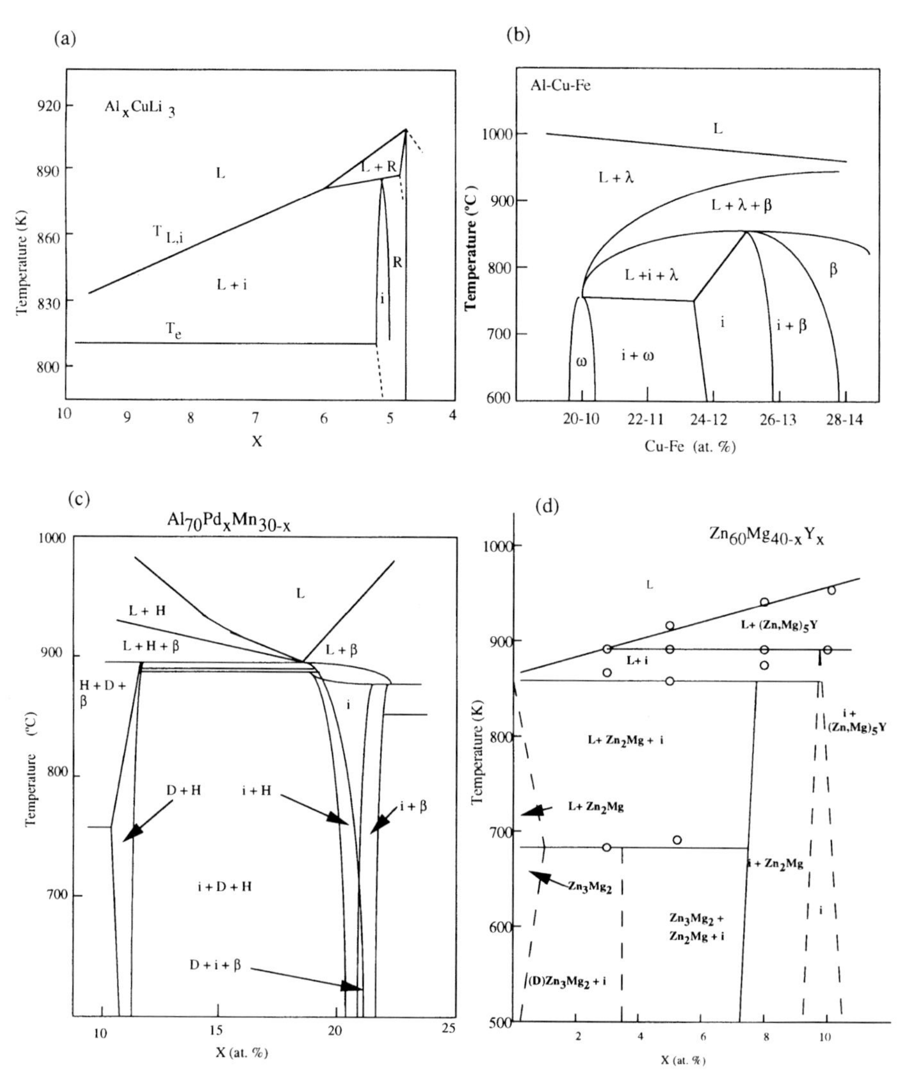

Fig. 2.15. Partial phase diagrams involving i phase along composition-temperature section for (a) Al-Cu-Li (after Chen et al. 1987b), (b) Al-Cu-Fe (after Faudot et al. 1991), (c) Al-Pd-Mn (after Gödecke and Lück 1995), and (d) Zn-Mg-Y systems.

results of Hardy and Silcock (1955-56). The R phase $Al_{4.8}Li_3Cu$ melts congruently at 910 K, while the i phase $Al_{5.1}Li_3Cu$ melts peritectically at ~884 K. The grains of the i phase once formed are very stable and melt directly upon heating, even at a rate of 20 K/min.

2.6.2 Al-Cu-Fe

The Al-Cu-Fe system phase diagram was first investigated some time ago (Bradley and Goldschmidt 1939, Phillips 1953-54). The i phase discovered by Tsai et al. (1987a) in the Al-Cu-Fe system is close to the composition of a ternary phase ψ found by Bradley and Goldschmidt (1939), the structure of which was left unknown. Recently, the phase diagram has been carefully studied again in the vicinity of the i phase (Faudot et al. 1991, Gayle et al. 1992). The i phase is surrounded by three crystalline phases, denoted by β, λ, and ω. Here β is a CsCl-type structure with a composition close to $Al_{50}(Cu,Fe)_{50}$, λ is isostructural to the monoclinic $Al_{13}Fe_4$ structure, and ω is the $Al_{70}Cu_{20}Fe_{10}$-based tetragonal structure. The β and λ phases melt at temperatures above 1373 K. In detailed studies, a single i phase has been found in the vicinity of the composition $Al_{63}Cu_{25}Fe_{12}$, as shown in Fig. 2.15b. The following reactions for solidification at $Al_{63}Cu_{25}Fe_{12}$ can be observed:

$$\begin{array}{llll} L & \rightarrow & L+\lambda & 1213\text{ K} - 1253\text{ K}, \\ L+\lambda & \rightarrow & L+\lambda+\beta & 1123\text{ K} - 1173\text{ K}, \\ L+\lambda+\beta & \rightarrow & L+i & 1073\text{ K} - 1123\text{ K}, \\ L+i & \rightarrow & i & \sim 1073\text{ K}. \end{array}$$

During solidification, the i phase first solidifies close to the composition $Al_{65.5}Cu_{21.5}Fe_{13}$. Its composition region lies between $Al_{64.5}Cu_{23}Fe_{12.5}$ and $Al_{62}Cu_{26.5}Fe_{11.5}$, while at 873 K it lies between $Al_{62}Cu_{26.5}Fe_{11.5}$ and $Al_{60.5}Cu_{29.5}Fe_{10}$ (Grushko et al. 1996). A detailed study in the vicinity of the i phase has shown that at 953 K the composition of the i phase extends over a triangle with vertices of Al, Cu, and Fe compositions of respectively 62.4-24.4-13.2, 65-23-12, and 61-28.4-10.6 (Gratias et al. 1993). The correlation between phason strains and the phase transition among the i phases and approximants has been discussed elsewhere (Bancel 1991).

2.6.3 Al-Pd-Mn

Several types of QCs can be obtained in the Al-Pd-Mn system either at equilibrium via normal casting methods or in quenched states via melt-spinning techniques. Therefore, the phase diagram of this system has been well investigated. A partial phase diagram was first produced along the $Al_{80-x}Pd_{20}Mn_x$ section (Yokoyama et al. 1991). The i phase was found to be formed via peritectic reactions and to be homogeneous from 9 to 10 at.% of Mn at 1143–1113 K. Figure 2.15c shows a partial phase diagram along the section with fixed

70 at.% of Al, where the i and d phases are both involved (Gödecke and Lück 1995, Audier et al. 1993). The i phase region shrinks with decreasing temperature. Around the composition $Al_{70}Pd_{20}Mn_{10}$, which is supposedly the stoichiometry of the i phase,

$$\begin{array}{llll} L & \rightarrow & L+\beta\,\mathrm{Al_{50}(Pd,Mn)_{50}} & \text{at } 1183\ \mathrm{K} \\ L+\beta\,\mathrm{Al_{50}(Pd,Mn)_{50}} & \rightarrow & i & \text{at } 1163\ \mathrm{K}, \end{array}$$

where β phase is a simple cubic structure analogous to CsCl. At the composition $Al_{70}Pd_{13}Mn_{17}$, which is supposedly the composition with mostly single d phase,

$$\begin{array}{llll} L & \rightarrow & L+\mathrm{Al_{11}(Pd,Mn)_3} & \text{at } 1225\ \mathrm{K} \\ L+\mathrm{Al_{11}(Pd,Mn)_3} & \rightarrow & L+\mathrm{Al_{11}(Pd,Mn)_3} & \\ & & +\beta\,\mathrm{Al_{50}(Pd,Mn)_{50}} & \text{at } 1196\ \mathrm{K} \\ L+\mathrm{Al_{11}(Pd,Mn)_3}+ & & & \\ \beta\,\mathrm{Al_{50}(Pd,Mn)_{50}} & \rightarrow & D & \text{at } 1169\ \mathrm{K}. \end{array}$$

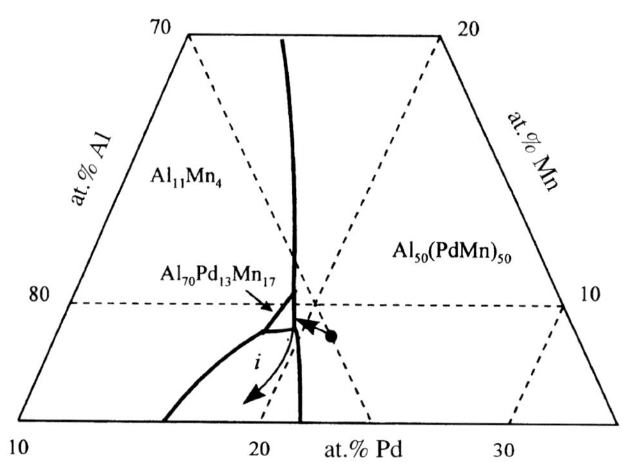

Fig. 2.16. Solidification path and sequence of formation of an $Al_{70}Pd_{21.5}Mn_{8.5}$ i phase. After Delaney et al. (1997).

Based on the results of the microstructural examination and chemical analysis, the sequence of solidification events and the order of precipitation of the phases have been mapped out qualitatively on the liquidus surface, as shown in Fig. 2.16. For example, for the $Al_{70}Pd_{21.5}Mn_{8.5}$ alloy, the initial composition is such that the $Al_{50}(Pd,Mn)_{50}$ phase is the first solidification product. The starting composition would be the composition of the liquid at the peritectic reaction

$$L + Al_{50}(Pd, Mn)_{50} + Al_{70}Pd_{13}Mn_{17} \rightarrow i.$$

However, the formation of the high temperature crystalline phase can generally be suppressed by solidification through a temperature gradient.

In the phase diagram (Tsai et al. 1991) for liquid-quenched Al-Pd-Mn alloys there are four quasicrystalline phases: F-type i phase, P-type i phase, and d phases with periods of 1.2 and 1.6 nm. All these four phases are formed in different composition regions, implying that a chemical factor plays a critical role in determining the quasicrystalline structure.

2.6.4 Zn-Mg-Y

The phase diagram of the Zn-Mg-Y system has been partially studied (Padezhnova et al. 1979, 1982). An unknown phase denoted as a Z phase, with a composition Zn_6Mg_3Y, has been identified to be the i phase from the published x-ray diffraction patern (Niikura et al. 1994, Tsai et al. 1997a, Langsdorf et al. 1997). It should be noted that a hexagonal Z phase also exists at a composition close to that of the i phase in the Zn-Mg-RE system and is shown to be structurally related to the i phase (Singh et al. 1998). Icosahedral phases have also been reported in the alloys $Zn_{56}Mg_{36}Y_8$ (Kondo et al. 1995) and $Zn_{56.8}Mg_{34.5}RE_{8.7}$ (Fisher et al. 1998). Thus it appears that the composition of the i phase is rigid in Y content (8~10 at.%) but flexible in Mg and Zn contents. Since the high vapor pressure of molten Zn creates a difficulty in the compositional control during the alloy preparation, details in the region (Zn-rich side) involving i phase are not well known. A partial phase diagram involving the i phase along the section $Zn_{60}Mg_{30-x}Y_x$ is shown in Figure 2.15d. The reactions to form the i phase at the stoichiometry of $Zn_{60}Mg_{30}Y_{10}$ are:

$$\begin{array}{lcll} L & \rightarrow & L + (\mathrm{Zn, Mg})_5Y & \text{at } 960\ \mathrm{K} \\ L + (\mathrm{Zn, Mg})_5Y & \rightarrow & i & \text{at } 820\ \mathrm{K}, \end{array}$$

where the $(Zn,Mg)_5Y$ alloy is possibly a hexagonal phase. One should note that at lower Y content, e.g., $Zn_{60}Mg_{37}Y_3$, the i phase solidifies as the primary phase. Annealing of the $Zn_{60}Mg_{37}Y_3$ alloy at 700 K or 620 K for 48 h produces the d phase at the interface between the i phase and the Zn_3Mg_2 crystalline phase. Based on the analysis of these phase diagrams, a common feature emerges: all the stable i phases solidify from the melt through peritectic reactions.

2.6.5 Al-Ni-Co

The constitution of the Al-rich alloys of Al-Ni-Co was studied some time ago (Raynor and Pfeil 1947). More recently, a stable d phase was obtained by conventional casting methods in a wide composition range from $Al_{70}Ni_{20}Co_{10}$ to $Al_{70}Ni_{10}Co_{20}$ (Tsai et al. 1989b). The d phase in this system is easily obtained

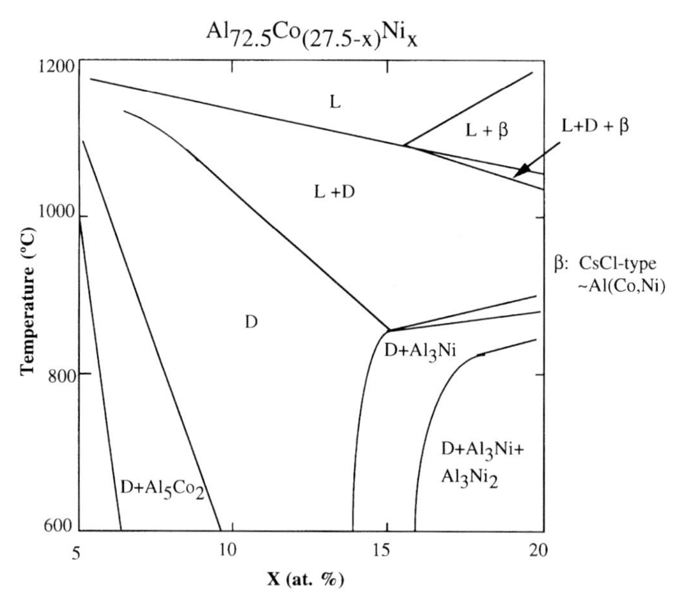

Fig. 2.17. Partial phase diagram at $Al_{72.5}Ni_xCo_{27.5-x}$ section involving a *d* phase in the Al-Ni-Co system. After Gödecke and Ellner (1996, 1997), Scheffer et al. (1998).

and its structure has been investigated extensively. It was shown earlier that the stoichiometric composition of the *d* phase is close to $Al_{70}Ni_{15}Co_{15}$. It has been found recently that at higher temperatures the *d* phase $Al_{70}Ni_{15}Co_{15}$ decomposes into another *d* phase with a composition close to $Al_{72}Ni_{12}Co_{16}$ (Fujiwara et al. 1998) and a β phase with higher Ni and Co contents. Thus, the strict stoichiometry of the *d* phase in the Al-Ni-Co system is close to $Al_{72}Ni_{12}Co_{16}$. A detailed phase diagram cut along the composition line $Al_{72.5}Ni_xCo_{27.5-x}$ is given in Fig. 2.17 (Gödecke and Ellner 1996, 1997, Scheffer et al. 1998). The *d* phase exists as a single phase from $x = 9$ to $x = 14$ and this phase field expands at high temperatures ($x = 7$ to $x = 15$). It should be noted that the *d* phase is the primarily solidified phase for all these compositions. The reactions forming the *d* phase in $Al_{72.5}Ni_{12}Co_{15.5}$ are

$$\begin{array}{llll} L & \rightarrow & L + D & \text{at } 1403\text{ K} \\ L + D & \rightarrow & D & \text{at } 1263\text{ K.} \end{array}$$

There are several variants of the *d* phase in the Al-Ni-Co system which are strongly dependent on both temperature and composition (Fig. 2.6). The distribution of four variants of the *d* phase corresponding to Fig. 2.6 has been investigated in detail and is shown in Fig. 2.18 (Fujiwara et al. 1998). Hiraga et al. (1991a) observed two different structures in an as-cast $Al_{70}Ni_{15}Co_{15}$ sample. Edagawa et al. (1992) observed a temperature dependent phase transition between two *d* structures S1 and S2 and described it as an order-disorder transition. Later, more variants of the *d* phase were found in samples with different compositions which had been treated at different temperatures (Grushko et al. 1994).

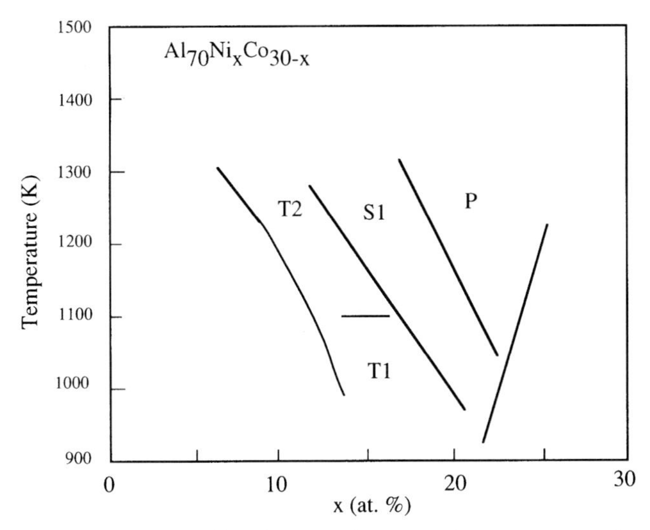

Fig. 2.18. Phase distribution of four variants of the *d* phase shown in Fig. 2.6.

2.7 Growth of Quasicrystals

As QCs are produced by both rapid solidification, their morphologies and growth mechanism can be interpreted in terms of the solidification theory. The nucleation and growth of rapidly solidified *i* phase have been studied in detail in the Al-Mn system and have been reviewed by Schaefer and Bendersky (1988). Here, the morphologies of stable QCs and the solidification mechanism and growth of single-grain samples are mainly described.

2.7.1 Morphologies of Quasicrystals

In most of the early experiments, QCs synthesized had highly dendritic morphologies. In 1986, however, crude pentagonal dodecahedra up to 0.01 mm were observed in Al-Mn-Si (Ishimasa and Nissen 1986, Robertson et al. 1986). Later, those faceted dendrites were observed to resemble stellated dodecahedra (Nishitani et al. 1986, Nissen et al. 1988). The most beautiful morphologies, however, have been observed in the stable phases. The Al-Li-Cu QC exhibits a form of rhombic triacontahedron (Lang et al. 1987, Kortan et al. 1989), as shown in Fig. 2.19. This polyhedron with 30 diamond faces, 32 vertices, and 60 edges, has the symmetry of the icosahedron. The overall morphology can be described as an assembly of either 12 triacontahedra or 12 rhombic icosahedra surrounding a stellate polyhedron with 20 branches. These triacontahedra are generally 0.1 mm across and usually form coarse dendrites in shrinkage cavities of the as-cast samples.

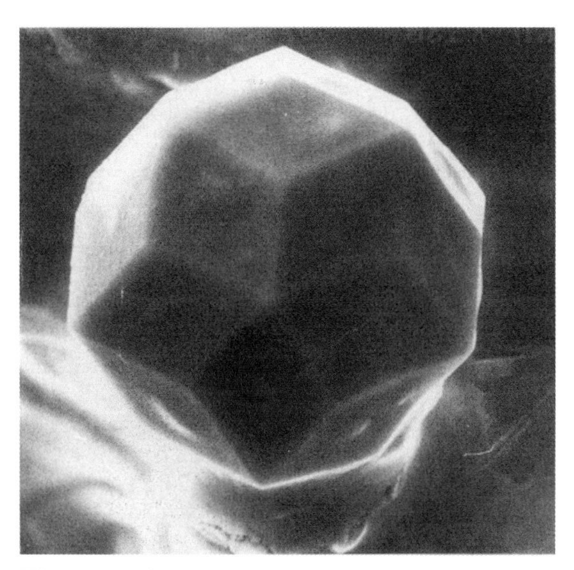

Fig. 2.19. An isolated single i phase single grain with a rhombic triacontrahedral form in Al-Li-Cu alloy. After Kortan et al. (1989).

The formation of the i phase follows a peritectic reaction involving the R phase and the liquid phase (Fig. 2.15b). The coexistence of the R phase and the i phase is observed frequently. Figure 2.20a shows a cubic grain of the R phase with three i QCs coherently nucleated and grown from the (111) edges of the cube. A representative schematic is shown in Fig. 2.20b where there is a coherent epitaxy between the R and the i phase grains. Three mutually perpendicular twofold planes of the i phase grains (denoted A, B, and C) are aligned parallel to the three <100> planes of the cubic grain. All the i grains have one of their threefold directions common with that of the cube (Kortan et al. 1989).

The i phases in the Al-Cu-Fe, Zn-Mg-Ga, Al-Pd-Mn, and Zn-Mg-Y alloys exhibit a nearly regular pentagonal dodecahedral form with a size of $\sim$100 μm. Among these, the morphological forms in the Al-Cu-Fe system are the most interesting. As shown in Fig. 2.15b, the Al-Cu-Fe i phase forms through a complicated process associated with the β phase or the λ phase over a large temperature range ($\sim$200 K). Various forms corresponding to the i phase and the crystalline phases are anticipated. Three morphological forms, dendritic, dodecahedral, and faceted spheres, are observed in solidification in the Al-

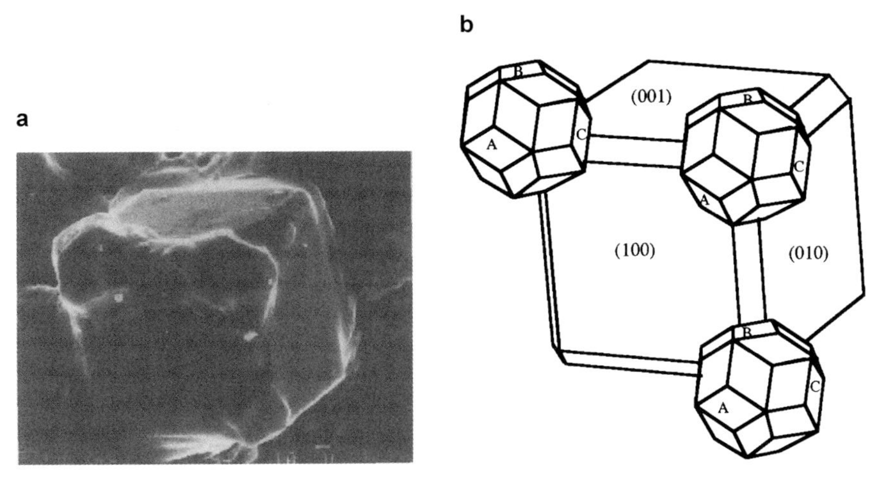

Fig. 2.20. (a) A single-crystal R-phase grain and three isolated but coherent i grains nucleated and grown on the (111) vertices of the R phase. (b) Illustration showing the relationship between the cubic phase and the three icosahedra. The i grains are not only epitaxial with the cubic phase but all have the same orientation. After Kortan et al. (1989).

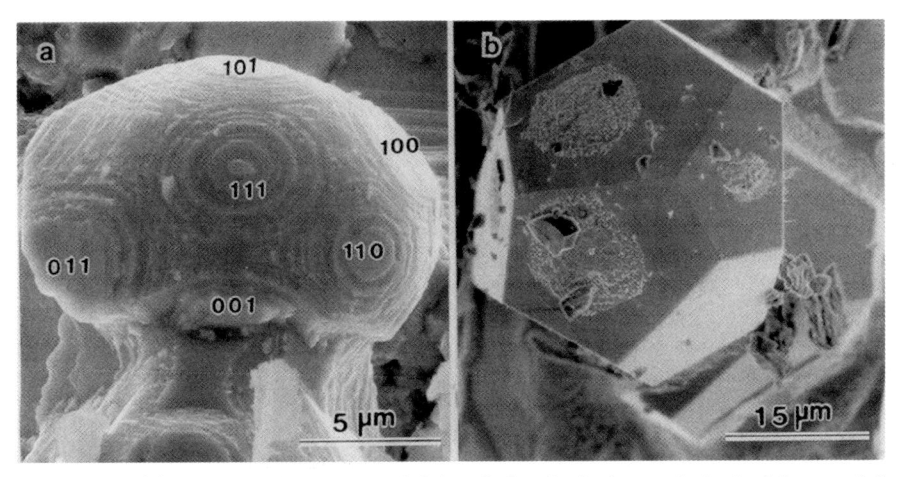

Fig. 2.21. (a) A facetted sphere and (b) a dodecahedral morphological forms of β and i phases in an as-solidified $Al_{65}Cu_{20}Fe_{15}$ alloy.

Cu-Fe system (Balzuweit et al. 1993). The structure of these three forms were verified to be the λ, i , and β phases, respectively. They exist in the phase diagram (Fig. 2.15b). The λ and β phases are the nucleation sites of the i phase. Thus, at a sufficiently slow cooling rate, the λ and β phases would be fully consumed to form the i phase. A faceted sphere can be considered as the primary crystallite. Figure 2.21 shows a faceted sphere and the dodecahedral morphological forms in an as-solidified $Al_{65}Cu_{20}Fe_{15}$ alloy. The faceted sphere is the β phase in which the facet planes can be recognized and indexed. The [110] axis has been found to be coincident with one of the fivefold axes of the i phase. Thus, one can expect that the [110] planes would grow and the [100] and [111] planes would degenerate.

The d phase in the Al-Ni-Co and Al-Cu-Co alloys appear to be thermodynamically stable. By employing conventional techniques, the single d QC can be grown to a size of several mm. Figure 2.22 shows a columnar d prismatic solidification morphology of a single grain in an $Al_{70}Ni_{15}Co_{15}$ alloy. The d phase possesses a 2D quasilattice plane stacked periodically. The d prism morphology brings out the crystallographic symmetry of the d phase. The elongated prismatic axis demonstrates that the growth is fastest along the tenfold axis and slowest along the twofold directions. The d phase in the Al-Pd-Mn system is stable at high temperatures and precipitates from a supersaturated i phase through a solid-state reaction. Figure 2.23 constitutes direct evidence for the precipitation of the d phase from the i phase. It is worth noting that the tenfold axis of the d phase coincides with the fivefold axis of the i phase.

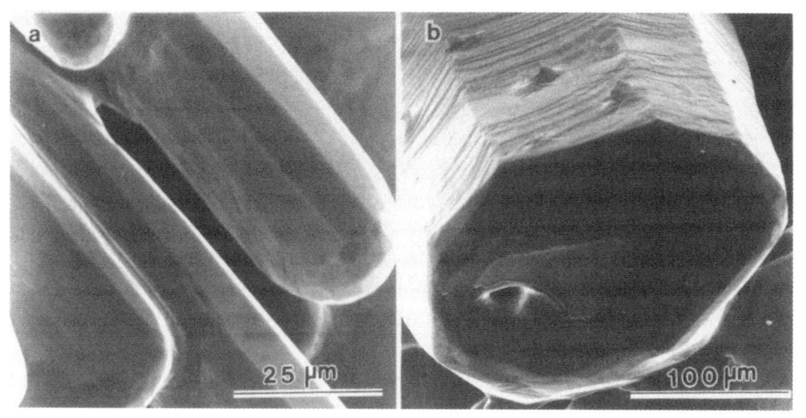

Fig. 2.22. A columnar d prismatic solidification morphology of a single grain of the d phase in an $Al_{70}Ni_{15}Co_{15}$ alloy.

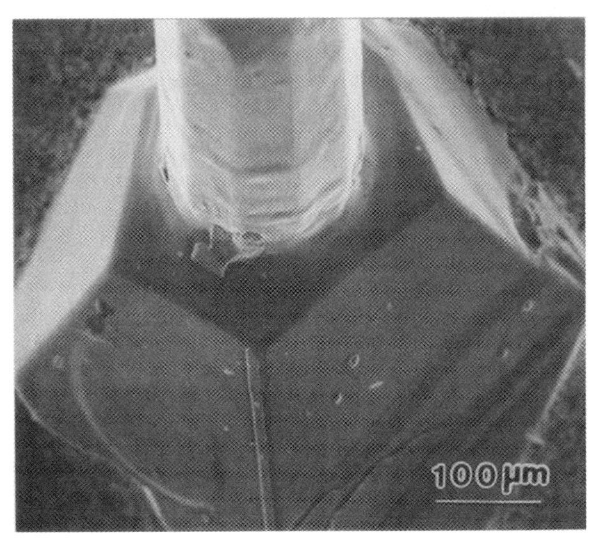

Fig. 2.23. Decagonal phase precipitated from the *i* phase in an Al-Pd-Mn alloy (courtesy of Y. Matsuo).

2.7.2 Solidification of Quasicrystals

With the exception of the Al-Li-Cu *i* phase, all the other *i* phases exhibit a pentagonal dodecahedral morphology. Thus it was believed that the fastest growth direction of this phase is along threefold axes, since the vertices in the pentagonal dodecahedron have threefold symmetry. Moreover, as in the Al-Mn system, the grains of the *i* phase embedded in a matrix of fcc Al have rounded outlines with dendrite arms radiating along the threefold axes (Schaefer et al. 1986). However, an x-ray diffraction study of the rapidly solidified Al-Mn and Al-Cu-Fe alloys with a near single-phase composition revealed that the grains of the *i* phase have a high degree of texture with a fivefold axis along the growth direction (Edagawa et al. 1991).

The disagreement among these observations has been resolved (Tsai et al. 1993b). A fast growth along fivefold directions can lead to an *i* morphology, since the vertices in an icosahedron are fivefold. However, if the growth front becomes planar, this would lead to a truncated icosahedron, which would look like a pentagonal dodecahedron. Various microstructures obtained under different conditions of growth are discussed here.

Figure 2.24 shows a TEM micrograph of a melt-quenched $Al_{72}Pd_{25}Cr_3$ consisting of an *i* phase and a *d* phase. The diffraction patterns along twofold directions are taken from the regions close to the interface between the *i* (A and C) and *d* (B and D) phase grains. As shown by the *i* twofold diffraction patterns from the grains A and C, both these grains have the same orientation. One of the fivefold directions coincides with the planar growth direction of the columnar structure. One of the fivefold axes of the *i* phase

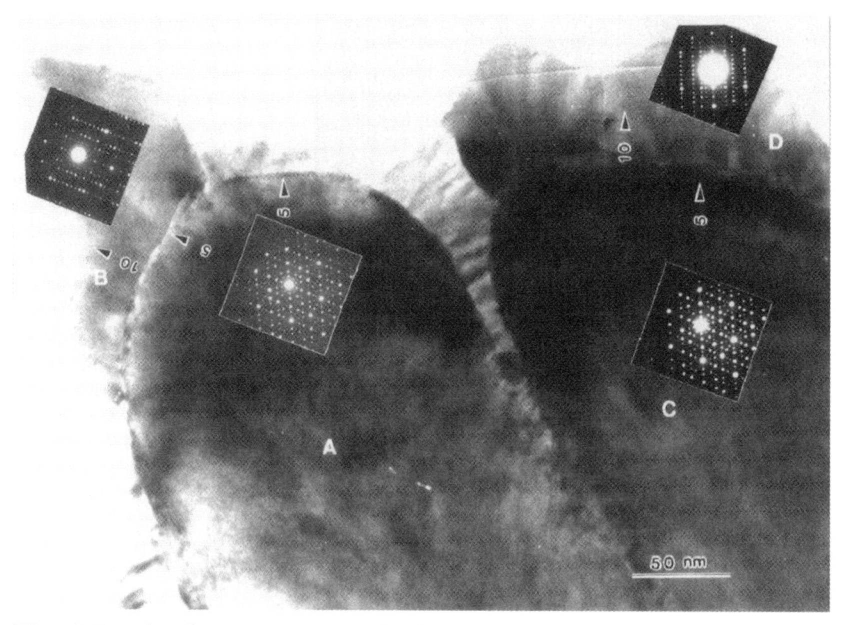

Fig. 2.24. An electron micrograph of a melt-quenched $Al_{72}Pd_{25}Cr_3$ alloy taken along the coincident two fold axes of the *d* and the *i* phases. Each *d* phase grain is observed to be growing with its tenfold axis along one of the fivefold axis of the parent *i* phase. After Tsai et al. (1993b).

coincides with the tenfold axis of a *d* phase grain. The tenfold axis of the other *d* phase grain is parallel to another fivefold axis of the *i* phase. Thus the orientation between the two grains of the *d* phase is about 63.43°, which is equal to the angle between two fivefold axes in an *i* phase. The tenfold axis is the fastest growth direction of the *d* phase, which coincides with the solidification direction.

Figure 2.25 shows a solid-liquid interface during the growth of a single *d* QC using an unidirectional solidification method (floating-zone technique). In the lower part of the micrograph is the grown single grain of the *d* phase, while above it is the part which was liquid during the growth. The interface is almost flat and perpendicular to the tenfold axis. This micrograph was taken in a back-scattered electron mode in a scanning electron microscope. In the image, therefore, different contrasts represent differences in composition. The bright and the dark areas represent areas respectively with lower and higher Al content. The concentration of the liquid just next to the interface fluctuates due to quenching. As a result, the planar growth is replaced by a cellular growth (Guo et al. 1998).

The growth phenomena are interpreted in terms of the classical solidification theory in which the growth morphology is determined by the stability of the solid-melt interface. When the interface changes from stable to unstable,

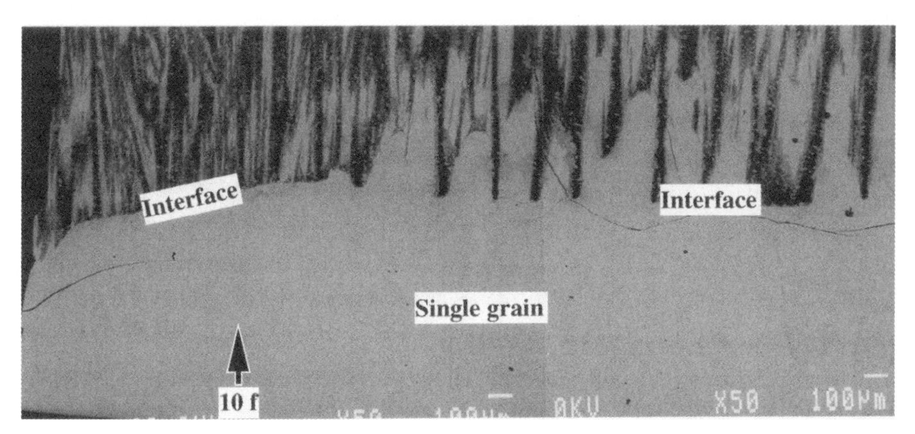

Fig. 2.25. A scanning electron micrograph of a solid-liquid interface of a $Al_{72}Ni_{12}Co_{16}$ alloy, quenched during the growth of the the d phase. The oriented dendritic growth region is the region which was liquid before quenching. After Guo et al. (1998).

the morphology of the crystalline grains changes from equiaxed grains with a planar interface to cellular grains, and then to dendrites. The constitutional supercooling is expressed in an usual form (Fleming 1974)

$$G/R > mC_0(k-1)/Dk, \tag{2.1}$$

where G is the interface temperature gradient, m is the liquidus slope, R is the rate of interface movement, C_0 is the initial solute concentration (here it is the excess Al concentration for the composition of QCs), k is the distribution coefficient (the ratio of the solute concentration to that of the liquid), and D is the diffusion coefficient. If the ratio G/R is smaller than $mC_0(k-1)/Dk$, the interface instability will occur. In the case shown in Fig. 2.25, R was 0.5 mm/h, which is much smaller than G. The term k is very close to unity. The left term of the inequality is thus much larger than the right term. As a result, the interface is stable and a planar growth is favorable. However, if the interface is quenched, R becomes very large, leading to an unstable interface, and thus a cellular growth.

For melt-quenched samples (Fig. 2.24), a stable planar growth can occur at concentrations far greater than those predicted by constitutional superscooling. By taking account of the interface energy, an absolute interface stability criterion has been derived by Mullins and Sekerka (1964) in a dilute alloy:

$$A = k^2 T_m G R/(k-1)mC_0 D, \tag{2.2}$$

where A is an interface stability parameter (being 1 for a stable interface), $G = \sigma/\Delta H$ is the surface tension constant, σ is the interfacial energy, and ΔH is the latent heat. $T_m G$ was estimated to be $\sim 1\times10^{-7}$ mK. Because of

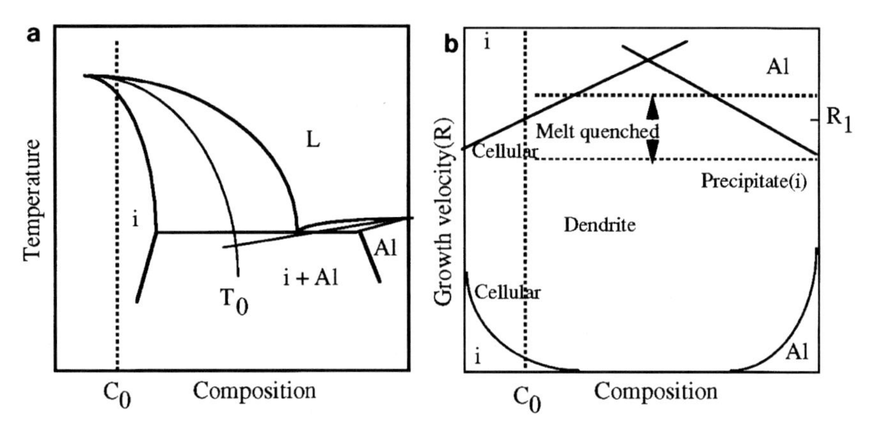

Fig. 2.26. A schematic of (a) the metastable phase equilibria between the i and fcc Al phases and (b) the dependence of microstructure on the growth velocity, R, and the composition C_0. After Tsai et al. (1993b).

the structural similarity between the solid and the melt, the magnitude of σ of the i phase is smaller than that for the crystalline phases. Therefore, the solid-melt interface instability will be higher. As the growth velocity of a dendritic structure approaches that for absolute stability, it is anticipated that the microstructure will change from dendritic to cellular before becoming a planar front.

The constitutional supercooling and absolute criterion are available for conventional cooling and rapid quenching, respectively (Coriel and Sekerka 1980). As an illustration, a metastable equilibrium pseudobinary phase diagram of the i phase and α-Al (fcc) phase in the Al-Mn system is shown in Fig. 2.26a. A schematic microstructure dependence on growth velocity, R, and composition, C_0, is constructed and shown in Fig. 2.26b. At small C_0, when R is small, the interface is stable and planar. This regime is represented by the lower left corner of the diagram. An example of this is the planar interface on slow solidification shown in Fig. 2.25. At larger growth rates, the interface becomes unstable and breaks down into a dendritic morphology. However, at very large R (and $A > 1$), this instablity is suppressed, and a planar front is retained, corresponding to the upper left corner of the diagram in Fig. 2.26b. An example of this is shown in the TEM micrograph of a rapidly solidified sample in Fig. 2.24. Although the i grain is divided into two regions A and C (Fig. 2.24), the growth front (perpendicular to the fivefold directions) is essentially planar. Thus the planar front, which was originally stable, has broken down due to a change in the solidification rate and the composition (this change in the solidification rate and the composition has eventually resulted in the formation of the d phase). This analysis is also in good agreement with the observation of texture in the Al-Mn and the Al-Cu-Fe i phases (Edagawa et al. 1991). With the alloy composition deviated from that of the single phase $Al_{80}Mn_{20}$, a dendritic structure is developed. In the $Al_{86}Mn_{14}$ alloy, a coral-like morphology with dendritic growth along the threefold axis was observed.

2.7.3 Growing Large Single Grains

Several QCs have been found to be thermodynamically stable and this led to the application of conventional techniques to grow large single grains. The first single-grain QC with a size of a cm was produced in the $Al_{5.1}Li_3Cu$ alloy by the Bridgman method (Parsey et al. 1987). In its growth an Al-rich Al_xCuLi_3 melt with $x > 5.1$ was used and critical temperature gradients and growth velocities were employed. This led to the growth of the *i* phase in preference to the R phase. The large *i* single grain produced was surrounded by eutectic phases. Although the largest single-grain regions were found to be greater than 1 cm, a lot of pin holes and precipitates were dispersed inside the *i* grain. These *i* single grains gave broad diffraction peaks, indicating a highly disordered structure.

Great progress in growing single-grain QCs came after the discovery of QCs in the Al-Pd-Mn system. A single-grain *i* QC, 1 cm in diameter and 5 cm long, was grown using the composition $Al_{70}Pd_{20}Mn_{10}$ by the Czochralski method (Yokoyama et al. 1992). Following this work, single QCs in this system have been grown using the Bridgman method (de Boissieu et al. 1992) and by using near equilibrium growth conditions of a 0.5 K/h cooling rate (Kortan et al. 1993). The structural quality of these single grains was evident in the back-scattered Laue x-ray patterns. Dynamical x-ray diffraction, which is very sensitive to structural imperfections, has further demonstrated that the single grains in this system are almost defect-free (Kycia et al. 1993).

Recently, single-grain *i* QCs with a size of about 3–5 mm have been grown by slow cooling methods in the Al-Cu-Fe (Lograsso and Delaney 1996) and the Zn-Mg-Y (Fisher et al. 1998) systems. A single cm-size grain in the Zn-Mg-Ho system has been produced by the Bridgman method (Sato et al. 1998c). These single-grain QCs reveal structural quality as good as that of the Al-Pd-Mn system. Figure 2.27 shows the transmission x-ray Laue patterns from a Zn-Mg-Y *i* QC produced by the Bridgman method. The patterns reveal fivefold, threefold, and twofold symmetries originating from the structure.

In the case of the *d* phase, the composition of the QC is incongruent to the melt. Therefore the Bridgman and Czochralski methods cannot be used for growing a large single grain in the Al-Ni-Co system. Single grains of ~5 mm were grown under equilibrium conditions (slow cooling), with cooling rates of a few K/h in the $Al_{70}Ni_{15}Co_{15}$ alloy (Kortan et al. 1993). A floating-zone method was employed to grow a single grain with a cm size on the optimal composition of $\sim Al_{72}Ni_{12}Co_{16}$, at which the *d* phase is a primarily solidified phase. The single-grain *d* QC produced by this method reveals a high structural quality as characterized by the back Laue scattering (Fig. 2.28) and neutron diffraction (Sato et al. 1998b). Remarkably, the neutron diffraction

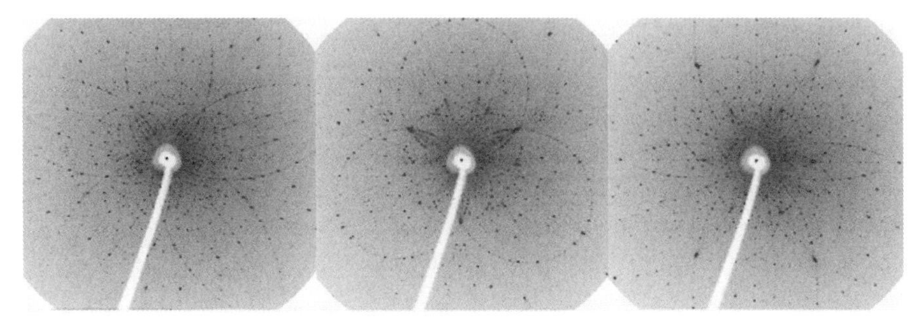

Fig. 2.27. X-ray transmission Laue patterns of the Zn-Mg-Y *i* QC.

exhibits a mosaic spread of as small as ~0.2°. The morphology of this *d* QC confirms that the natural growth direction is parallel to the tenfold axis. The two twofold diffraction patterns shown in Figs. 2.28b and 2.28c were obtained at every 36° around the tenfold axis, confirming the *d* symmetry of this large single-grain QC.

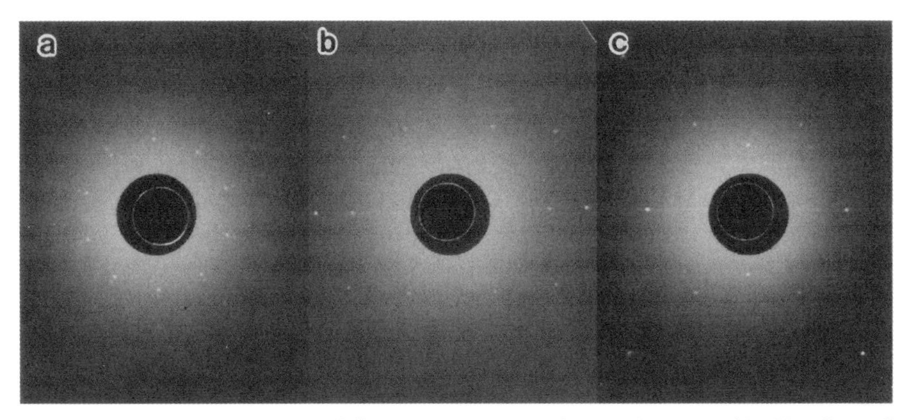

Fig. 2.28. X-ray back Laue diffraction patterns of a single-grain $Al_{72}Ni_{12}Co_{16}$ *d* QC. After Sato et al. (1998b).

2.8 Summary

Recent studies on the metallurgy of QCs have been reviewed. QCs can be formed in several alloy systems by several routes, such as melt solidification, solid state reactions, and vapor condensation. Quasicrystalline phases can be metastable or stable, depending on the alloy system and the composition.

There are 3D (icosahedral) and 2D QCs. Electron diffraction patterns show that the *i* QCs exist either as F-type or P-type lattices, corresponding to a face-centered or primitive hypercubic lattice in 6D space. 2D QCs exhibiting *d*, octagonal, and dodecagonal symmetries have been observed in various alloys. Several variants of the *d* phase have been observed in the Al-Ni-Co system and these are very sensitive to the composition and temperature.

Almost all stable *i* QC-forming alloys have the e/a ratio of about 1.75 in Al-TM class and 2.1 in Zn-Mg-Al class. Thus, the stable *i* QCs can be empirically described as Hume-Rothery electronic compounds.

The tranformation kinetics of *a* phase to QCs has been described. The transformation is a polymorphous reaction which is dominated by the interface in the Al-Cu-V alloy, whereas it is a diffusion-controlled reaction in the Zr-Al-Cu-Ni alloy.

Phase diagrams have been redrawn to include QCs. None of the *i* QCs crystallizes from melt as the primary phase. They are formed through peritectic reactions. The Al-Ni-Co system shows primary crystallization of the *d* phase at a composition around $Al_{72}Ni_{12}Co_{16}$.

Large ($\sim$1 cm in size) and high structural quality single-grain QCs have been grown by conventional crystal growth techniques in the Al-Pd-Mn and the Al-Ni-Co alloys. The availability of large single grains of other QCs made by crystal growth techniques is also expected.

Acknowledgments

Special thanks are due to Dr. Alok Singh for a critical reading of the manuscript and arranging its layout. I am grateful to Drs. T.J. Sato, H. Takakura, E. Abe, and J. Guo for help and stimulating discussions. Thanks are also due to Drs. K. Saito, K. Tsuda, M. Tanaka, K.H. Kuo, A.R. Kortan, H.S. Chen, B. Harbrecht, and Y. Matsuo for making available original micrographs used in this article. The author is grateful to "CREST", Japan Science and Technology Cooperation, for financial support.

References

Akiyama, H., Honda, Y., Hashimoto, T., Edagawa, K., Takeuchi, S. (1993): Jpn. J. Appl. Phys. **32**, L1003

Alisowa, S.P., Kovneristyi, Yu. K., Lasarewa, J.E., Budberg, P.B. (1990): Dokl. Akad. Nauk. SSSR **315**, 116

Audier, M., Durand-Charre, M., de Boissieu, M. (1993): Philos. Mag. B **68**, 607

Avrami, M. (1939): J. Chem. Phys. **7**, 1103

Ball, M.D., Lloyd, D.J. (1985): Scr. Metall. **19**, 1065

Balzuweit, K., Meeks, H., Van Tendeloo, G., De Boer, J.L. (1993): Philos. Mag. B **67**, 513

Bancel, P.A. (1991): in Quasicrystals: State of the Art, Vincenzo, D.P., Steinhardt, P.J. (eds). World Scientific, Singapore, p 17
Bancel., P.A., Heiney, P.A. (1986): Phys. Rev. B **33**, 7917
Beeli, C., Nissen, H.-U., Robadey, J. (1991): Philos. Mag. Lett. **63**, 87
Bendersky, L.A. (1985): Phys. Rev. Lett. **55**, 1461
Bendersky, L.A., Kaufman, M.J. (1986): Philos. Mag. B **53**, L75
Bendersky, L.A., Biancaniello, F.S., (1987): Scr. Metall. **21**, 531
Bergman, G., Waugh, J.L.T., Pauling, L. (1957): Acta Crystallogr. **10**, 254
Blanpain, B., Mayer, J.W. (1990): in Thin Film Structures and Phase Stability, Clemens, B.M., Johnson, W.L. (eds), Mater. Res. Soc. Symp. Proc. **187**, 21
Bradley, A.J., Goldschmidt, H.J. (1939): J. Inst. Met. **65**, 403
Cahn, J.W., Shechtman, D., Gratias, D. (1986): J. Mater. Res. **1**, 13
Calvayrac, Y., Quivy, A., Bessiere, M., Lefebvre, S., Cornier-Quiquandon, M., Gratias, D. (1990): J. Phys. (Paris) **51**, 417
Cao, W., Ye, H.Q., Kuo, K.H. (1988): Phys. Status Solidi A **107**, 511
Chattopadhyay, K., Lele, S., Ranganathan, S., Subbanna, G.N., Thangaraj, N. (1985): Current Science **54**, 895
Chen, H.S., Chen, C.H. (1986): Phys. Rev. B **33**, 2814
Chen, H.S., Phillips, J.C., Villars, P., Kortan, A.R., Inoue, A. (1987a): Phys. Rev. B **35**, 9326
Chen, H.S., Koskenmaki, D., Chen, C.H. (1987b): Phys. Rev. B **35**, 3715
Chen, H.S., Kortan, A.R., Parsey Jr., J.M. (1987c): Phys. Rev. B **36**, 7681
Chien, C.L., Lu, M. (1992): Phys. Rev. B **45**, 12 793
Cooper, M., Robinson, K. (1966): Acta Crystallogr. **20**, 614
Coriell, S.R., Sekerka, R.F. (1980): in Rapid Solidification and Technology II, Mehrabian, R., Kear, B.H., Cohen, M. (eds). Claitors, Baton Rouge, Los Angeles, p 35
de Boissieu, M., Rurand-Charre, M., Bastie, P., Cacabelli, A., Boudard, M., Bessiere, M., Lefebvre, S., Janot, C., Audier, M. (1992): Philos. Mag. Lett. **65**, 147
Delaney, D.W., Bloomer, T.E., Lograsso, T.A. (1997): in New Horizons in Quasicrystals: Research and Applications, Goldman, A.I., Sordelet, D.J., Thiel, P.A., Dubois, J.M. (eds). World Scientific, Singapore, p 45
Dong, C., Hei, Z.K., Wang, L.B., Song, Q.H., Wu, Y.K., Kuo, K.H. (1986): Scr. Metall. **20**, 1155
Dubost, B., Lang, J.M., Tanaka, M., Sainfort, P., Audier, M. (1986): Nature **324**, 48
Dunlap, R.A., Dini, K. (1986): J. Mater. Res. **1**, 415
Ebalard, S., Spaepen, F. (1989): J. Mater. Res. **4**, 39
Eckert, J., Schultz, L., Urban, K. (1991): Mater. Sci. Eng. A **133**, 393
Edagawa, K., Sugawara, T., Tokoh, S., Seki, F., Oda, K., Ito, K., Ino, H. (1991): J. Jpn. Inst. Met. **55**, 607
Edagawa, K., Ichihara, M., Suzuki, K., Takeuchi, S. (1992): Philos. Mag. Lett. **66**, 19
Elser, V., Henley, C.L. (1985): Phys. Rev. Lett. **55**, 2883
Faudot, F., Quivy, A., Calvayrac, Y., Gratias, D., Harmelin, M. (1991): Mater. Sci. Eng. A **133**, 383
Fisher, I.R., Islam, Z., Panchula, A.F., Cheon, K.O., Kramer, M.J., Canfield, P.C., Goldman, A.I. (1998): Philos. Mag. B **77**, 1601
Fleming, M.C. (1974): Solidification Process. McGraw-Hill, New York, p 273
Follstaedt, D.M., Knapp, J.A. (1986): Phys. Rev. Lett. **56**, 1827
Frank, F.C. (1952): Proc. R. Soc. London, Ser. A **215**, 43

Fujiwara, A., Inoue, A., Tsai, A.P. (1998): in Proceedings of the 6th International Conference on Quasicrystals, Takeuchi, S., Fujiwara, T. (eds). World Scientific, Singapore, p 341
Gayle, F.W., Shapiro, A.J., Biancaniello, F.S., Boettinger, W.J. (1992): Metall. Trans. A **23**, 2409
Gödecke, T., Lück, T. (1995): Z. Metallk. **86**, 109
Gödecke, T., Ellner, M. (1996): Z. Metallk. **87**, 854
Gödecke, T., Ellner, M. (1997): Z. Metallk. **88**, 382
Gratias, D., Calvayrac, Y., Devaud-Rzepski, Q., Faudot, F., Hermelin, M., Quivy, A., Bancel, P. (1993): J. Non-Cryst. Solids **153-154**, 482
Grushko, B., Urban, K. (1994): Philos. Mag. B **70**, 1063
Grushko, B., Wittenberg, R., Holland-Moritz, D. (1996): J. Mater. Res. **11**, 2177
Guo, J., Sato, T.J., Hirano, T., Tsai, A.P. (1998): unpublished
Guryan, C.A., Goldman, A.I., Stephens, P.W., Hiraga, K., Tsai, A.P., Inoue, A., Masumoto, T. (1989): Phy. Rev. Lett. **62**, 2409
Hardy, H.K., Silcock, J.M. (1955-56): J. Inst. Met. **84**, 423
He, L.X., Li, X.Z., Zhang, Z., Kuo, K.H. (1988a): Phys. Rev.Lett. **61**, 1116
He, L.X., Wu, Y.K., Kuo, K.H. (1988b): J. Mater. Sci. Lett. **7**, 1284
Heiney, P.A., Bancel, P.A., Horn, P.M., Jordan, J.L., Laplaca, S., Angilello, J., Gayle, F.W. (1987): Science **238**, 660
Henley, C.L. (1988): Philos. Mag. Lett. **58**, 87
Henley, C.L., Elser, V. (1986): Philos. Mag. B **53**, 59
Hiraga, K., Lincoln, F.J., Sun, W. (1991a): Mater. Trans., Jpn. Inst. Met. **32**, 308
Hiraga, K., Sun, W., Lincoln, F.J., Kaneko, M., Matsuo, Y., (1991b): Jpn. J. Appl. Phys. **30**, 2028
Holzer, J.C., Kelton, K.F. (1991): Acta Metall. Mater. **39**, 1833
Hume-Rothery, W. (1926): J. Inst. Met. **36**, 295
Ichikawa, N., Matsumoto, O., Hara, T., Kitahara, T., Yamauchi, T., Matsuda, T., Takeuchi, T., Mizutani, U. (1994): Jpn. J. Appl. Phys. **33**, L736
Inoue, A., Kimura, H.M., Masumoto, T. (1987): J. Mater. Sci. Lett. **22**, 1758
Inoue, A., Bizen, Y., Masumoto, T. (1988): Metall. Trans. A **19**, 383
Ishimasa, T., Nissen, H.U. (1986): Phys. Scr. T **13**, 291
Ishimasa, T., Nissen, H.U., Fukano, Y. (1985): Phys. Rev. Lett. **55**, 511
Ishimasa, T., Fukano, Y., Tsuchimori, M., (1988): Philos. Mag. Lett. **58**, 157
Johnson, W.A., Mehl, R.F. (1939): Trans. Am. Inst. Min. Met. Mining Eng. **134**, 416
Jones, H. (1937): Proc. Phys. Soc., London **49**, 250
Kafolt, D., Nanao, S., Egami, T., Wong, K.M., Poon, S.J. (1986): Phys. Rev. Lett. **57**, 114
Kelton, K.F., Kim, Y.J., Stroud, R.M. (1997): Appl. Phys. Lett. **70**, 3230
Kimura, K., Hashimoto, T., Suzuki, K., Nagayama, K., Ino, H., Takeuchi, S. (1985): J. Phys. Soc. Jpn. **54**, 3217
Kittel, C. (1995): Introduction to Solid State Physics, 7th ed. John Wiley and Sons, New York
Klein, T., Symko, O.C. (1994): Appl. Phys. Lett. **64**, 431
Kondo, R., Honda, Y., Hashimoto, T., Edagawa, K., Takeuchi, S. (1995): in Proceedings of the 5th International Conference on Quasicrystals, Janot, C., Mosseri, R. (eds). World Scientific, Singapore, p 476
Kortan, A.R., Chen, H.S., Parsey Jr., J.M., Kimerling, L.C. (1989): J. Mater. Sci. **24**, 1999
Kortan, A.R., Thiel, F.A., Kopylov, N., Chen, H.S. (1993): J. Cryst. Growth **128**, 1086

Koshikawa, N., Sakamoto, S., Edagawa, K., Takeuchi, S. (1992): Jpn. J. Appl. Phys. **31**, L966
Köster, U., Meinhardt, J., Roos, S., Liebertz, H. (1996): Appl. Phys. Lett. **69**, 179
Krumeich, F., Conrad, M., Harbrecht, B. (1994): in Electron Microscopy 1994, Proceedings of the 13th International Congress on Electron Microscopy, Vol. 2A, Jouffrey, B., Colliex, C. (eds). Les Editions de Physique, Les Ulis, p 751
Kycia, S.W., Goldman, A.I., Lograsso, T.A., Delaney, D.W., Black, D., Sutton, M., Dufresne, E., Brüning, R., Rodricks, B. (1993): Phys. Rev. B **48**, 3544
Lang, J.M., Audier, M., Dubost, B., Sainfort, P. (1987): J. Cryst. Growth **83**, 456
Langsdorf, A., Ritter, F., Assmus, W. (1997): Philos. Mag. Lett. **75**, 381
Lilienfeld, D.A., Nastasi, M., Johnson, H.H., Ast, D.G., Mayer, J.W. (1985): Phys. Rev. Lett. **55**, 1587
Lilienfeld, D.A., Nastasi, M., Johnson, H.H., Ast, D.G., Mayer, J.W. (1986): J. Mater. Res. **1**, 237
Lograsso, T.A., Delaney, D.W. (1996): J. Mater. Res. **11**, 2125
Luo, Z., Zhang, S., Tang, Y., Zhao, D. (1993): Scr. Metall. **28**, 1513
Matsubara, E., Waseda, Y., Tsai, A.P., Inoue, A., Masumoto, T. (1988): Z. Naturforsch. A **43**, 505
Mizutani, U., Sakabe, Y., Matsuda, T. (1990a): J. Phys. Condens. Matter. **2**, 6153
Mizutani, U., Sakabe, Y., Shibuya, T., Kishi, K., Kimura, K., Takeuchi, S. (1990b): J. Phys. Condens. Matter **2**, 6169
Mizutani, U., Takeuchi, T., Fukunaga, T. (1993): Mater. Trans., Jpn. Inst. Met. **34**, 102
Monokanov, V.V., Chebotnikov, V.N. (1990): J. Non-Cryst. Solids **117-118**, 789
Mukhopadhyay, N.K., Subbanna, G.N., Ranganathan, S., Chattopadhyay, K. (1986): Scr. Metall. **20**, 525
Mukhopadhyay, N.K., Ranganathan, S., Chattopadhyay, K. (1987): Philos. Mag. Lett. **56**, 121
Mullins, W.W., Sekerka, R.F., (1964): J. Appl. Phys. **35**, 444
Nelson, D.R., Halperin, B.I. (1985): Science **229**, 233
Niikura, A., Tsai, A.P., Inoue, A., Masumoto,T. (1994): Philos. Mag. Lett. **69**, 351
Nishitani, S.R., Kawaura, H., Kobayashi, K.F., Shingu, P.H. (1986): J. Cryst. Growth **76**, 209
Nissen, H.U., Wessicken, R., Beeli, C., Csanady, A., (1988): Philos. Mag. B **57**, 587
Ohashi, W., Spaepen, F. (1987): Nature **330**, 555
Padezhnova, E.M., Mel'nik, E.V., Dobatkina, T.V. (1979): Russ. Metall. (Engl. Transl.) No. 1 179
Padezhnova, E.M., Mel'nik, E.V., Miliyevskiy, R.A., Dobatkina, T.V., Kinzhibalo, V.V. (1982): Russ. Metall. (Engl. Transl.) No. 4 185
Parsey Jr., J.M., Chen, H.S., Kortan, A.R., Thiel, F.A., Miller, A.E., Farrow, R.C. (1987): J. Mater. Res. **3**, 233
Pauling, L. (1938): Phys. Rev. **54** 899
Phillips, H.W.L. (1953-54): J. Inst. Met. **82**, 197
Poon, S.J., Drehman, A.J., Lawless, K.R. (1985): Phys. Rev. Lett. **55**, 2324
Ramachandrarao, P., Sastry, G.V.S. (1985): Pramāna **25**, L225
Raynor, G.V.S. (1949): Prog. Met. **1**, 1
Raynor, G.V.S., Pfeil, P.C.L. (1947): J. Inst. Met. **73**, 609
Ritsch, S. (1996): PhD Thesis, Swiss Federal Institute of Technology Zürich
Ritsch, S., Beeli, C., Nissen, H.-U., Luck, R. (1995): Philos. Mag. Lett. **71**, 671
Robertson, J.L., Mesenheimer, M.E., Moss, M.E., Bendersky, L.A. (1986): Acta Metall. **34**, 2177
Rokhsar, D.S., Wright, D.C., Mermin, N.D. (1987): Phys. Rev. B **35**, 5487
Saito, Y., Chen, H.S., Mihama, K., (1986): Appl. Phys. Lett. **48**, 581

Saito, M., Tanaka, M., Tsai, A.P., Inoue, A., Masumoto, T. (1992): Jpn. J. Appl. Phys. A **31**, L109
Sastry, G.V.S., Rao, V.V., Ramachandrarao, P., Anantharaman, T.R. (1986): Scr. Metall. **20**, 191
Sato, T.J., Abe, E., Tsai, A.P. (1997): Jpn. J. Appl. Phys. A **36**, L1038
Sato, T.J., Abe, E., Tsai, A.P. (1998a): Philos. Mag. Lett. **77**, 213
Sato, T.J., Hirano, T., Tsai, A.P. (1998b): J. Cryst. Growth, in press
Sato, T.J., Takakura, H., Tsai, A.P. (1998c): Jpn. J. Appl. Phys. **37**, L663
Scheffer, M., Gödecke, T., Lück, R., Ritsch, S., Beeli, C. (1998): Z. Metallk. **89**, 270
Schaefer, R.J., Bendersky, L.A. (1988): in Introduction to Quasicrystals, Vol. 1, M.V. Jarić (ed). Academic Press, Boston, p 111
Schaefer, R.J., Bendersky, L.A., Shechtman, D., Boetinger, W.J., Biancaaiello, F.S., (1986): Metall. Trans. A **17**, 2117
Shechtman, D., Blech, I.A. (1985): Metall. Trans. A **16**, 1005
Shechtman, D., Blech, I., Gratias, D., Cahn, J.W. (1984): Phy. Rev. Lett. **53**, 195
Shen, Y., Poon, S.J., Shiflet, G. J. (1986): Phys. Rev. B **34**, 3516
Singh, A., Abe, E., Tsai, A.P. (1998): Philos. Mag. Lett. **77**, 95
Steinhardt, P.J., Nelson, D.R., Ronchetti, M. (1983): Phys. Rev. B **28**, 784
Stephens, W., Goldman, I. (1986): Phys. Rev. Lett. **56**, 1168
Steurer, W. (1996): in Physical Metallurgy, Vol. 1, Cahn, R.W., Haasen (eds). Elsevier Science BV, Amsterdam, p 408
Tsai, A.-P., Inoue, A., Masumoto, T. (1987a): Jpn. J. Appl. Phys. **26**, 1505
Tsai, A.-P., Inoue, A., Masumoto, T. (1987b): Jpn. J. Appl. Phys. **26**, 1994
Tsai, A.-P., Inoue, A., Masumoto, T. (1988): Jpn. J. Appl. Phys. **27**, L1587
Tsai, A.-P., Inoue, A., Masumoto, T. (1989a): Mater. Trans., Jpn. Inst. Met. **30**, 150
Tsai, A.-P., Inoue, A., Masumoto, T., (1989b): Mater. Trans., Jpn. Inst. Met. **30**, 463
Tsai, A.P., Inoue, A., Bizen, Y., Masumoto, T. (1989c): Acta Metall. **37**, 1443
Tsai, A.P., Yokoyama, Y., Inoue, A., Masumoto., T. (1990a): Philos. Mag. Lett. **61**, 9
Tsai, A.P., Inoue, A., Yokoyama, Y., Masumoto., T. (1990b): Mater. Trans., Jpn. Inst. Met. **31**, 98
Tsai, A.-P., Yokoyama, Y., Inoue, A., Masumoto, T (1990c): Jpn. J. Appl. Phys. **29**, L1161
Tsai, A.P., Yokoyama, Y., Inoue, A., Masumoto, T. (1991): J. Mater. Res. **6**, 2646
Tsai, A.P., Sato., A., Yamamoto, A., Inoue, A., Masumoto, T. (1992a): Jpn. J. Appl. Phys. **31**, L970
Tsai, A.P., Yamamoto, A., Masumoto, T. (1992b): Philos. Mag. Lett. **66**, 203
Tsai, A.P., Tsurui, T., Memezawa, A., Aoki, K., Inoue, A., Masumoto, T. (1993a): Philos. Mag. Lett. **67**, 393
Tsai, A.P., Chen, H.S., Inoue, A. Masumoto, T., (1993b): J. Non-Cryst. Solids **153-154**, 513
Tsai, A.P., Niikura, A., Inoue, A., Masumoto, T., Nishida, Y., Tsuda, K., Tanaka, M. (1994a): Philos. Mag. Lett. **70**, 169
Tsai, A.P., Hiraga, K., Inoue, A., Masumoto, T., Chen, H.S. (1994b): Phys. Rev. B **49**, 3569
Tsai, A.P., Niikura, A., Inoue, A., Masumoto, T. (1997): J. Mater. Res. **12**, 1468
Tsuda, K., Saito, M., Terauchi, M., Tanaka, M., Tsai, A.P., Inoue, A., Masumoto, T. (1993): Jpn. J. Appl. Phys. A **32**, 129
Villars, P., Phillips, J.C., Chen, H.S. (1986): Phys. Rev. Lett. **57**, 3085
Wanger, J.L., Biggs, B.D., Wong, K.M., Poon, S.J. (1988): Phys. Rev. B **38**, 7436
Wang, N., Chen, H., Kuo, K.H. (1987): Phys. Rev. Lett. **59**, 1010

Yokoyama, Y., Tsai, A.-P., Inoue, A., Masumoto, T., Chen, H.S. (1991a): Mater. Trans., Jpn. Inst. Met. **32**, 421
Yokoyama, Y., Tsai, A.-P., Inoue, A., Masumoto, T. (1991b): Mater. Trans., Jpn. Inst. Met. **32**, 1089
Yokoyama, Y., Miura, T., Tsai, A.-P., Inoue, A., Masumoto, T. (1992): Mater. Trans., Jpn. Inst. Met. **33**, 97
Yoshida, K., Taniguchi, Y. (1991): Philos. Mag. Lett. **63**, 127
Yoshioka, A., Edagawa, K., Kimura, K., Takeuchi, S. (1995): Jpn. J. Appl. Phys. **34**, 1606
Zhang, H., Kuo, K.H. (1990): Phys. Rev. B **41**, 3482

3. Crystallography of Quasicrystals

Walter Steurer and Torsten Haibach

3.1 Introduction

Quasiperiodic structures (QSs), incommensurately modulated structures (IMSs) and composite structures (CSs) are the main types of aperiodic crystal structures (Cummins 1990, van Smaalen 1995, Axel and Gratias 1995, Yamamoto 1996, Steurer and Haibach 1998). Diffraction patterns of aperiodic crystals consist of sharp Bragg reflections like those of regular periodic crystals. In contrast, aperiodic crystal structures lack lattice periodicity (Burzlaff et al. 1992, Wondratschek 1992) in d-dimensional (dD) physical space. It was first shown by de Wolff (1974) that diffraction patterns of IMSs can be described as projections of appropriate n-dimensional (nD) reciprocal lattices onto physical space. Consequently, the direct space IMSs result from cutting nD periodic hypercrystals with physical space. All types of aperiodic crystals (excluding almost periodic structures) can be described entirely by the same nD approach. Crystallography of quasicrystals (QCs), therefore, corresponds mainly to crystallography extended to n dimensions. Hermann (1949) was the first to study the number of dimensions necessary for the existence of m-fold rotational symmetry preserving translation symmetry. For instance, a lattice being invariant under five-, eight- or twelvefold rotational symmetry has to be at least four-dimensional. In general, any aperiodic structure with a Fourier module corresponding to a $\mathbb{Z}$-module of rank n, can be embedded in n dimensions to achieve a lattice periodic structure. Symmetry restrictions of the dD physical space drastically limit the number of symmetry groups needed in the corresponding nD symmetry analysis. A comprehensive description of the symmetry operations for nD periodic and quasiperiodic structures has been given, e.g., by Janssen (1992a).

The purpose of this chapter is to provide crystallographic background for analysis, description, and interpretation of QSs. In the first section the concept of nD crystallography is briefly introduced and illustrated with simple examples. The subsequent sections focus on the crystallography of three main types of QCs: planar QCs, i.e., QSs with a quasiperiodic stacking of lattice periodic layers; axial QCs, i.e., QS with a periodic stacking of quasiperiodic layers; and icosahedral QCs. The discussion of axial QCs will be restricted to the decagonal case, as most of the experimentally observed axial QS belong to this type. For each type embedding, metrics and symmetry in direct and reciprocal space are discussed in a way to provide experimentalists with practical information. The relationships of these types of QS to their rational approximants, i.e., their corresponding periodic crystal structures, and their periodic average structures will be shown in detail. The 1D Fibonacci sequence as well as 2D and 3D PT's are used as basic examples of QSs.

3.2 *N*-Dimensional Description of Quasicrystals

Aperiodic crystals lack lattice periodicity in physical space. Their Fourier spectrum $M_{\mathrm{F}}^* = \{F(\boldsymbol{H})\}$ consists of δ-peaks on

$$M^* = \left\{ \boldsymbol{H} = \sum_{i=1}^{n} h_i \boldsymbol{a}_i^* \middle| h_i \in \mathbb{Z} \right\}, \tag{3.1}$$

a $\mathbb{Z}$-module (an additive group) of rank n ($n > d$) with basis vectors $\boldsymbol{a}_i^*$, $i = 1, \ldots, n$. In the embedding approach n determines the minimal dimension of the embedding space and d the one of the aperiodic crystal. In our considerations the dimension of the aperiodic crystal always equals the dimension of 3D physical space $V^{\parallel}$.

3.2.1 Embedding of Direct and Reciprocal Space

The nD embedding space can be separated into two orthogonal subspaces both preserving the point symmetry according to the nD space group

$$V = V^{\parallel} \oplus V^{\perp} \tag{3.2}$$

with $V^{\parallel} = \mathrm{span}(\boldsymbol{v}_1, \boldsymbol{v}_2, \boldsymbol{v}_3)$ and $V^{\perp} = \mathrm{span}(\boldsymbol{v}_4, \ldots, \boldsymbol{v}_n)$. If not indicated explicitly the V-basis will refer to a Cartesian coordinate system. The n-star of rationally independent vectors defining the $\mathbb{Z}$-module M^* can be considered as appropriate *projection* $\boldsymbol{a}_i^* = \pi^{\parallel}(\boldsymbol{d}_i^*)$ ($i = 1, \ldots, n$) of the basis vectors $\boldsymbol{d}_i^*$ of an nD reciprocal lattice Σ^* with

$$M^* = \pi^{\parallel}(\Sigma^*) \; . \tag{3.3}$$

The decomposition [Eq. (3.2)] has to keep the orthogonal subspaces invariant under the symmetry operations $\Gamma(R)$ of the nD point group $K^{n\mathrm{D}}$ of Σ^*. These restrictions have the important consequence that only a small subset of all nD symmetry groups is necessary to describe the symmetry of aperiodic crystals in the nD approach.

The two invariant subspaces are defined by the eigenvectors of the symmetry operators. The reduced symmetry operators are obtained by the similarity transformation:

$$W\Gamma(R)W^{-1} = \Gamma^{\mathrm{red}}(R) = \Gamma^{\parallel}(R) \oplus \Gamma^{\perp}(R); \; R \in K^{n\mathrm{D}} \; . \tag{3.4}$$

The reduced symmetry matrix is blockdiagonal consisting of the symmetry operators of each subspace. The columns of W are the Euclidean basis vectors, while the blocks of rows can be considered as projectors $\pi^{\parallel}$ and $\pi^{\perp}$ onto $V^{\parallel}$ and $V^{\perp}$, respectively. In direct space the aperiodic crystal structure results from a *cut* of a periodic nD *hypercrystal* with dD physical (parallel) space $V^{\parallel}$ (cf. Janssen 1988). An nD hypercrystal corresponds to an nD lattice Σ decorated with nD *hyperatoms*. The basis vectors of Σ are obtained via the orthogonality condition of direct and reciprocal space

$$\boldsymbol{d}_i \boldsymbol{d}_j^* = \delta_{ij} \ . \tag{3.5}$$

The atomic positions in physical space thus depend on the embedding and the shape of the *atomic surfaces*. Atomic surfaces are the components of hyperatoms in $(n-d)$D complementary (perpendicular) space $V^{\perp}$ (Fig. 3.1). Cutting a hypercrystal structure with physical space at different perpendicular space components will result in different physical space structures. This is a consequence of the irrational slope of the physical space section with respect to the n-dimensional lattice. All sections with different perpendicular space components belong to the same *local isomorphism class* (i.e., they are homometric) and will show identical diffraction patterns. Consequently, only QCs belonging to different local isomorphism classes can be distinguished by diffraction experiments.

The various types of aperiodic crystals differ from each other by the characteristics of their atomic surfaces. QCs show *discrete* atomic surfaces (which may also be of fractal shape) while those of IMSs and CSs are essentially *continuous*. Essentially continuous means that they may consist also of discrete segments in the presence of a density modulation. However, their atomic surfaces can always be described by modulation functions. With the amplitudes of the modulation function going to zero, a continuous transition to a periodic structure will be performed. Composite structures consist of two or more substructures which themselves may be modulated. In reciprocal space the characteristics of IMSs and CSs are the crystallographic point symmetry of their Fourier modules M^* and the existence of large Fourier coefficients on a distinct subset $\Lambda^* \subset M^*$ related to the reciprocal lattice of their *average structures*. The characteristic features of QCs are the possibilities of non-crystallographic point symmetry (as is observed in most cases) and of scaling symmetry $SM^* = sM^*$ (Fig. 3.2, Janssen 1992b). S denotes a scaling symmetry matrix acting on a Fourier module and s its eigenvalue. In the case of QSs with crystallographic point symmetry the structures may be described either as QCs or as IMSs or CSs, respectively (see Sect. 3.3.3). In practice, the embedding technique applied will depend on the intensity distribution. If large Fourier coefficients exist on a subset Λ^*, the description as IMSs may be preferable. However, if the major Fourier coefficients are related by scaling the QC will be the more appropriate description.

The periodic average structure of an IMS can be obtained by projecting its nD hypercrystal structure along the perpendicular space upon the physical space. In the case of a QS this would give a dense structure. Discrete periodic average structures of (at least) some QSs can be obtained by

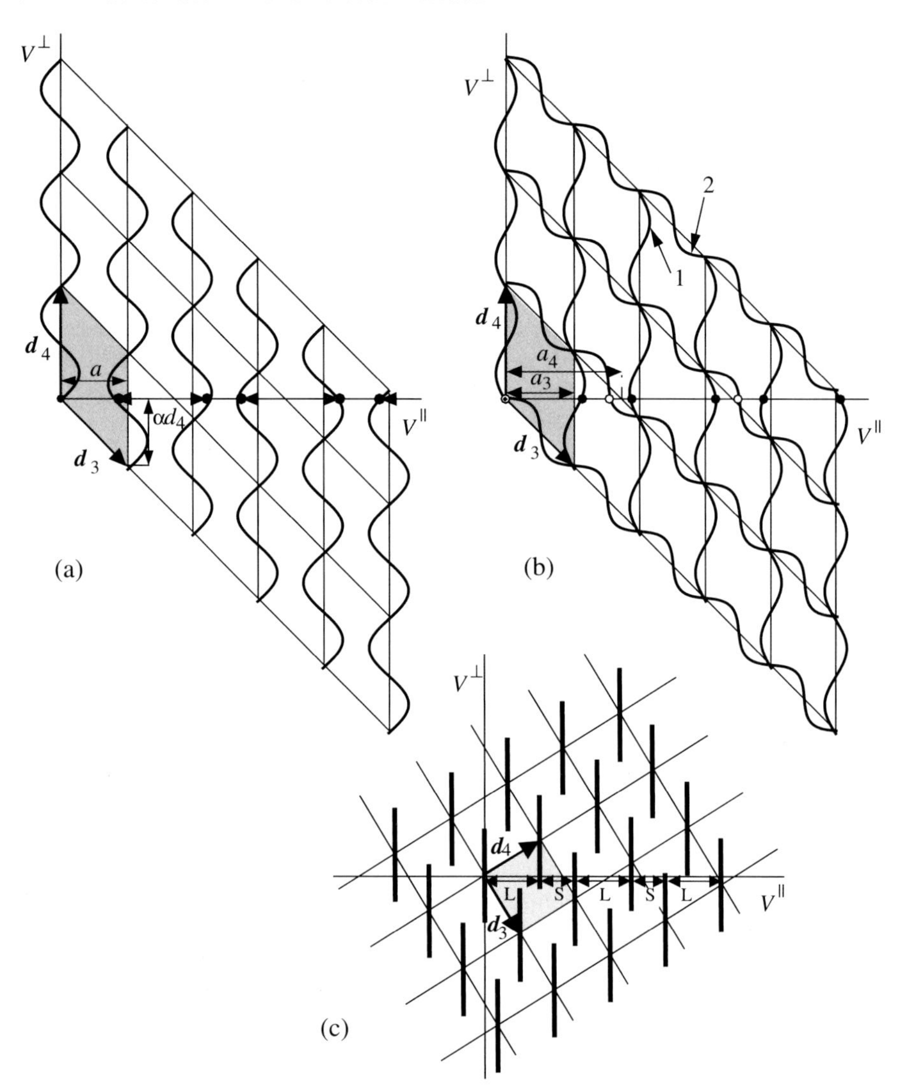

Fig. 3.1. N-dimensional embedding of the three fundamental types of dD aperiodic structures: (a) modulated structure, (b) composite structure with modulated subsystems (marked 1 and 2), and (c) quasiperiodic sequence. Vectors $\boldsymbol{d}_i$ mark the nD basis vectors while a and a_i refer to the lattice parameters of the average structures. L and S denote the two unit tiles of the Fibonacci chain.

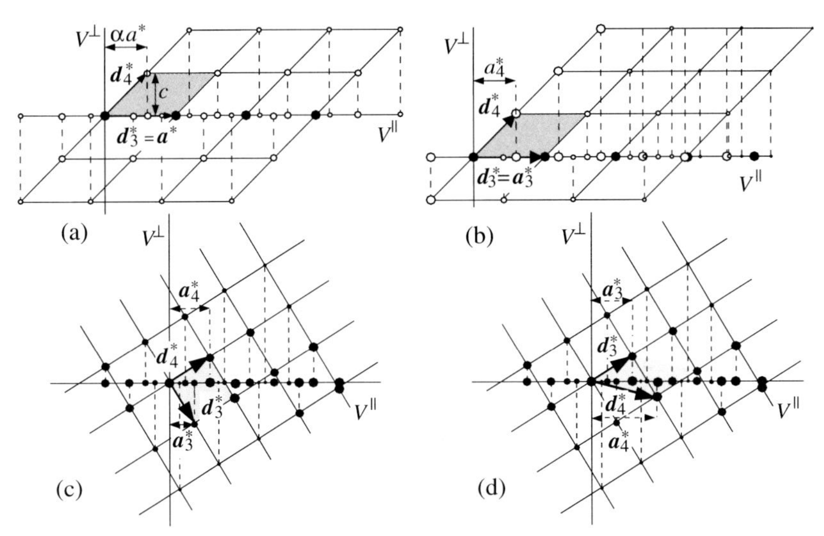

Fig. 3.2. (a)–(c) N-dimensional reciprocal space embedding of the dD aperiodic structures shown in Fig. 3.1. (d) Scaled unit cell of (c) resulting in τ-times larger lattice parameters in physical space. Dashed lines indicate the projections, vectors $\boldsymbol{d}_i^*$ refer to the nD reciprocal basis, a^* and a_i^* are the lattice parameters in reciprocal physical space.

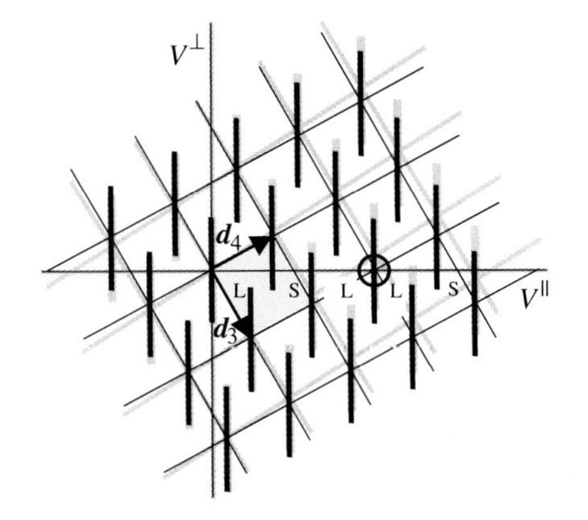

Fig. 3.3. N-dimensional embedded Fibonacci chain (light gray) and its rational approximant (black). The encircled lattice node is mapped upon physical space by shearing the dD lattice.

oblique projection (Steurer and Haibach 1997, and references therein). They may serve for a better understanding of the QS-to-crystal phase transformation. Additionally, they help determining the Jones-zones, i.e., the Brillouin zones of QS.

The nD approach allows an illustrative representation of the relationships between aperiodic crystals and their *rational approximants* (Goldman and Kelton 1993, Gratias et al. 1995). The analogue to the lock-in transition of an IMS to a commensurately modulated structure (superstructure) is the transition of a QC to a rational approximant (Fig. 3.3). While in the case of an IMS the modulation vector changes from an irrational to a rational value, for a QS the irrational slope of the cut of the nD lattice with physical space changes to a rational one. This means, that the corresponding lattice nodes lie exactly in the physical world and determine the lattice parameters of the periodic approximant. This transition can be described by a shear deformation (phason strain) of the hypercrystal parallel to $V^{\perp}$ (cf. Gratias et al. 1995). Thereby a position vector $\boldsymbol{r}$ of the nD hypercrystal is transformed to the vector $\boldsymbol{r}'$ of the approximant

$$\boldsymbol{r}' = A\boldsymbol{r} \tag{3.6}$$

with the shear matrix (subscript V denotes that the matrix elements refer to the V-basis)

$$A = \left(\begin{array}{ccc|ccc} 1 & & 0 & 0 & \cdots & 0 \\ & \ddots & & \vdots & \ddots & \vdots \\ 0 & & 1 & 0 & \cdots & 0 \\ \hline A_{41} & \cdots & A_{43} & 1 & & 0 \\ \vdots & \ddots & \vdots & & \ddots & \\ A_{n1} & \cdots & A_{n3} & 0 & & 1 \end{array}\right)_V = \left(\begin{array}{c|c} 1 & 0 \\ \hline (\tilde{A}^{-1})^T & 1 \end{array}\right)_V . \tag{3.7}$$

The determinant of A is equal to one. Thus the volume of the nD unit cell does not change during the transition. However, due to the rational slope of physical space the atomic surfaces will not be dense anymore but discrete. Consequently, the point density of approximant and QC differ and shifting physical space parallel to $V^{\perp}$ will change the structure. The symmetry group of the approximant is a subgroup of the QC symmetry group. The eliminated symmetry elements may serve as twin laws (cf. Koch 1992), as observed, e.g., in tenfold twinned orthorhombic approximants of decagonal $Al_{70}Co_{15}Ni_{15}$ (cf. Kalning et al. 1997).

In reciprocal space the phason strain leads to a shift of the scattering vectors as a function of their perpendicular space components:

$$\boldsymbol{H}^{\|'} = \boldsymbol{H}^{\|} + \tilde{A}\boldsymbol{H}^{\perp} . \tag{3.8}$$

The nD reciprocal lattice vectors transform according to

$$\boldsymbol{H}' = (A^{-1})^T \boldsymbol{H} \tag{3.9}$$

with

$$\begin{aligned}(A^{-1})^T &= \left(\begin{array}{ccc|ccc} 1 & & 0 & -A_{41} & \cdots & -A_{n1} \\ & \ddots & & \vdots & \ddots & \vdots \\ 0 & & 1 & -A_{43} & \cdots & -A_{n3} \\ \hline 0 & \cdots & 0 & 1 & & 0 \\ \vdots & \ddots & \vdots & & \ddots & \\ 0 & \cdots & 0 & 0 & & 1 \end{array}\right)_V \\ &= \left(\begin{array}{c|c} 1 & \tilde{A} \\ \hline 0 & 1 \end{array}\right)_V .\end{aligned} \tag{3.10}$$

Since the approximant structure results from a rational cut of the nD lattice with physical space, its diffraction pattern corresponds to a projection of nD reciprocal space along rational reciprocal lattice lines. Consequently, the Fourier coefficients of the approximant correspond to the sum of the Fourier coefficients (structure factors) that project onto one and the same $\boldsymbol{H}^{\mathrm{Ap}}$ in physical reciprocal space.

3.2.2 Structure Factor

The structure factor $F(\boldsymbol{H})$ of a periodic structure is defined as the Fourier transform of the electron density distribution function $\rho(\boldsymbol{r})$ of its unit cell (UC)

$$F(\boldsymbol{H}) = \int_{\mathrm{UC}} \rho(\boldsymbol{r})\, \mathrm{e}^{2\pi \mathrm{i} \boldsymbol{H}\boldsymbol{r}} \mathrm{d}\boldsymbol{r} = \sum_{k=1}^{n} T_k(\boldsymbol{H})\, f_k(\|\boldsymbol{H}\|)\, \mathrm{e}^{2\pi \mathrm{i} \boldsymbol{H}\boldsymbol{r}_k} . \tag{3.11}$$

$T_k(\boldsymbol{H})$ is the temperature factor and $f_k(\|\boldsymbol{H}\|)$ the atomic scattering factor of the k-th of the n atoms in the unit cell. In case of a dD QC, the Fourier transform of the electron density distribution function $\rho(\boldsymbol{r})$ of the nD unit cell can be separated into the product of dD parallel and $(n-d)$D perpendicular space components, and we obtain

$$F(\boldsymbol{H}) = \sum_{k=1}^{n} T_k\left(\boldsymbol{H}^{\parallel}, \boldsymbol{H}^{\perp}\right) f_k\left(\|\boldsymbol{H}^{\parallel}\|\right) g_k\left(\boldsymbol{H}^{\perp}\right) \mathrm{e}^{2\pi \mathrm{i} \boldsymbol{H}\boldsymbol{r}_k} . \tag{3.12}$$

In parallel space one obtains the conventional atomic scattering factors $f_k\left(\|\boldsymbol{H}^{\parallel}\|\right)$ and temperature factors $T_k\left(\boldsymbol{H}^{\parallel}\right)$. In perpendicular space, the Fourier transform of the atomic surfaces, called geometrical form factor $g_k\left(\boldsymbol{H}^{\perp}\right)$, is

$$g_k\left(\boldsymbol{H}^{\perp}\right) = \frac{1}{A_{\mathrm{UC}}^{\perp}} \int_{A_k} \mathrm{e}^{2\pi \mathrm{i} \boldsymbol{H}^{\perp}\boldsymbol{r}^{\perp}} \mathrm{d}\boldsymbol{r}^{\perp} , \tag{3.13}$$

where $A^{\perp}_{\mathrm{UC}}$ is the volume of the nD unit cell projected onto $V^{\perp}$ and A_k is the volume of the k-th atomic surface. The perpendicular space component $T_k\left(\boldsymbol{H}^{\perp}\right)$ of the temperature factor describes random phason fluctuations related to random phason flips in physical space. Assuming harmonic (static or dynamic) displacements in nD space one obtains in analogy to the usual expression given by Willis and Pryor (1975)

$$T_k\left(\boldsymbol{H}\right) = T_k\left(\boldsymbol{H}^{\parallel}, \boldsymbol{H}^{\perp}\right) = \mathrm{e}^{-2\pi^2 \boldsymbol{H}^{\parallel T} \langle \boldsymbol{u}^{\parallel} \boldsymbol{u}^{\parallel T} \rangle \boldsymbol{H}^{\parallel}} \times \mathrm{e}^{-2\pi^2 \boldsymbol{H}^{\perp T} \langle \boldsymbol{u}^{\perp} \boldsymbol{u}^{\perp T} \rangle \boldsymbol{H}^{\perp}} \tag{3.14}$$

with

$$\langle \boldsymbol{u}^{\parallel} \boldsymbol{u}^{\parallel T} \rangle = \begin{pmatrix} \langle u_1^2 \rangle & \langle u_1 u_2 \rangle & \langle u_1 u_3 \rangle \\ \langle u_2 u_1 \rangle & \langle u_2^2 \rangle & \langle u_2 u_3 \rangle \\ \langle u_3 u_1 \rangle & \langle u_3 u_2 \rangle & \langle u_3^2 \rangle \end{pmatrix}_V \tag{3.15}$$

and

$$\langle \boldsymbol{u}^{\perp} \boldsymbol{u}^{\perp T} \rangle = \begin{pmatrix} \langle u_4^2 \rangle & \cdots & \langle u_4 u_n \rangle \\ \vdots & \ddots & \vdots \\ \langle u_n u_4 \rangle & \cdots & \langle u_n^2 \rangle \end{pmatrix}_V . \tag{3.16}$$

The elements of type $\langle u_i u_j \rangle$ represent the mean displacements of the hyperatoms along the i-th axis times the displacements of the atoms along the j-th axis on the V-basis. This model excludes phonon-phason interactions as no coupling is defined.

3.3 One-Dimensional Quasicrystals

Structures with 1D quasiperiodic order and 2D lattice periodicity (1D QCs) are the simplest representatives of QCs. A few phases of this structure type have been identified experimentally. Structures with a linear quasiperiodic ordering of only a subset of atoms (density or vacancy ordering; cf. Steurer 1996, and references therein) may loose any lattice symmetry in physical space and should better be dealt with as IMSs or CSs.

A fundamental model of a 1D QS is the Fibonacci sequence. It will be used as an example to describe the quasiperiodic direction of 3D structures with 1D quasiperiodic stacking of periodic atomic layers. 1D QSs are on the borderline between QSs and IMSs. They can be described in either of the two approaches. The description as QS is advantageous if some kind of scaling symmetry is present or if there is a close structural relationship with 2D or 3D QCs. This is the case for 1D QCs occurring as intermediate states during QC-to-crystal transformations. The description as IMSs may be helpful in the course of structure analysis. The diffraction pattern can then be separated into a set of main reflections and a set of satellite reflections. The main reflections are related to the 3D periodic average structure, which can be determined with conventional methods. However, indexing a typical 1D QC as an IMS may be difficult as the intensity distribution does not allow main reflections to be determined easily (see Sect. 3.2.1).

In the following a simple example of a 1D QC structure that can be embedded in $(d+1)$D space is described (all 1D QCs observed so far belong to this class). There exists, however, an infinite number of different 1D QSs that need an embedding space of dimension $n > d+1$. These are, for instance, all quasiperiodic sequences formed by substitution rules based on n letters with $n > d+1$ (Luck et al. 1997).

3.3.1 Indexing

The electron density distribution function $\rho(\boldsymbol{r})$ of a 1D QC is given by the Fourier series

$$\rho(\boldsymbol{r}) = \frac{1}{V}\sum_{\boldsymbol{H}} F(\boldsymbol{H})\mathrm{e}^{-2\pi\mathrm{i}\boldsymbol{H}\boldsymbol{r}} \,. \tag{3.17}$$

The Fourier coefficients (structure factors) $F(\boldsymbol{H})$ are functions of the scattering vectors $\boldsymbol{H} = \sum_{i=1}^{3} h_i^{\parallel}\tilde{\boldsymbol{a}}_i^*$ with $h_1^{\parallel}, h_2^{\parallel} \in \mathbb{Z}$, $h_3^{\parallel} \in \mathbb{R}$. Introducing four reciprocal basis vectors, all scattering vectors can be indexed with integer components: $\boldsymbol{H} = \sum_{i=1}^{4} h_i\boldsymbol{a}_i^*$ with $\boldsymbol{a}_4^* = \alpha\boldsymbol{a}_3^*$, where α is an irrational algebraic number and $h_i \in \mathbb{Z}$. The set M^* of all diffraction vectors $\boldsymbol{H}$ forms a vector module ($\mathbb{Z}$-module) of rank four. The vectors $\boldsymbol{a}_i^*$, $i = 1, \ldots, 4$, can be considered as physical space projections of the basis vectors $\boldsymbol{d}_i^*$, $i = 1, \ldots, 4$, of the corresponding 4D reciprocal lattice Σ^* with

$$\boldsymbol{d}_1^* = \boldsymbol{a}_1^*, \boldsymbol{d}_2^* = \boldsymbol{a}_2^*, \boldsymbol{d}_3^* = \|\boldsymbol{a}_3^*\| \begin{pmatrix} 0 \\ 0 \\ 1 \\ -\alpha c \end{pmatrix}_V, \boldsymbol{d}_4^* = \|\boldsymbol{a}_3^*\| \begin{pmatrix} 0 \\ 0 \\ \alpha \\ c \end{pmatrix}_V \,. \tag{3.18}$$

Here c is an arbitrary constant usually set to unity and the subscript V denotes components referring to a 3D crystallographic (symmetry adapted) coordinate system with an orthogonal fourth dimension (V-basis). The direct 4D basis results from the orthogonality condition [Eq. (3.5)]. For physical space vectors the reciprocity relationship $\boldsymbol{a}_i\boldsymbol{a}_j^* = \delta_{ij}$ is valid only for $i, j = 1, 2, 3$.

According to the scaling symmetry (see Sect. 3.3.2) the indexing of the quasiperiodic axis is not unique. Even if all Bragg peaks can be indexed, a set of α-times enlarged basis vectors again will describe the locations equivalently. A first attempt to solve the problem of indexing was given by Elser (1985). In the case of a primitive QS having a simple atomic surface the intensity distribution is a simple function of the geometrical form factor [Eq. (3.13)] and consequently a monotonically decreasing function of $\|\boldsymbol{H}^{\perp}\|$. If the intensity of scaled scattering vectors decreases monotonically in the same way as predicted, the proper basis has been selected. However, given a more complicated structure this approach may fail. It has been shown by Cervellino et al. (1998) that a detailed analysis of the Patterson function (autocorrelation function) depending on perpendicular space components allows the basis vectors of more complex structures to be determined properly.

3.3.2 Symmetry

The possible Laue symmetry group $K^{3\mathrm{D}}$ of the intensity weighted Fourier module (diffraction pattern)

$$M_I^* = \left\{ I(\boldsymbol{H}) = \|F(\boldsymbol{H})\|^2 \,\middle|\, \boldsymbol{H} = \sum_{i=1}^{4} h_i a_i^*, h_i \in \mathbb{Z} \right\} \tag{3.19}$$

results from the direct product $K^{3\mathrm{D}} = K^{2\mathrm{D}} \otimes K^{1\mathrm{D}} \otimes \bar{1}$. Here $K^{2\mathrm{D}}$ is one of the ten crystallographic 2D point groups and $K^{1\mathrm{D}} = 1$ or $\bar{1}$. Consequently, all 3D crystallographic Laue groups except the two cubic ones (they would mix periodic and aperiodic directions) are permitted: $\bar{1}$, $2/m$, mmm, $4/m$, $4/mmm$, $\bar{3}$, $\bar{3}m$, $6/m$, and $6/mmm$. If one distinguishes between symmetry operators $R \in K^{2\mathrm{D}}$ and $R' \in K^{1\mathrm{D}}$, the Laue group $2/m$ can occur in two different orientations with regard to the unique axis $[0010]_V$: $2'/m$ and $2/m'$. Thus, there are 10 different Laue groups.

Thirty-one point groups result from the direct products $K^{3\mathrm{D}} = K^{2\mathrm{D}} \otimes K^{1\mathrm{D}}$ and their subgroups of index 2. These are all 27 3D crystallographic point groups except the five cubic point groups. Four additional point groups are obtained by considering the different settings in 2, $2'$, m, m', $2/m'$, $2'/m$, $2'mm'$, and $2mm$. The necessity to distinguish between primed and non-primed operators is based on reduced tensor symmetries of physical properties. A table of the 80 3D space groups compatible with 1D quasiperiodicity has been derived by Wang et al. (1997). These space groups contain no symmetry operations with translation components along the unique direction $[0010]_V$. The 80 symmetry groups leaving the 4D hypercrystal structure invariant are a subset of the (3+1)D space groups (superspace groups) given by Janssen et al. (1992). This subset corresponds to all superspace groups with the basis space group being one of the 80 3D space groups mentioned above marked by the bare symbols (00γ), $(\alpha\beta0)$ or $(\alpha\beta\gamma)$. In the last two cases only one of the coefficients α, β, γ is allowed to be irrational.

3.3.3 Example of a One-Dimensional Quasicrystal: Fibonacci Phase

If the Fibonacci sequence is chosen for the quasiperiodic direction of a 1D QC, it may simply be called a Fibonacci phase. We define the Fibonacci phase geometrically as layer structure: layers with 2D lattice periodicity are stacked quasiperiodically. The distances between the layers follow the Fibonacci sequence ...LSLLS... . The Fibonacci sequence can be obtained from the substitution rule $\sigma : \mathrm{S} \mapsto \mathrm{L}, \mathrm{L} \mapsto \mathrm{LS}$. In case of structural ordering, the letters L and S may denote long and short interatomic distances or atoms with large and small scattering cross sections. The substitution rule can be rewritten with the substitution matrix S

$$\sigma : \begin{pmatrix} \mathrm{S} \\ \mathrm{L} \end{pmatrix} \mapsto \underbrace{\begin{pmatrix} 0 & 1 \\ 1 & 1 \end{pmatrix}}_{=S} \begin{pmatrix} \mathrm{S} \\ \mathrm{L} \end{pmatrix} = \begin{pmatrix} \mathrm{L} \\ \mathrm{LS} \end{pmatrix} . \tag{3.20}$$

Assigning the eigenvector $\binom{1}{\tau}$ of S to $\binom{\mathrm{S}}{\mathrm{L}}$ results in an infinite Fibonacci sequence invariant under scaling with its corresponding eigenvalue τ ($\tau = \frac{1+\sqrt{5}}{2} = 2\cos\frac{\pi}{5} = 1.618\ldots$). Furthermore, the ratios of two subsequent substitutions (called inflation operations, as the number of tiles is inflated) are given by the ratio of the eigenvector components

$$\frac{\mathrm{L}}{\mathrm{S}} = \frac{\mathrm{LS}}{\mathrm{L}} = \frac{\mathrm{LSL}}{\mathrm{LS}} = \frac{\mathrm{LSLLS}}{\mathrm{LSL}} = \cdots = \frac{\tau}{1} . \tag{3.21}$$

Based on the scaling symmetry matrix in Eq. (3.20), the 4D reciprocal lattice Σ^* is spanned by basis vectors according to Eq. (3.18) with $\alpha = \tau$. Consequently, the embedding matrix W [cf. Eq. (3.4)] is given by

$$W = \begin{pmatrix} 1 & 0 & 0 & 0 \\ 0 & 1 & 0 & 0 \\ 0 & 0 & 1 & \tau \\ 0 & 0 & -\tau & 1 \end{pmatrix} . \tag{3.22}$$

The physical space structure ("quasilattice") of the Fibonacci phase is a subset of the vector module $M = \left\{ \boldsymbol{r} = \sum_{i=1}^{4} n_i \pi^{\parallel}(\boldsymbol{d}_i), n_i \in \mathbb{Z} \right\}$:

$$M_{Fibonacci} =$$
$$\left\{ \boldsymbol{r} = \pi^{\parallel}(\sum_{i=1}^{4} m_i \boldsymbol{d}_i) \middle| m_i \in \mathbb{Z}, \|\pi^{\perp}(\sum_{i=1}^{4} m_i \boldsymbol{d}_i)\| \le \frac{1+\tau}{2\|\boldsymbol{a}_3^*\|(2+\tau)} \right\} . \tag{3.23}$$

It can be obtained from cutting a decorated 4D hyperlattice Σ spanned by the basis vectors according to Eq. (3.5)

$$\begin{aligned} \boldsymbol{d}_1 &= \boldsymbol{a}_1, \boldsymbol{d}_2 = \boldsymbol{a}_2, \\ \boldsymbol{d}_3 &= \frac{\|\boldsymbol{a}_3\|}{(2+\tau)} \begin{pmatrix} 0 \\ 0 \\ 1 \\ -\tau \end{pmatrix}_V , \boldsymbol{d}_4 = \frac{\|\boldsymbol{a}_3\|}{(2+\tau)} \begin{pmatrix} 0 \\ 0 \\ \tau \\ 1 \end{pmatrix}_V \end{aligned} \tag{3.24}$$

with physical space. The corresponding atomic surfaces are line segments. The basis vectors determine the 4D metric tensor G defined as

$$\mathsf{G} = \begin{pmatrix} \boldsymbol{d}_1\boldsymbol{d}_1 & \boldsymbol{d}_1\boldsymbol{d}_2 & \boldsymbol{d}_1\boldsymbol{d}_3 & \boldsymbol{d}_1\boldsymbol{d}_4 \\ \boldsymbol{d}_2\boldsymbol{d}_1 & \boldsymbol{d}_2\boldsymbol{d}_2 & \boldsymbol{d}_2\boldsymbol{d}_3 & \boldsymbol{d}_2\boldsymbol{d}_4 \\ \boldsymbol{d}_3\boldsymbol{d}_1 & \boldsymbol{d}_3\boldsymbol{d}_2 & \boldsymbol{d}_3\boldsymbol{d}_3 & \boldsymbol{d}_3\boldsymbol{d}_4 \\ \boldsymbol{d}_4\boldsymbol{d}_1 & \boldsymbol{d}_4\boldsymbol{d}_2 & \boldsymbol{d}_4\boldsymbol{d}_3 & \boldsymbol{d}_4\boldsymbol{d}_4 \end{pmatrix} \tag{3.25}$$

and the volume of the 4D unit cell is $V = \sqrt{\det \mathsf{G}}$. The point density $\mathrm{D_p}$ in physical space (the reciprocal of the mean atomic volume) is determined by the size of the atomic surfaces A_i

$$\mathrm{D_p} = \frac{1}{V}\sum_{i=1}^{n} A_i \,. \tag{3.26}$$

Weighting each atomic surface in Eq. (3.26) with its atomic weight $\mathrm{M}_{\mathrm{A}i}$, the mass density D can be expressed as

$$\mathrm{D} = \frac{1}{V}\sum_{i=1}^{n} A_i \mathrm{M}_{\mathrm{A}i} \,. \tag{3.27}$$

3.3.3.1 Structure Factor. The 4D hyperlattice is decorated with 4D hyperatoms. The Fourier transform of their parallel space components corresponds to the usual 3D atomic scattering factors. The atomic surfaces along the 1D perpendicular space are line segments of length $\frac{1+\tau}{\|\boldsymbol{a}_3^*\|(2+\tau)}$. They are centered at positions $(x_1 x_2 00)$. Consequently, the structure factor for the Fibonacci phase is obtained by substituting

$$g_k\left(\boldsymbol{H}^{\perp}\right) = \frac{2+\tau}{\pi\tau^2\left(-\tau h_3 + h_4\right)} \sin\left(\frac{\pi\tau^2\left(-\tau h_3 + h_4\right)}{2+\tau}\right) \tag{3.28}$$

in Eq. (3.12). The geometrical form factor $g_k\left(\boldsymbol{H}^{\perp}\right)$ is of the form $\frac{\sin x^{\perp}}{x^{\perp}}$. The upper and lower envelopes of this function are hyperbolae $\pm\frac{1}{x^{\perp}}$. Hence, the envelope of the diffracted intensity is proportional to $\frac{1}{x^{\perp 2}}$ and is convergent.

The atomic surface can be decomposed into sections, which show the same local environment (Voronoi domains) in physical space. Projecting all nearest neighbors of the hyperatom of interest onto $V^{\perp}$ encodes all different environments as shown in Fig. 3.4. If physical space $V^{\parallel}$ cuts the hyperatom, e.g., in the region marked **a**, the central atom has a coordination of one atom at distance S on the left side and another at distance L on the right side. Consequently, all hyperatoms that share a distinct region of the atomic surface in the projection onto perpendicular space determine all bond distances and angles in physical space.

As a consequence of the local isomorphism the point density has to be invariant for any shift of physical space parallel to perpendicular space. This leads to the *closeness condition*: when the atomic surfaces are projected onto perpendicular space each boundary of an atomic surface has to fit exactly to another one (the uppermost and lowest hyperatoms in Fig. 3.4 fit exactly to the central one).

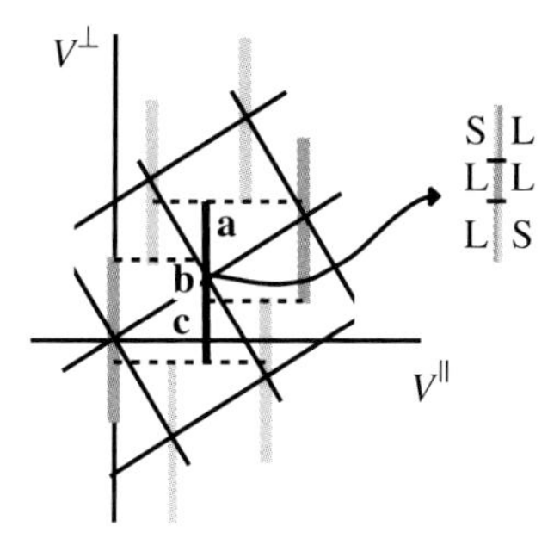

Fig. 3.4. Projecting all nearest neighbors of the hyperatom onto $V^\perp$ results in three different coordinations. Cutting the hyperatom in (a) leads to atoms at S and L, in (b) at L and L, and in (c) at L and S for the left and right neighboring atoms. The nearest neighbors of the hyperatom show the closeness condition.

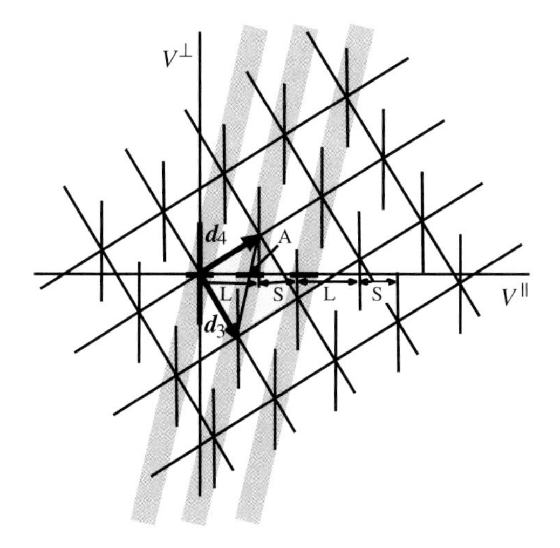

Fig. 3.5. Oblique projection (marked gray) onto reciprocal space leads to the average structure of the Fibonacci chain. The bold horizontal bars represent the projected atomic surfaces. The unit cell length of the average structure is marked with A.

3.3.3.2 Periodic Average Structure. As mentioned in Sect. 3.3, 1D QCs can equivalently be treated as IMSs showing a periodic average structure. The periodic average structure of a Fibonacci phase can be derived by an oblique projection onto physical space $V^{\parallel}$ (Fig. 3.5) as demonstrated by Steurer and Haibach (1998). Based on the projection with

$$\pi^{\parallel}(\boldsymbol{r}) = \begin{pmatrix} 1 & 0 & 0 & 0 \\ 0 & 1 & 0 & 0 \\ 0 & 0 & 1 & 3-2\tau \end{pmatrix}_V \boldsymbol{r}_V = \begin{pmatrix} 1 & 0 & 0 & 0 \\ 0 & 1 & 0 & 0 \\ 0 & 0 & \frac{1}{\tau^2} & \frac{1}{\tau^2} \end{pmatrix}_D \boldsymbol{r}_D, \quad (3.29)$$

the basis vectors of the average periodic structure are $\bar{\boldsymbol{a}}_1 = \boldsymbol{a}_1$, $\bar{\boldsymbol{a}}_2 = \boldsymbol{a}_2$, $\bar{\boldsymbol{a}}_3 = \tau^{-2}\boldsymbol{a}_3$ and $\bar{\boldsymbol{a}}_1^* = \boldsymbol{a}_1^*$, $\bar{\boldsymbol{a}}_2^* = \boldsymbol{a}_2^*$, $\bar{\boldsymbol{a}}_3^* = \tau^2\boldsymbol{a}_3^*$. The oblique projection in physical space results in an oblique section in reciprocal space (Fig. 3.6). Consequently, all reflections of type $(h_1h_2h_3h_3)_D$ are main reflections.

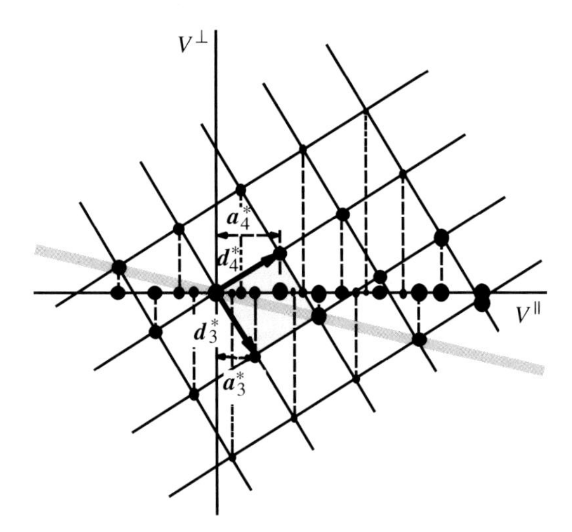

Fig. 3.6. Oblique section (marked gray) of physical space leads to the diffraction pattern of the average structure of the Fibonacci chain.

3.3.3.3 Approximant Structures. The $\langle m,n\rangle$-approximant ($m,n \in \mathbb{N}$) of a Fibonacci phase can be obtained applying the shear matrix [Eq. (3.7)] with $A_{43} \neq 0$ upon $\boldsymbol{r}$

$$\boldsymbol{r} = m\boldsymbol{d}_3 + n\boldsymbol{d}_4 = \frac{\|\boldsymbol{a}_3^*\|}{2+\tau}\begin{pmatrix} 0 \\ 0 \\ m+n\tau \\ n-m\tau \end{pmatrix}_V . \tag{3.30}$$

From the condition that the perpendicular space component of the approximant basis vector has to vanish

$$A\boldsymbol{r} = \frac{\|\boldsymbol{a}_3^*\|}{2+\tau}\begin{pmatrix} 0 \\ 0 \\ m+n\tau \\ A_{43}(m+n\tau)n-m\tau \end{pmatrix}_V \stackrel{!}{=} \frac{\|\boldsymbol{a}_3^*\|}{2+\tau}\begin{pmatrix} 0 \\ 0 \\ m+n\tau \\ 0 \end{pmatrix}_V \tag{3.31}$$

the shear matrix coefficient is

$$A_{43} = \frac{m\tau - n}{n\tau + m} . \tag{3.32}$$

The basis vectors $\boldsymbol{a}_i^{\mathrm{Ap}}, i = 1,2,3$, of the $\langle m,n\rangle$-approximant are $\boldsymbol{a}_i^{\mathrm{Ap}} = \boldsymbol{a}_i, i = 1,2$, $\boldsymbol{a}_3^{\mathrm{Ap}} = \frac{(m+n\tau)}{2+\tau}\boldsymbol{a}_3$.

All peaks are shifted according to Eq. (3.8). Projecting the 4D reciprocal space onto physical space results in a periodic reciprocal lattice. Thus, all diffraction vectors $\boldsymbol{H} = (h_1h_2h_3h_4)$ of the QC are transformed to $\boldsymbol{H}^{\mathrm{Ap}} = \left(h_1h_2(mh_3+nh_4)\right) = \left(h_1^{\mathrm{Ap}}h_2^{\mathrm{Ap}}h_3^{\mathrm{Ap}}\right)$ on the basis of the $\langle m,n\rangle$-approximant. Consequently, all structure factors $F(\boldsymbol{H})$ with $\boldsymbol{H} = \left(h_1h_2(h_3-on)(h_4+om)\right)$, $o \in \mathbb{Z}$ are projected onto each other.

3.4 Decagonal Quasicrystals

QCs which can be seen as periodic stackings of quasiperiodic layers and which exhibit decagonal diffraction symmetry are called decagonal phases. Many stable and metastable representatives of this class of QCs have been observed experimentally (Steurer 1996, and references therein). The Penrose tiling will be used as an example for the 2D quasiperiodic atomic layers in a decagonal phase.

3.4.1 Indexing

The electron density distribution function $\rho(\boldsymbol{r})$ of a 2D QC can be represented by the Fourier series given in Eq. (3.17). All Fourier coefficients, i.e., the structure factors $F(\boldsymbol{H})$, can be indexed with reciprocal space vectors $\boldsymbol{H} = \sum_{i=1}^{3} h_i^{\parallel} \tilde{\boldsymbol{a}}_i^*$ with $h_1^{\parallel}, h_2^{\parallel} \in \mathbb{R}$, $h_3^{\parallel} \in \mathbb{Z}$. Introducing five reciprocal basis vectors, all possible reciprocal space vectors can be indexed with integer components: $\boldsymbol{H} = \sum_{i=1}^{5} h_i \boldsymbol{a}_i^*$ with $\boldsymbol{a}_i^* = a^* \left(\cos \frac{2\pi i}{5}, \sin \frac{2\pi i}{5}, 0\right), i = 1, \ldots, 4$, $a^* = \|\boldsymbol{a}_1^*\| = \|\boldsymbol{a}_2^*\| = \|\boldsymbol{a}_3^*\| = \|\boldsymbol{a}_4^*\|, \boldsymbol{a}_5^* = \|\boldsymbol{a}_5^*\| (0, 0, 1)$, and $h_i \in \mathbb{Z}$ (Fig. 3.7). The vector components refer to a Cartesian coordinate system in physical space $V^{\parallel}$. The set of all diffraction vectors $\boldsymbol{H}$ forms a $\mathbb{Z}$-module M^* of rank five. The vectors $\boldsymbol{a}_i^*$, $i = 1, \ldots, 5$, can be considered as physical space projections of the basis vectors $\boldsymbol{d}_i^*$, $i = 1, \ldots, 5$, of the 5D reciprocal lattice Σ^* with

$$\boldsymbol{d}_i^* = a^* \begin{pmatrix} \cos \frac{2\pi i}{5} \\ \sin \frac{2\pi i}{5} \\ 0 \\ c \cos \frac{6\pi i}{5} \\ c \sin \frac{6\pi i}{5} \end{pmatrix}_V, i = 1, \ldots, 4; \boldsymbol{d}_5^* = a_5^* \begin{pmatrix} 0 \\ 0 \\ 1 \\ 0 \\ 0 \end{pmatrix}_V. \tag{3.33}$$

Here c is an arbitrary constant which is usually set to 1 (as it is also done in the following). The subscript V denotes components referring to a 5D Cartesian coordinate system (V-basis), while subscript D refers to the 5D crystallographic basis (D-basis). The embedding matrix W [cf. Eq. (3.4)] is

$$W = \begin{pmatrix} \cos \frac{2\pi}{5} & \cos \frac{4\pi}{5} & \cos \frac{6\pi}{5} & \cos \frac{8\pi}{5} & 0 \\ \sin \frac{2\pi}{5} & \sin \frac{4\pi}{5} & \sin \frac{6\pi}{5} & \sin \frac{8\pi}{5} & 0 \\ 0 & 0 & 0 & 0 & 1 \\ \cos \frac{6\pi}{5} & \cos \frac{2\pi}{5} & \cos \frac{8\pi}{5} & \cos \frac{4\pi}{5} & 0 \\ \sin \frac{6\pi}{5} & \sin \frac{2\pi}{5} & \sin \frac{8\pi}{5} & \sin \frac{4\pi}{5} & 0 \end{pmatrix}. \tag{3.34}$$

The direct 5D basis is obtained from the orthogonality condition [Eq. (3.5)]

$$\boldsymbol{d}_i = \frac{2}{5a^*} \begin{pmatrix} \cos \frac{2\pi i}{5} - 1 \\ \sin \frac{2\pi i}{5} \\ 0 \\ \cos \frac{6\pi i}{5} - 1 \\ \sin \frac{6\pi i}{5} \end{pmatrix}_V, i = 1, \ldots, 4; \boldsymbol{d}_5 = \frac{1}{a_5^*} \begin{pmatrix} 0 \\ 0 \\ 1 \\ 0 \\ 0 \end{pmatrix}_V. \tag{3.35}$$

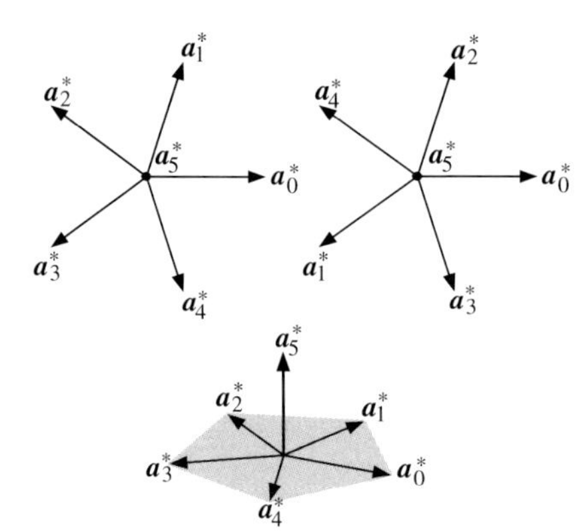

Fig. 3.7. Reciprocal basis of the decagonal phase. The projections upon the parallel (left) and the perpendicular space (right) are shown. The additional vector $\boldsymbol{a}_0^* = -(\boldsymbol{a}_1^* + \boldsymbol{a}_2^* + \boldsymbol{a}_3^* + \boldsymbol{a}_4^*)$ is linear dependent. In the lower drawing, a perspective view is given of the physical space reciprocal basis.

The metric tensors G and G^* are of type

$$\begin{pmatrix} A & C & C & C & 0 \\ C & A & C & C & 0 \\ C & C & A & C & 0 \\ C & C & C & A & 0 \\ 0 & 0 & 0 & 0 & B \end{pmatrix} \tag{3.36}$$

with $A = 2a^{*2}$, $B = a_5^{*2}$, $C = -\frac{1}{2}a^{*2}$ for reciprocal space and $A = \frac{4}{5a^{*2}}$, $B = \frac{1}{a_5^{*2}}$, $C = \frac{2}{5a^{*2}}$ for direct space. Therefrom the direct and reciprocal lattice parameters can be derived as

$$d_i^* = a^*\sqrt{2}, d_5^* = a_5^*, \alpha_{ij} = 104.5^\circ, \alpha_{i5} = 90^\circ, i,j = 1,\ldots,4 \tag{3.37}$$

and

$$d_i = d = \frac{2}{a^*\sqrt{5}}, d_5 = \frac{1}{a_5^*}, \alpha_{ij} = 60^\circ, \alpha_{i5} = 90^\circ, i,j = 1,\ldots,4\,. \tag{3.38}$$

The volume of the 5D unit cell is

$$V = \sqrt{\det(\mathsf{G})} = \frac{4}{5\sqrt{5}a^{*4}a_5^*} = \frac{\sqrt{5}d^4 d_5}{4}\,. \tag{3.39}$$

3.4.2 Symmetry

The diffraction symmetry of decagonal phases, i.e., the point symmetry group leaving the intensity weighted Fourier module (diffraction pattern) M_I^* invariant, is one of the two Laue groups $10/mmm$ or $10/m$. The 18 space groups leaving the 5D hypercrystal structure invariant are that subset of the 5D space groups of which the point groups are isomorphous to the 6 possible 3D decagonal point groups (Table 3.1). The orientation of the symmetry elements of the 5D space groups is fixed by the isomorphism of the 3D and 5D point groups. The tenfold axis defines the unique direction $[00100]_V$ or

Table 3.1. 3D point groups of order k and corresponding 5D decagonal space groups with reflection conditions (cf. Rabson et al. 1991).

3D point group	k	5D space group	reflection condition
$\frac{10}{m}\frac{2}{m}\frac{2}{m}$	40	$P\frac{10}{m}\frac{2}{m}\frac{2}{m}$	no condition
		$P\frac{10}{m}\frac{2}{c}\frac{2}{c}$	$h_1h_2h_2h_1h_5 : h_5 = 2n$
			$h_1h_2\bar{h}_2\bar{h}_1h_5 : h_5 = 2n$
		$P\frac{10_5}{m}\frac{2}{m}\frac{2}{c}$	$h_1h_2\bar{h}_2\bar{h}_1h_5 : h_5 = 2n$
		$P\frac{10_5}{m}\frac{2}{c}\frac{2}{m}$	$h_1h_2h_2h_1h_5 : h_5 = 2n$
$\frac{10}{m}$	20	$P\frac{10}{m}$	no condition
		$P\frac{10_5}{m}$	$0000h_5 : h_5 = 2n$
$10\,2\,2$		$P10\,2\,2$	no condition
		$P10_j\,2\,2$	$0000h_j : jh_5 = 10n$
$10mm$	20	$P10mm$	no condition
		$P10cc$	$h_1h_2h_2h_1h_5 : h_5 = 2n$
			$h_1h_2\bar{h}_2\bar{h}_1h_5 : h_5 = 2n$
		$P10_5mc$	$h_1h_2\bar{h}_2\bar{h}_1h_5 : h_5 = 2n$
		$P10_5cm$	$h_1h_2h_2h_1h_5 : h_5 = 2n$
$\overline{10}m2$	20	$P\overline{10}m2$	no condition
		$P\overline{10}c2$	$h_1h_2h_2h_1h_5 : h_5 = 2n$
		$P\overline{10}2m$	no condition
		$P\overline{10}2c$	$h_1h_2\bar{h}_2\bar{h}_1h_5 : h_5 = 2n$
10	10	$P10$	no condition
		$P10_j$	$0000h_j : jh_5 = 10n$

$[00001]_D$, which is the periodic direction. The reflection and inversion operations $\Gamma(m)$ and $\Gamma(\bar{1})$ are equivalent in both subspaces $V^{\parallel}$ and $V^{\perp}$. $\Gamma(10)$, a $\pi/5$ rotation in $V^{\parallel}$ around the tenfold axis corresponds to a $3\pi/5$ rotation in $V^{\perp}$ (cf. Fig. 3.7), is given by

$$\Gamma(10) = \left(\begin{array}{ccc|cc} \cos\frac{\pi}{5} & -\sin\frac{\pi}{5} & 0 & 0 & 0 \\ \sin\frac{\pi}{5} & \cos\frac{\pi}{5} & 0 & 0 & 0 \\ 0 & 0 & 1 & 0 & 0 \\ \hline 0 & 0 & 0 & \cos\frac{3\pi}{5} & -\sin\frac{3\pi}{5} \\ 0 & 0 & 0 & \sin\frac{3\pi}{5} & \cos\frac{3\pi}{5} \end{array}\right)_V = \left(\begin{array}{c|c} \Gamma^{\parallel}(10) & 0 \\ \hline 0 & \Gamma^{\perp}(10) \end{array}\right)_V, \tag{3.40}$$

$$\Gamma(m) = \begin{pmatrix} 1 & 0 & 0 & 0 & 0 \\ 0 & \bar{1} & 0 & 0 & 0 \\ 0 & 0 & 1 & 0 & 0 \\ 0 & 0 & 0 & 1 & 0 \\ 0 & 0 & 0 & 0 & \bar{1} \end{pmatrix}_V, \Gamma(\bar{1}) = \begin{pmatrix} \bar{1} & 0 & 0 & 0 & 0 \\ 0 & \bar{1} & 0 & 0 & 0 \\ 0 & 0 & \bar{1} & 0 & 0 \\ 0 & 0 & 0 & \bar{1} & 0 \\ 0 & 0 & 0 & 0 & \bar{1} \end{pmatrix}_V. \tag{3.41}$$

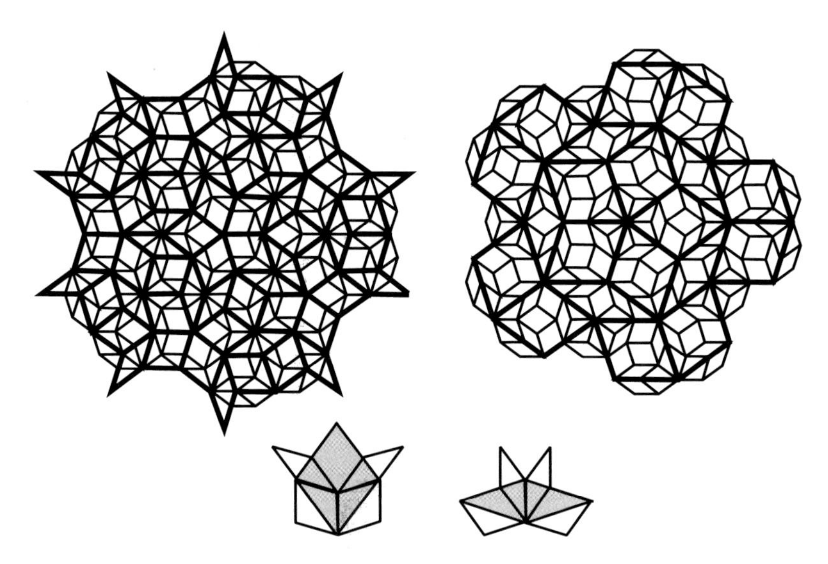

Fig. 3.8. Scaling properties of the PT. In the left drawing a PT (thin lines) is superposed by a PT (thick lines) scaled by S. In the right drawing the scaling by S^2 is shown. The lower drawing demonstrates the inflation rule of the PT.

The translation components of the tenfold screw axis and the c-glide planes are along the periodic direction.

3.4.3 Example of a Decagonal Phase: Layers of Penrose Tilings

The Penrose Tiling (PT) (Penrose 1974, Pavlovitch and Kléman 1987) can be constructed from two unit tiles: a skinny (acute angle $\alpha = \frac{\pi}{5}$) and a fat rhomb (acute angle $\alpha = \frac{2\pi}{5}$) with equal edge lengths a_r and areas $a_r^2 \sin\frac{\pi}{5}$ and $a_r^2 \sin\frac{2\pi}{5}$, respectively. Their areas and frequencies in the PT are both in a ratio $1 : \tau$. The construction has to obey matching rules, which can be derived from the scaling properties of the PT (Fig. 3.8). The set of vertices of the PT, M_{PT}, is a subset of the vector module $M = \left\{\boldsymbol{r} = \sum_{i=0}^{4} n_i a_r \boldsymbol{e}_i \middle| \boldsymbol{e}_i = \left(\cos\frac{2\pi i}{5}, \sin\frac{2\pi i}{5}, 0\right)\right\}$. M_{PT} consists of five subsets $M_{PT} = \cup_{k=0}^{4} M_k$ with

$$M_k = \left\{\pi^{\parallel}(\boldsymbol{r}_k) \middle| \pi^{\perp}(\boldsymbol{r}_k) \in T_{ik}, i = 0, \ldots, 4\right\} , \tag{3.42}$$

where $\boldsymbol{r}_k = \sum_{j=0}^{4} \boldsymbol{d}_j \left(n_j + \frac{k}{5}\right)$ and $n_j \in \mathbb{Z}$. The i-th triangular subdomain T_{ik} of the k-th pentagonal atomic surface corresponds to

$$T_{ik} = \left\{\boldsymbol{t} = x_i \boldsymbol{e}_i + x_{i+1} \boldsymbol{e}_{i+1} \middle| x_i \in [0, \lambda_k], x_{i+1} \in [0, \lambda_k - x_i]\right\} \tag{3.43}$$

with λ_k the radius of a pentagonally shaped atomic surface: $\lambda_0 = 0$, for $\lambda_{1,\cdots,4}$ [see Eq. (3.48)]. Performing the scaling operation SM_{PT} with the matrix

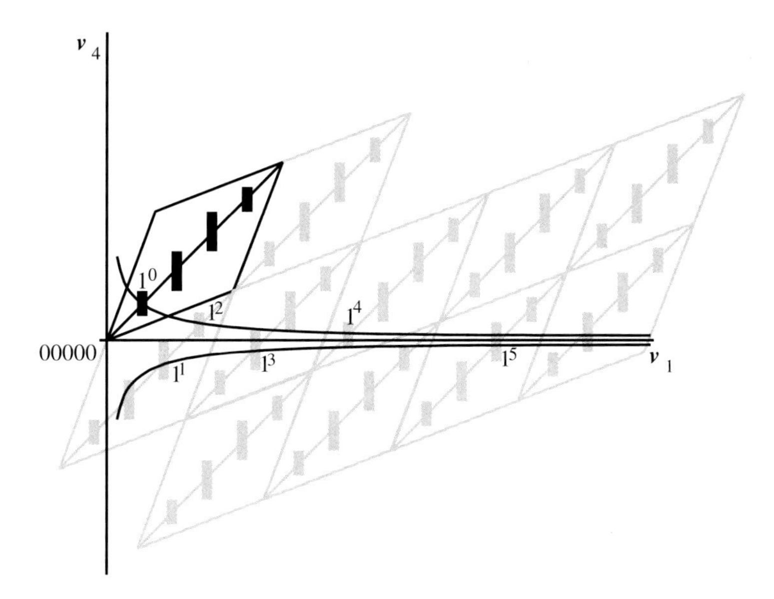

Fig. 3.9. Hyperbolic rotation in superspace demonstrated on the example of atomic surface 1. A given point 1^0 of the AS is successively mapped upon the sites marked by 1^1, 1^2, 1^3, 1^4, 1^5. In each step its v_4-component is decreased by a factor $\frac{-1}{\tau}$ and its v_1-component is increased by a factor τ. The drawing corresponds to the characteristic $(10010)_V$ section of the PT.

$$S = \begin{pmatrix} 0 & 1 & 0 & \bar{1} & 0 \\ 0 & 1 & 1 & \bar{1} & 0 \\ \bar{1} & 1 & 1 & 0 & 0 \\ \bar{1} & 0 & 1 & 0 & 0 \\ 0 & 0 & 0 & 0 & 1 \end{pmatrix}_D = \left(\begin{array}{ccc|cc} \tau & 0 & 0 & 0 & 0 \\ 0 & \tau & 0 & 0 & 0 \\ 0 & 0 & 1 & 0 & 0 \\ \hline 0 & 0 & 0 & -\frac{1}{\tau} & 0 \\ 0 & 0 & 0 & 0 & -\frac{1}{\tau} \end{array}\right)_V$$

$$= \left(\begin{array}{c|c} S^{\parallel} & 0 \\ \hline 0 & S^{\perp} \end{array}\right)_V \tag{3.44}$$

yields a tiling dual to the original PT and enlarged by a factor τ. Only scaling by S^{4n} results in a PT (increased by a factor τ^{4n}) of original orientation. Then the relationship $S^{4n}M_{PT} = \tau^{4n}M_{PT}$ holds. S^2 maps the vertices of an inverted and by a factor τ^2 enlarged PT upon the vertices of the original PT. This operation corresponds to a hyperbolic rotation in superspace shown by Janner (1992) (Fig. 3.9). The rotoscaling operation $\Gamma(10)S^2$ leaves the subset of vertices of a PT forming a pentagram invariant (Janner 1992, Fig. 3.10).

3.4.3.1 Structure Factor. The structure factor of a decagonal phase with PT's as layers can be calculated according to Eq. (3.12). The geometrical form factors g_k for the PT correspond to the Fourier transforms of four pentagonally shaped atomic surfaces [Eq. (3.13)] with the volume of the projected unit cell

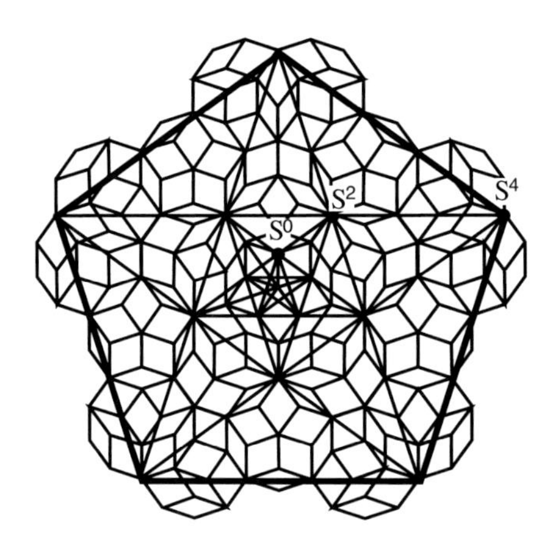

Fig. 3.10. Scaling symmetry of the vertices of a PT forming a pentagram. The rotoscaling operation $\Gamma(10)S^2$, for instance, maps each vertex of the pentagram upon another vertex of the pentagram. S^0 is mapped upon S^2 and S^4 subsequently.

$$A^{\perp}_{\mathrm{UC}} = \frac{4}{25a^{*2}}\left[(7+\tau)\sin\frac{2\pi}{5} + (2+\tau)\sin\frac{4\pi}{5}\right]. \tag{3.45}$$

Integrating the pentagons by triangularisation yields

$$\begin{aligned} g_k\left(\boldsymbol{H}^{\perp}\right) &= \frac{1}{A^{\perp}_{\mathrm{UC}}}\sin\left(\frac{2\pi}{5}\right) \\ &\quad\times\sum_{j=0}^{4}\frac{A_j\left(\mathrm{e}^{\mathrm{i}A_{j+1}\lambda_k}-1\right)-A_{j+1}\left(\mathrm{e}^{\mathrm{i}A_j\lambda_k}-1\right)}{A_jA_{j+1}\left(A_j-A_{j+1}\right)} \end{aligned} \tag{3.46}$$

with j running over five triangles of a pentagon with radius λ_k, $A_j = 2\pi\boldsymbol{H}^{\perp}\boldsymbol{e}_j$ and

$$\boldsymbol{H}^{\perp} = \pi^{\perp}(\boldsymbol{H}) = a^*\sum_{j=0}^{4}h_j\begin{pmatrix}0\\0\\0\\\cos\frac{6\pi j}{5}\\\sin\frac{6\pi j}{5}\end{pmatrix}_V . \tag{3.47}$$

The radii of the pentagons are

$$\lambda_0 = 0,\ \lambda_{1,4} = \frac{2}{5\tau^2a^*},\ \lambda_{2,3} = \frac{2}{5\tau a^*}. \tag{3.48}$$

According to Eq. (3.26), the point density $\mathrm{D_p}$ of the PT in physical space is

$$\mathrm{D_p} = \frac{1}{V}\sum_{i=1}^{n}A_i = \frac{5}{2}a^*(2-\tau)^2\tan\frac{2\pi}{5}. \tag{3.49}$$

The atomic surfaces of the PT can be decomposed into sections showing the same local coordination in physical space. Projecting all nearest neighbors of a hyperatom onto $V^{\perp}$ determines all different Voronoi polyhedra in physical space analogously to the Fibonacci chain (Fig. 3.11). Any point within a

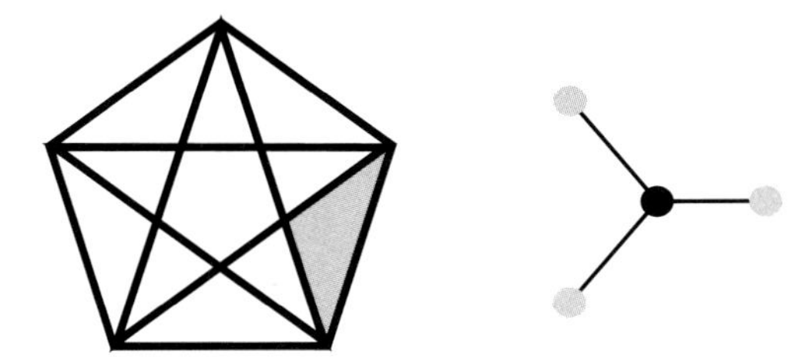

Fig. 3.11. Projecting all nearest neighbors onto the hyperatom results in 3 different local coordinations. All atoms belonging to the central pentagon exhibit local fivefold symmetry in parallel space. The local environment corresponding to the obtuse triangle is shown on the right (cf. Pavlovitch and Kléman 1987).

special region is determined by the neighboring hyperatoms that share this region. The central small pentagon is related to atoms in physical space with five neighbors located at the vertices of a pentagon.

3.4.3.2 Periodic Average Structure. A discrete periodic average structure of a decagonal phase with PT layers can be derived by an oblique projection along $[11110]_D$ and $[\bar{4}1110]_D$ onto $V^{\parallel}$ (Fig. 3.12) (Steurer and Haibach 1997). Based on the projection

$$\begin{aligned}\pi^{\parallel}(\boldsymbol{r}) &= \begin{pmatrix} 1 & 0 & 0 & \bar{1} & -\tau\sqrt{3-\tau} \\ 0 & 1 & 0 & 0 & -\tau \\ 0 & 0 & 1 & 0 & 0 \end{pmatrix}_V \boldsymbol{r}_V \\ &= \frac{2\sqrt{5}}{5a^*}\begin{pmatrix} 0 & \frac{\tau-1}{2} & -\frac{\tau+1}{2} & 1 & 0 \\ 0 & \cos\frac{\pi}{10} & -\cos\frac{\pi}{10} & 0 & 0 \\ 0 & 0 & 0 & 0 & 1 \end{pmatrix}_D \boldsymbol{r}_D, \end{aligned} \tag{3.50}$$

the basis vectors of the monoclinic periodic average structure are $\|\bar{\boldsymbol{a}}_1\| = \|\bar{\boldsymbol{a}}_2\| = \frac{2}{5a^*}(2\tau-1)$, $\bar{\boldsymbol{a}}_i = \frac{3-\tau}{\tau}\boldsymbol{a}_i, i = 1,2$, and $\bar{\boldsymbol{a}}_3 = \boldsymbol{a}_3$, $\alpha_3 = \frac{2\pi}{5}$ (Fig. 3.13). With the constraint of equal densities of the QS and its average structure, an occupancy factor of $\frac{3-\tau}{\tau}$ results for the averaged atoms, i.e., the distorted pentagons (Fig. 3.13). atomic surface. The frequency of averaged hyperatoms containing exactly one vertex of the PT is 0.7236, of those with two vertices - 0.0652, and of empty ones - 0.2112. This is similar to an average structure of an IMS with displacive and density modulation. The fraction of the monoclinic unit cell covered by projected AS is $\frac{2}{3\tau+1} = 0.342$. The reciprocal lattice of the average structure is spanned by the vectors

$$\begin{aligned}\bar{\boldsymbol{a}}_1^* &= a^*\sqrt{3-\tau}\begin{pmatrix} \cos\frac{\pi}{10} \\ -\sin\frac{\pi}{10} \\ 0 \end{pmatrix}_V, \bar{\boldsymbol{a}}_2^* = a^*\sqrt{3-\tau}\begin{pmatrix} 0 \\ 1 \\ 0 \end{pmatrix}_V, \\ \bar{\boldsymbol{a}}_3^* &= \boldsymbol{a}_3^* \,. \end{aligned} \tag{3.51}$$

All reflections of type $\boldsymbol{H} = \left(0h_2 - (h_1+h_2)h_1\right)_D$ are main reflections.

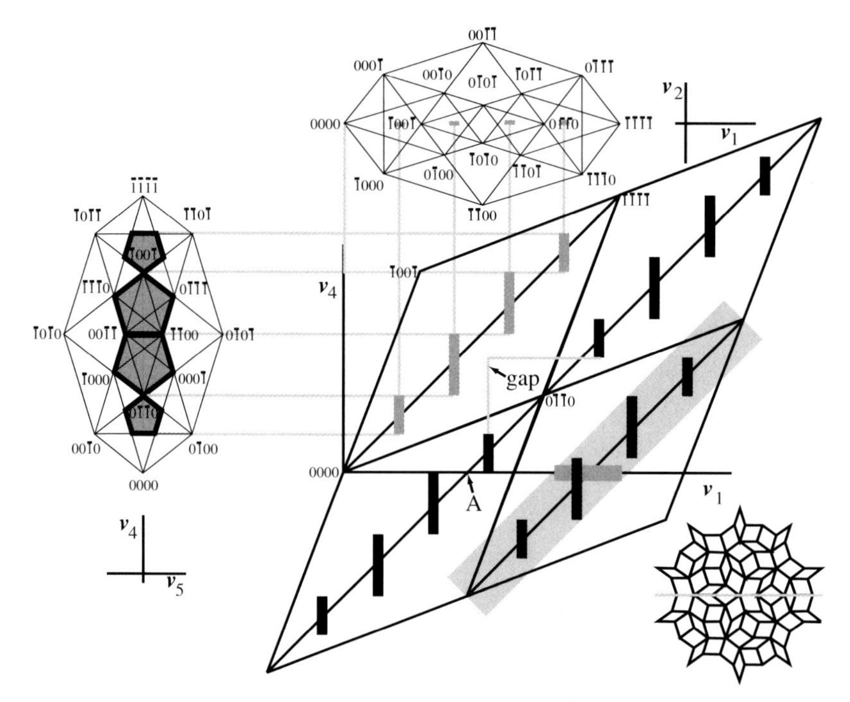

Fig. 3.12. Characteristic $(10010)_V$ section of the PT together with the parallel (above) and perpendicular space (left) projections of one 5D unit cell. In the lower left unit cell, the oblique projection direction is highlighted. The PT in the lower right corner indicates the orientation of the characteristic section.

3.4.3.3 Approximant Structures. The symmetry and metrics of rational approximants of 2D decagonal phases with rectangular symmetry have been discussed in detail by Niizeki (1991), and for some concrete 3D approximants by Zhang and Kuo (1990) and Edagawa et al. (1991). However, the authors use different approaches. In the sequel we will derive the shear matrix on the settings and nomenclature introduced in Sect. 3.4.3.

According to the group-subgroup symmetry relationship between a QC and its rational approximants, the approximants of the decagonal phase may exhibit orthorhombic, monoclinic, or triclinic symmetry. Since only orthorhombic rational approximants of the decagonal phase have been observed so far, we will focus on that special case. Preserving two mirror planes orthogonal to each other allows only matrix coefficients A_{41} and A_{53} besides the diagonal coefficients $A_{ii} = 1, i = 1, \ldots, 5$, in the shear matrix [Eq. (3.7)] to differ from zero. The action of the shear matrix is to deform the 5D lattice in a way to bring two selected lattice vectors into the physical space. If we define these lattice vectors along two orthogonal directions (P- and D-direction, respectively; Fig. 3.14) according to

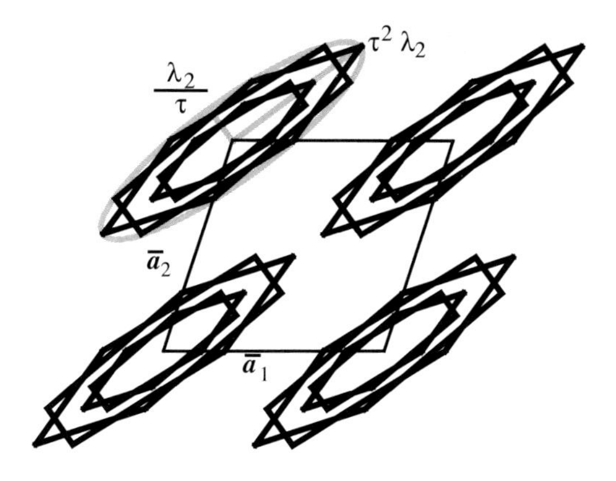

Fig. 3.13. Unit cell of the average structure of the PT. All vertices of a PT project into the projected atomic surfaces (distorted pentagons).

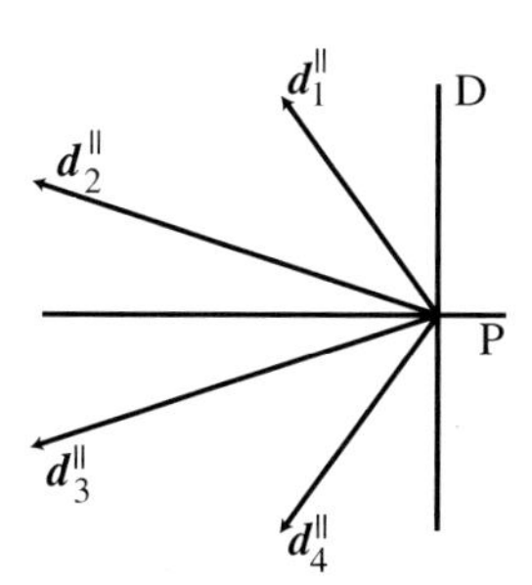

Fig. 3.14. Basis vectors in direct physical space of a decagonal QS. Pairwise combination defines the P and D direction.

$$\begin{aligned} \boldsymbol{r}_P &= -\left\{p\left(\boldsymbol{d}_2+\boldsymbol{d}_3\right)+q\left(\boldsymbol{d}_1+\boldsymbol{d}_4\right)\right\} \\ &= \frac{2\left(3-\tau\right)}{5a^*}\begin{pmatrix} \tau^2 p+q \\ 0 \\ 0 \\ p+\tau^2 q \\ 0 \end{pmatrix}_V \end{aligned} \tag{3.52}$$

and

$$\begin{aligned} \boldsymbol{r}_D &= \left\{r\left(\boldsymbol{d}_1-\boldsymbol{d}_4\right)+s\left(\boldsymbol{d}_2-\boldsymbol{d}_3\right)\right\} \\ &= \frac{2\sqrt{3-\tau}}{5a^*}\begin{pmatrix} 0 \\ \tau r+s \\ 0 \\ 0 \\ -r+\tau s \end{pmatrix}_V \end{aligned} \tag{3.53}$$

with $p, q, r, s \in \mathbb{Z}$, the $mm2$ point group symmetry is retained. From the condition that the perpendicular space components of the approximant basis vectors have to vanish

$$\pi^{\perp}\left(\boldsymbol{r}_P\right) = \pi^{\perp}\left(-\left\{p\left(\boldsymbol{d}_2+\boldsymbol{d}_3\right)+q\left(\boldsymbol{d}_1+\boldsymbol{d}_4\right)\right\}\right) = \boldsymbol{0} \tag{3.54}$$

$$\pi^{\perp}\left(\boldsymbol{r}_D\right) = \pi^{\perp}\left(\left\{r\left(\boldsymbol{d}_1-\boldsymbol{d}_4\right)+s\left(\boldsymbol{d}_2-\boldsymbol{d}_3\right)\right\}\right) = \boldsymbol{0} \tag{3.55}$$

we obtain with Eq. (3.35)

$$\frac{2\,(3-\tau)}{5a^*}\begin{pmatrix} 1 & 0 & 0 & 0 & 0 \\ 0 & 1 & 0 & 0 & 0 \\ 0 & 0 & 1 & 0 & 0 \\ A_{41} & 0 & 0 & 1 & 0 \\ 0 & A_{52} & 0 & 0 & 1 \end{pmatrix}_V \begin{pmatrix} \tau^2 p + q \\ 0 \\ 0 \\ p + \tau^2 q \\ 0 \end{pmatrix}_V$$

$$= \frac{2(3-\tau)}{5a^*}\begin{pmatrix} \tau^2 p + q \\ 0 \\ 0 \\ A_{41}\left(\tau^2 p + q\right) + p + \tau^2 q \\ 0 \end{pmatrix}_V \stackrel{!}{=} \frac{2(3-\tau)}{5a^*}\begin{pmatrix} \tau^2 p + q \\ 0 \\ 0 \\ 0 \\ 0 \end{pmatrix}_V \tag{3.56}$$

and

$$\frac{2\sqrt{3-\tau}}{5a^*}\begin{pmatrix} 1 & 0 & 0 & 0 & 0 \\ 0 & 1 & 0 & 0 & 0 \\ 0 & 0 & 1 & 0 & 0 \\ A_{41} & 0 & 0 & 1 & 0 \\ 0 & A_{52} & 0 & 0 & 1 \end{pmatrix}_V \begin{pmatrix} 0 \\ \tau r + s \\ 0 \\ 0 \\ -r + \tau s \end{pmatrix}_V$$

$$= \frac{2\sqrt{3-\tau}}{5a^*}\begin{pmatrix} 0 \\ \tau r + s \\ 0 \\ 0 \\ A_{52}\left(\tau r + s\right) - r + \tau s \end{pmatrix}_V \stackrel{!}{=} \frac{2\sqrt{3-\tau}}{5a^*}\begin{pmatrix} 0 \\ \tau r + s \\ 0 \\ 0 \\ 0 \end{pmatrix}_V . \tag{3.57}$$

Therefrom, the coefficients A_{41} and A_{52} are

$$A_{41} = -\frac{p + \tau^2 q}{\tau^2 p + q},\, A_{52} = \frac{r - \tau s}{\tau r + s} \tag{3.58}$$

and the basis vectors spanning the unit cell of the $\langle p/q, r/s\rangle$-approximant are given by

$$\begin{aligned} \boldsymbol{a}_1^{\mathrm{Ap}} &= \pi^{\parallel}\left(\boldsymbol{r}_P\right) = \frac{2\,(3-\tau)}{5a^*}\begin{pmatrix} \tau^2 p + q \\ 0 \\ 0 \end{pmatrix}_V , \\ \boldsymbol{a}_2^{\mathrm{Ap}} &= \pi^{\parallel}\left(\boldsymbol{r}_D\right) = \frac{2\sqrt{3-\tau}}{5a^*}\begin{pmatrix} 0 \\ \tau r + s \\ 0 \end{pmatrix}_V , \\ \boldsymbol{a}_3^{\mathrm{Ap}} &= \pi^{\parallel}\left(\boldsymbol{d}_5\right) = \frac{1}{a_5^*}\begin{pmatrix} 0 \\ 0 \\ 1 \end{pmatrix}_V . \end{aligned} \tag{3.59}$$

For the most common approximants the coefficients p, q, r, s correspond to Fibonacci numbers F_n defined as

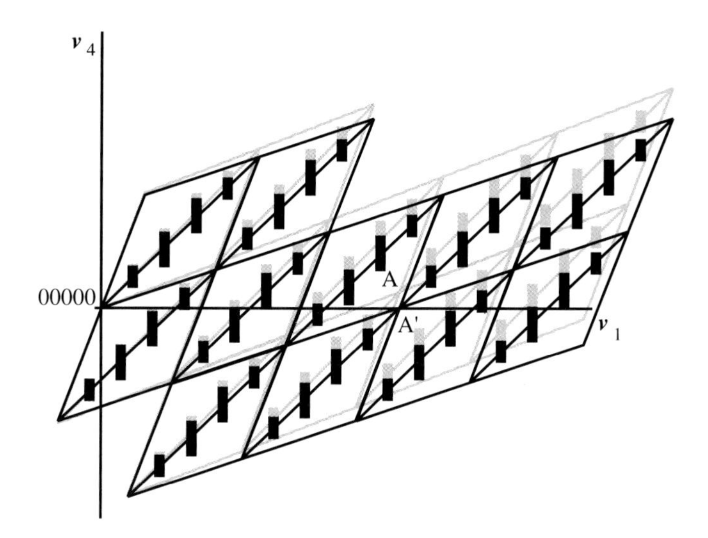

Fig. 3.15. Characteristic $[10010]_V$ section of the PT (light gray) superimposed on its rational approximant (black) with $p = 3, q = -1$. The lattice point A is mapped upon A′ by shearing the 5D lattice.

$$F_{n+1} = F_n + F_{n-1}, F_0 = 0, F_1 = 1 \,. \tag{3.60}$$

If we set $p = F_{n+2}, q = -F_n, r = F_{n'+1}, s = F_{n'}$, then we obtain the $\langle -F_{n+2}/F_n, F_{n'+1}/F_{n'} \rangle$- or, for short, $\langle n, n' \rangle$-approximants (Fig. 3.15) with lattice parameters

$$\begin{aligned} \left\| \boldsymbol{a}_1^{\mathrm{Ap}} \right\| &= \frac{2\,(3-\tau)}{5a^*}\tau^{n+2} = a_r\,(3-\tau)\,\tau^n, \\ \left\| \boldsymbol{a}_2^{\mathrm{Ap}} \right\| &= \frac{2\sqrt{3-\tau}}{5a^*}\tau^{n'+1} = a_r\sqrt{3-\tau}\tau^{n'-1}, \left\| \boldsymbol{a}_3^{\mathrm{Ap}} \right\| = \frac{1}{a_5^*} \end{aligned} \tag{3.61}$$

using the equality $\tau F_{n+1} + F_n = \tau^{n+1}$ and $a_r = \frac{2\tau^2}{5a^*}$. The approximants of this type are centered orthorhombic if $n \bmod 3 = (n'+1) \bmod 3$. In this case not only $\boldsymbol{r}_P$ and $\boldsymbol{r}_D$ are lattice vectors but also $\frac{\boldsymbol{r}_P + \boldsymbol{r}_D}{2}$, as shown by Edagawa et al. (1991).

All Bragg peaks are shifted according to Eq. (3.8). Projecting the 5D reciprocal space onto physical space results in a periodic reciprocal lattice. All reflections $\boldsymbol{H} = (h_1 h_2 h_3 h_4 h_5)$ are transformed to

$$\boldsymbol{H}^{\mathrm{Ap}} = \big([-p(h_2+h_3) - q(h_1+h_4)]\ [r(h_1-h_4) + s(h_2+h_3)]\ h_5 \big) \,.$$

3.5 Icosahedral Quasicrystals

QCs exhibiting icosahedral diffraction symmetry are called icosahedral QCs. The most perfect quasiperiodic phases known belong to this class (cf. Steurer 1996, and references therein). The Ammann tiling or 3D PT will be used as an example of a 3D QS.

3.5.1 Indexing

The set of diffraction vectors M^* forms a $\mathbb{Z}$-module of rank six. Sextuplets of integers are needed, therefore, to describe the diffraction vectors $\boldsymbol{H} = \sum_{i=1}^{6} h_i \boldsymbol{a}_i^*$, $h_i \in \mathbb{Z}$. Since there are several different indexing schemes in use, the indices may refer to different reciprocal bases. The generic indexing scheme (*setting 1*) uses six reciprocal basis vectors $\boldsymbol{a}_i^*$ directed towards the corners of an icosahedron: $\boldsymbol{a}_1^* = a^*(0,0,1), \boldsymbol{a}_i^* = a^*\left(\sin\theta\cos\frac{2\pi i}{5}, \sin\theta\sin\frac{2\pi i}{5}, \cos\theta\right), i = 2,\ldots,6$, and $\tan\theta = 2$. Here θ is the angle between two adjacent fivefold axes, $a^* = \|\boldsymbol{a}_i^*\|$, and $h_i \in \mathbb{Z}$ (Fig. 3.16). The vectors $\boldsymbol{a}_i^*$, $i = 1,\ldots,6$, can be considered as physical space projections of the basis vectors $\boldsymbol{d}_i^*$, $i = 1,\ldots,6$, of the 6D reciprocal lattice Σ^* with

$$\boldsymbol{d}_1^* = a^* \begin{pmatrix} 0 \\ 0 \\ 1 \\ 0 \\ 0 \\ c \end{pmatrix}_V, \boldsymbol{d}_i^* = a^* \begin{pmatrix} \sin\theta\cos\frac{2\pi i}{5} \\ \sin\theta\sin\frac{2\pi i}{5} \\ 1 \\ -c\sin\theta\cos\frac{4\pi i}{5} \\ -c\sin\theta\sin\frac{4\pi i}{5} \\ -c\cos\theta \end{pmatrix}_V, i = 2,\ldots,6\,. \quad (3.62)$$

Here c is an arbitrary constant usually set equal to one. The direct 6D basis results from the orthogonality condition [Eq. (3.5)] and we obtain

$$\begin{aligned} \boldsymbol{d}_1 &= \frac{1}{2ca^*} \begin{pmatrix} 0 \\ 0 \\ c \\ 0 \\ 0 \\ 1 \end{pmatrix}_V, \boldsymbol{d}_i = \frac{1}{2ca^*} \begin{pmatrix} c\sin\theta\cos\frac{2\pi i}{5} \\ c\sin\theta\sin\frac{2\pi i}{5} \\ c \\ -\sin\theta\cos\frac{4\pi i}{5} \\ -\sin\theta\sin\frac{4\pi i}{5} \\ -\cos\theta \end{pmatrix}_V, \\ i &= 2,\ldots,6\,. \end{aligned} \quad (3.63)$$

The metric tensors G and G^* are of type

$$\begin{pmatrix} A & B & B & B & B & B \\ B & A & B & -B & -B & B \\ B & B & A & B & -B & -B \\ B & -B & B & A & B & -B \\ B & -B & -B & B & A & B \\ B & B & -B & -B & B & A \end{pmatrix} \quad (3.64)$$

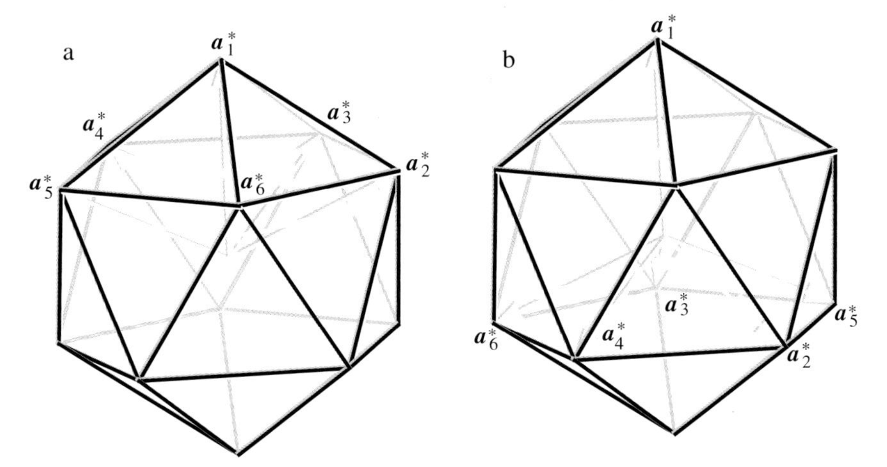

Fig. 3.16. Perspective view of the reciprocal basis of the icosahedral phase: (a) parallel and (b) perpendicular space components (setting 1).

with $A = \left(1 + c^2\right) a^{*2}$, $B = \frac{\sqrt{5}}{5}\left(1 - c^2\right) a^{*2}$ for reciprocal space, and $A = \frac{1+c^2}{4(ca^*)^2}$, $B = \frac{\sqrt{5}\left(c^2-1\right)}{20(ca^*)^2}$ for direct space. Thus, for $c = 1$ hypercubic lattices result. The direct and reciprocal lattice parameters are

$$\|\boldsymbol{d}_i^*\| = a^*\sqrt{2}, \alpha_{ij}^* = 90°, i, j = 1, \ldots, 6 \tag{3.65}$$

and

$$\|\boldsymbol{d}_i\| = \frac{1}{a^*\sqrt{2}}, \alpha_{ij} = 90°, i, j = 1, \ldots, 6\,. \tag{3.66}$$

The volume of the 6D direct lattice unit cell is

$$V = \sqrt{\det\left(\mathsf{G}\right)} = \left(\frac{1}{a^*\sqrt{2}}\right)^6 = \|\boldsymbol{d}_i\|^6\ . \tag{3.67}$$

Additionally, there exists another common setting of the icosahedral QCs. The same six-star of reciprocal basis vectors in different orientation (*setting 1'*) is referred to a Cartesian coordinate system (C-basis) oriented along twofold axes (Bancel et al. 1985)

$$\begin{pmatrix} \boldsymbol{a}_1^* \\ \boldsymbol{a}_2^* \\ \boldsymbol{a}_3^* \\ \boldsymbol{a}_4^* \\ \boldsymbol{a}_5^* \\ \boldsymbol{a}_6^* \end{pmatrix} = \frac{a^*}{\sqrt{2+\tau}} \begin{pmatrix} 0 & 1 & \tau \\ -1 & \tau & 0 \\ -\tau & 0 & 1 \\ 0 & -1 & \tau \\ \tau & 0 & 1 \\ 1 & \tau & 0 \end{pmatrix} \begin{pmatrix} \boldsymbol{c}_1 \\ \boldsymbol{c}_2 \\ \boldsymbol{c}_3 \end{pmatrix}\,. \tag{3.68}$$

The C-basis is related to the V-basis by the rotation

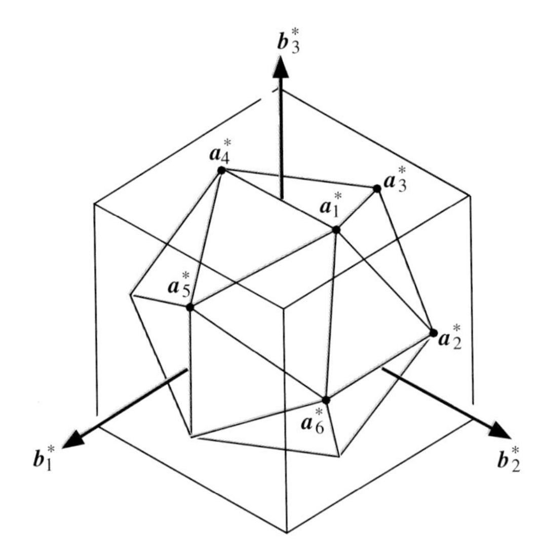

Fig. 3.17. Perspective parallel space view of the two alternative reciprocal bases of the icosahedral phase: the cubic and the icosahedral setting, represented by the bases $b_i^*, i = 1, 2, 3$ (setting 2) and $a_i^*, i = 1, \ldots, 6$ (setting 1′), respectively.

$$\begin{pmatrix} \boldsymbol{c}_1 \\ \boldsymbol{c}_2 \\ \boldsymbol{c}_3 \end{pmatrix} = \begin{pmatrix} \cos\frac{\pi}{10} & \sin\frac{\pi}{10} & 0 \\ -\cos\frac{\theta}{2}\sin\frac{\pi}{10} & \cos\frac{\theta}{2}\cos\frac{\pi}{10} & \sin\frac{\theta}{2} \\ \sin\frac{\theta}{2}\sin\frac{\pi}{10} & -\sin\frac{\theta}{2}\cos\frac{\pi}{10} & \cos\frac{\theta}{2} \end{pmatrix} \begin{pmatrix} \boldsymbol{v}_1 \\ \boldsymbol{v}_2 \\ \boldsymbol{v}_3 \end{pmatrix} . \quad (3.69)$$

Though both bases are represented on different Cartesian bases, the 6D description is equivalent and the 6D indices are identical.

A different way of indexing is based on a cubic basis (*setting 2*) (Fig. 3.17)

$$\begin{pmatrix} \boldsymbol{b}_1^* \\ \boldsymbol{b}_2^* \\ \boldsymbol{b}_3^* \end{pmatrix} = \frac{1}{2}\begin{pmatrix} 0 & \bar{1} & 0 & 0 & 0 & 1 \\ 1 & 0 & 0 & \bar{1} & 0 & 0 \\ 0 & 0 & 1 & 0 & 1 & 0 \end{pmatrix} \begin{pmatrix} \boldsymbol{a}_1^* \\ \boldsymbol{a}_2^* \\ \boldsymbol{a}_3^* \\ \boldsymbol{a}_4^* \\ \boldsymbol{a}_5^* \\ \boldsymbol{a}_6^* \end{pmatrix}$$

$$= \frac{a^*}{\sqrt{2+\tau}} \begin{pmatrix} \boldsymbol{c}_1 \\ \boldsymbol{c}_2 \\ \boldsymbol{c}_3 \end{pmatrix} . \quad (3.70)$$

The indices $(h_1h_2h_3h_4h_5h_6)$ of setting 1 are related to those of setting 2 $(h/h' \;\; k/k' \;\; l/l')$ by the transformation

$$\begin{pmatrix} h \\ h' \\ k \\ k' \\ l \\ l' \end{pmatrix}_C = \begin{pmatrix} 0 & \bar{1} & 0 & 0 & 0 & 1 \\ 0 & 0 & \bar{1} & 0 & 1 & 0 \\ 1 & 0 & 0 & \bar{1} & 0 & 0 \\ 0 & 1 & 0 & 0 & 0 & 1 \\ 0 & 0 & 1 & 0 & 1 & 0 \\ 1 & 0 & 0 & 1 & 0 & 0 \end{pmatrix} \begin{pmatrix} h_1 \\ h_2 \\ h_3 \\ h_4 \\ h_5 \\ h_6 \end{pmatrix}_D$$

$$= \begin{pmatrix} h_6 - h_2 \\ h_5 - h_3 \\ h_1 - h_4 \\ h_6 + h_2 \\ h_5 + h_3 \\ h_1 + h_4 \end{pmatrix}_D . \quad (3.71)$$

The primed indices refer to τ-times enlarged basis vectors $\boldsymbol{b}_i^*$.

3.5.2 Symmetry

The diffraction symmetry of icosahedral phases, i.e., the point symmetry group of the intensity weighted Fourier module (diffraction pattern) M_I^* can be described by the Laue group $m\bar{3}\bar{5}$. The 11 symmetry groups leaving the 6D hypercrystal structure invariant are that subset of the 6D space groups of which the point groups are isomorphous to the two possible 3D icosahedral point groups (Table 3.2). Besides primitive 6D Bravais lattice symmetry (P, primitive hypercubic, phc) also body centered (I, body centered hypercubic, bchc) and all-face centered (F, all-face centered hypercubic, fchc) Bravais lattices occur. The centering translations are $\frac{1}{2}(111111)$ for I and of type $\frac{1}{2}(110000)$ for F. A 6D hypercube has 64 corners and 240 faces. Each corner belongs to 64 and each face to 16 hypercubes. Thus the multiplicities, i.e., the number of symmetrically equivalent positions (equipoint positions), for the P, I, and F Bravais lattices are 1,2, and 16, respectively.

The orientation of the symmetry elements of the 6D space groups is fixed by the isomorphism of the 3D and 6D point groups. The reducible matrix representations of the generating symmetry operators are

$$\Gamma_D(5) = \begin{pmatrix} 1 & 0 & 0 & 0 & 0 & 0 \\ 0 & 0 & 0 & 0 & 0 & 1 \\ 0 & 1 & 0 & 0 & 0 & 0 \\ 0 & 0 & 1 & 0 & 0 & 0 \\ 0 & 0 & 0 & 1 & 0 & 0 \\ 0 & 0 & 0 & 0 & 1 & 0 \end{pmatrix}_D , \quad (3.72)$$

$$\Gamma_D(3) = \begin{pmatrix} 0 & 1 & 0 & 0 & 0 & 0 \\ 0 & 0 & 0 & 0 & 0 & 1 \\ 0 & 0 & 0 & \bar{1} & 0 & 0 \\ 0 & 0 & 0 & 0 & \bar{1} & 0 \\ 0 & 0 & 1 & 0 & 0 & 0 \\ 1 & 0 & 0 & 0 & 0 & 0 \end{pmatrix}_D , \quad (3.73)$$

$$\Gamma_D(m) = \begin{pmatrix} 1 & 0 & 0 & 0 & 0 & 0 \\ 0 & 0 & 0 & 0 & 0 & 1 \\ 0 & 0 & 0 & 0 & 1 & 0 \\ 0 & 0 & 0 & 1 & 0 & 0 \\ 0 & 0 & 1 & 0 & 0 & 0 \\ 0 & 1 & 0 & 0 & 0 & 0 \end{pmatrix}_D , \quad (3.74)$$

Table 3.2. 3D point groups of order k and corresponding 6D hypercubic space groups with their reflection conditions (cf. Levitov and Rhyner 1988, Rokhsar et al. 1988).

3D point group	k	6D space group	reflection condition
$\frac{2}{m}\bar{3}\bar{5}$	120	$P\frac{2}{m}\bar{3}\bar{5}$	no condition
		$P\frac{2}{n}\bar{3}\bar{5}$	$h_1h_2\bar{h}_1\bar{h}_2h_5h_6 : h_5 - h_6 = 2n$
		$I\frac{2}{m}\bar{3}\bar{5}$	$h_1h_2h_3h_4h_5h_6 : \sum_{i=1}^{6} h_i = 2n$
		$F\frac{2}{m}\bar{3}\bar{5}$	$h_1h_2h_3h_4h_5h_6 : \sum_{i\neq j=1}^{6} h_i + hj = 2n$
		$F\frac{2}{n}\bar{3}\bar{5}$	$h_1h_2h_3h_4h_5h_6 : \sum_{i\neq j=1}^{6} h_i + hj = 2n$
			$h_1h_2\bar{h}_1\bar{h}_2h_5h_6 : h_5 - h_6 = 2n$
235	60	$P235$	no condition
		$P235_1$	$h_1h_2h_2h_2h_2h_2 : h_1 = 5n$
		$I235$	$h_1h_2h_3h_4h_5h_6 : \sum_{i=1}^{6} h_i = 2n$
		$I235_1$	$h_1h_2h_3h_4h_5h_6 : \sum_{i=1}^{6} h_i = 2n$
			$h_1h_2h_2h_2h_2h_2 : h_1 = 5n$
		$F235$	$h_1h_2h_3h_4h_5h_6 : \sum_{i\neq j=1}^{6} h_i + hj = 2n$
		$F235_1$	$h_1h_2h_3h_4h_5h_6 : \sum_{i\neq j=1}^{6} h_i + hj = 2n$
			$h_1h_2h_2h_2h_2h_2 : h_1 = 5n$

$$\Gamma_D(\bar{1}) = \begin{pmatrix} \bar{1} & 0 & 0 & 0 & 0 & 0 \\ 0 & \bar{1} & 0 & 0 & 0 & 0 \\ 0 & 0 & \bar{1} & 0 & 0 & 0 \\ 0 & 0 & 0 & \bar{1} & 0 & 0 \\ 0 & 0 & 0 & 0 & \bar{1} & 0 \\ 0 & 0 & 0 & 0 & 0 & \bar{1} \end{pmatrix}_D . \tag{3.75}$$

The block-diagonalisation of these matrices with the matrix

$$W = a^* \begin{pmatrix} 0 & s\cos\frac{4\pi}{5} & s\cos\frac{6\pi}{5} & s\cos\frac{8\pi}{5} & s & s\cos\frac{2\pi}{5} \\ 0 & s\sin\frac{4\pi}{5} & s\sin\frac{6\pi}{5} & s\sin\frac{8\pi}{5} & 0 & s\sin\frac{2\pi}{5} \\ 1 & c & c & c & c & c \\ 0 & -s\cos\frac{8\pi}{5} & -s\cos\frac{2\pi}{5} & s\cos\frac{6\pi}{5} & -s & s\cos\frac{4\pi}{5} \\ 0 & -s\sin\frac{8\pi}{5} & -s\sin\frac{2\pi}{5} & s\sin\frac{6\pi}{5} & 0 & s\sin\frac{4\pi}{5} \\ 1 & -c & -c & -c & -c & -c \end{pmatrix} \tag{3.76}$$

with $s = \sin\theta$ and $c = \cos\theta$ gives the irreducible representations of the symmetry operations in the orthogonal subspaces. The reflection and inversion operations $\Gamma_V(m)$ and $\Gamma_V(\bar{1})$ are equivalent in both subspaces $V^{\parallel}$ and $V^{\perp}$. $\Gamma_V(5)$, a $2\pi/5$ rotation in $V^{\parallel}$ around the fivefold axis, however, corresponds to a $4\pi/5$ rotation in $V^{\perp}$

$$
\Gamma_V(5) = \left(\begin{array}{ccc|ccc}
\cos\frac{2\pi}{5} & -\sin\frac{2\pi}{5} & 0 & 0 & 0 & 0 \\
\sin\frac{2\pi}{5} & \cos\frac{2\pi}{5} & 0 & 0 & 0 & 0 \\
0 & 0 & 1 & 0 & 0 & 0 \\
\hline
0 & 0 & 0 & \cos\frac{4\pi}{5} & -\sin\frac{4\pi}{5} & 0 \\
0 & 0 & 0 & \sin\frac{4\pi}{5} & \cos\frac{4\pi}{5} & 0 \\
0 & 0 & 0 & 0 & 0 & 1
\end{array}\right)_V
= \left(\begin{array}{c|c}
\Gamma^{\parallel}(5) & 0 \\
\hline
0 & \Gamma^{\perp}(5)
\end{array}\right)_V . \tag{3.77}
$$

The same holds for the threefold rotation operation.

The Fourier module in physical reciprocal space M^* of icosahedral QCs with primitive Bravais hyperlattice is invariant under the action of the scaling matrix S^3

$$
S = \frac{1}{2}\begin{pmatrix}
1 & 1 & 1 & 1 & 1 & 1 \\
1 & 1 & 1 & \bar{1} & \bar{1} & 1 \\
1 & 1 & 1 & 1 & \bar{1} & \bar{1} \\
1 & \bar{1} & 1 & 1 & 1 & \bar{1} \\
1 & \bar{1} & \bar{1} & 1 & 1 & 1 \\
1 & 1 & \bar{1} & \bar{1} & 1 & 1
\end{pmatrix}_D ,
$$

$$
S^3 = \begin{pmatrix}
2 & 1 & 1 & 1 & 1 & 1 \\
1 & 2 & 1 & \bar{1} & \bar{1} & 1 \\
1 & 1 & 2 & 1 & \bar{1} & \bar{1} \\
1 & \bar{1} & 1 & 2 & 1 & \bar{1} \\
1 & \bar{1} & \bar{1} & 1 & 2 & 1 \\
1 & 1 & \bar{1} & \bar{1} & 1 & 2
\end{pmatrix}_D
$$

and we obtain $S^3 M^* = \tau^3 M^*$. In the case of centered Bravais hyperlattices the respective scaling operations correspond to the matrix S. By similarity transformation with the matrix W the components of the scaling operation in the two subspaces can be obtained

$$
S_V = \left(\begin{array}{ccc|ccc}
\tau & 0 & 0 & 0 & 0 & 0 \\
0 & \tau & 0 & 0 & 0 & 0 \\
0 & 0 & \tau & 0 & 0 & 0 \\
\hline
0 & 0 & 0 & -\frac{1}{\tau} & 0 & 0 \\
0 & 0 & 0 & 0 & -\frac{1}{\tau} & 0 \\
0 & 0 & 0 & 0 & 0 & -\frac{1}{\tau}
\end{array}\right)_V
= \left(\begin{array}{c|c}
S^{\parallel} & 0 \\
\hline
0 & S^{\perp}
\end{array}\right)_V . \tag{3.78}
$$

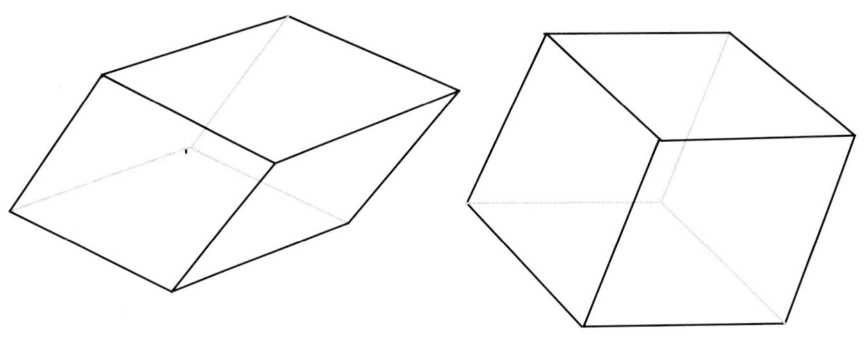

Fig. 3.18. The two unit tiles of the AT: a prolate (left) and an oblate (right) rhombohedron with equal edge lengths a_r.

3.5.3 Example of a Three-Dimensional Quasilattice: Ammann Tiling

The 3D analogue to the PT is called Ammann tiling (AT) or 3D PT (cf. Janssen 1988, Levine and Steinhardt 1986, Socolar and Steinhardt 1986, Steurer and Haibach 1998). It consists of two kinds of unit tiles: a prolate and an oblate rhombohedron with equal edge lengths a_r (Fig. 3.18). The acute angles of the rhombs covering these rhombohedra amount to $\alpha_r = \theta = \arctan(2) = 63.44°$. The volumes of the unit tiles are given by

$$V_p = \frac{4}{5}a_r^3 \sin\frac{2\pi}{5},\, V_o = \frac{4}{5}a_r^3 \sin\frac{\pi}{5} = \frac{V_p}{\tau} \tag{3.79}$$

and their relative frequencies in the AT are $\tau : 1$. Therefrom the point density $\mathrm{D_p}$ is

$$\mathrm{D_p} = \frac{\tau+1}{\tau V_p + V_o} = \frac{\tau}{a_r^3}\sin\frac{2\pi}{5}\,. \tag{3.80}$$

The set of vertices of the AT, M_{AT}, is

$$M_k = \left\{\pi^{\parallel}(\boldsymbol{r})\middle\| \pi^{\perp}(\boldsymbol{r}) \in T_i, i = 1, \ldots, 60\right\} \tag{3.81}$$

with $\boldsymbol{r} = \sum_{j=1}^{6} n_j \boldsymbol{d}_j, n_j \in \mathbb{Z}$. The 60 trigonal pyramidal subdomains T_i of the triacontahedron correspond to

$$T_i =$$

$$\left\{\boldsymbol{t} = \sum_{j=1}^{3} x_j \boldsymbol{e}_j \middle\| x_1 \in [0,\lambda], x_2 \in [0, \lambda - x_1], x_3 \in [0, \lambda - x_1 - x_2]\right\}, \tag{3.82}$$

where $\lambda = \tau a_r$ is the central distance of the vertices and $\boldsymbol{e}_j$ are vectors pointing to adjacent vertices of the triacontahedron.

In the 6D description the AT is obtained by an irrational cut of a hypercubic lattice decorated with triacontahedral atomic surfaces (Fig. 3.19) at the

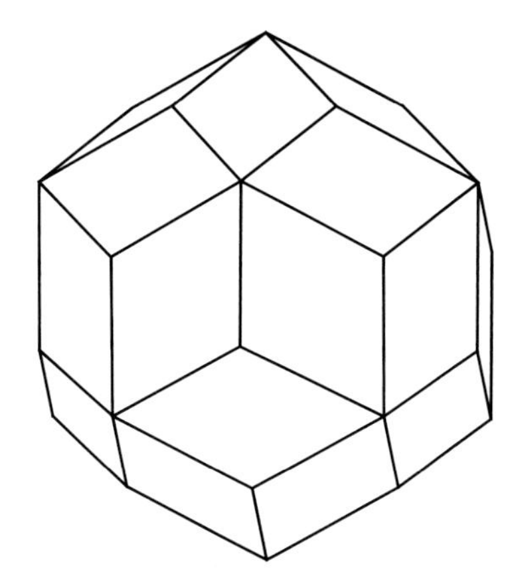

Fig. 3.19. Atomic surface of the AT in the 6D description. It results from the projection of one 6D unit cell upon $V^{\perp}$.

hyperlattice nodes. The AT is a canonical tiling, i.e., the shape of the AS corresponds to the 6D unit cell projected upon the perpendicular space. Thus, the edge length of the rhombs covering the atomic surface is equal to the perpendicular space component of the basis vectors $\left\|\pi^{\perp}\left(\boldsymbol{d}_i\right)\right\| = \frac{1}{2a^*} = a_r$.

3.5.3.1 Structure Factor. The structure factor of the AT can be calculated according to the general formula [Eq. (3.12)]. The geometrical form factor g for the AT corresponds to the Fourier transform of one triacontahedral atomic surface at the origin of the 6D unit cell. The volumes of the projected unit cell and of the atomic surface are in the case of the canonical AT equal and amount to

$$A_{\mathrm{UC}}^{\perp} = 8a_r^3\left(\sin\frac{2\pi}{5} + \sin\frac{\pi}{5}\right) . \tag{3.83}$$

Integrating the triacontahedron by decomposition into trigonal pyramids (directed from the center of the triacontahedron to three of its corners with the vectors $\boldsymbol{e}_i, i = 1,2,3;\ \|\boldsymbol{e}_i\| = \frac{1}{2a^*}$) yields

$$g\left(\boldsymbol{H}^{\perp}\right) = \frac{1}{A_{\mathrm{UC}}^{\perp}}\sum_R g_k\left(R^T\boldsymbol{H}^{\perp}\right) \tag{3.84}$$

with $k = 1,\ldots,60$, running over all symmetry operations R of the icosahedral point group

$$g_k\left(\boldsymbol{H}^{\perp}\right) =$$

$$-\mathrm{i}V_r\frac{A_2A_3A_4\mathrm{e}^{\mathrm{i}A_1} + A_1A_3A_5\mathrm{e}^{\mathrm{i}A_2} + A_1A_2A_6\mathrm{e}^{\mathrm{i}A_3} + A_4A_5A_6}{A_1A_2A_3A_4A_5A_6} \tag{3.85}$$

with $A_j = 2\pi\boldsymbol{H}^{\perp}\boldsymbol{e}_j, j = 1,2,3, A_4 = A_2 - A_3, A_5 = A_3 - A_1, A_6 = A_1 - A_2$, and the volume of the parallelepiped $V_r = \boldsymbol{e}_1\cdot(\boldsymbol{e}_2\times\boldsymbol{e}_3)$ defined by the vectors $\boldsymbol{e}_j, j = 1,2,3$ (Yamamoto 1992).

3.5.3.2 Periodic Average Structure. An all-face centered periodic cubic average structure of the AT can be obtained by oblique projection of the 6D hypercrystal structure along $[\bar{1}11010]_D$, $[01\bar{1}10\bar{1}]_D$, and $[\bar{1}001\bar{1}1]_D$ onto $V^\parallel$ (Fig. 3.20) with the projector

$$\pi^\parallel = \begin{pmatrix} 1 & 0 & 0 & 0 & 0 & -(2\tau-3) \\ 0 & 1 & 0 & 0 & 2\tau-3 & 0 \\ 0 & 0 & 1 & 2\tau-3 & 0 & 0 \end{pmatrix}_V = \frac{1}{2a^*} \tag{3.86}$$

$$\times \begin{pmatrix} -(2\tau-3) & -(\tau-1) & -(\tau-1) & 2-\tau & 1 & 2-\tau \\ 0 & \tan\frac{\pi}{5} & -\tan\frac{\pi}{5} & -\tan\frac{\pi}{5} & 0 & \tan\frac{\pi}{5} \\ 1 & 2-\tau & 2-\tau & \tau-1 & 2\tau-3 & \tau-1 \end{pmatrix}_D .$$

The lattice parameter is

$$\begin{aligned} \|\bar{\boldsymbol{a}}\| &= \left\| \pi^\parallel \begin{pmatrix} 0 \\ 0 \\ 0 \\ \bar{1} \\ 0 \\ 1 \end{pmatrix}_D \right\| = \left\| \frac{1}{a^*} \begin{pmatrix} 0 \\ \tan\frac{\pi}{5} \\ 0 \end{pmatrix}_V \right\| \\ &= \frac{\tan\frac{\pi}{5}}{a^*} = 2a_r \tan\frac{\pi}{5} . \end{aligned} \tag{3.87}$$

The projected atomic surfaces are still of regular triacontahedral shape and by a factor $\cos\phi = 0.230$, $\phi = \arctan\left(\tau^3\right)$ smaller than the original ones (Fig. 3.21). The occupancy factor of $\frac{5}{2\tau+1} = 1.180$ results from the fact that

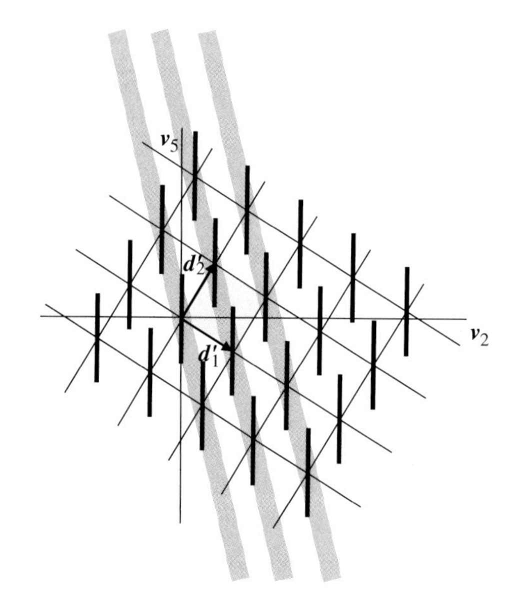

Fig. 3.20. Characteristic twofold section of the Ammann tiling in the 6D description. The vectors $\boldsymbol{d}_1'$ and $\boldsymbol{d}_2'$ correspond to the vectors $(000\bar{1}01)_D$ and $(01\bar{1}000)_D$. The oblique projection is indicated by gray strips.

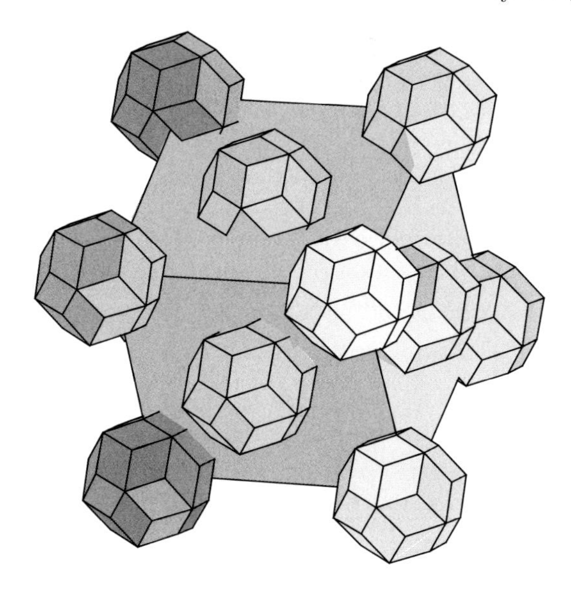

Fig. 3.21. Perspective view of one 3D unit cell of the periodic average structure of the Ammann tiling. The all-face centered unit cell is decorated by undistorted but shrunk triacontahedra resulting from the oblique projection.

a fraction of $\frac{2}{5\tau+3} = 0.180$ of all averaged hyperatoms contains an additional vertex of the AT. This is comparable to the average structure of an IMS with both displacive and substitutional modulation. The fraction of the volume occupied by the projected atomic surfaces in the average structure amounts to

$$\frac{4\tau\cos\phi\sin\frac{2\pi}{5}}{\left(\tan\frac{\pi}{5}\right)^3} = 0.195\,. \tag{3.88}$$

The reciprocal lattice Λ^* corresponding to the average structure is spanned by the vectors

$$\begin{aligned}
\bar{\boldsymbol{a}}_1^* &= a^*\tan\frac{3\pi}{10}\begin{pmatrix}\cos\frac{\theta}{2}\\ 0\\ \sin\frac{\theta}{2}\end{pmatrix}_V, \bar{\boldsymbol{a}}_2^* = a^*\tan\frac{3\pi}{10}\begin{pmatrix}0\\ 1\\ 0\end{pmatrix}_V,\\
\bar{\boldsymbol{a}}_3^* &= a^*\tan\frac{3\pi}{10}\begin{pmatrix}\sin\frac{\theta}{2}\\ 0\\ \cos\frac{\theta}{2}\end{pmatrix}_V.
\end{aligned} \tag{3.89}$$

They are enlarged by a factor τ^2 compared with the reciprocal basis vectors of the setting 2 discussed above. Thus all reflections of the type

$$\begin{aligned}
\boldsymbol{H} &= \Big(\frac{1}{2}(-h_1+h_3)\ \frac{1}{2}(-h_1+h2)\ \frac{1}{2}(-h_1-h_2)\\
&\quad \frac{1}{2}(-h_2+h_3)\ \frac{1}{2}(h_1+h_3)\ \frac{1}{2}(h_2+h_3)\Big)
\end{aligned}$$

are main reflections.

3.5.3.3 Approximant Structures. The symmetry and metrics of rational approximants of 3D icosahedral phases with pentagonal, cubic, and trigonal symmetry have been discussed in detail by Gratias et al. (1995) and for orthorhombic approximants by Niizeki (1992). In the following, we will demonstrate the derivation of shear matrix and lattice parameters on the example of cubic rational approximants consistent with our settings and nomenclature.

Preserving a particular subset of threefold axes of the icosahedral point group results in cubic approximants. The action of the shear matrix is to deform the 6D lattice Σ defined by the basis matrix

$$(\boldsymbol{d}_1\boldsymbol{d}_2\boldsymbol{d}_3\boldsymbol{d}_4\boldsymbol{d}_5\boldsymbol{d}_6) = \frac{1}{a^*2\sqrt{2+\tau}}\begin{pmatrix} 0 & -1 & -\tau & 0 & \tau & 1 \\ 1 & \tau & 0 & -1 & 0 & \tau \\ \tau & 0 & 1 & \tau & 1 & 0 \\ 0 & 0 & -1 & \tau & -\tau & 1 \\ 1 & 1 & -\tau & 0 & 0 & -\tau \\ \tau & -\tau & 0 & -1 & -1 & 0 \end{pmatrix}_C \quad (3.90)$$

in a way to bring three selected lattice vectors into the physical space. If we define these lattice vectors along the cubic axes of setting 2 according to

$$\begin{aligned} \boldsymbol{r}_1 &= \{p\,(\boldsymbol{d}_6 - \boldsymbol{d}_2) + q\,(\boldsymbol{d}_5 - \boldsymbol{d}_3)\} \\ &= \frac{1}{a^*2\sqrt{2+\tau}}\begin{pmatrix} 2\,(p+\tau q) \\ 0 \\ 0 \\ p-(\tau-1)\,q \\ -\tau^2 p+\tau q \\ \tau p - q \end{pmatrix}_C \end{aligned} \quad (3.91)$$

$$\begin{aligned} \boldsymbol{r}_2 &= \{r\,(\boldsymbol{d}_1 - \boldsymbol{d}_4) + s\,(\boldsymbol{d}_2 + \boldsymbol{d}_6)\} \\ &= \frac{1}{a^*2\sqrt{2+\tau}}\begin{pmatrix} 0 \\ 2\,(r+\tau s) \\ 0 \\ -\tau r + s \\ r-(\tau-1)\,s \\ \tau^2 r - \tau s \end{pmatrix}_C \end{aligned} \quad (3.92)$$

$$\begin{aligned} \boldsymbol{r}_3 &= \{t\,(\boldsymbol{d}_3 + \boldsymbol{d}_5) + u\,(\boldsymbol{d}_1 + \boldsymbol{d}_4)\} \\ &= \frac{1}{a^*2\sqrt{2+\tau}}\begin{pmatrix} 0 \\ 0 \\ 2\,(t+\tau u) \\ -\tau^2 t + \tau u \\ -\tau t + u \\ -t+(\tau-1)\,u \end{pmatrix}_C \end{aligned} \quad (3.93)$$

with $p, q, r, s, t, u \in \mathbb{Z}$, the $m\bar{3}$ point group symmetry is retained. From the condition that the perpendicular space components of the approximant basis vectors have to vanish, we obtain

$$A \frac{1}{a^* 2\sqrt{2+\tau}} \begin{pmatrix} 2(p+\tau q) \\ 0 \\ 0 \\ p-(\tau-1)q \\ -\tau^2 p+\tau q \\ \tau p - q \end{pmatrix}_C$$

$$= \frac{1}{a^* 2\sqrt{2+\tau}} \begin{pmatrix} 2(p+\tau q) \\ 0 \\ 0 \\ A_{41} 2(p+\tau q) + p-(\tau-1)q \\ A_{51} 2(p+\tau q) - \tau^2 p+\tau q \\ A_{61} 2(p+\tau q) + \tau p - q \end{pmatrix}_C$$

$$\stackrel{!}{=} \frac{2(p+\tau q)}{a^* 2\sqrt{2+\tau}} \begin{pmatrix} 1 \\ 0 \\ 0 \\ 0 \\ 0 \\ 0 \end{pmatrix}_C , \tag{3.94}$$

$$A \frac{1}{a^* 2\sqrt{2+\tau}} \begin{pmatrix} 0 \\ 2(r+\tau s) \\ 0 \\ -\tau r + s \\ r-(\tau-1)s \\ \tau^2 r - \tau s \end{pmatrix}_C$$

$$= \frac{1}{a^* 2\sqrt{2+\tau}} \begin{pmatrix} 0 \\ 2(r+\tau s) \\ 0 \\ A_{42} 2(r+\tau s) - \tau r + s \\ A_{52} 2(r+\tau s) + r-(\tau-1)s \\ A_{62} 2(r+\tau s) + \tau^2 r - \tau s \end{pmatrix}_C$$

$$\stackrel{!}{=} \frac{1}{a^* 2\sqrt{2+\tau}} \begin{pmatrix} 0 \\ 2(r+\tau s) \\ 0 \\ 0 \\ 0 \\ 0 \end{pmatrix}_C , \tag{3.95}$$

$$A \frac{1}{a^* 2\sqrt{2+\tau}} \begin{pmatrix} 0 \\ 0 \\ 2(t+\tau u) \\ -\tau^2 t + \tau u \\ -\tau t + u \\ -t + (\tau-1)u \end{pmatrix}_C$$

$$= \frac{1}{a^* 2\sqrt{2+\tau}} \begin{pmatrix} 0 \\ 0 \\ 2\,(t+\tau u) \\ A_{43} 2\,(t+\tau u) - \tau^2 t + \tau u \\ A_{53} 2\,(t+\tau u) - \tau t + u \\ A_{63} 2\,(t+\tau u) - t + (\tau - 1)\,u \end{pmatrix}_C$$

$$\stackrel{!}{=} \frac{1}{a^* 2\sqrt{2+\tau}} \begin{pmatrix} 0 \\ 0 \\ 2\,(t+\tau u) \\ 0 \\ 0 \\ 0 \end{pmatrix}_C . \tag{3.96}$$

In the case of cubic symmetry we have the equalities $p = r = t$ and $q = s = u$. Therewith, the submatrix $(\tilde{A}^{-1})^T$ is

$$(\tilde{A}^{-1})^T = \frac{q - \tau p}{2\,(p + \tau q)} \begin{pmatrix} \frac{1}{\tau} & 1 & \tau \\ \tau & \frac{1}{\tau} & 1 \\ 1 & \tau & \frac{1}{\tau} \end{pmatrix}_C . \tag{3.97}$$

The basis vectors spanning the unit cell of the cubic $\langle n, n' \rangle$-approximant are given by

$$\boldsymbol{a}_1^{\mathrm{Ap}} = \left\| \pi^{\parallel}(\boldsymbol{r}_1) \right\| = \frac{p + \tau q}{a^* \sqrt{2+\tau}} \begin{pmatrix} 1 \\ 0 \\ 0 \end{pmatrix}_C ,$$

$$\boldsymbol{a}_2^{\mathrm{Ap}} = \left\| \pi^{\parallel}(\boldsymbol{r}_2) \right\| = \frac{p + \tau q}{a^* \sqrt{2+\tau}} \begin{pmatrix} 0 \\ 1 \\ 0 \end{pmatrix}_C ,$$

$$\boldsymbol{a}_3^{\mathrm{Ap}} = \left\| \pi^{\parallel}(\boldsymbol{r}_3) \right\| = \frac{p + \tau q}{a^* \sqrt{2+\tau}} \begin{pmatrix} 0 \\ 0 \\ 1 \end{pmatrix}_C . \tag{3.98}$$

For the most common approximants the coefficients p, q, r, s, t, u correspond to Fibonacci numbers F_n. Setting $p = r = t = F_n, q = s = u = F_{n+1}$ we obtain the $\langle n, n+1 \rangle$-approximants with lattice parameters

$$\left\| \boldsymbol{a}_1^{\mathrm{Ap}} \right\| = \frac{\tau^{n+1}}{a^* \sqrt{2+\tau}} = \left\| \boldsymbol{a}_2^{\mathrm{Ap}} \right\| = \left\| \boldsymbol{a}_3^{\mathrm{Ap}} \right\| \tag{3.99}$$

by using the equality $\tau F_{n+1} + F_n = \tau^{n+1}$.

All Bragg peaks are shifted according to Eq. (3.8). Projecting the 6D reciprocal space onto physical space results in a periodic reciprocal lattice. All reflections $\boldsymbol{H} = (h_1 h_2 h_3 h_4 h_5 h_6)$ are transformed to

$$\begin{aligned} \boldsymbol{H}^{\mathrm{Ap}} \;\; = \;\; & \big([p(h_6 - h_2) + q(h_5 - h_3)] \;\; [r(h_1 - h_4) + s(h_2 + h_6)] \\ & [t(h_3 + h_5) + u(h_1 + h_4)] \big) . \end{aligned}$$

References

Axel, F., Gratias, D. (1995): Beyond Quasicrystals. Springer-Verlag, Berlin
Bancel, P.A., Heiney, P.A., Stephens, P.W., Goldman, A.I., Horn, P.M. (1985): Phys. Rev. Lett. **54**, 2422
Burzlaff, H., Zimmermann, H., de Wolff, P.M. (1992): in International Tables for Crystallography, Vol. A, Hahn, T. (ed). Kluwer Academic Publishers, Dordrecht, p 736
Cervellino, A., Haibach, T., Steurer, W. (1998): Phys. Rev. B **57**, 11 223
Cummins, H.Z. (1990): Phys. Rep. **185**, 211
de Wolff, P.M. (1974): Acta Crystallogr. A **30**, 777
Edagawa, K., Suzuki, K., Ichihara, M., Takeuchi, S., Shibuya, T. (1991): Philos. Mag. B **64**, 629
Elser, V. (1985): Phys. Rev. B **32**, 4892
Goldman, A.I., Kelton, R.F. (1993): Rev. Mod. Phys. **65**, 213
Gratias, D., Katz, A., Quiquandon, M. (1995): J. Phys. Condens. Matter **7**, 9101
Hermann, C. (1949): Acta Crystallogr. **2**, 139
Janner, A. (1992): Acta Crystallogr. A **48**, 884
Janssen, T. (1988): Phys. Rep. **168**, 55
Janssen, T. (1992a): Z. Kristallogr. **198**, 17
Janssen, T. (1992b): Philos. Mag. B **66**, 125
Janssen, T., Janner, A., Looijenga-Vos, A, de Wolff, P.M. (1992): in International Tables for Crystallography, Vol. C, Wilson, A.J.C. (ed). Kluwer Academic Publishers, Dordrecht, p 797
Kalning, M., Kek, S., Krane, H.G., Dorna, V., Press, W., Steurer, W. (1997): Phys. Rev. B **55**, 187
Koch, E. (1992): in International Tables for Crystallography, Vol. C, Wilson, A.J.C. (ed). Kluwer Academic Publishers, Dordrecht, p 10
Levine, D., Steinhardt, P.J. (1986): Phys. Rev. B **34**, 596
Levitov, L.S., Rhyner, J. (1988): J. Phys. (Paris) **49**, 1835
Luck, J.M., Godrèche, C., Janner, A., Janssen, T. (1997): J. Phys. A **26**, 1951
Niizeki, K. (1991): J. Phys. A **24**, 3641
Niizeki, K. (1992): J. Phys. A **25**, 1843
Pavlovitch, A., Kléman, M. (1987): J. Phys. A **20**, 687
Penrose, R. (1974): Bull. Inst. Math. Appl. **10**, 266
Rabson, D.A., Mermin, N.D., Rokhsar, D.S., Wright, D.C. (1991): Rev. Mod. Phys. **63**, 699
Rokshar, D.S., Mermin, N.D., Wright, D.C. (1988): Acta Crystallogr. A **44**, 197
Socolar, J.E.S., Steinhardt, P.J. (1986): Phys. Rev. B **34**, 617
Steurer, W. (1996): in Physical Metallurgy, Vol. 1, Cahn, R.W., Haasen, P. (eds). Elsevier Science BV, Amsterdam, p 371
Steurer, W., Haibach, T. (1998): in International Tables for Crystallography, Vol. B, Shmueli, U. (ed). Kluwer Academic Publishers, Dordrecht, in press
Steurer, W., Haibach, T. (1998): Acta Crystallogr. A, in press
van Smaalen, S. (1995): Cryst. Rev. **4**, 79
Wang, R., Wenge, Y., Hu, C., Ding, D. (1997): J. Phys. Condens. Matter **9**, 2411
Willis, B.T.M., Pryor, A.W. (1975): Thermal Vibrations in Crystallography. Cambridge University Press, Cambridge
Wondratschek, H. (1992): in International Tables for Crystallography, Vol. A, Hahn, T. (ed). Kluwer Academic Publishers, Dordrecht, p 711
Yamamoto, A. (1992): Phys. Rev. B **45**, 5217
Yamamoto, A. (1996): Acta Crystallogr. A **52**, 509
Zhang, H., Kuo, K. H. (1990): Phys. Rev. B **42**, 8907

4. Experimental Determination of the Structure of Quasicrystals

Michel Boudard and Marc de Boissieu

4.1 Introduction

The atomic structure of the matter can be studied by means of diffraction techniques, such as electron, x-ray, and neutron diffraction (Baruchel et al. 1993, Janot 1994). Indeed, the diffraction pattern of a given system depends directly on its structure. In a very schematic way Fig. 4.1 shows general results expected for a gas, liquid, amorphous solid, and crystals. Each pattern of the distribution of intensities is characteristic of the different systems, and thus information on the structure can be obtained from the analysis of these diffraction spectra. In the case of periodic crystals, the structure determination is now a routine task, even for complex structures, owing to the application of direct methods.

Quasicrystalline systems exhibit diffraction patterns very similar to those of crystalline systems in that they consist of sharp diffraction peaks (Bragg reflections) which are characteristic for a long-range order present in these materials. However, quasicrystals (QCs) have a diffraction pattern whose symmetry is not compatible with periodicity. This was the most startling point when these materials were discovered in 1984 (Shechtman et al. 1984). For a while it was even suggested that the quasicrystalline order could not exist and that QCs would be either a random assembly of clusters with only orientational order (Shechtman et al. 1984, Stephens and Goldman 1986, Goldman and Stephens 1988) or the result of twining of large unit cell crystals (Pauling 1985). A comparison with the early available data was not completely consistent with these models. Moreover, the discovery of perfect icosahedral (i) QCs in the Al-Cu-Fe and Al-Pd-Mn systems (Tsai et al. 1987, Tsai et al. 1990a, Tsai et al. 1990b) allowed to rule out both of these hypotheses (Guryan et al. 1989, Bessière et al. 1991, de Boissieu et al. 1992, Kycia et al. 1993). Indeed, QCs can now be grown as equilibrium phases, with large single grains and their diffraction pattern shows extremely sharp Bragg reflections, which correspond to a correlation length larger than a few μm (see Chap. 2).

The atomic structure of QCs is best described in the high-dimensional picture where the periodicity is recovered (see Chap. 3). We will use the cut method (Janner and Janssen 1977, 1979, 1980, Bak 1986a, 1986b, Janssen 1986), which is well suited for data analysis (Gratias et al. 1988). In this

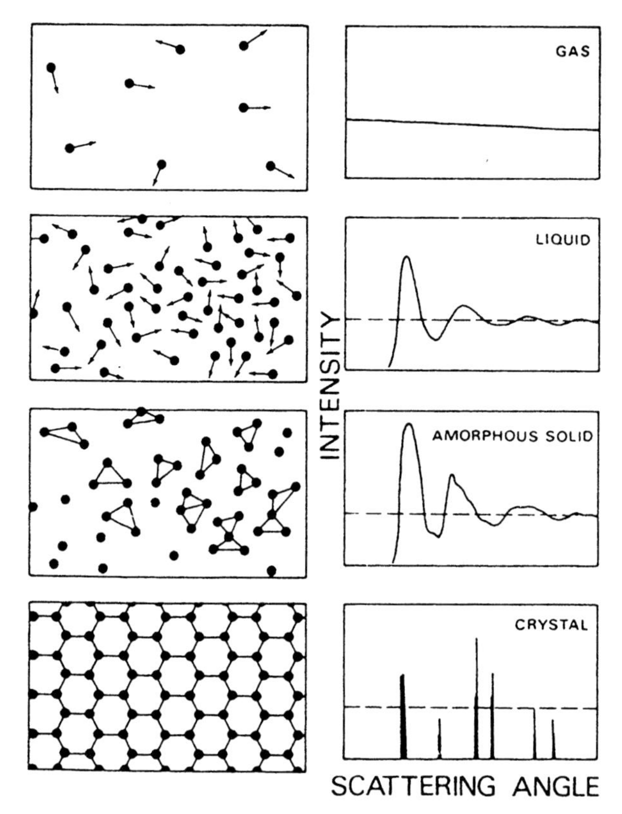

Fig. 4.1. Pictorial representation of atom distributions in matter and expected typical diffraction patterns (Janot 1994).

method, the structure is described by a set of atomic surfaces (ASs) inside a periodic unit cell. Solving the structure thus involves determining both the position and the shape of the ASs. Owing to periodicity, tools such as the Patterson analysis can be used. However, such an analysis may not lead to a unique solution and a final step involving modeling is necessary.

We recall briefly below, using a 1D example, the basic ingredients of the cut method. The 1D quasiperiodic structure can be given a periodic image in a 2D space. The periodic lattice is decorated with segment lines called ASs. The 1D structure is obtained as a section through this periodic lattice. In the most general case, the structure is defined by a set of several ASs inside the periodic 2D unit cell.

Figure 4.2 represents an example built of 2 segments in each unit cell: a segment Sperp(A) decorating the point of the square lattice and generating atoms of chemical species A, and a segment Sperp(B) decorating the middle point of each square cell and generating atoms of chemical species B. The periodic image is then determined by 1) the periodic lattice, i.e., the unit cell,

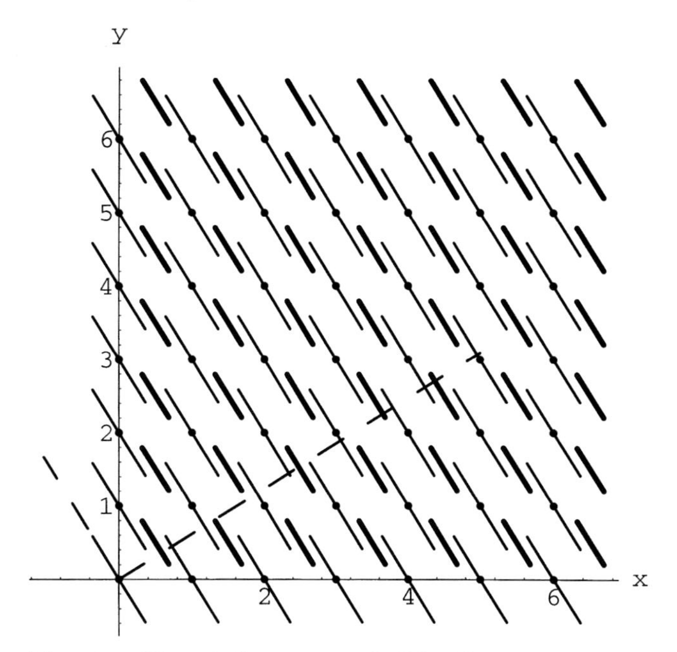

Fig. 4.2. 2D periodic structure build with two segments.

2) the sites occupied in the unit cell (e.g., the origin and center in Fig. 4.2), and 3) the shape, dimension, and chemical species of the different segments [segments Sperp(A) and Sperp(B)].

The shape of the AS is a crucial parameter which is directly related to the local atomic environment in the physical space. Indeed, if segment lines are made longer it is easy to realize that a new local environment will be generated.

The Fourier transform (FT) of the decorated lattice is also a square lattice with a side of $2\pi/a$. This lattice has Fourier coefficients, called structure factors, which are labeled with two integer indices. In the most general case, the amplitude of a Bragg reflection (or a structure factor) is given by

$$F(Q_{n_1,n_2}) = \frac{1}{a^2} \sum_i e^{i\boldsymbol{Q}\cdot\boldsymbol{R}_i} G_i(Q_{\mathrm{perp}}), \tag{4.1}$$

where the summation is over all the ASs inside the unit cell, R_i is the position of the i-th AS in the unit cell, and $G_i(Q_{\mathrm{perp}})$ is its FT. This is an oscillating function which decreases rapidly with Q_{perp}. In the computation, there is a phase factor and interferences between the different FTs $G_i(Q_{\mathrm{perp}})$. This induces a strong variation of the scattered intensity.

The diffraction pattern has an intensity distribution equal to the square of the structure factor. The 2D diffraction pattern corresponding to Fig. 4.2 is shown in Fig. 4.3. The square of the modulus of the structure factor cor-

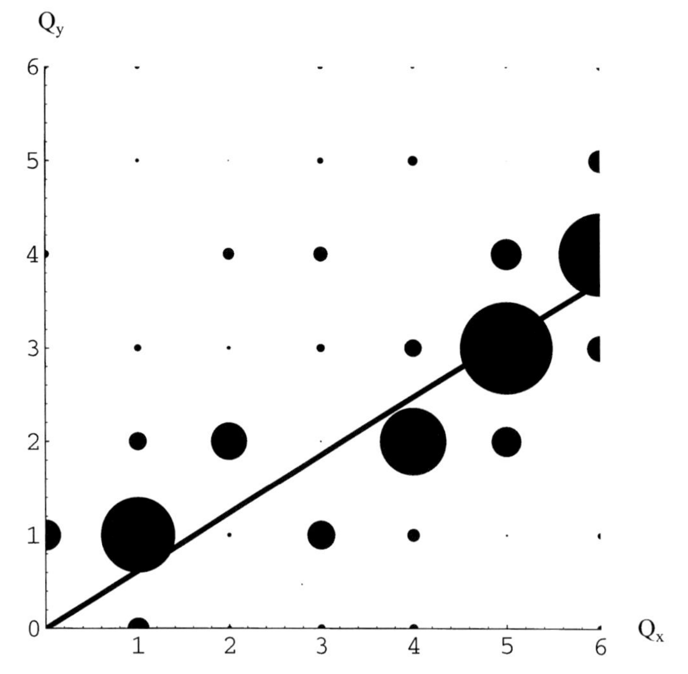

Fig. 4.3. Intensities corresponding to the model of Fig. 4.2.

responding to each Bragg peak is represented by the area of the circles. The points j, which have a small j component of Q_{perp}, have larger areas due to the behavior of the G function (this function has its maximum at zero and decays rapidly for increasing Q_{perp} values). The experimental 1D diffraction pattern is obtained as a projection of this periodic 2D pattern along the parallel (physical) direction.

4.2 X-ray and Neutron Diffraction

When performing a diffraction experiment, an incoming x-ray or neutron beam is directed on the sample. The outgoing beam will have noticeable intensity only for special directions. These directions are related to the positions in the reciprocal space and can be indexed with three indices in the case of crystals and five or six integer indices in the case of QCs (see Chap. 3).

The intensity of the diffracted beam is proportional to the square of the structure factor F. Had it been possible to measure directly the scattered amplitude (both with modulus and phases), the atomic structure would have been obtained as a direct inverse FT. However, only the intensity (and thus the modulus of the scattered amplitude) is accessible through a diffraction experiment. Thus the main problem is to determine the phases in order to obtain structural information. This problem was solved in the case of periodic crystals by applying the so-called direct methods (Hauptman 1991), which allows a routine determination of the structure of even complex periodic systems. Such a tool does not exist for a QC and one has to use another approach.

4.2.1 Patterson Analysis

Some information about the structure can be obtained by performing the FT of the intensity, called the Patterson function, which is defined as

$$P(R) = \sum_Q |F(Q)|^2 e^{i\boldsymbol{Q}\cdot\boldsymbol{R}} , \tag{4.2}$$

where $|F(Q)|^2 = F(Q)F^*(Q)$ and $F(Q)$ is the FT of the density $\rho(\boldsymbol{r})$. The FT of a product of functions is given by a convolution product

$$P(\boldsymbol{r}) = \rho(\boldsymbol{r}) * \rho(-\boldsymbol{r}) = \int_{\text{all space}} \rho(\boldsymbol{u})\rho(\boldsymbol{u}+\boldsymbol{r})d^3\boldsymbol{u}. \tag{4.3}$$

This function is the density density correlation function, or the pair correlation function. It takes non-zero values only if there is density both at $\boldsymbol{u}$ and $\boldsymbol{u}+\boldsymbol{r}$, i.e., at points separated by $\boldsymbol{r}$. Thus, this function gives a distribution of all interatomic distances. In the case of periodic structures, the Patterson function has the same periodicity as the structure and is always centrosymmetrical.

An example of the Patterson function of a simple 1D structure is shown in Fig. 4.4. For very simple structures, it is possible to get all the information about the atomic structure in the unit cell from the Patterson function. However, when one deals with several atoms per unit cell and with different chemical species, the problem can be difficult to solve from a Patterson analysis. Figure 4.4 indicates the complexity which can arise in such a case.

The procedure is similar for QCs when they are described in the periodic space, but with a further complication arising from the extended size of the ASs The carrier of the Patterson function is no longer a distribution of points. Let us take a simple 2D structure and compute the corresponding Patterson function. As shown in Fig. 4.5, the square unit cell is decorated by two different atomic surfaces located at the origin and at the center of the square, with respective lengths L_1 and L_2, which correspond to the structure in Fig 4.2. Each AS has a different length (for simplicity, we consider here that there is no difference in the chemical species). To construct the Patterson function, we follow the same procedure as that in the 1D periodic case. The autocorrelation function is obtained by translating a copy of the structure upon itself. When translating the structure onto itself, only two correlations will show up: a trivial one at the origin and the other at the

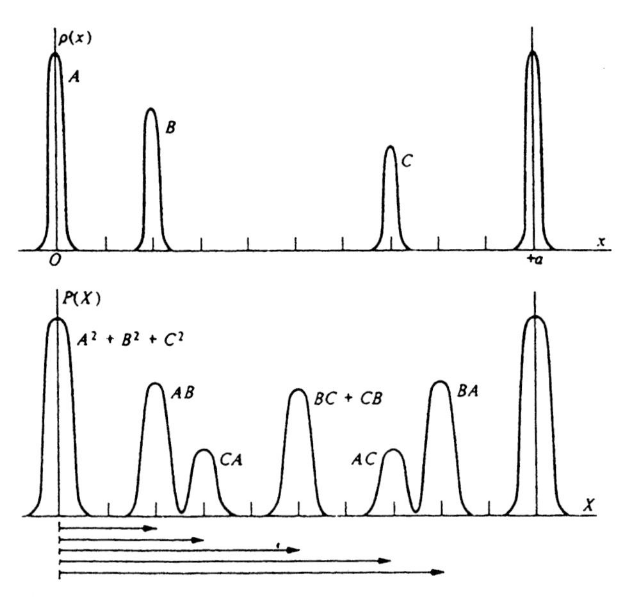

Fig. 4.4. Construction of the Patterson function for a 1D periodic structure having three atoms A, B, and C. Top panel: the electron density function. Lower panel: the corresponding Patterson function. After Warren (1990).

center of the square. Because the periodic structure is decorated with segments lines instead of points, the Patterson function will have an extended size in the perpendicular space. The profile in the perpendicular space of the Patterson function can be computed in two steps. First, the copy is translated along the diagonal of the square until it reaches the center, and then it is displaced in the perpendicular direction. For a certain length there is an intersection between the two segment lines and the Patterson function takes a constant value equal to $2L_1 \times L_2$. Then, when the displacement in the perpendicular direction reaches the value equal to $L_1 - L_2$, the overlap decreases continuously and reaches zero when $R_{\text{perp}} = L_1 + L_2$. The profile in the perpendicular direction is shown in Fig. 4.5c.

In the above example there is a one-to-one correspondence between the Patterson function and the position of the ASs: there is only one solution with the ASs at (0,0) and (1/2,1/2). In principle, the shape of the Patterson function also contains information on the shape of different ASs. In practice, because of the limited number of measured Bragg reflections, the Patterson function is smeared out and the detailed information on the shape of the ASs is difficult to extract (de Boissieu et al. 1988).

4.2.2 Contrast Variation

The Patterson analysis becomes much simplified if the different atomic contributions can be disentangled. This can be achieved by the so-called contrast

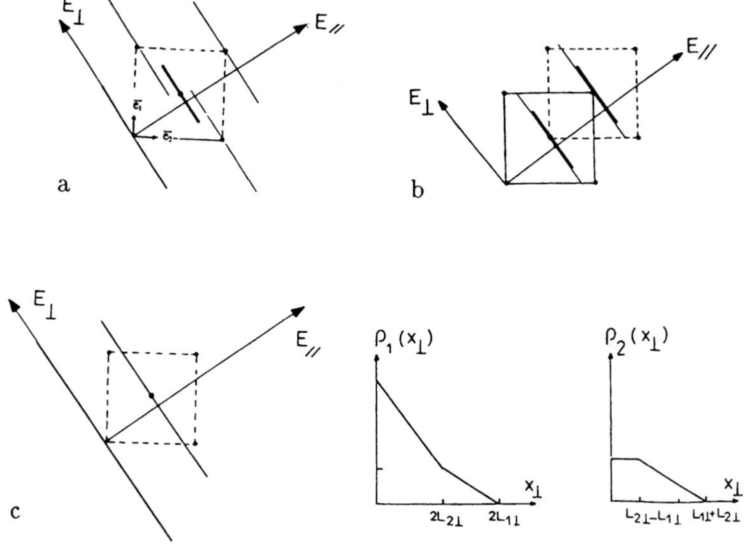

Fig. 4.5. Construction of a 2D Patterson function. (a) The initial structure contains two segment lines located at the origin and at the center, with respective lengths L1 and L2. (b) Translation (1/2,1/2) of the copy of the structure. (c) Resulting Patterson function and corresponding profile (see text).

variation technique. The principle of this method is to vary either the neutron scattering length or the x-ray form factor of a given element. This can be achieved by isotopic (or isomorphic) substitution in the case of neutron diffraction, or by anomalous x-ray scattering. The variation of the scattering power of one element produces intensity variations in the diffraction pattern. In the case of a binary alloy A-B, it can be shown that the measure of four diffraction patterns allows an extraction of the partial structure factors F_A and F_B, where the partial structure factors correspond to the FT of only the atomic densities of atoms A or B (Roth et al. 1984).

Figure 4.6 (Boudard et al. 1991) shows the results for a pseudo-binary alloy A-T in which the scattering length of the element T has been varied by isomorphic substitution. Note the strong relative intensity variation of some Bragg reflections. This makes it possible to separate the contributions from atoms A and T.

4.3 Structure of the Al-Pd-Mn Icosahedral Phase

We will now demonstrate how the basic tools of quasicrystallography described above apply when solving the structure of the *i*-Al-Pd-Mn phase (Boudard et al. 1991, Boudard et al. 1992). In the case of *i* phases, the 6D periodic space is decomposed in two 3D subspaces: the physical space and the

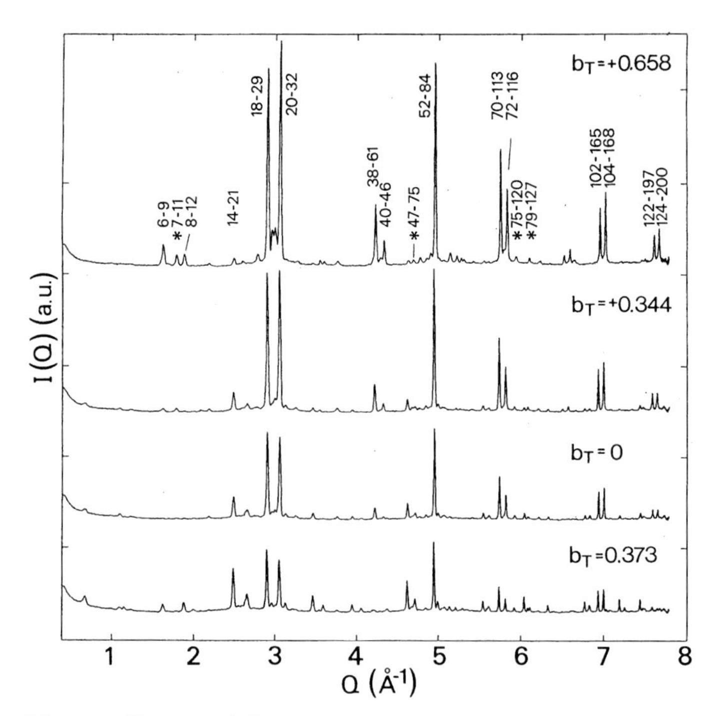

Fig. 4.6. Neutron diffraction pattern measured in a contrast variation experiment. The different scattering lengths corresponding to T are indicated. The principal peaks are indexed with the N,M indices following Cahn et al. (1986).

perpendicular space. The ASs are now 3D objects and one has to determine their shape to completly specify the structure.

4.3.1 Space Group Determination

The space group determination is very similar to what is done in 3D crystallography. There are three possible 6D hypercubic Bravais lattices in the i phase: primitive (P), face centered (F) and body centered (I). They correspond respectively to P, I, and F 6D hypercubic lattices in the reciprocal space. Once the equivalent of the screw axis or glide plane operations are added to the point group symmetry (see Chap. 3), we end up with 16 space groups (Janssen and Fasolino (1998), Rokshar et al. 1987, Levitov and Rhyner 1988). Each of these space groups leads to specific extinctions in the diffraction pattern. Finally, note that there are two possible i point groups, 235 and m$\bar{3}\bar{5}$, with or without an inversion center. Because of the Friedel rule, it is not possible to determine whether the inversion center exists from the diffraction pattern only.

A typical twofold x-ray single crystal diffraction pattern is shown in Fig. 4.7 (Boudard et al. 1992). The twofold plane goes through the origin of

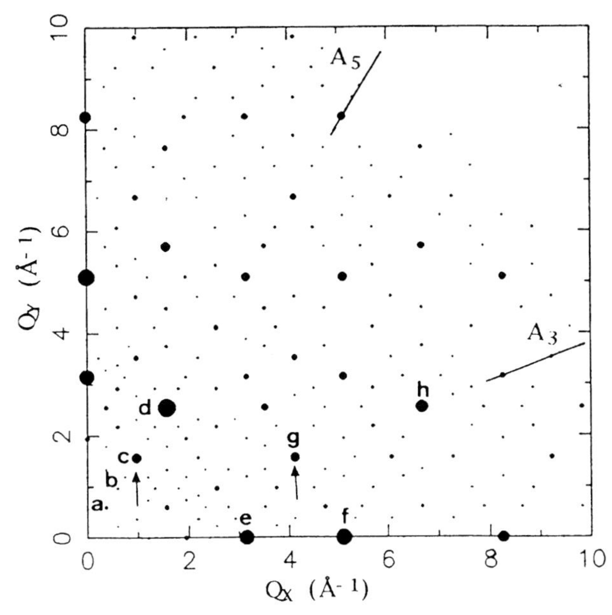

Fig. 4.7. Twofold plane x-ray diffraction pattern of the Al-Pd-Mn *i* phase.

the reciprocal space and is such that $Q_z = 0$. It contains all the principal symmetry axes (twofold, threefold, and fivefold). A simple way to determine the Bravais lattice is by looking at the inflation properties of the diffraction pattern. Indeed, along a fivefold axis we should have a τ^3 scaling for the P lattice and a τ scaling for the F or I lattices. Reflections labeled a, b, c, and d are obviously related by a τ inflation, which means that we have either the F or I lattice. A careful indexing of the diffraction pattern leads to the I reciprocal lattice (Ebalard and Spaepen 1989, Tsai et al. 1990b). Because of this scaling property, the 6D cell parameter is determined only within a τ factor. Following the convention proposed by Cahn et al. (1986), one has to ensure that the strongest diffraction peaks do correspond to a small Q_{perp} component. In other words, they must correspond to the N and M values such that M is the largest possible. All Bragg reflections are indexed when considering a 6D cubic cell with the lattice constant $a_{\text{F}} = 12.9$ Å.

As will be seen in Sect. 4.3.2, it is more convenient to consider the F direct lattice as resulting from an ordering on the P lattice (Calvayrac et al. 1989) with the 6D lattice constant $a_P = a_{\text{F}}/2 = 6.45$ Å. The I reciprocal lattice is such that all 6D indices are even or odd. In fact, the experimental diffraction pattern is such that all odd reflections are weak when compared to all even reflections. This is the signature of a superstructure ordering on a primitive lattice in the direct space.

It is not easy to visualize the 6D space and it is very convenient to produce rational sections of the 6D space containing two orthogonal directions: one in the physical space and the other in the perpendicular space. Figure 4.8 shows such a section in a plane containing two fivefold directions: one in the physical space $[1, \tau, 0]_{\text{par}}$ and another in the perpendicular space $[-\tau, 1, 0]_{\text{perp}}$. The segment lines correspond to the trace of the 3D ASs which lie in the perpendicular space. The picture is analogous to a simple 2D example (Fig. 4.2).

The 6D superstructure is illustrated in Fig. 4.8. The nodes of the primitive lattice are decorated with two different ASs in an alternative way. The real periodic 6D unit cell has the lattice constant $2a_P$, but it is convenient to consider this lattice as resulting from two P lattices shifted by a [100000] vector. The description of the atomic structure is done in the primitive unit cell (a_P) and considering two classes of atomic coordinates: their sum is either even (as [000000] or [200000]) or odd (as [100000]). In the reciprocal space, the diffraction pattern is described in a unit cell of side $2\pi/a_P$ and with the indices which are either all integers or all half integers. The intensity of the Bragg peaks corresponding to the structure presented in Fig. 4.8 is given by

$$F(Q) = G_0(Q_{\text{perp}}) + G_1(Q_{\text{perp}}), \tag{4.4}$$

when Q has all 6D integer indices, and by

$$F(Q) = G_0(Q_{\text{perp}}) - G_1(Q_{\text{perp}}), \tag{4.5}$$

when Q has all 6D half-integer indices. $G_0(Q_{\text{perp}})$ and $G_1(Q_{\text{perp}})$ are the FTs of the ASs attached to, respectively, even and odd nodes. As a direct consequence of this relation, reflections with all half-integer indices are sensitive to the difference between even and odd sites. They are rather weak and are called superstructure reflections. In the following, we shall consider four different sites with their 6D cordinates N_0 (000000), N_1 (100000), BC_0 $\frac{1}{2}$(111111), and BC_1 $\frac{1}{2}$(11111$\bar{1}$).

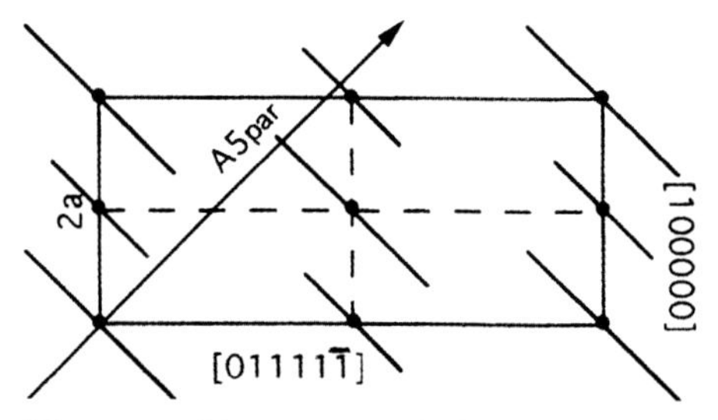

Fig. 4.8. Illustration of the superstructure ordering in direct space. The figure shows a slice of the 6D cube containing a fivefold axis in both the physical and perpendicular space. The F-type unit cell has the lattice constant $2a_P = a_F$, but the description of the structure can be done in the primitive unit cell a_P (dashed line).

4.3.2 Patterson Analysis

Two different experiments were carried out to extract information on the different atomic sublattices. In the first step, a contrast variation carried out with neutron scattering gave information on the Mn and AlPd sublattices. However, these experiments were performed on powder samples, which has some drawbacks. In the second step, anomalous x-ray scattering was used on a single-grain sample. Working close to the Pd edge, it is possible to vary the Pd x-ray form factor. Although the contrast variation is much smaller than what is obtained with neutrons, the partial Pd contribution could be extracted (de Boissieu et al. 1994).

4.3.2.1 Mn Sublattice. We have already shown in Fig. 4.6 the four different neutron diffraction patterns obtained by varying the scattering length of Mn (Boudard et al. 1991). This is done by replacing a certain amount of Mn atoms (with a neutron scattering length $b = -0.373 \times 10^{-15}$ m by a (Fe,Cr) mixture whose scattering length is $b = 0.658 \times 10^{-15}$ m [the Mn-(Fe,Cr) mixture will be called hereafter T]. The (Fe,Cr) atoms substitute the Mn atoms randomly and this kind of substitution is called isomorphic. By varying the amount of (Fe,Cr) in T one varies its corresponding scattering length and it is possible to extract the modulus of the partial structure factors $|F_{\mathrm{Al,Pd}}|$, $|F_{\mathrm{Mn}}|$, and also their phase difference $\delta\phi$ (Boudard et al. 1991). $|F_{\mathrm{Al,Pd}}|$ and $|F_{\mathrm{Mn}}|$ correspond, respectively, to the contribution of the Al and Pd atoms grouped together and to the Mn atoms. The regroupment of the Al and Pd atoms is due to the fact that we are performing the substitution on Mn atoms, which allows a separation of the contribution of this element from the rest. Thus, the ternary alloy Al-Pd-Mn is treated as a binary alloy A-Mn.

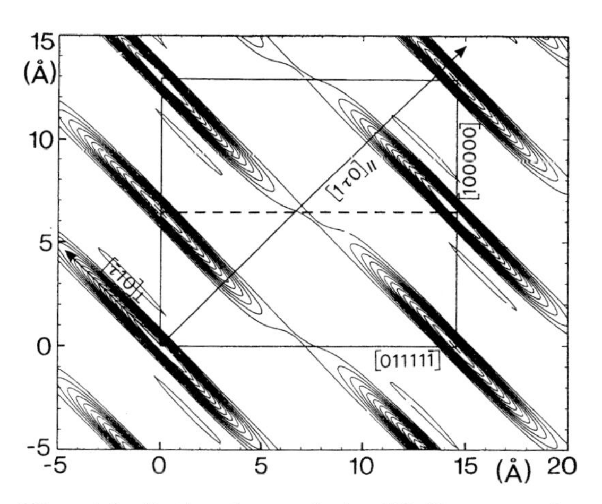

Fig. 4.9. Rational cut of the 6D Patterson function corresponding to the Mn sublattice.

Once all the Bragg peaks have been indexed and the partial structure factor extracted, we can compute the partial Patterson functions. This is simply obtained by the FT of the square of the partial structure factor $|F_{\mathrm{AlPd}}|$ and $|F_{\mathrm{Mn}}|$. Namely, we compute the function

$$P(R) = \sum_{Q} |F_{\mathrm{Mn}}|^2(Q)e^{i\boldsymbol{Q}\cdot\boldsymbol{R}} , \tag{4.6}$$

where the summation extends over all measured Q vectors, and R is a vector in the 6D space. A systematic search procedure looks for correlations in the 6D unit cell. It turns out that the Patterson function of the Mn sublattice is very simple. Correlations are only located on the lattice nodes of the 6D space. The 6D Patterson function is visualized using rational sections of the 6D space containing two orthogonal directions, one in the physical space and the other in the perpendicular space. Figure 4.9 shows the Patterson function of the Mn sublattice in a plane containing two fivefold directions: one in the physical space $[1, \tau, 0]_{\mathrm{par}}$ and another in the perpendicular space $[-\tau, 1, 0]_{\mathrm{perp}}$ (the unit cell is outlined). The lines correspond to iso-intensity contours of the Patterson function. Two observations can be made. First, it is very striking that all the features are extended in the perpendicular space as cigar-like objects. This implies that, to a first approximation, the ASs describing the Mn quasiperiodic structure are confined in the perpendicular space. In other words, the parallel components of the ASs, if any, must be small. Second, the correspondence between the Patterson function and the density function is straightforward. This is a case where the combined use of the contrast variation and 6D crystallography shows all its power.

The density correlations of the Patterson map are only located at the two sites $N_0 = [000000]$ and $N_1 = [100000]$, as seen in Fig. 4.9. The trace of the 6D unit cell has been indicated in this figure, showing that, for instance, there are no correlations on the two body-centered sites BC_0 and BC_1 or on the mid-edge site $\frac{1}{2}[100000]$. More information can be obtained when looking at the profiles of the Patterson features in the perpendicular space, in a way similar to what had been explained in the 1D case (Fig. 4.10). The intensity of these two correlations is slightly different, as shown in Fig. 4.10a. The N_0 correlation is slightly stronger than the N_1 one. Because of the very simple features, there is only one solution compatible with this Patterson map. There are two ASs which describe the Mn sublattice located at N_0 and N_1. They are, to a first approximation, spheres with radii R_0 and R_1. To be in agreement with the Patterson map, R_0 must be larger than R_1. An estimate of the relative sizes of these two ASs is deduced from the Q_{perp} dependence of the partial structure factor, once corrected for the thermal Debye-Waller factor. Taking into account the atomic density and chemical composition, it is then possible to have an estimate of the volume of the two atomic surfaces. Assuming that they are spheres, the final volumes retained correspond to two spheres with the radii $R_0 = 5.68$ Å and $R_1 = 3.55$ Å ($R_0/R_1 = 1.6$).

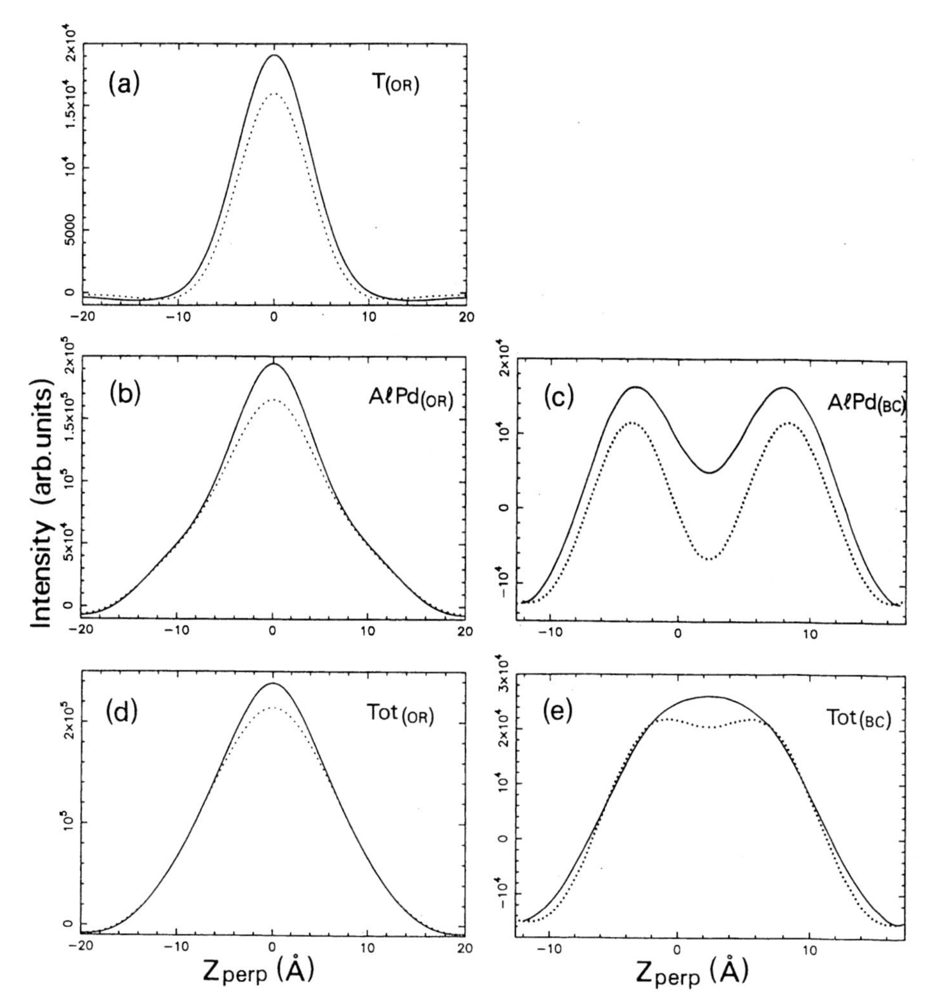

Fig. 4.10. Profile of the Patterson functions. Full (dashed) lines correspond to even (odd) $[N_0(N_1)]$ sites. (a) T origin node. (b) A origin node. (c) A body-centered node. (d) Total Patterson function (T+A) at origin node. (e) and (d) are for body-centered nodes.

4.3.2.2 Pd Sublattice. In a second step, anomalous x-ray scattering experiments on a single grain were performed. The main advantage of this procedure is that the contrast experiment is performed on the same and unique sample, as opposed to what is achieved with the isomorphic substitution. In this case the incident energy of the x-ray beam is tuned close to an absorption edge, namely the Pd K edge. For energies just below the absorption edge, anomalous scattering takes place and the x-ray form factor of Pd atoms, f_{Pd}, is decreased by six electrons. If a second experiment is carried out far from the edge, it is then possible to extract the partial Pd structure factor. The drawback of this technique is that the contrast variation is small: six electrons to be compared to $f_{\rm Pd} = 46$ electrons. Nevertheless, the Pd contribution could be extracted.

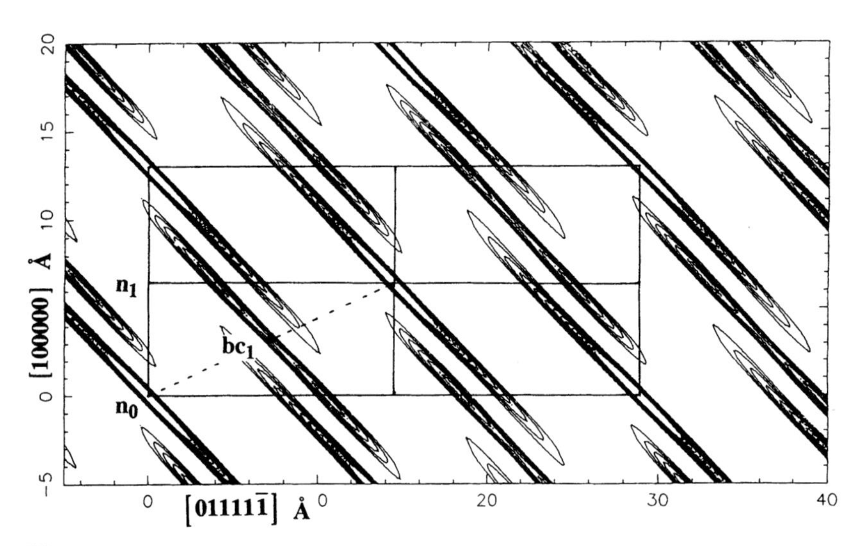

Fig. 4.11. Rational cut of the partial Pd 6D Patterson map in a plane containing a fivefold axis in both parallel and perpendicular space. The primitive underlying cell is outlined. Only contributions at N_0 and BC_1 have a significant intensity.

Figure 4.11 shows the resulting partial Pd map. Again, the structure of the Patterson map is very simple: there are significant contributions only at N_0 and BC_1 sites. In particular, at variance to what was observed for the Mn sublattice, there is no contribution at the N_1 site. This Patterson map is very simple to interpret and leads to a unique solution: Pd ASs have to be located at N_0 and BC_1 sites. A closer look at the profiles of the Patterson map give also some clues about the average shape and size of the ASs. If ASs are modeled with spheres, the N_0 AS has to be a spherical shell and the BC_1 site - a small ball. Their radii can be estimated by comparison with the available data.

Figure 4.12 shows a comparison between the experimental partial structure factors and the ones calculated with a spherical model (de Boissieu et al. 1994). Because we are using spheres as ASs, the $Q_{\rm perp}$ dependence of the structure factor is a continuous function corresponding to the FT of the ASs. Because we have two ASs located at two different sites, the $Q_{\rm perp}$ dependence of the partial structure factor splits into two branches. The size of the spherical shell located at the $n0$ site is limited on the inner side by the N_0 Mn AS. The best agreement between the data and the model is then obtained when ASs have the following sizes: - a spherical shell centered at the sites N_0 with internal and external radii equal to $0.8a_P$ and $1.25a_P$ ($a_P = 6.45$ Å), -

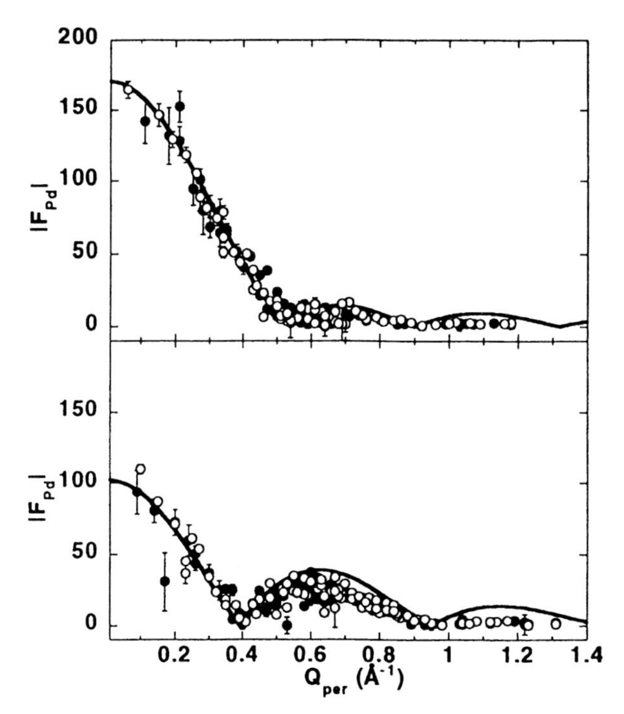

Fig. 4.12. Q_{perp} dependence of the partial structure factor F_{Pd} extracted from the data. Partial structure factors are corrected for a parallel Debye-Waller factor. Upper and lower panels correspond respectively to reflections having N and M of the same parity and different parity. The filled (open) circles are primitive (superstructure) reflections. The solid line corresponds to the calculation with a spherical model.

a small ball of Pd centered at the sites bc_1 with the radius of $0.71a_P$. This model compares quite well with the experimental data, as shown by the solid line in Fig. 4.12.

4.3.3 First-Order Model

Further information on the remaining Al atoms can be obtained by computing the total Patterson maps. Complete data sets have been measured by neutron and x-ray diffraction on single grains (Boudard et al. 1992). Both data sets are complementary because of the large difference in the scattering power of different chemical species when going from neutron to x-rays. Neutrons are sensitive to Mn sublattices whereas x-rays are more sensitive to Pd atoms. Locations of Mn and Pd ASs are already known from the contrast variation analysis and only Al ASs have to be determined. The analysis of the total Patterson maps showed that Al atoms have to be located at N_0 and N_1 sites. Based on this, a spherical model, which considers, as a first trial, ASs as spheres or spherical shells, can be constructed. This is only a very crude first-order approximation and certainly ASs do have a much more complex structure. However, such a model compares already well with the low Q_{perp} data and gives some clues on the local order.

The resulting model has the following characteristics: - a core of Mn with a radius of $0.8a_P$ is centered at the origin sites N_0 and is surrounded successively by a shell of Pd with an external radius of $1.26a_P$ and a shell of Al with an external radius of $1.55a_P$; - a core of Mn with a radius of $0.52a_P$ is centered at the origin sites N_1 and is surrounded by a shell of Al with an external radius of $1.64a_P$; - a small ball of Pd with a radius of $0.71a_P$ is centered at the sites BC_1; - the BC_0 site is empty. The resulting model is shown in Fig. 4.13.

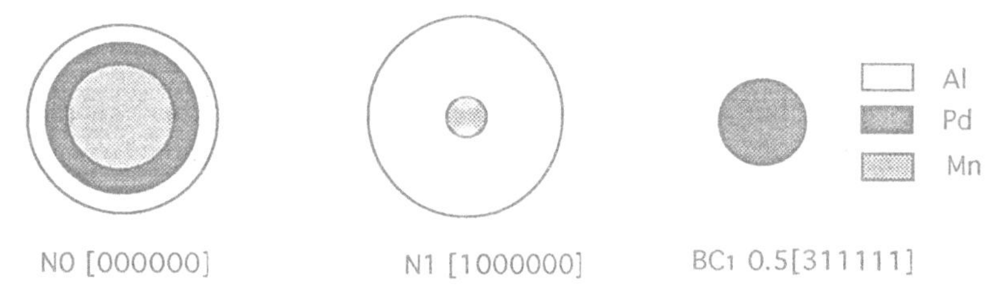

Fig. 4.13. Model built with spherical atomic surfaces.

The spherical model fits the data relatively well and, as expected from a spherical model, is better for low-Q_{perp} values. The residual factor R, defined as

$$R = \sum \frac{|I_{\text{obs}} - I_{\text{calc}}|}{\sum I_{\text{obs}}}, \tag{4.7}$$

is 9% for x-rays and 16% for neutrons when considering values of Q_{perp} lower than 0.5 (in $\frac{2\pi}{a_P}$units), and is 14% (x-rays) and 24% (neutrons) for all the data.

Although we have learned a lot about the structure, we are still far away from what is usually achieved for typical crystalline materials (R factor of the order of 5% for all the data). This is due to the fact that we have very few constraints and free parameters in our model and the final details are not accessible. After describing in the next sections the resulting 3D structure and the limitations of the direct approach, we shall present ideas that are being developed to improve the model.

4.3.4 About the Resulting Atomic Structure

From the 6D spherical model presented above, it is easy to generate a list of atomic coordinates which are required for calculations of many physical properties of quasicrystals. Such a list is difficult to handle and it is more useful to describe the 3D structure either in terms of local i clusters or alternatively in terms of dense atomic planes. The nature of i clusters in the 3D structure can be found following the procedure proposed by Duneau and Oguey (1989) in their Al-Mn model (this model is described in the next section). Comparing the sphere sizes of the ASs related to the origin nodes (sites N_0 and N_1) with the ASs proposed by Duneau and Oguey (Figs. 4.13 and 4.18), it can be shown that the external shell of the Mackay icosahedron is actually present in the structure. This atomic cluster has been found in the cubic α-Al-Mn-Si phase (Cooper and Robinson 1966) and has been the basic ingredient of various modelings of the i-Al-Mn-Si phase (Guyot and Audier 1985, Elser and Henley 1985). In the α-Al-Mn-Si phase the Mackay icosahedron (Fig. 4.14) decomposes into three successive shells: a small Al icosahedron, a Mn icosahedron which is twice as large, and an Al icosidodecahedron obtained by placing 30 atoms on the edges of the Mn icosahedron (along the twofold directions). In the spherical model of the Al-Pd-Mn i phase, only the exter-

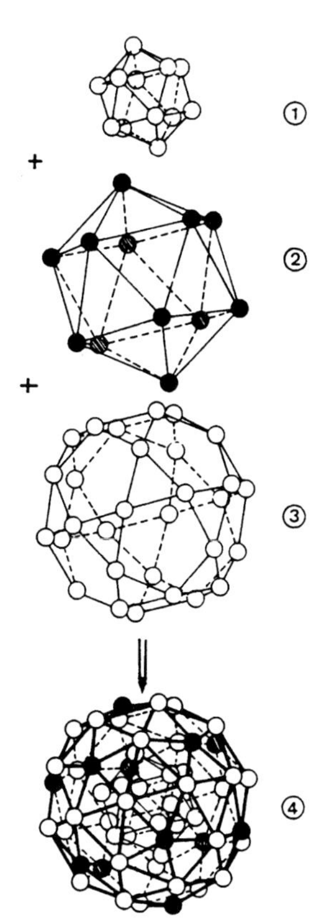

Fig. 4.14. The Mackay icosahedron in the α-Al-Mn-Si phase decomposed into three different shells. White and black dots represent Al and Mn atoms.

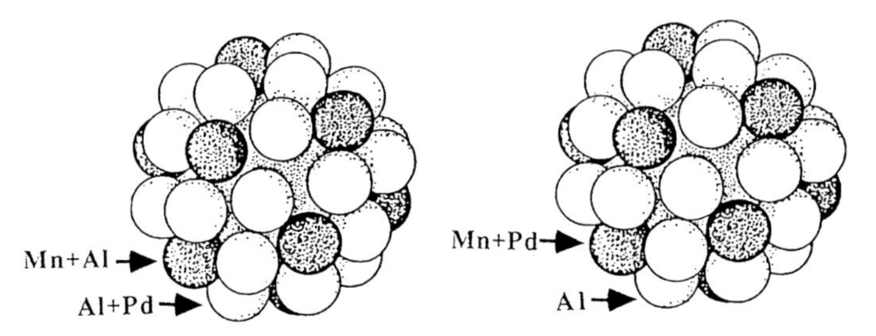

Fig. 4.15. Two kinds of Mackay clusters in the Al-Pd-Mn *i* phase.

nal shell of the cluster, i.e., the large icosahedron and the icosidodecahedron, exists. The internal small icosahedron is replaced by part of a dodecahedron. Two kinds of external shells are found, depending on the parity of the high-dimensional lattice node where the cluster center is located. In the 6D description, the Mn inner core has a radius smaller than the standard TR; this implies that the resulting 3D large icosahedron is occupied by Mn atoms plus a small number of Al or Pd atoms. The external icosidodecahedron is made of either Al atoms alone, or of Al and Pd atoms. In summary, two types of clusters are present in the 3D structure which are related to two 6D origin nodes (even and odd) (Fig. 4.15). Namely, a pseudo-Mackay cluster type 1 with a large icosahedron of Mn+Al and an icosidodecahedron of Pd+Al, and a pseudo-Mackay cluster type 2 with a large icosahedron of Mn+Pd and an icosidodecahedron of Al. About 60% of the atoms belong to one of these two clusters. There is also another type of cluster (Katz and Gratias 1994) that can be considered as the glue atoms between the Mackay-type clusters. This cluster is built of 33 atoms, which correspond to the first two successive shells of the Bergman cluster (Bergmann (1984)) in the following way: a central Pd atom is surrounded by an icosahedron of 12 atoms (first shell) and by a dodecahedron of 20 atoms (second shell). About 80% of atoms belong to this cluster.

The existence of dense planes in the 3D structure may be found from planar projections of the structure. This structure can be obtained either analytically using the 6D image or by computing the FT of all structure factors (with their phases) lying in a reciprocal plane (since the FT of a cut is a projection). Structure factors deduced from x-ray data, together with phases of the spherical model, were used for the computation of the twofold projection shown Fig. 4.16a. When looking at this projection at glancing angle, series of lines corresponding to atomic planes are clearly visible. The densest planes are characterized by larger spacing. They are found to be orthogonal to a fivefold axis. Twofold planes are also visible, but they are closer to each other and contain a smaller density of points. Since the x-ray structure factors were used, the plots in Fig. 4.16 may be compared, in a very crude approximation, to the high-resolution electron microscopy (HREM) data. A projection along a fivefold axis is shown Fig. 4.16b. 2D cuts of the 3D structure perpendicular

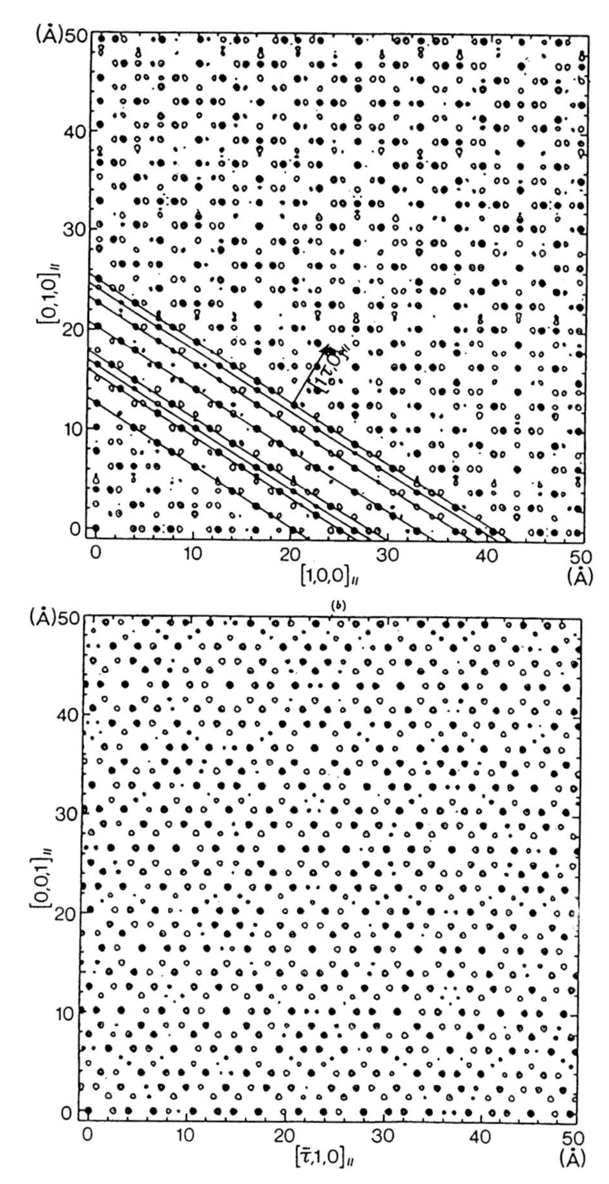

Fig. 4.16. (a) Twofold projection of the *i*-Al-Pd-Mn structure as obtained by Fourier transform of the $Q_{\|}(z) = 0$ layer of the reciprocal space. The map shows contour plots of the electronic density. Dense contour regions correspond to high electronic density projection. The densest planes show up perpendicularly to the fivefold direction. (b) Fivefold projection.

to a fivefold axis and made at different levels along this axis are shown in Fig. 4.17. Some planes have no thickness, whereas the other are corrugated and consist of one dense plane sandwiched between layers 0.5 Å apart. As a result of quasiperiodicity, all planes are, in principle, different. However, within very small changes in composition and local order, there is only a finite number of different dense atomic planes to be considered. Each type of a plane has a similar local order and a chemical composition lying between two similar values. Obviously, the position of such planes is not random but quasiperiodic.

The traces of some shells of the Mackay icosahedron have been outlined in Fig. 4.17. This also gives some idea about the quasiperiodic distribution of the Mackay clusters (see, for instance, the cut at $z_{A5} = 4.56$ Å). The hierarchical packing of cluster is also visible: tenfold wheels are packed together to form a new ring of ten wheels, and so on. This also illustrates two basic properties of any quasiperiodic structure: the long-range orientational order and the quasiperiodic long-range translational order.

4.3.5 Limitations of the Direct Approach

One of the main drawbacks of the direct approach is that the precise shape of the ASs is not specified. This results, for instance, in distances in the 3D structure which are too small to be physically realistic. This problem can be solved by tailoring properly the external shape of the ASs, but it would be useful to have some guide from the experimental results. In standard 3D crystallography, once an approximate solution is found, a model can be extracted from the density map obtained by the FT of the structure factors with their correct phases. This route is almost impossible in the quasicrystalline case because of the infinite number of parameters necessary for the shape specification of ASs. Moreover, truncation effects arising from the computation of the density map with a finite number of reflections may have dramatic consequences (de Boissieu et al. 1988).

4.3.6 Modeling

A modeling with spheres allows an easy analytical calculation of the structure factors and gives a good agreement for low values of Q_{perp}. It allows a good estimate of the occupied sites in the 6D unit cell and an evaluation of the size of the corresponding ASs for different chemical species. This approach for modeling is only a first step toward a realistic model and it will be compared to two other 6D models based on more complicated ASs. We will illustrate how additional constraints arising from density, chemical composition, and minimum distances between atoms in the physical space can be introduced into the models.

Figure 4.18 compares the ASs at the origin nodes for the spherical model with the one proposed by Duneau and Oguey (1989) for the Al-Mn-Si i

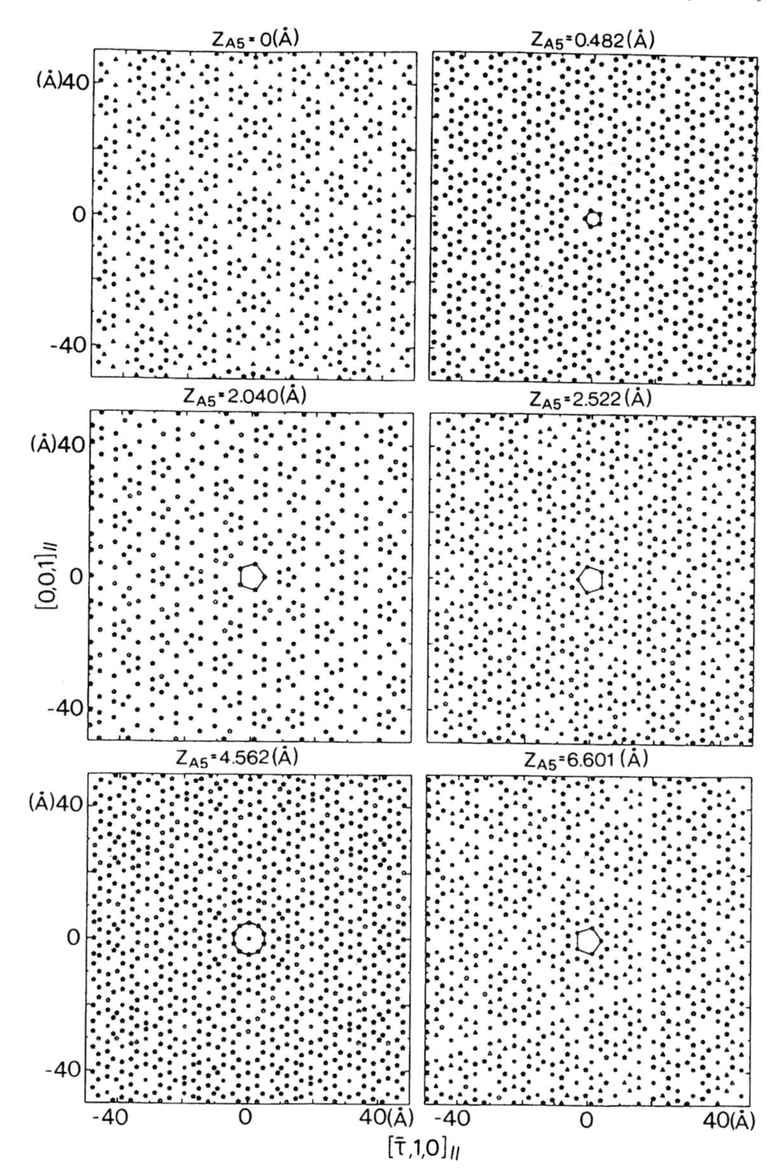

Fig. 4.17. A series of six successive dense atomic planes perpendicular to a fivefold axis. The level of each plane is indicated. Atomic species are indicated: (*) Al, (△) Pd, and (•) Mn atoms. Some fivefold rings corresponding to external shells of a pseudo-Mackay cluster have been outlined.

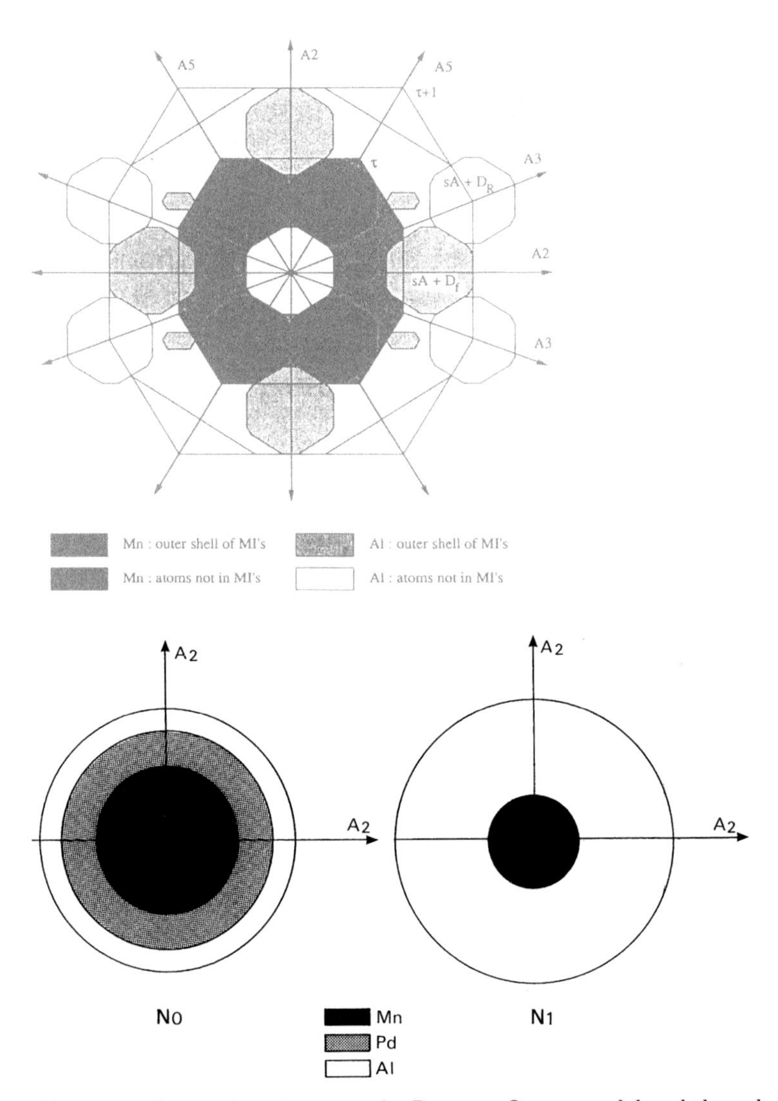

Fig. 4.18. Comparison between the Duneau–Oguey model and the spherical model.

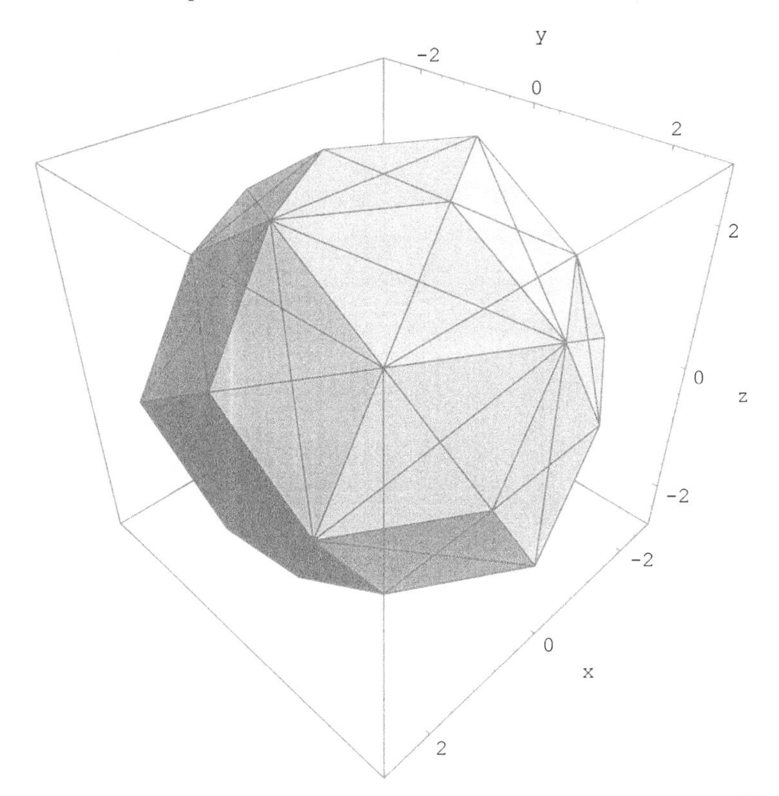

Fig. 4.19. Triacontahedron generating the standard 3D Penrose tiling.

phase (DO model). The ASs proposed in this model are built in order to maximize the number of generated Mackay icosahedra (Mackay 1962, Cooper and Robinson 1966). Besides the condition of generating a lot of Mackay icosahedra, the DO model does not create short distances. This model can be built up starting from a unique AS located at the nodes of the 6D unit cell. This AS is a triacontahedron (TR) which is the AS corresponding to the standard Penrose tiling (Fig. 4.19).

It is possible to increase the density of the Penrose tiling by taking a TR τ times larger than the canonical one, i.e., a large triacontahedron (LTR). In particular, a new distance is generated along the twofold axis (see Fig. 4.20). There are no short distances along this direction and neither along the threefold direction (see Fig. 4.21). By a short distance we mean a distance which is too short to be physically realistic. In the case of metallic alloys, the shortest interatomic distance is equal to 2.5 Å. However, there is a problem of short distances along the fivefold direction, as can be seen in Fig. 4.22: the two ASs centered at [2,0,0,0,0,0] and [0,1,1,1,1,$\bar{1}$] will lead in the parallel space to an unphysical short distance of 1 Å. This can be seen in the figure in the region where the dashed lines are closeby.

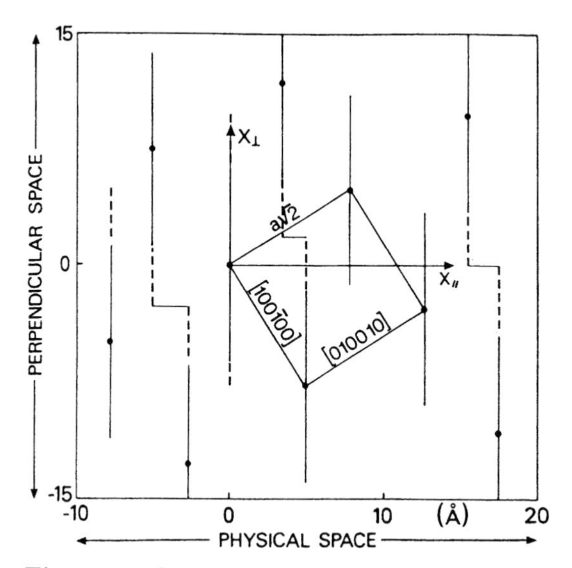

Fig. 4.20. 2D cut containing a twofold direction in both the physical and perpendicular space. The atomic surfaces corresponding to the canonical TR are drawn with a solid line and the inflation by τ is drawn with a dashed line.

To avoid these short distances one has thus to tailor correctly the ASs This can be carried out in the perpendicular space where the projection of the two polyhedra have been represented in Fig. 4.23. To avoid short distances, there should be no intersection between the two polyhedra. This problem is shown for the LTR in Fig. 4.23 where it can be seen that there is an intersection generating short distances. They can be avoided by cutting both ASs by planes perpendicular to the fivefold axes (only one is represented in Fig. 4.23). These planes define a dodecahedron. The AS located at the origin is thus a LTR truncated by a dodecahedron. It is possible to add a small polyhedron at the BC site whose maximum diameter along the fivefold direction is represented in the figure. In the DO model the polyhedron is a small dodecahedron. This model does not present a superstructure, i.e., no difference is considered between the two origin sites N_0 and N_1 (respectively, BC_0 and BC_1). However, it can be generalized to the *i*-Al-Pd-Mn phase by considering different chemical occupation of the sites according to their parity. Without additional modifications, the LTR AS truncated by a dodecahedron will generate a pseudo-Mackay cluster which has a central Mn atom surrounded by an incomplete dodecahedron with six or seven atoms, instead of the inner Al icosahedron.

In the original DO model, the central part of ASs at the origin is removed. Moreover, some external parts are removed along the threefold axes. This is

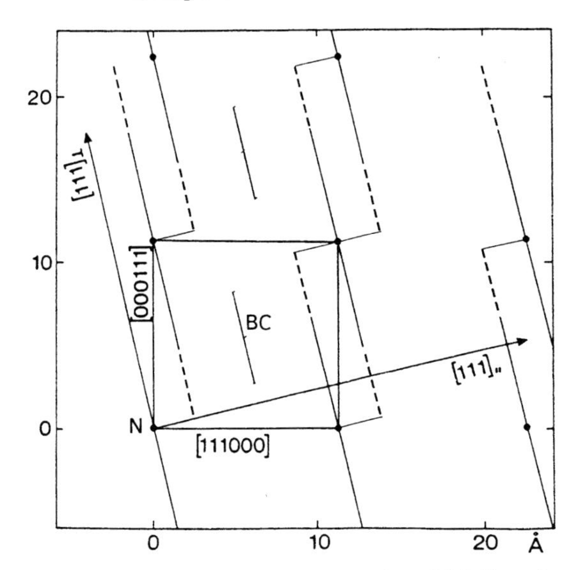

Fig. 4.21. 2D cut containing a threefold direction in both the physical and perpendicular space. The atomic surface corresponding to the canonical TR is drawn with a solid line and the inflation by τ is drawn with a dashed line.

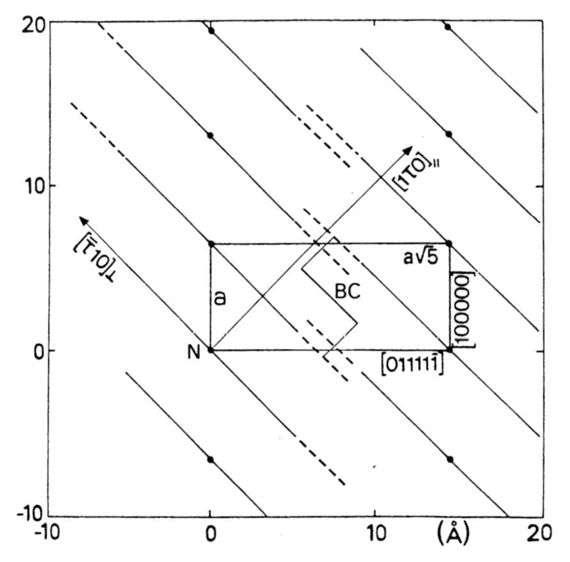

Fig. 4.22. 2D cut containing a fivefold direction in both the physical and perpendicular space. The atomic surfaces corresponding to the canonical TR is drawn with a solid line and the inflation by τ is drawn with a dashed line.

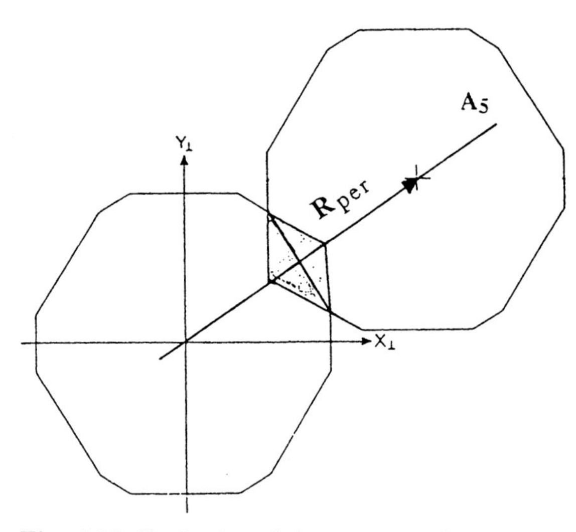

Fig. 4.23. Projection of the atomic surfaces corresponding to the LTR.

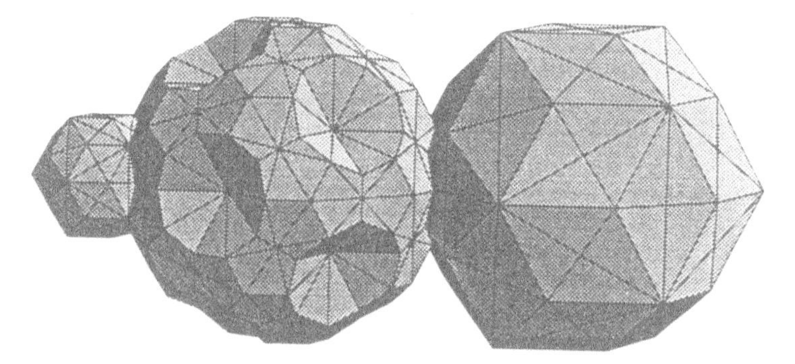

Fig. 4.24. The atomic surfaces corresponding to the Katz–Gratias model.

done to avoid short distances due to the additional small ASs which are added in order to generate the inner Al icosahedron of the Mackay icosahedron. This modification allows an increase in the density by 5%.

Another solution which avoids short distances along fivefold axes has been proposed by Katz and Gratias and will be called K model hereafter (Cornier-Quiquandon 1991, Katz and Gratias 1994). The AS centered at the node equivalent to [2,0,0,0,0,0] is the LTR, while the AS retained at the node equivalent to $[0,1,1,1,1,\bar{1}]$ is now a similar LTR but with holes corresponding to the intersection with the LTR at [2,0,0,0,0,0] (see Figs. 4.22, 4.23, and 4.24). As shown in Fig. 4.25, with these two ASs located at the nodes it is possible to have a supplementary AS only at one of the two BC positions, but with a diameter along the fivefold axis two times larger than in the previous

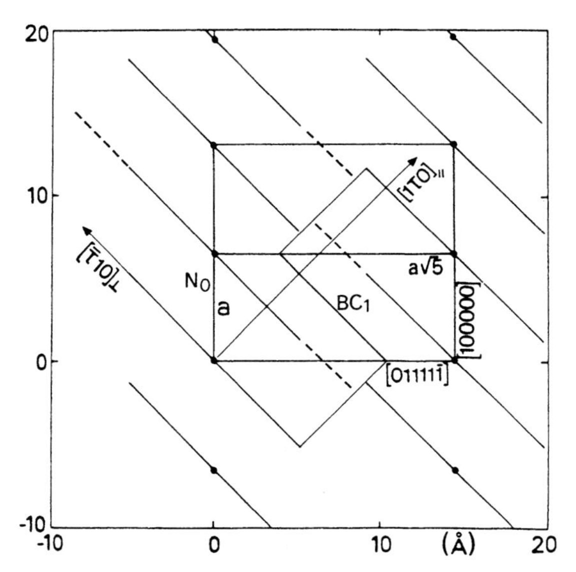

Fig. 4.25. 2D cut containing a fivefold direction in both the physical and perpendicular space. The atomic surfaces correspond to the Katz–Gratias model.

DO model. This AS is taken as a small TR (see Fig. 4.24). This model is in qualitative agreement with the spherical model. The DO and K models fit well the measured density ($0.064 < \rho < 0.07$ at/Å^3). Similarly to the DO model, it is possible to increase the density of the K model up to 0.067 at/Å^3 by adding ASs for generating a small icosahedron in the Mackay icosahedron.

Katz and Gratias have built their model by considering two hypotheses: the closeness condition (Katz 1990) and the existence of matching rules. The closeness condition is related to phason disorder. If the cut space is moved in the perpendicular direction, in some places an atom will occupy a new position with respect to the initial structure; the atom "displacement" is called a phason jump. However, such a situation is true only if the ASs are connected or obey the so-called closeness condition. This is, for instance, the case for the 2D representation of the Fibonnacci sequence shown by a thin segment lines in Fig. 4.2. The thick segment lines do not obey the closeness condition - this implies that a shift of the cut space in the perpendicular direction will lead to the "creation" or the "annihilation" of an atom. Such a situation is unphysical. The closeness condition implies constraints on the possible shape of the ASs. Together with a necessary condition for the existence of matching rules, only 10 ASs bounded by planes of high symmetry have to be considered (Cornier-Quiquandon et al. 1991). Using these ASs, there is only a limited set of solutions with the proper atomic density and chemical composition. One should note, however, that there is no experimental evidence to support these hypotheses. In particular, other tiling models can be described by fractal ASs Although the K model represents an improvement by taking into account the physical constraints (no short distances and closeness condition), the agreement of the model with the diffraction data is of the same order of magnitude as for the spherical model.

A slight improvement was obtained by Yamamoto et al. (1994b). The first stage of their model corresponds to the spherical model. Then from Fourier synthesis, the deviation from spherical ASs is calculated. Based on this information, a cluster model is considered which has occupation domains consistent with the spherical model. The final model has 15 positional, seven thermal, and 28 occupational parameters, and the reliability factor is similar to that obtained with the spherical model. This factor can be improved by allowing shifts from ideal positions.

Another theoretical model was recently proposed by Elser (1996). It is based on a decoration of the τ inflated oblate and prolate rhomboedra which build the 3D Penrose tiling. This model is essentially the same as the K model, but allows an introduction of random tiling by considering different rearrangements of the rhomboedra.

So far we have considered perfectly ordered quasicrystalline models. An alternative to the perfect QC model is the random tiling model. In such a model disorder is introduced in a subtle way. Although they have been named "random tiling", these structures remain long-range ordered and have a diffraction pattern with sharp Bragg reflections. Disorder is introduced by the so-called phason defects. In a simple 1D Fibonnacci chain, such a defect corresponds to the exchange of a sequence SL by LS. For the 2D Penrose tiling, a phason defect corresponds to a position exchange inside a hexagonal configuration. This will result in violations of the matching rules. The complete discussion of phason defects and their energetics can be found elsewhere (Henley 1991).

In the high-dimensional picture of QCs, phason disorder introduced in random tiling models can be viewed as bounded fluctuations of the cut surface along the perpendicular direction. Such bounded fluctuations do not destroy the long-range order, i.e., Bragg peaks remain delta functions but they lead to the changes in the intensity distribution and to the occurence of the diffuse scattering. This is analogous to the effect of thermal vibrations which give rise to a decrease of Bragg-peak intensities by the Debye-Waller term together with thermal diffuse scattering located beneath Bragg peaks. Similarly, to a first approximation, bounded fluctuations of the cut space result in a decrease of the Bragg peak intensity which is proportional to $\exp(-2B_{\text{perp}}Q_{\text{perp}})$, where B_{perp} is the perpendicular Debye-Waller factor. This intensity loss is distributed into diffuse scattering located closely to the Bragg peaks.

In a random tiling model one has to define the tiling and the atomic decoration. The starting point is the existence of atomic clusters, such as the Mackay icosahedron cluster. From the first order model or from the structure of periodic approximant, the linkages between clusters can be deduced: these linkages have specific directions (for instance, along two- and threefold directions) and length. This is a guide to define an underlying tiling, although one should note that there is no unique solution. Henley proposed a set of four cells, called canonical cells (Henley 1991), which have been used in the modeling of Al-Mn-Si QCs Elser proposed to use a 3D Penrose tiling decorated by Bergman clusters (Elser 1996).

We have seen that various models have been proposed. They all share common building blocks, but still do not reproduce correctly the available diffraction data, and in particular, the weak reflections.

4.4 Structure of the Al-Ni-Co Decagonal Quasicrystal

2D QCs consist of quasiperiodically ordered atomic layers which are stacked periodically. These systems are particularly interesting for the study of their physical properties since they share both periodic and quasiperiodic characteristics in the same sample. Three types of 2D QCs have been obtained experimentally so far: octogonal, dodecagonal, and decagonal (d). Their diffraction pattern have eight-, twelve-, and tenfold symmetry. Most of the structural studies have been carried out on d phases because single grains suitable for x-ray diffraction experiments could be obtained in the Al-Mn, Al-Cu-Co, and Al-Ni-Co systems (Steurer 1989, 1990, Steurer and Kuo 1990, Steurer et al. 1993, Yamamoto and Ishihara 1988). The structure determination of a d QC follows the same route as the one described for the i-Al-Pd-Mn phase. However, due to the periodicity along one direction, the interpretation of the HREM images is easier. This has proven to be a very interesting and complementary approach to the one based on the x-ray or neutron diffraction.

In all d phases the periodicity has been found to be equal to a multiple of 4 Å. In the case of the Al-Ni-Co d phase, whose structure will be discussed below, the periodicity is 4 Å. However, a lot of disorder is present in those phases. In particular, their diffraction pattern presents a set of diffuse layers, orthogonal to the periodic axis and halfway between two quasiperiodic diffraction planes. For convenience, the 3D structure can be described in an orthonormal basis (a, b, c), where (a, b) is the quasiperiodic plane and c is the periodic direction.

The atomic structure of the d-Al-Ni-Co QC has been studied by x-ray diffraction on a single-grain sample (Steurer et al. 1993, Yamamoto and Ishihara 1988). The structure has been analyzed in the 5D space necessary to recover periodicity in the case of d symmetry. The symmetry of the diffraction pattern of the Al-Ni-Co phase presents extinctions compatible with the P105/mmc space group (see Chap. 3). As in the icosahredral case, the first step in the analysis is the computation of a Patterson map. The interpretation of the Patterson map is done in a straightforward way. There are only two ASs in the asymmetric unit of the 5D unit cell whose 5D coordinates are (2/5,2/5, 2/5,2/5,1/4) and (4/5,4/5,4/5,4/5,1/4). In this notation the fifth

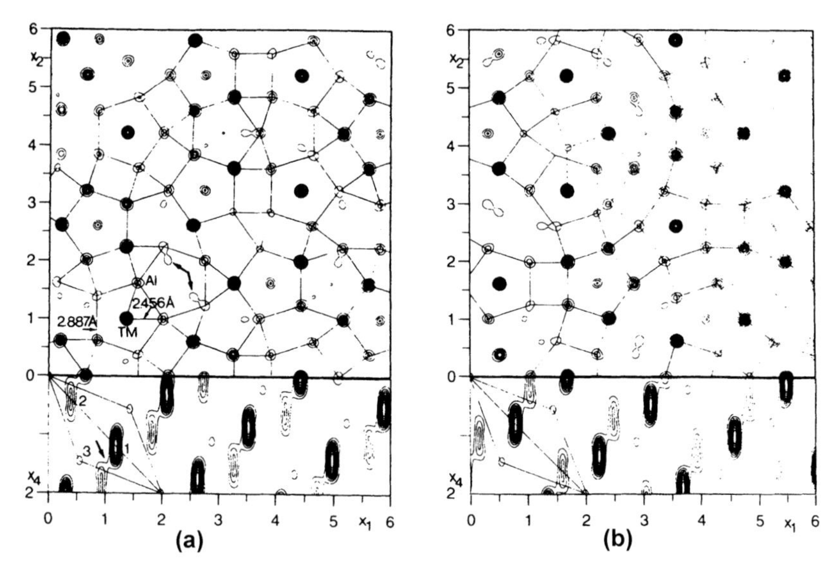

Fig. 4.26. Section of the 3D structure obtained by the Fourier transform of the phased structure factors. The c axis is vertical, and the two different planes locted at z=1/4 (a) and z=3/4 (b) are related by a screw axis operation. The bottom part of the figure shows the corresponding section of the periodic 5D space in a plane containing a direction in both the parallel and perpendicular space. The 5D unit cell is outlined.

index corresponds to the c axis. These are specific positions of the 5D unit cell, similar to those obtained in generating a 2D Penrose tiling (see Chap. 3). Starting from these ASs, Steurer and coworkers have carried out a refinement in order to characterize the shape of the ASs. To a first approximation, ASs are modeled as regular decagons or pentagons. The characteristic of each AS (its radius, chemical content, and Debye-Waller factor) can be adjusted in a fitting procedure which compares the diffraction data with those obtained with the model. As in the i-Al-Pd-Mn case, generally ASs have not a single atomic content but are made of several concentric shells with different atomic contents.

Having obtained the structure factors with phases from the model, the 3D atomic structure can be obtained by a FT. Figure 4.26 shows the two layers obtained (Steurer et al. 1993). The bottom part shows the corresponding section of the high-dimensional space, illustrating the complete equivalence between the two descriptions. Several chains of pentagonal clusters are outlined. It is striking that the same kind of local atomic order can be found in various crystalline approximant phases, in particular, in the $Al_{13}Co_4$ monoclinic phase.

Further information can be obtained through the maximum entropy method. This method is now a well-established technique to avoid the truncation effects which arise when a FT is performed. This technique was used by

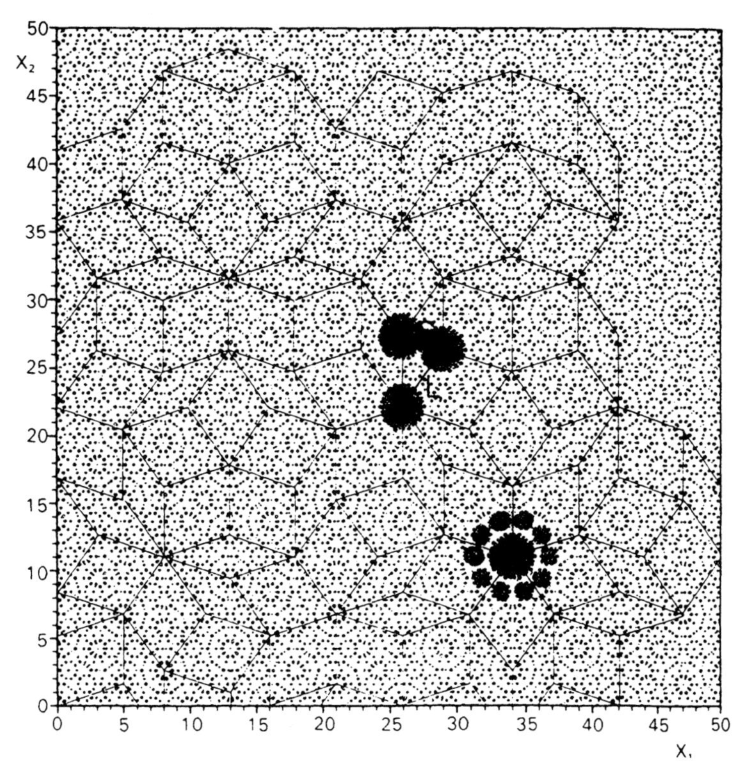

Fig. 4.27. Projection of the electron density calculated by the maximum entropy method. The size of the box is 190×190 Å^2. The shaded area corresponda to 12 Å characteristic columnar clusters. A large Penrose tiling with an edge length of 20 Å is superimposed on the map.

Steurer et al. (1993) in the case of *d* QCs and by Yamamoto et al. (1995) for *i* QCs. Starting from the measured structure factor with the phases obtained by the model, this technique allows to find the most probable density map compatible with the diffraction data. The main advantage of the procedure is that the positivity of the electronic density is imposed. This produces density maps where most of the spurious wiggles due to truncation effect are removed. Figure 4.27 shows a projection of the structure obtained by the maximum entropy method on a square 190×190 Å^2. A large number of *d* "wheels" are clearly visible in the figure (Steurer et al. 1993). Because we are looking at a projection down the periodic axis, these wheels are in fact atomic columns. The diameter of the columns is about 12 Å. Around each decagon, we can also notice ten pentagons surrounding them, forming a larger wheel with a diameter of 20 Å. The packing of these atomic columns is quasiperiodic and can be described as a decorated 2D Penrose tiling with an edge length equal to about 19 Å. The columns are sitting on the vertices of the tiling and on the diagonal of the wide rhombic tiles (Fig. 4.27).

Because of the periodicity along one direction, HREM images are easier to interpret. In particular, it is possible to compute the image of a cluster and compare it with the experimental one. To a first crude approximation, an HREM image corresponds to the projection of the electronic density and can be compared with Fig. 4.27, for instance. More realistic calculations can also be carried out using the dynamical theory for the interaction between electrons and solid state matter. Such calculations are possible when the periodicity is present (Hiraga et al. 1994).

Figure 4.28 shows, as an example, the results of calculations carried out on the 20 Å columnar cluster (Hiraga et al. 1994a). A characteristic contrast is predicted which should be seen on experimentally obtained images. Such a local arrangement is effectively observed in images, as shown in Fig. 4.29. The network of atom columnar clusters is outlined and is very similar to the one obtained from the x-ray diffraction study. This has been the basis for various models based on decorated tiling. The most difficult task in this procedure is to evidence the possible underlying tiling and the decoration.

In the case of the decorated Penrose tiling of Fig. 4.29, a 5D model has been proposed by Yamamoto (1997) (Fig. 4.30). In this case, the shape of the AS is no longer a simple decagon or pentagon, but presents a rather complicated border. Each AS has only a mirror symmetry element. This model generates a superstructure with a larger unit cell and this is observed experimentally for a slowly cooled sample. Indeed, the phase diagram of the Al-Ni-Co system is extremely complex. Depending on the annealing temperature and on the precise chemical composition, different *d* phases have been observed.

Fig. 4.28. Structure model and calculated HREM image of an atomic cluster. (a) projection of the two layers related by a screw axis symmetry operation. Large circles are for the z=1/4 layer and small circles for the z=3/4 layer. Open and closed circles are Al and transition metals. Dark circle are considered to be a mixture of Al and transition metals. (b) Calculated image: there is a ring contrast surrounded by ten wheels. The diameter of the image is 20 Å.

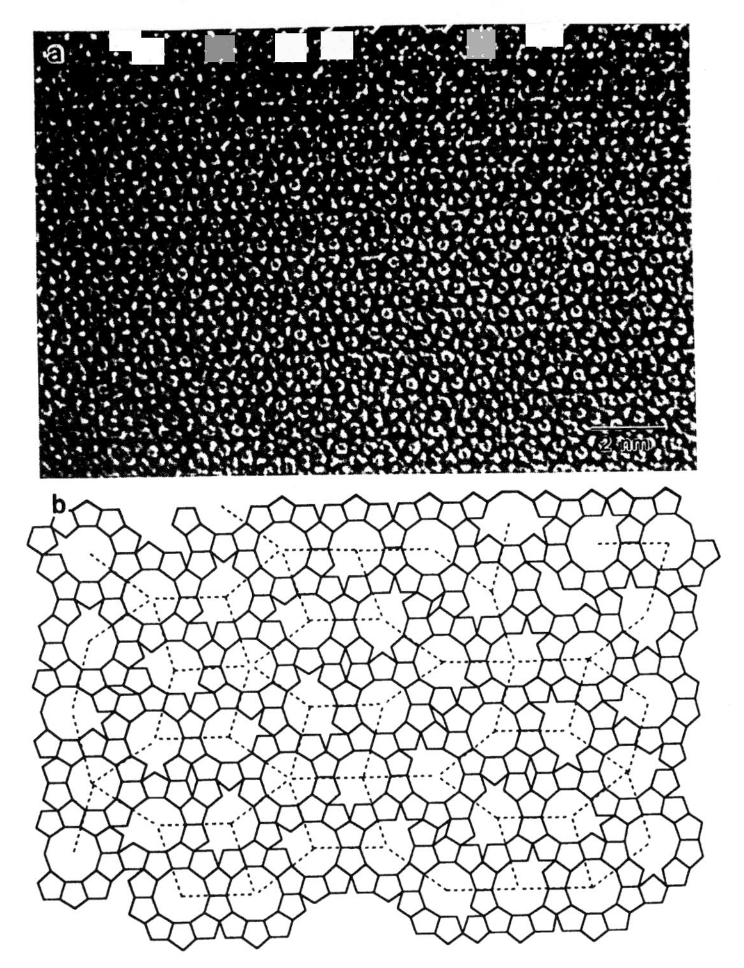

Fig. 4.29. HREM image of the Al-Ni-Co *d* phase. Ring contrasts similar to the one of Fig. 4.28 are observed (indicated with black circles). The framework of the arrangement of the columnar clusters is shown in (b). A tiling is shown with dotted lines.

Generally, when cooling down the sample, the high-temperature *d* phase transforms into a super-ordered *d* phase (Edagawa et al. 1992, Ritsch 1996) and this shows up as supplementary Bragg reflections in the diffraction pattern. The satellite reflections can be indexed when considering a larger unit cell in the 5D space and it can be shown that the resulting structure is a superstructure ordering of the high-temperature phase. The precise nature of this superstructure is still not completely understood.

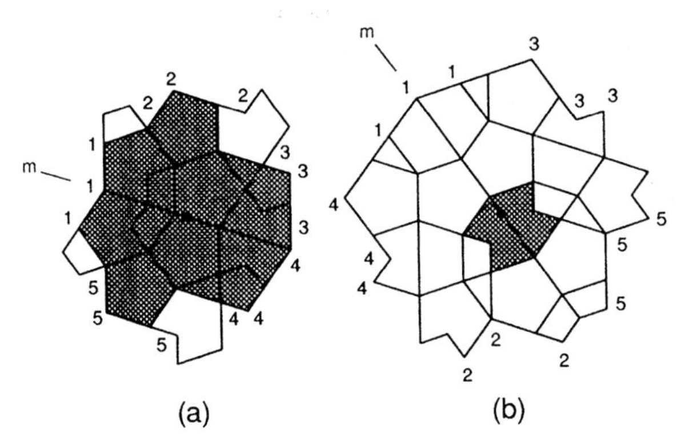

Fig. 4.30. Two independent atomic surfaces corresponding to the network of Fig. 4.29. The shaded part corresponds to transition metals.

4.5 Conclusions

The atomic structure of QCs is now well understood. We have seen that the 6D approach together with contrast variation are powerful tools. In the corresponding 3D structure, basic clusters and dense atomic planes have been identified. Various routes and atomic models have been proposed to go further in the structure determination of i and d QCs. All models have in common a "hard core" which represents about 80% of the atomic position. However, the comparison of various models with the available experimental diffraction data is far from reaching the accuracy which is currently achieved in the case of periodic materials. This means that the details of the structure still have to be determined. The structure determination of QCs is thus still a challenging problem (Hippert and Gratias 1994, Steurer 1996, Yamamoto 1996). Nevertheless, atomic models are realistic enough to allow a theoretical calculation of the physical properties of QCs (see Chaps. 6 and 7).

References

Bak, P. (1986a): Phys. Rev. Lett. **56**, 861

Bak, P. (1986b): Scripta Metall. **20**, 1199

Baruchel, J., Hodeau, J.L., Lehmann, M.S., Regnard, J.R., Schlenker, C. (eds) (1993): Neutron and Synchrotron Radiation for Condensed Matter Studies. Springer-Verlag, Berlin

Bergman, G., Waugh, J.L.T., Pauling, L. (1957): Acta Crystallogr. **10**, 254

Bessière, M., Quivy, A., Lefebvre, S., Devaud-Rzepski, J., Calvayrac, Y., (1991) J. Phys. (Paris) I **1**, 1823

Boudard, M., de Boissieu, M., Janot, C., Dubois, J.M., Dong, C. (1991): Philos. Mag. Lett. **64**, 197

Boudard, M., de Boissieu, M., Janot, C., Heger, G., Beeli, C., Nissen, H. U., Vincent, H., Ibberson, R., Audier, M., Dubois, J.M. (1992): J. Phys. Condens. Matter **4**, 10 149
Cahn, J.W., Shechtman, D., Gratias, D. (1986): J. Mater. Res. **1**, 13.
Calvayrac Y., Devaud-Rzepski, J., Bessière, M., Lefebvre, S., Quivy, A., Gratias, D. (1989): Philos. Mag. B **59**, 439
Cooper, M., Robinson, K. (1966): Acta Crystallogr. **20**, 614
Cornier-Quiquandon, M., Quivy, M., Lefebvre, S., Elkaim, E., Heger, G. (1991): Phys. Rev. B **44**, 2071
Cornier-Quiquandon, M., Bellissent, R., Calvayrac, Y., Cahn, J. W., Gratias, D. and Mozer, B. (1993): J. Non-Cryst. Solids **153-154**, 10
de Boissieu, M., Janot, C., Dubois, J.M. (1988): Europhys. Lett. **7**, 593
de Boissieu, M., Durand-Charre, M., Bastie, P., Carabelli, A., Boudard, M., Bessiere, M., Lefebvre, S., Janot, C., Audier, M. (1992): Philos. Mag. Lett. **65**, 147
de Boissieu, M., Stephens, M., Boudard, M., Janot, C., Chapman, D.L., Audier, M. (1994): J. Phys. Condens. Matter **6**, 10 725
Duneau, M., Oguey, C. (1989): J. Phys. (France) **50**, 135.
Ebalard, S., Spaepen, F. (1989): J. Mater. Res. **4**, 39
Edagawa, K., Ichihara, M., Suzuki, K., Takeuchi, S. (1992): Philos. Mag. Lett. **61**, 19
Elser, V. (1985): Phys. Rev. B **32**, 4892
Elser, V. (1996): Philos. Mag. B **73**, 641
Elser, V., Henley, C. L. (1985): Phys. Rev. Lett. **55**, 2883
Goldman A. I., Stephens P. W. (1988) Phys. Rev. B **37**, 2828
Gratias, D., Cahn J.W., Mozer, B. (1988): Phys. Rev. B **38**, 1643
Guryan, C.A., Goldman, A.I., Stephens, P.W., Hiraga, K., Tsai A.-P., Inoue, A., Masumoto, T. (1989): Phys. Rev. Lett. **62**, 2409
Guyot, P., Audier, M. (1985): Philos. Mag. B **52**, L15
Hauptman, H.A. (1991): Rep. Prog. Phys. **54**, 1427
Henley, C.L. (1991): in Quasicrystals, The State of the Art, DiVincenzo, D.P., Steinhardt, P.J. (eds). World Scientific, Singapore, p 429
Hippert, F., Gratias, D. (eds) (1994): Lectures on Quasicrystals. Les Editions de Physique, Les Ulis
Hiraga, K., Sun, W., Yamamoto, A. (1994): Mater. Trans., Jpn. Inst. Met. **35**, 657
Janner, A., Janssen, T. (1977): Phys. Rev. B **15**, 643
Janner, A., Janssen, T. (1979): Physica A **99**, 47
Janner, A., Janssen, T. (1980): Acta Crystallogr. A **36**, 408
Janot, C. (1994): Quasicrystals, A Primer, 2nd ed. Oxford University Press, New York
Janssen, T. (1986): Acta Crystallogr. A **42**, 261
Katz, A. (1990): Number Theory and Physics, Luck, J.-M., Moussa, P., Waldschmidt, M. (eds). Springer-Verlag, Berlin, p 100
Katz, A, Gratias, D. (1994): in Lectures on Quasicrystals, Hippert, F., Gratias, D. (eds). Les Editions de Physique, Les Ulis, p 187
Kycia, S.W., Goldman, A.I., Lograsso, T.A., Delancy, D.W., Sutton, M., Dufresne, E., Brüning, R., Rodricks B. (1993): Phys. Rev. B **48**, 3544
Levitov, L.S., Rhyner, J. (1988): J. Phys. (Paris) **49**, 1835
Mackay, A.L. (1962): Acta Crystallogr. **15**, 916
Pauling, L. (1985): Nature **312**, 512
Ritsch, S. (1996): PhD Thesis, Swiss Federal Institute of Technology Zürich
Rokshar, D.S., Mermin, N.D., Wright, D.C. (1987): Phys. Rev. B **35**, 5487
Roth, M., Lewit-Bentley, A., Bentley, G.A. (1984): J. Appl. Cryst. **17**, 77

Shechtman, D., Blech, I., Gratias, D., Cahn, J.W. (1984) Phys. Rev. Lett. **53**, 1951
Stephens P. W., Goldman A. I. (1986): Phys. Rev. Lett. **56**, 1168
Steurer, W. (1989): Acta Crystallogr. B **45**, 534
Steurer, W. (1990): Z. Kristallogr. **190**, 179
Steurer, W. (1996): in Physical Metallurgy, Vol 1, Cahn, R.W., Haasen, P. (eds). Elsevier Science BV, Amsterdam, p 371
Steurer, W., Kuo, K.H. (1990): Acta Crystallogr. B **46**, 703
Steurer, W., Haibach, T., Zhang, B., Kek, S., Lück, R. (1993): Acta. Crystallogr. B **49**, 661
Tsai, A.-P., Inoue, A., Masumoto, T. (1987): Jpn. J. Appl. Phys. **26**, L1505
Tsai, A.-P., Inoue, A., Yokoyama, Y., Masumoto, T. (1990a): Mater. Trans., Jpn. Inst. Met. **31**, 98
Tsai, A.-P., Inoue, A., Masumoto, T. (1990b) Philos. Mag. Lett. **62**, 95
Warren, B.E. (1990): X-Ray Diffraction. Dover Publ., New York
Yamamoto, A. (1990): in Quasicrystals, Fujiwara, T., Ogawa, W. (eds). Springer-Verlag, Berlin, p 57
Yamamoto, A. (1992): Phys Rev B **45**, 5217
Yamamoto, A. (1996): Acta. Crystallogr. A **52**, 509
Yamamoto, A., Ishihara, K.N. (1988): Acta Crystallogr. A **44**, 707
Yamamoto, A., Weber, S. (1997): Phys. Rev. Lett. **79**, 861
Yamamoto, A., Matsuo, Y., Yamanoi, T., Tsai, A.-P., Hiraga, K., Masumoto, Y. (1994): in Aperiodic'94, Proceedings of the International Conference on Aperiodic Crystals, Chapuis, G., Paciorek, W. (eds). World Scientidic, Singapore, p 393
Yamamoto, A., Weber, S., Sato, A., Ohshima, K.-I., Tsai, A.P., Hiraga, K., Inoue, A., Masumoto, T. (1995): in Proceedings of the 5th International Conference on Quasicrystals, Janot, C., Mosseri, R. (eds). World Scientidic, Singapore, p 124

5. Electronic Transport Properties of Quasicrystals – Experimental Results

Östen Rapp

5.1 Introduction

5.1.1 Background

A first challenging property of quasicrystals (QCs) is their structure which combines long-range order with absence of translational symmetry. One might then conjecture that a metal in this remarkable state of condensed matter would be described as a material in between crystals with long-range periodicity and amorphous (a) metals where long-range order is absent. In the first years of the brief history of QCs it was accordingly expected that also the physical properties of QCs should be in between those of crystalline and a metals of the corresponding chemical composition. This was in fact born out by the early experiments. For instance, the electrical resistivity, ρ, and the magnitude of the negative temperature derivative, $d\rho/\mathrm{d}T$, were both found to be smaller for a QC of Pd-U-Si than for the same material in the a state (Poon et al. 1985) and a similar result was observed also in Al-Mn-Si (Kimura et al. 1987) for much larger resistivities in both the a and quasicrystalline phases. This point is also clearly illustrated by the results of Graebner and Chen (1987). Values for several properties were determined or estimated from specific heat and resistivity measurements of superconducting $Mg_8Zn_3Al_2$ in crystalline, a, and quasicrystalline states. The width of the superconducting transition, ΔT_c, was comparable in the crystalline and quasicrystalline phases, and normally broadened for the a state. Since ΔT_c is usually an excellent measure of sample homogeneity, this result indicates that measured properties were intrinsic, and not due to defects which otherwise hampered the interpretation of many early results on QCs Graebner and Chen (1987) found that the results for the quasicrystalline phase were all in between those of the crystalline and a modifications of the same alloy, including the resistivity, the average temperature derivative of resistivity, the density of states (DOS), the Debye temperature, the sound velocity, the superconducting transition temperature, and the electron-phonon interaction.

This situation changed drastically with the discovery of stable icosahedral ($\imath$) QCs and the subsequent completely unexpected results for the physical properties, in particular for the resistivity. The initial incredulity in the physics community can perhaps be illustrated by the reactions to the first resistivity measurements on *i*-Al-Li-Cu (Wagner et al. 1988), where the results for ρ of a phase pure sample of 500–1000 $\mu\Omega\,\mathrm{cm}$ were not believed to be intrinsic to the quasicrystalline phase and were instead attributed to cracks and other sample imperfections. Such an interpretation was quite natural at the time, considering on the one hand the constituting elements of only simple non-transition metals, and on the other hand the brittle samples. It was not until somewhat later that a large value of ρ for *i*-Al-Li-Cu of 870 $\mu\Omega\,\mathrm{cm}$ at 4 K was confirmed (Kimura et al. 1989). Clearly the quasicrystalline structure has a profound influence on the physical properties. Recurrent new surprises since 1990 have further emphasized this point. Yet our understanding of these properties and their relation to the structure is at the very best fragmentary. This is the challenge which has nurtured the sustained excitement and vitality of the field.

5.1.2 Resistance Anomalies

As a brief overview of the phenomena to be described, some anomalous resistance properties are briefly summarized.

First, the resistivity of several quasicrystalline systems is remarkably large. In *i*-Al-Cu-Fe, $\rho(4\mathrm{K})$ was found to be 4.3 $\mathrm{m}\Omega\,\mathrm{cm}$ (Klein et al. 1990a), and 11 $\mathrm{m}\Omega\,\mathrm{cm}$ with improved control of sample stoichiometry (Klein et al. 1991), and in *i*-Al-Cu-Ru, values of $\rho(4\mathrm{K})$ up to 30 $\mathrm{m}\Omega\,\mathrm{cm}$ were reported at about the same time (Biggs et al. 1990). A few years later even more anomalous results were found for *i*-Al-Pd-Re with $\rho(4\mathrm{K})$ above 200 $\mathrm{m}\Omega\,\mathrm{cm}$ (Akiyama et al. 1993a, Pierce et al. 1993c). Such values of the resistivities are larger than those observed in some systems on the insulating side of a metal-insulator transition (Mott 1990), and have given rise to an intensive discussion if there can be a metal-insulator transition in QCs composed only of metallic elements.

Second, the temperature dependence of the electrical resistivity is anomalous with values of $\mathrm{d}\rho/\mathrm{d}T$ which are negative for high resistivity samples and often of quite large magnitude. It is convenient to describe an average temperature dependence of ρ by the resistance ratio R, with

$$R = \frac{\rho(4\mathrm{K})}{\rho(295\mathrm{K})} \; . \tag{5.1}$$

As discussed below, in addition to describing an interesting physical property, R is also a convenient parameter for characterizing large resistivity quasicrystalline samples. R values for the samples mentioned are 1.1 for *i*-Al-Li-Cu, (Kimura et al. 1989), up to about 2 for *i*-Al-Cu-Fe (Klein et al. 1991), about 4 for *i*-Al-Cu-Ru (Biggs et al. 1990), and in a range of values reaching 100 and above for *i*-Al-Pd-Re (Pierce et al. 1994, Gignoux et al. 1997). Amorphous metals, e.g., of transition metal-transition metal (TM-TM) type, typically have R values around 1.05 (Howson and Gallagher 1988). It is interesting to recall that the reason for a negative temperature derivative with this magnitude of R, which by today standards must be characterized as a tiny anomaly, was nevertheless hotly debated for more than a decade from the mid seventies, without reaching any satisfying solution.

Third, with improved sample quality the resistivity of a QC increases (Klein et al. 1990a, Mizutani et al. 1990b). This property is counterintuitive and violates conventional thinking about the resistivity.

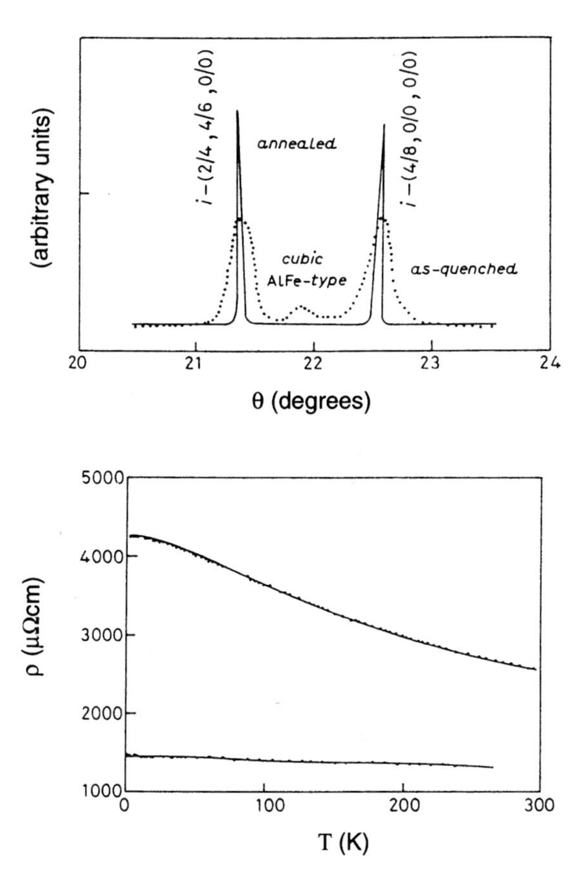

Fig. 5.1. Increased sample quality of QCs leads to enhanced anomalous properties. Top panel: x-ray diffraction pattern of i-$Al_{63}Cu_{25}Fe_{12}$ before and after annealing. Bottom panel: The temperature dependence of the electrical resistivity for the same sample. Lower curve: as-quenched Upper curve: annealed sample. After Klein et al. (1990a).

Figure 5.1 from Klein et al. (1990a) shows an early result on i-Al-Cu-Fe, where all these properties are illustrated: the large resistivity, the large negative value of $\mathrm{d}\rho/\mathrm{d}T$, and the increased ρ and R with improved sample quality as seen from the smaller widths of the peaks in x-ray diffraction.

A fourth anomalous transport property is the large positive magnetoresistance that can be observed in some QCs, with values of $\Delta\rho/\rho$ in excess of 100% at low temperatures for samples of *i*-Al-Pd-Re in a certain range of resistivities (Ahlgren et al. 1997a). Such values are otherwise known only in pure metals at low temperatures or in giant (or colossal) magnetoresistance materials.

5.1.3 Outline

There are several review articles on the subject of QCs. Examples where electronic transport properties have been treated from an experimental point of view include articles by Kimura and Takeuchi (1991a), Poon (1992), Berger (1994), and Takeuchi, S. (1994), some of which are rather short. For much of the earlier results on transport properties of QCs the reader is referred to this work. Emphasis in this review is on the development since 1990. The outline is as follows. In Sect. 5.2 an overview of published results on electrical resistivity and other transport properties of QCs will be given. Materials studied are *i* and decagonal (*d*) QCs and approximants to both these structures. Some attempts to understand or obtain an overview of these properties will be described in Sect. 5.3. Brief concluding remarks are made in Sect. 5.4.

5.2 Experimental Results

5.2.1 Overview

From the point of view of electronic transport properties the most interesting QCs are the stable *i* systems, and most of the experimental interest has accordingly been devoted to this group. If we describe Cu as a late TM, a well-known generalization both from work on *a* metals and empirical rules for the occurrence of superconductivity, these systems are of similar type: Al-(late TM)-(TM close to a half filled *d* band).

Tables 5.1 and 5.2 give an extensive, yet partial, list of published work on transport properties of these materials. The resistivity and residual resistivity are given, and other transport measurements reported are indicated. The magnetoresistance, Hall effect, thermoelectric power, high temperature resistance, the electronic part of the thermal conductivity, $\kappa_{\rm el}$, and also a few references to results for the optical conducitvity, $\sigma(\omega)$, have been included.

These tables give ample evidence of the abundant and well-documented anomalously large resistivities and temperature dependences of ρ in different *i* QCs. Detailed results have been included to illustrate that ρ(4K) and R vary substantially not only between different QCs, but also within each alloy system, and depend crucially on alloy composition and on the detailed metallurgical state of the sample.

Table 5.1. Summary of transport properties on *i*-Al-Cu-Fe QCs *R* is the residual resistance ratio. *Other* includes: *M* (magnetoresistance), *H* (Hall effect), and *S* (thermoelectric power)

Sample	ρ(4K) [mΩ cm]	R	Other exp.	Reference
$Al_{63.5}Cu_{24.5}Fe_{12}$	4.3	1.7	M, H	Klein et al. (1990a)
$Al_{63.5}Cu_{24.5}Fe_{12}$	4.2	1.5	H, S	Biggs et al. (1991)
$Al_{63}Cu_{25}Fe_{12}$	3	1.8	H, S	Srinivas and Dunlap (1991)
$Al_{63}Cu_{25}Fe_{12}$	4.3	1.7	M	Klein et al. (1990b, 1991)
$Al_{63}Cu_{24.5}Fe_{12.5}$	4.5	1.7		Klein et al. (1991)
$Al_{62}Cu_{25.5}Fe_{12.5}$	7.7	2.1		Klein et al. (1991)
$Al_{62.5}Cu_{25}Fe_{12.5}$	11	2.3		Klein et al. (1991) and
			M	Klein et al. (1992)
$Al_{63.5}Cu_{24.5}Fe_{12}$	4.4	1.6	$\sigma(\omega)$	Homes et al (1991)
Al-Cu-Fe	$4.6 - 9.7$[a]	1.66–2.2	M	Sahnoune et al. (1992, 1993)
$Al_{62.5}Cu_{26.5}Fe_{11}$	3.7	1.6	H, S	Pierce et al. (1993a)
$Al_{63.5}Cu_{24.5}Fe_{12}$	4.3	1.5	H, S	Pierce et al. (1993a)
$Al_{62.5}Cu_{24.5}Fe_{13}$	6.7	2.0	H, S	Pierce et al. (1993a)
Al-Cu-Fe	3.1–10	1.5–2.4	high T	Mayou et al. (1993)
$Al_{63}Cu_{25}Fe_{12}$	1.8	1.5	M	Matsuo et al. (1993)
$Al_{65}Cu_{25}Fe_{10}$	7.8	1.15		Takeuchi et al. (1993)
Al-Cu-Fe	4.3–10	1.7–2.3	M, H	Berger et al. (1993a),
				Klein et al. (1993)
$Al_{63.2}Cu_{24.8}Fe_{12.0}$	4.3	1.8	M, H	Haberkern et al. (1993b, 1993c)
$Al_{62.0}Cu_{25.5}Fe_{12.5}$	9.1	2.1	H	Haberkern et al. (1993b)
$Al_{62.5}Cu_{25.0}Fe_{12.5}$	6.4–9.1	2.1–2.5	M, H, S	Haberkern et al. (1993b, 1993c)
Al-Cu-Fe	$4.2 - 9.6$[a]		M, H	Haberkern et al. (1993a)
[b]$Al_{100-x-y}Cu_xFe_y$	3.8–10		H	Lindqvist et al. (1993)
Al-Cu-Fe	3.7–8.3		S	Pierce et al. (1993b)
$Al_{63}Cu_{25.0}Fe^{c}_{12}$	2.6	1.4		Tamura et al. (1994a)
$Al_{62.5}Cu_{25.5}Fe_{12}$	3.8[c]	1.7[c]	H	Tamura et al. (1994a)
$Al_{62.5}Cu_{25}Fe_{12.5}$	5.7[c]	2.2[c]		Tamura et al. (1994a)
Al-Cu-Fe thin film	2.9, 4.4	1.4, 1.6		Klein et al. (1995)
Al-Cu-Fe	4.5, 10	1.61, 2.15		Rodmar et al. (1995),
			M	Ahlgren et al. (1995, 1996a)
Al-Cu-Fe thin films	3–6	1.5–2		Klein and Symko (1994)
Al-Cu-Fe	4.8–10.5	1.75–2.45	M	Lindqvist et al. (1995),
				Grenet et al. 1995)
Al-Cu-Fe	6.4–9.1	2.1–2.5	H, S	Haberkern et al. (1995)
Al-Cu-Fe thin films	3.6–4.1	1.5–1.6		Yoshioka et al. (1995)
$Al_{62.5}Cu_{25.05}Fe_{12.45}$	4.3	1.7	H	Berger et al. (1995a)
$Al_{62.3}Cu_{24.9}Fe_{12.8}$	6.5	1.8	H	Berger et al. (1995a)
$Al_{62.8}Cu_{24.8}Fe_{12.4}$	7.2	1.9	H	Berger et al. (1995a)
$Al_{62.8}Cu_{26}Fe_{11.2}$	7.5	2.1	H	Berger et al. (1995a)
$Al_{62.5}Cu_{25}Fe_{12.5}$	10.5	2.2	H	Berger et al. (1995a),
				Giroud et al. (1996)
Al-Cu-Fe	4.5, 10	1.61, 2.15	M	Ahlgren et al. (1997c)

[a] Obtained from analyses of the magnetoresistance.
[b] $24.4 \le x \le 26.0, 12.0 \le y \le 13.0$
[c] Evaluated at 30 K, the lowest measuring temperature.

Table 5.2. Al-TM(late)-TM(half-filled): Al-Cu-(Ru,Os,Cr), Al-Pd-(Fe,Mn,Re)

Sample	ρ(4K) [mΩ cm]	R	Other exp.	Reference
$Al_{68}Cu_{17}Ru_{15}$	2.1	1.3		Mizutani et al. (1990b)
$Al_{68}Cu_{17}Ru_{15}$	5.5	2.0	H, S	Biggs et al. (1990)
$Al_{70}Cu_{15}Ru_{15}$	10	2.3	H, S	Biggs et al. (1990)
$Al_{65}Cu_{20}Ru_{15}$	30	4.2	M, H, S	Biggs et al. (1990), Poon
(1993a, 1993b, 1994),				
$Al_{68}Cu_{17}Ru_{15}$	2.5	1.3	M	Kimura et al. (1991b)
Al-Cu-Ru	1.7, 1.9	1.2, 1.3	H	Nakamura and Mizutani (1994)
Al-Cu-Ru	2.3–7.2	1.2–2.3	H	Tamura et al. (1994a)
[a]$Al_{65}Cu_{20+x}Ru_{15-x}$	3.2–51	1.2–3.4	M	Lalla et al. (1995a, 1995b)
Al-Cu-Ru	2.3–7.1	1.4–2.2	H	Tamura et al. (1995a, 1995b)
$Al_{65}Cu_{20}Os_{15}$	77–143	4.0–4.5		Honda et al. (1995)
$Al_{65}Cu_{20}Cr_{15}$	1.1–2.2[c]	$\simeq$ 1.04	M	Banerjee et al. (1995, 1997)
$Al_{70}Pd_{20}Fe_{10}$	6.7	1.2	Mn subst.	Wang et al. (1994)
$Al_{70.5}$(PdMn)	3.4,10	1.1, 2.2	high T	Lanco et al. (1992)
[b]$AlPd_xMn_y$	1.5–10	1.08–2.4		Lanco et al. 1993)
Al-Pd-Mn	1.5–9.2		M	Akyiama et al. (1993b)
$Al_{72}Pd_{20}Mn_8$	9	1.2	M	Takeuchi et al. (1993)
$Al_{70}Pd_{21}Mn_9$	7		M	Chernikov et al. (1993a, 1993b)
$Al_{70}Pd_{21}Mn_9$	2.6	0.96	M	Saito et al. (1994)
$Al_{70}Pd_{21}Mn_9$	7		M,$\sigma(\omega)$	Chernikov et al. (1994)
$Al_{70}Pd_{21}Mn_9$	2.5	0.96	M	Matsuo et al. (1994, 1995)
$Al_{70}Pd_{21}Mn_9$			κ_{el}	Chernikov et al. (1995)
$Al_{70.5}Pd_{22}Mn_{7.5}$	10	2.45	M	Lindqvist et al. (1995)
Al-Pd-Mn			S	Giroud et al. (1996)
$Al_{70.5}Pd_{21.1}Mn_{8.4}$	2	0.9	M	Rodmar et al. (1996)
$Al_{70}Pd_{20}Re_{10}$	35–300	6–20		Akiyama et al. (1993a)
Al-Pd-Re	38–280	5–28		Pierce et al. (1993c)
Al-Pd-Re	3.5–25	1.7–5	M	Berger et al. (1993b)
$Al_{70.5}Pd_{21}Re_{8.5}$	240–1500	20,100	H	Pierce et al. 1994 and
			S	Poon et al. (1995a)
Al-Pd-Re	1–1900	1.4–51	M	Honda et al. (1994a, 1994b)
$Al_{70.5}Pd_{21}Re_{8.5}$	$\simeq$ 1000	35-100	M	Poon et al. (1995b, 1996)
Al-Pd-Re	40,50	7,11	H	Tamura et al. (1995a, 1995b)
$Al_{70.5}Pd_{21}Re_{8.5}$	$\simeq$ 1500	190	M	Gignoux et al. (1995)
Al-Pd-Re	1–35 (2K)	1.2–46		Takeuchi et al. (1995)
$Al_{70.5}Pd_{20}Re_{7.5}$	20	4	M	Lindqvist et al. (1995)
$Al_{70.5}Pd_{21.0}Re_{8.5}$	3–220	2–45	M	Ahlgren et al. (1996b, 1997a)
Al-Pd-Re	$\simeq$ 1500		M	Bianchi et al. (1996)
$Al_{70}Pd_{22.5}Re_{7.5}$	150, 400	10, 50		Lin C.R. et al. (1996)
$Al_{70}Pd_{21.4}Re_{8.6}$			κ_{el}	Chernikov et al. (1996)
$Al_{70}Pd_{22.5}Re_{7.5}$	60–440	7–47	H	Lin C.R. et al. (1997)
Al-Pd-Re	30, 200	3, 7	M,$\sigma(\omega)$	Bianchi et al. (1997)
$Al_{70.5}Pd_{21.0}Re_{8.5}$	3–600	2–94	M	Ahlgren et al. (1997b)
$Al_{70.5}Pd_{21.0}Re_{8.5}$	3–830	3–190	M,high T	Gignoux et al. (1997)

[a] x=2,1,0, -1. [b] $18 \leq x \leq 23, 7.5 \leq y \leq 11$. [c] entrapped cryst. phase in QC grain.

For Al-Cu-Fe the stoichiometry and heat treatment conditions can be extremely well controlled (Lindqvist et al. 1993, Berger et al. 1995a, Giroud et al. 1996). Consistent variations of $\rho(4\text{K})$ and R can be obtained within variations of nominal compositions of a few tenths of an atomic percent.

A similar precision has not been achieved in i-Al-Pd-Mn and i-Al-Pd-Re. The reason is not clear. Samples of i-Al-Pd-Mn close to 70 at.% Al have a high quality already in the as-quenched condition in contrast, e.g., to i-Al-Cu-Fe (Lanco et al. 1992). This could be due to the absence of phason strain as inferred from the narrow linewidths in x-ray diffraction at this composition (Tsai et al. 1991). Yet the resistivity is quite sensitive to annealing also in i-Al-Pd-Mn (Lanco et al. 1992). The state of the Mn atoms in the QC and the role of possible minute precipitates of Mn remain to be clarified. A correlation between increasing ρ and a successive transition from a paramagnetic to a diamagnetic susceptibility, χ, has been observed by Lanco et al. (1993), and is in contrast to i-Al-Cu-Fe where χ remains diamagnetic for samples of strongly varying resistivity, and of magnitude apparently uncorrelated to $\rho(4\text{K})$ (Rapp et al. 1993).

One can note also that R for i-Al-Pd-Mn is often smaller than expected from the ρ values and the trend for the other quasicrystalline systems (except Al-Cu-Cr). In many cases $\rho(T)$ displays a maximum between room temperature and 4.2 K, and sometimes in addition a minimum at lower temperatures. R is therefore less informative for these samples. Such behavior could be due to a combination of quantum interference effects (QIE) and a Kondo effect from Mn precipitates.

For i-Al-Pd-Re, and presumably also for i-Al-Cu-Os, the problems of correct composition and heat treatment are further accentuated. For a given nominal composition, the resistivity varies quite strongly with heat treatment (Pierce et al. 1994). It varies also between samples of the same nominal composition annealed at the same temperature (Honda et al. 1994b). At one nominal composition, and varying heat treatments, a variation in $\rho(4\text{K})$ by a factor close to 300 was achieved by Gignoux et al. (1997) in a series of samples which all were a single i phase in standard x-ray diffraction. In addition, the difficulty to make accurate resistivity measurements on small and brittle samples gives rise to a substantial uncertainty in this description.

However, ρ and R in Tables 5.1 and 5.2 are generally well correlated; an increasing ρ also leads to a larger R. This trend is consistent through most of the data and gives an experimental handle to conveniently characterize i samples, even though the details of compositional and metallurgical problems have not been fully understood. Conversely, when $\rho(4\text{K})$ is larger than expected from R, the possibility of cracks and other sample imperfections must be given careful attention.

An interesting line of work is alloying in QCs. One example of such work in i samples is the substitution of Re by Mn in i-Al-Pd-Re by Guo and Poon (1996). In this way a continous transition in the direction of decreasing

ρ and R could be monitored. Another example is the work by Wang et al. (1994) where Mn was substituted for Fe in *i*-Al-Pd-Fe. Non-monotonous ρ(T) relations, similar to those observed in *i*-Al-Pd-Mn, were found already when 20% of the Fe atoms were substituted by Mn.

5.2.2 High-Temperature Electrical Resistivity

The increasing electrical conductivity with increasing temperature, up to room temperature, is in itself a remarkable property, as commented on above. The results by Mayou et al. (1993) show that the conductivity continues to increase at higher temperatures and that this increase accelerates at least up to 1000 K. These results are shown in Fig. 5.2 together with measurements on some approximants.

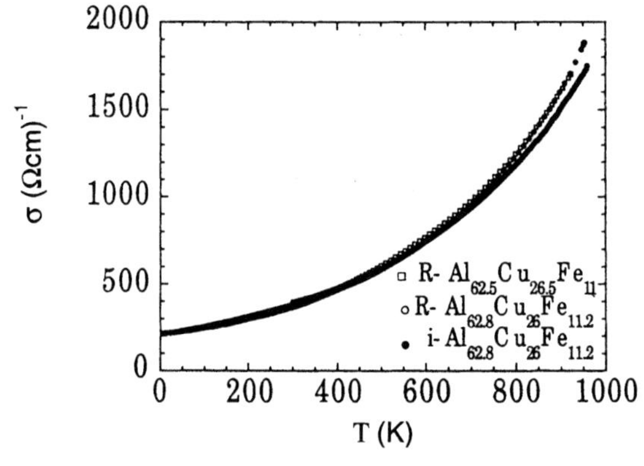

Fig. 5.2. Temperature dependence of the electrical conductivity σ for *i*-$Al_{62.8}Cu_{26}Fe_{11.2}$ and the approximants $Al_{62.8}Cu_{26}Fe_{11.2}$ and $Al_{62.5}Cu_{26.5}Fe_{11}$. After Mayou et al. (1993).

The conductivity and its temperature dependence are closely similar for the i and approximant phases over most of the measurement range. Since the resistivity is dominated by properties on the length scale of a mean free path, where the electronic environment in these materials is similar, this agreement may not be surprising.

Gignoux et al. (1997) considered the temperature dependent part of the conductivity, $\delta\sigma(T) = \sigma(T) - \sigma(0)$. A simple power law $\delta\sigma(T) \simeq T^{1.37}$ could describe data for *i*-Al-Pd-Re (R ~120) over two orders of magnitude in temperature from 7 to 700 K. This result could indicate that a single mechanism for electronic transport dominates over a large temperature range in this material. If so, it must also be understood why the temperature exponent varies strongly between different materials. Perrot and Dubois (1993), e.g., observed that a similar, but weaker accelerated increase of $\sigma(T)$ for *i*-Al-Pd-Mn up to

900 K, could be well described as a linear decrease of $\rho(T)$ over the same temperature range. It is a challange to incorporate these results into a viable theory for the conductivity of QCs.

5.2.3 Hall Effect and Thermoelectric Power

Some features of the Hall effect and thermoelectric power in *a* metals will first be briefly recalled. In these metals the Seebeck coeeficient S is often dominated by the diffusion term, with a linear T dependence as in the Drude model (Naugle 1984). For the non-transition *a* metals the Hall coefficient, R_{H}, is negative, and of the magnitude expected from a free-electron model (Mizutani and Yoshida 1982). For *a* metals composed of TMs, R_{H} is sometimes positive, at first a surprising observation for metals expected to be free-electron like. The most likely explanation seems to be based on a model of *s*-*d* hybridization. A small temperature dependence of R_{H} at low temperatures has been successfully described by QIE. The corresponding identification in the thermoelectric power is more uncertain. In this case most of the discussion of the temperature dependence has centered on the low-temperature enhancement related to the electron-phonon interaction, which can often be observed when data are plotted as S/T vs T. The Hall effect and thermoelectric power of *a* metals have been described in review articles on *a* metals, e.g., by Naugle (1984) and by Howson and Gallagher (1988). In comparison, the corresponding properties of high-resistivity QCs display rather different behavior.

A negative Hall coefficient observed in *i*-Al-Cu-Fe and *i*-Al-Cu-Ru was about two orders of magnitude larger than in *a* metals. The low-temperature value would correspond to an effective carrier density of 2–5$\times 10^{20}$ cm^{-3} (Klein et al. 1990a, Biggs et al. 1990). In addition, there can be a significant temperature dependence, including a sign change, with increasing temperature (Biggs et al. 1990). A strong concentration dependence of R_{H} in *i*-Al-Cu-Fe, also including a sign change, has been observed by Lindqvist et al. (1993), and will be discussed in Sect. 5.3.1. The composition appears to be similarly crucial in *i*-Al-Cu-Ru, where Tamura et al. (1995b) observed a positive R_{H} in $Al_{63}Cu_{25}Ru_{12}$ decreasing with temperature up to room temperature, and a negative R_{H} in $Al_{62}Cu_{25}Ru_{13}$ at low temperatures changing sign to positive values above about 150 K, in qualitative agreement with results for R_{H} of $Al_{65}Cu_{20}Ru_{15}$ as reported by Biggs et al. (1990).

In *i*-Al-Pd-Re, R_{H} at low temperatures is positive and another order of magnitude larger (Pierce et al. 1994). In this case R_{H} changed sign to negative values with increasing temperature for the sample with the largest $\rho(4\mathrm{K})$ and R, while it remained positive for samples of smaller ρ and R (Pierce et al. 1994, Tamura et al. 1995b). Sign changes in $R_{\mathrm{H}}(T)$ similar to those observed in *i* samples are well known in crystalline alloys and compounds (Hurd 1972), but are not observed in *a* metals, unless they are strongly paramagnetic.

The thermoelectric power also displays strong temperature dependent features. An example is shown in Fig. 5.3 from Biggs et al. (1990). For low resistivity *i* alloys, the Seebeck coefficient is dominated by the diffusion term, as in *a* metals of comparable resistivities. For *i*-Al-Cu-Ru, S varies strongly with temperature and may change sign. Such features are reminiscent of band structure effects in the thermoelectric power of elements and crystalline alloys (Barnard 1972). For instance, in elemental Rh, S has a (negative) minimum at 30 K and changes sign to positive values above 60 K.

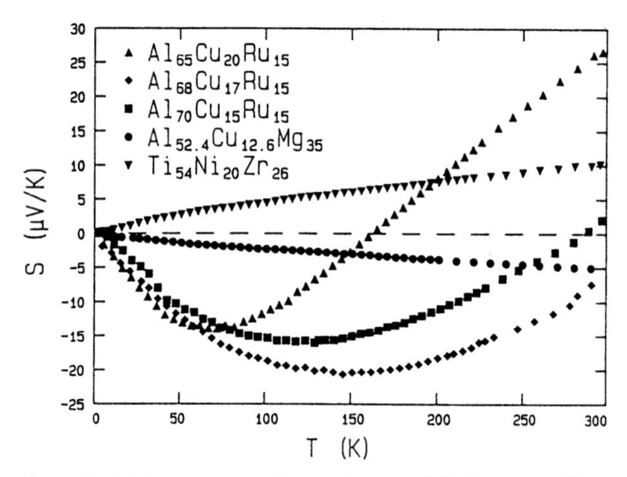

Fig. 5.3. Temperature dependence of Seebeck coefficient of *i*-Al-Cu-Ru Data for *i*-Al-Cu-Mg and *i*-Ti-Ni-Zr are also shown, where the resistivities and $S(T)$ are both similar to observations in *a* metals. After Biggs et al. (1990).

Thus, for the more resistive QCs, simplifications similar to those for *a* metals do not occur when interpreting observations for the Hall effect and the thermoelectric power. These phenomena may instead be at least as complex as in crystalline metals and compounds. Apparently band-structure effects are important in understanding R_H and S of QCs. From a general point of view these difficulties may not be surprising since the conductivity depends on energy dependent properties at the Fermi surface, while the thermoelectric power and Hall effect depend on energy derivatives.

In spite of these problems, quantitative information can sometimes be obtained from experiments, which is useful in the study of other problems. The temperature dependence of the Hall effect at low temperatures is one example. A $\sqrt{T}$ dependence was observed in R_H of *i*-Al-Cu-Fe (Klein et al. 1990), which is characteristic for electron-electron interaction (EEI) effects. Haberkern et al. (1993a) used this fact to determine the Coulomb interaction parameter in the enhanced EEI in order to reduce the number of free parameters when analyzing the magnetoresistance.

Descriptions of $\sigma(T)$ and $R_{\mathrm{H}}(T)$ in conjunction have been attempted in different empirical models (Haberkern et al. 1993b, Haberkern and Fritsch 1995, Tamura et al. 1995a, 1995b). Excellent descriptions could be obtained of both $\sigma(T)$ as well as $R_{\mathrm{H}}(T)$. However, a strong temperature dependence of the carrier densities was assumed for holes and electrons, with details obtained from fits to observed data. It is not clear why the carrier density should increase with temperature similar to an activated behavior in semiconductors or as a power law, and this approach would seem to require justification.

A correlation between the temperature dependence of the Hall coefficient and the resistivity of *i*-Al-Cu-Fe has been found (Berger et al. 1995a, Giroud et al. 1996). $R_{\mathrm{H}}(T)$ versus $\rho(T)$ was found to be linear from helium temperatures up to room temperature for samples of varying resistivities and Hall coefficients of both signs. This relation is interesting but not understood.

5.2.4 Icosahedral Approximants

The fact that a perfect QC can be described by crystalline compounds to arbitrary precision by a series of rational approximants of increasing order has provided a powerful tool in the studies of QCs. One can ask, e.g., what is the size of the cell (order of the approximant) required to reproduce certain properties observed in the corresponding QC? An example of this approach is given by Krajčí et al. (1995), who studied a series of approximants to *i*-Al-Pd-Mn from 1/1 with 128 atoms in the unit cell to 8/5 with about 41 000 atoms in the unit cell. It was found that the lower order approximants could not reproduce essential features of the band structure in contrast to the higher approximants. Experimental work may follow the opposite route towards the solution of the same problem: which is the lowest order approximant where quasicrystalline-like properties can be observed? An aim in both cases is to find a minimal number of characteristic elements of a structure which displays quasicrystalline-like properties.

In an early report on the electronic properties of an approximant to the high-resistivity phases Biggs et al. (1991) investigated the Al-Cu-Fe system. The resistivity, Hall coefficient, and the thermoelectric power were compared for the *i* phase, a 3/2 rhombohedral *r* approximant, and an *a* phase of compositions close to $Al_{63}Cu_{25}Fe_{12}$. Although the room temperature resistivity for the *a* phase was quite large, 1.3 mΩ cm, and about half of that observed for the *i* and *r* phases, R was only 1.08, while for the *i* and *r* phases R was 1.5–1.8. The Hall effect was small and positive with a weak temperature dependence for the *a* phase, and more than an order of magnitude larger, negative, and strongly temperature dependent for the *i* and *r* phases. S was small and almost linear in temperature for the *a* phase and much larger and strongly temperature dependent for the *i* and *r* phases. These results show that essential features of the quasicrystalline properties were obtained already for the 3/2 approximant. For samples of similar composition Berger et al. (1993a) observed a closely similar magnitude and temperature dependence of $\sigma(T)$

below 300 K for an i phase and the r approximant. The magnetoresistance of this r phase has also been studied (Berger et al. 1995b, Lindqvist et al. 1995) and was found to correspond to that observed in i crystals of comparable but slightly larger resistivities. A 1/1 approximant has also been studied in a cubic compound of Al-Si-Cu-Fe (Quivy et al. 1996). The resistivity and the R value of this approximant were both found to be smaller than for the orthorhombic, pentagonal, and rhombohedral approximants, which in turn were smaller than for different quasicrystalline samples in the i phase.

Icosahedral Al-Mn-Si and the crystalline 1/1 approximant α-Al-Mn-Si have been studied for a long time. Much of the early work is covered in the reviews by Kimura and Takeuchi (1991a) and by Poon (1992). Over a range of Si concentration in $Al_{82.6-x}Mn_{17.4}Si_x$ from $x = 4$ to 16, the resistivity in the approximant was found to be larger by a factor up to six at room temperature as compared to the i phase (Belin et al. 1994). After exploring various annealing treatments, Poon (1992) found low temperature resistivities of the approximant phase up to 6.7 mΩ cm. The results for the thermoelectric power and the Hall coefficient were in line with the expectations for a high-resistivity QC, with strongly temperature dependent and fairly large values of S and R_H (Biggs et al. 1992). However, the minimum in R_H observed at 50 K by Berger et al. (1995b) is not present in the data by Biggs et al. (1992) for the same concentration. Also for this QC the magnetoresistance indexmagnetoresistivity is of the sign and magnitude expected for a strong spin-orbit scattering material with ρ(4K) of about 4 mΩ cm (Berger et al. 1995b, Lindqvist et al. 1995).

Similar investigations have been made of the low-resistivity of a, metal-like, i crystals. The observations by Edagawa et al. (1992) on Mg-Ga-Al-Zn samples are illustrative. In the i phase, ρ(4K) was 120 μΩ cm and $R = 1.045$. In the 2/1 approximant the corresponding values were respectively 110 μΩ cm and 1.035, similar to the QC. For the 1/1 approximant, $\rho(4\ \mathrm{K}) = 60$ μΩ cm, $R = 0.89$, and the quasicrystalline character of the results was thus lost. Low-resistivity QCs and approximants of Mg-Al-Pd type have been studied by Hashimoto et al. (1994). In this case ρ at 300 K was somewhat larger for the 1/1 approximant than for the i sample.

5.2.5 Decagonal Quasicrystals

Decagonal QCs offer the possibility to study transport properties in a periodic and a quasicrystalline direction in the same sample. At a time when it was not yet clear if there was any particular influence of quasicrystallinity on physical properties, this possibility was particularly attractive. A number of interesting papers appeared around 1990.

As can be seen from Table 5.3, there are significant differences between different d crystals and sometimes between the results by different groups on the same material. However, some essential features common to all or

Table 5.3. Early results on d QCs. Indices q (p) denote current parallel to quasicrystalline (periodic) direction. Other notations as in Table 5.1

Sample	$\rho_q(4K)$ $\rho_p(4K)$ [μΩ cm]	R_q R_p	Other exp.	Reference
$Al_{62}Si_3Cu_{20}Co_{15}$	270[a]	1.07[a]	S_q	Lin et al. (1990)
	67[a]	0.91[a]	S_p	
$Al_{62}Si_3Cu_{20}Co_{15}$			H_q, H_p	Zhang et al. (1990)
$Al_{70}Ni_{15}Co_{15}$	$\sim$ 1200	$\sim$ 0.9		Shibuya et al. (1990)
	60	$\sim$ 0.8		
$Al_{65}Ni_{20}Co_{15}$	50, 60[b]	0.4, 0.5[b]	therm. cond.	Zhang et al. (1991)
	9, 12[b]	0.3, 0.5[b]		
$Al_{62}Si_3Cu_{20}Co_{15}$			therm. cond.	Zhang et al. (1991)
$Al_{65}Cu_{15}Co_{20}$	340	1.03		Martin et al. (1991
	30	0.8		
$Al_{70}Ni_{15}Co_{15}$	340	$\sim$ 1		Martin et al. (1991)
	38	0.8		
Al-Ni-Co	300	$\sim$ 1.05	M_q	Edagawa et al. (1996a)
	50	$\sim$ 0.9		

[a] at 77 K. [b] extrapolated.

most of these results can be identified: The resistivity in the periodic direction, ρ_p, is small, and $d\rho_p/dT$ is positive ($R_p < 1$), while in the quasiperiodic direction, ρ_q is comparatively large, albeit small on the scale of the i materials (Sect. 5.2.1), and $d\rho_q/dT$ is negative ($R_q > 1$). The anisotropy, $\rho_q(4K)/\rho_p(4K)$ varies in the range from about 4 to 20 for the materials listed in Table 5.3. Later investigations of d-$Al_{70}Ni_{15}Co_{15}$ (Markert et al. 1994, Naugle et al. 1996) have shown similar results. A surprising result is the positive $d\rho/dT$ in the quasicrystalline direction in d-Al-Ni-Co reported by Shibuya et al. (1990), which is anomalous considering the large resistivity (> 1 mΩ cm) as pointed out already by the authors.

Wang and Zhang (1994) examined published data for the resistivity of d QCs and found that $\rho_p(T)$ was linear over an extended temperature range, while it was necessary to include a T^2 term in the description of $\rho_q(T)$. They interpreted this observation as reflecting a low Debye temperature in the periodic direction, while in the quasicrystalline direction, a generalized Faber-Ziman model was suggested with an enhanced electron-phonon interaction.

The low-temperature magnetoresistance of d-Al-Co-Ni was studied by Edagawa et al. (1996a) with the current in the quasicrystalline plane and the magnetic field both parallel and perpendicular to the quasicrystalline plane. The magnetoresistance was positive and small, in both cases of order 5×10^{-4}, which is comparable to the usual results in a metals.

Recently some new and interesting results on d QCs have appeared. Edagawa et al. (1996b) measured the thermal conductivity, κ, of d-$Al_{65}Cu_{20}Co_{15}$

and derived estimates of the electron contribution κ_{el} for the periodic and quasicrystalline directions. It was found that $\kappa_{q,el}$ was smaller than $\kappa_{p,el}$ by about an order of magnitude at 100 K, and that this difference increased with decreasing temperature to two orders of magnitude at 1 K.

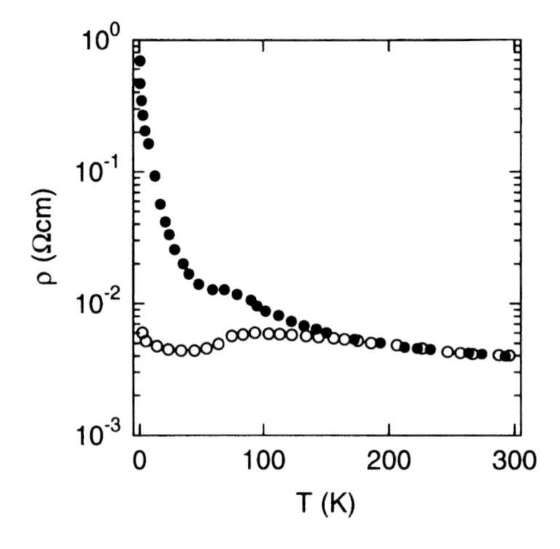

Fig. 5.4. Electrical resistivity of the d approximant Al_3Mn ($Al_{11}Mn_4$) along the pseudo-quasicrystalline plane (open circles) and along the crystal axis (filled circles). After Volkov and Poon (1995).

The ansisotropy in the thermoelectric power S of d-$Al_{73}Cu_{17}Co_{10}$ was studied by Lin, S.Y. et al. (1996). They found evidence for a strong enhancement at low temperatures of (the negative) S in the quasicrystalline direction. Although the electron-phonon mass enhancement, λ, cannot be accurately determined from such measurements, this result provides a strong indication that λ is significantly enhanced in the quasicrystalline plane.

Studies have also been made of d approximants (Volkov and Poon 1995). The results for Al_3Mn are particularly interesting and are illustrated in Fig. 5.4. The resistivity of this material was 5–7 $m\Omega\,cm$ at room temperature with small anisotropy. With decreasing temperature the anisotropy increases to above 200 at 0.45 K, and most remarkably, ρ_q along the pseudo-quasicrystalline plane has a small temperature dependence while ρ_p in the crystalline direction increases to several hundreds $m\Omega\,cm$. These observations are opposite to those found in other approximants by the same authors, and also in contrast to the results for d phases described above.

5.3 Towards Understanding Transport Properties

Different ideas have been used when trying to understand electronic transport in QCs. These efforts are often uncorrelated however, and represent, one could say, views on QCs from different angles, each one selecting a somewhat arbitrary subset of properties. Some such attempts will be reviewed, reflecting a wide range of ideas and methods explored.

5.3.1 Strong Sensitivity to Electron Concentration

It was realized early that a Hume-Rothery like mechanism could stabilize QCs with no occupied *d*-electron states (Friedel and Dénoyer 1987, Smith and Ashcroft 1987). In descriptions based on the nearly free-electron approximation, and in analogy with ordinary crystals, there is also an energy gap associated with each reciprocal lattice vector in QCs, although very few of these reciprocal lattice vectors should be of importance for the overall electronic structure. This leads to a reduction of the DOS, i.e., a pseudogap, and stabilization of the structure. In a construction corresponding to that of the Brillouin zone for crystals, the large multiplicity of the vectors giving the strongest diffraction peaks will form a quasi-Brillouin zone (quasi-BZ), with a large number of surfaces. For the proper electron concentration, the interaction between Fermi surface and quasi-zone boundaries can therefore be strongly enhanced.

The further developments of this picture play an important role in our understanding of QCs. Theoretical and experimental results on these topics are discussed elsewhere in this book. Here some examples will be given on the influence of electron concentration on transport properties.

When alloying in QCs, the sensitivity of observed properties to concentration changes was soon observed, as described in early reviews (Kimura and Takeuchi 1991a, Poon 1992). The concept of a quasi-BZ has proved to be quite useful in understanding these results.

The experiments on Mg doping of *i*-Al-Cu-Li by Kishi et al. (1990) in Fig. 5.5 illustrate clearly an interpretation along this line. The resistivity at 4 K of $Al_5(Mg_xLi_{1-x})_{3.5-4}Cu_1$ was found to decrease monotonously from the large value of 870 $\mu\Omega$ cm at $x = 0$ to 60 $\mu\Omega$ cm for $x = 1$, while the resistance ratio R decreased from 1.12 to 0.97. For $x = 0$, the electron-per-atom ratio, e/a, is 2.1. The DOS was found to drop drastically when this concentration was approached, indicating the appearence of a pseudo-gap. The radius of a free-electron Fermi surface at $e/a = 2.1$ is close to the distance from the quasi-zone center to the Bragg plane associated with the strong (111101) reflexion of the i phase. When electron concentration increases beyond this value, the quasi-BZ-Fermi surface interactions become less important, the free surface of the Fermi sphere grows, ρ drops, and the temperature dependence of ρ approaches that of a good metal.

Wagner et al. (1990) varied concentration in *i*-Ga-Zn-Mg and *i*-Al-Cu-Mg and observed striking examples of quasi-BZ-Fermi surface interactions. The thermoelectric power is illustrated in Fig. 5.6 as an example. In these alloys the free-electron concentration could be varied from 2.18 to 2.26 electrons per atom for *i*-Ga-Zn-Mg, and from 2.25 to 2.54 for *i*-Al-Cu-Mg. The free-electron Fermi sphere makes contact with the zone planes corresponding to (222100) and (311111)/(222110) for Z close respectively to 2.2 and 2.4, in the range covered by alloying.

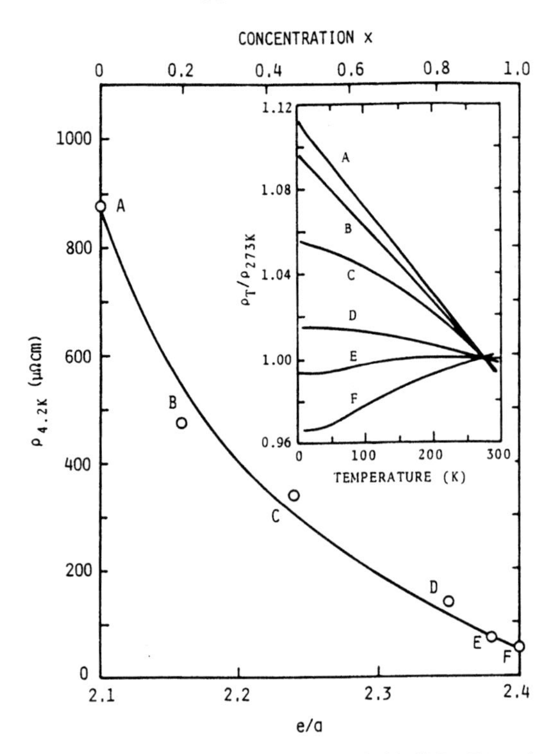

Fig. 5.5. Electrical resistivity of $Al_5(Mg_xLi_{1-x})_{3.5-4}Cu_1$ at 4.2 K vs electron-per-atom ratio (bottom) or Mg concentration (top). Inset shows $\rho(T)/\rho(273K)$ vs T for the six samples of the main panel. After Kishi et al. (1990).

At these electron concentrations there are clear anomalies in the Z-dependence of the thermoelectric power, as illustrated in Fig. 5.6, and also in the Hall coefficient, and in the electronic specific heat coefficient, although these last anomalies were somewhat less clear.

The results by Kishi et al. (1990) and by Wagner et al. (1990) illustrate elegantly the simple and yet powerful physical description made possible with concepts generalized in a straightforward way from conventional metal theory. However, this success is limited to the simple QCs composed of metals with s and p electrons at the Fermi surface.

When i QCs with unfilled d bands are studied the picture is more complicated. First, as mentioned above, these materials are often much more sensitive to variations in composition. Furthermore, the control of this composition is far from mastered. An exception to the last statement is the work on i-Al-Cu-Fe by Lindqvist et al. (1993), illustrated in Fig. 5.7.

In contrast to the results for i-Al-Cu-(LiMg) and i-Ga-Mg-Zn discussed above, graphs with ρ and R_H of i-Al-Cu-Fe vs electron-per-atom ratio displayed scattered data from which conclusive results could not be obtained.

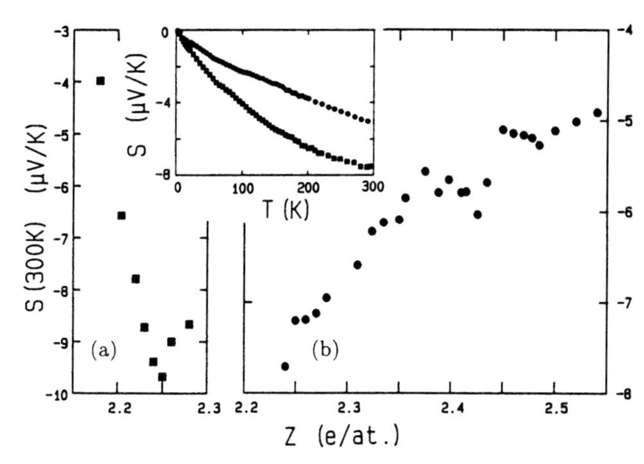

Fig. 5.6. The thermoelectric power , S, at room temperature vs electron-per-atom ratio Z for (a) i-Ga-Zn-Mg and (b) i-Al-Cu-Mg samples. Inset: typical results for $S(T)$ of i-Ga-Zn-Mg (bottom curve) and i-Al-Cu-Mg (top curve). After Wagner et al. (1990).

Figure 5.7 suggests that the crucial parameter in this case is instead the iron concentration and points to a particular significance of Fe concentrations around 12.5 at.% Fe in this system, with a maximum in ρ in conjunction with a sign change of R_{H}. It is interesting to note that the stability was also the highest at this concentration in an investigation of the thermal behavior of i-Al-Cu-Fe by Bessière et al. (1991). Such a correlation between high resistivity and stability is in agreement with the model discussed above.

However, the strong concentration dependence of ρ and R_{H} shown in Fig. 5.7 does not fit into a simple Hume-Rothery picture. If the valence of Fe is taken to be -2, the variation of charge concentration in Fig. 5.7 is 0.04 electrons per atom. This is an order of magnitude smaller than in Figs. 5.5 and 5.6, and would not appear to produce significant changes in quasi-BZ-Fermi surface interactions which could account for the observed variations. Furthermore, the choice of -2 for the Fe-valence is not well founded, and would appear to have been justified mainly by the fact that such a value is favorable for a free-electron like Fermi surface to make contact with the quasi-zone. Model calculations in related crystalline compounds can provide qualitative support for a negative valence of the TM atom, e.g., by *sp-d* hybridization (Trambly de Laissardière et al. 1993), but quantitative results have not been obtained. From calculations of the electronic structure in approximants to i-Al-Pd-Mn, charge transfer from the Al atom to the TM atoms was found to be small (Krajčí et al. 1995).

In analogy with *a* metals, where a sign change from negative to positive R_{H} has been observed as a function of increasing TM concentration and qualitatively described in a model based on *sp-d* hybridization (Nguyen-Manh et al. 1987), a similar mechanism was suggested for the sign change of R_{H} in

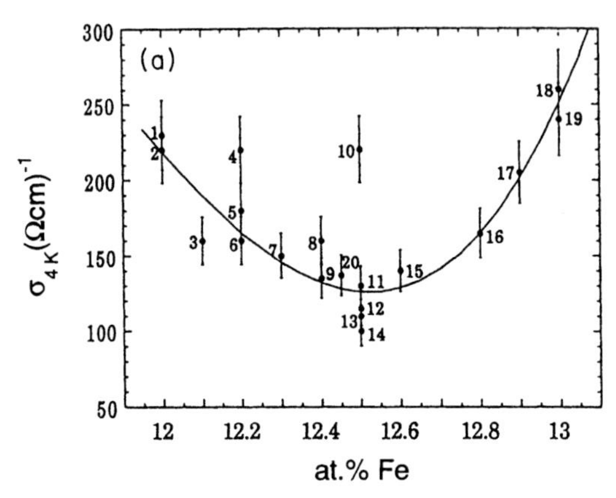

Fig. 5.7. The electrical conductivity $\sigma(4K)$ (a) and the Hall coefficient $R_H(4K)$ (b) vs nominal Fe concentration in *i*-Al-Cu-Fe. Results for 20 samples are shown with Al concentration in the range 61.8–63, Cu between 24.4–26, and Fe from 12-13 at.%, respectively. After Lindqvist et al. (1993).

i-Al-Cu-Fe (Lindqvist et al. 1993). It must be recalled however, that sign reversals in *a* metals and in *i*-Al-Cu-Fe occur on different scales. In *a*-Cu-Zr, e.g., R_H varies monotonously with concentration. The observation of the effect is made over a range of charge variation which corresponds to several tenths of one electron/atom, and from values of order -5×10^{-11} to $+5\times10^{-11}$ $m^3\,C^{-1}$ with increasing Zr concentration (Pavuna 1985). R_H of *i*-Al-Cu-Fe varies non-monotonously with Fe concentration, changes sign over a concentration range where the charge variation is two orders of magnitude smaller than for *a* metals, and occurs between values of R_H which are numerically larger by two orders of magnitude.

Therefore the observations for R_H in Fig. 5.7 represent a quite dramatic effect, which apparently is qualitatively different from *a* metals and *i* QCs with *sp* electrons at the Fermi surface. Unfortunately, as mentioned above, it has not yet been possible to achieve a similar control of concentration in the other Al-TM(late)-TM(half-filled) *i* QCs. The understanding of this exceptional concentration dependence therefore remains an experimental as well as a theoretical challenge.

5.3.2 Magnitude of the Electrical Resistivity

The large magnitude of the resistivity in several quasicrystalline systems is a major problem and the subject of an intensive theoretical interest. Four recent papers illustrate the breadth of these efforts. Janot (1996) considered a variable-range hopping (VRH) mechanism with hierarchially distributed hopping distances. Burkov et al. (1996) suggested a fractional multicomponent Fermi surface and focussed on phonon and impurity scattering in "dirty" QCs. Fujiwara et al. (1996) studied an approximant of *d*-Al-Cu-Ni with about

4400 atoms in the unit cell and calculated scaling properties for conductivity related parameters. Roche and Mayou (1997) made numerical studies of propagation modes in three-dimensional QCs from a real space method developed by them.

Experimentally, various empirical approaches have been explored to organize observations and extract essential features. As mentioned, it was early observed that a high resistivity appears to be correlated to a low electronic specific heat coefficient, γ. Such a relation was corroborated by plotting ρ as a function of γ for a number of *sp i* QCs (Kimura et al. 1989). Data for *i*-Al-Cu-Ru were later included, suggesting that this trend was continued to the higher resistivity of this QC (Mizutani et al. 1990b, Mizutani 1993). A different approach was based on comparisons of measurements of the specific heat and the Hall coefficient of *i*-Al-Cu-Ru, *i*-Al-Cu-Fe, and α-Al-Mn-Si (Biggs et al. 1992). It was concluded that the high resistivity of these materials was due to a reduction of the effective carrier concentration rather than a decrease of the mobility.

The interest of attempts to correlate ρ and γ of course lies in the information that can be obtained on the role of the DOS, which enters in both these quantities. Writing $N(\epsilon_F)$ for the bare, band structure DOS, these relations are:

$$\rho = \frac{1}{e^2 N(\epsilon_F) D} \tag{5.2}$$

and

$$N(\epsilon_F) = \frac{3\gamma}{\pi^2 k_B^2} \frac{1}{[1 + \lambda + \lambda_m]} \; . \tag{5.3}$$

The diffusion constant $D = v_F^2 \tau / 3$, v_F is the Fermi velocity, τ is the elastic relaxation time, λ is the electron phonon interaction parameter, and λ_m is a corresponding term for magnetic interactions such as spin fluctuations.

When a correlation between ρ and γ is extended to high-resistivity materials, where the temperature dependence of ρ is significant, the temperature taken to evaluate ρ is important. If helium temperatures are used (Mizutani 1993), QIE affecting $N(\epsilon_F)$ and D in Eq. (5.2), will influence this correlation. Taking ρ at room temperature reduces this problem, although some quantum corrections could still be present.

Figure 5.8 shows that a simple correlation between γ and ρ(295 K) extends over more than an order of magnitude in γ and two orders of magnitude in ρ, and comprises simple as well as TM-based *i* QCs. The only systems which have been intentionally excluded are *i*-Al-Pd-Mn and other Mn-based QCs. These samples will be discussed below.

The mass enhancement factors, λ and λ_m, which affect the calculation of the DOS from the observed γ are sample dependent and in general poorly known. However, *i*-Al-Mg-Zn is superconducting at 0.4 K (Graebner and Chen 1987) and *i*-Al-Cu-Li at 1.5 K (Wagner et al. 1988), and λ has then

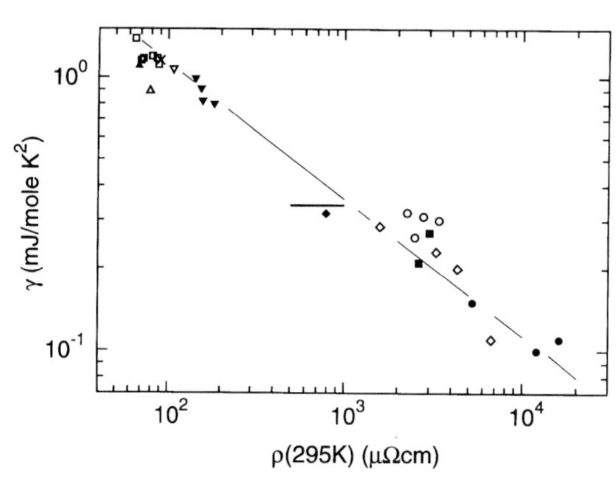

Fig. 5.8. γ vs ρ at room temperature for i QCs. The straight line summarizing data has the form $\gamma \simeq 1/\sqrt{\rho}$. Data sources: **Al-Mg-Cu**: open squares, 6 alloys from Mizutani et al. (1990a); filled up triangle, Wagner et al. (1989). **Al-Mg-Ag**: cross, Mizutani et al. (1990a). **Mg-Ga-Zn**: filled down triangles, four samples, Mizutani et al. (1990a). **Al-Zn-Mg**: unfilled up triangle, Graebner and Chen (1987). **Al-Cu-V**: unfilled down triangle, Wagner et al. (1989). **Al-Cu-Li**: bar, Wagner et al. (1988) with uncertain ρ; filled rhomboid, Kimura et al. (1989). **Al-Cu-Ru**: unfilled rhomboid, one sample from Mizutani et al. (1990b); three samples from Biggs et al. (1990). **Al-Cu-Fe**: unfilled circles, four samples from Klein et al. (1991), Biggs et al. (1991), Pierce et al. (1993a), and Wang, K. et al. (1993), respectively. In the latter case ρ was not given, and was taken from Klein et al. (1991) for a sample of the same composition and heat treatment. **Al-Cu-Ru(Si)**: filled squares, two samples from Pierce et al. (1993a). **Al-Pd-Re**: filled circles, two samples from Pierce et al. (1994), one sample from Gignoux et al. (1997).

been estimated with the result of about 0.4 in both cases. For the other samples displayed in Fig. 5.8, λ and λ_{m} can be expected to be small and a correction of the DOS by at most 20% would seem likely. This may imply some scatter for the DOS, the physical quantity of interest, but is a minor effect on a logarithmic scale.

Quasicrystals with Mn are likely exceptions. In i-Al-Pd-Mn the importance of magnetic interactions has been inferred from studies of the magnetic susceptibility and the magnetization in conjunction with low-temperature specific heat measurements (Chernikov et al. 1993a, Lasjaunias et al. 1995). Results for γ were difficult to evaluate and therefore uncertain, and furthermore λ_{m} may be significant. The first point is illustrated by the different results for γ with (0.41 ± 0.12) mJ/mole K^2 (Chernikov et al. 1993a) and 0.25 mJ/mole K^2 suggested by Lasjaunias et al. (1995) to be a possible value. In view of this uncertainty in γ as well as in λ_{m}, data for i-Al-Pd-Mn have been excluded. Similar considerations apply to Al-Mn QCs, where an enhanced DOS has been observed, presumably associated with the low temperature magnetic state of Mn (Lasjaunias et al. 1987, Berger et al. 1988).

It should also be mentioned that *a* metals in general do not fit into the correlation shown in Fig. 5.8. Typical results, e.g., for *a*-Cu-Zr (Samwer and von Löhneysen 1982), would cluster around 180 μΩ cm and 4 mJ/mole K^2, i.e., much above the data shown, which reflects the large carrier density of *a* metals. This is further illustrated by Mizutani et al. (1990b) from an extensive compilation of data for *a* metals.

Figure 5.8 clearly illustrates that the DOS decreases when the resistivity increases in *i* QCs. Furthermore, the straight line shown, which summarizes data well, within the limitations discussed, has the slope of $-1/2$. Thus, the decrease of the DOS is not sufficient to explain the increase of the resistivity. From Eq. (5.2) it follows that the diffusion constant also decreases. The straight line in Fig. 5.8 would suggest an additional $D \sim N(\epsilon_F)$ behavior. Within a conventional model of transport processes one can then infer that the elastic scattering rate increases when going to the high-resistivity *i* alloys, or that the charge concentration decreases leading to a decreased Fermi energy and Fermi velocity, or that some combination of these changes occur.

Related conclusions have been reached for selected systems from band structure calculations in the tight binding-LMTO scheme and recursive methods for large systems. Krajčí et al. (1995) concluded that the high resistivity of a series of approximants to *i*-Al-Pd-Mn arises both from a low effective number of carriers and from a low mobility of carriers close to the Fermi surface. Fujiwara et al. (1996) found for a series of approximants to *d*-Al-Cu-Co that in addition to the low DOS, the diffusion constant decreased with increasing systems size, i.e., approaching a perfect QC.

5.3.3 The Magnetoresistivity

The magnetoresistance is unique among the unusual electronic transport properties of QCs since only in this area most, but not all results can be quantitatively described within an existing theoretical framework, in this case QIE in weakly disordered metals. Main features of QIE will be briefly recalled. An overview of observations is then given and results and consequences of analyses in terms of QIE are discussed.

Background. In metals with strong electronic scattering, one must consider interference between scattered waves. Let τ be the elastic scattering time, the characteristic time the electron spends in an eigenstate of momentum, and τ_{ie} the inelastic scattering time, the corresponding time in an energy eigenstate. $\tau_{\mathrm{ie}}(\mathrm{T})$ increases with decreasing temperature since the energy available for inelastic events decreases. For a strong scattering material one can then reach a region, often limited to low temperatures, where

$$\tau << \tau_{\mathrm{ie}} \; . \tag{5.4}$$

When Eq. (5.4) is fulfilled, several elastic scattering events take place within one inelastic scattering event and different types of interference effects occur:

weak localization (WL) and enhanced EEI. WL is a one-electron effect where the elastically scattered electron interferes with itself during the time of preserved phase coherence, usually taken to be the inelastic scattering time. This leads to an increased tendency towards backscattering, from which the terminology weak localization has been coined. EEI results from interference between the scattered waves from two electrons. Two different EEI contributions are considered: from the diffusion channel (particle-particle) and from the Cooper channel (particle-hole), named according to the nature of the interacting particles, and the relation to superconducting interactions for the latter.

The contributions from these mechanisms to the magnetoconductivity, $\Delta\sigma(B,T) = \sigma(B,T) - \sigma(0,T)$, at magnetic field B, and temperature T, have been calculated in considerable quantitative detail. The two main contributions were obtained by Fukuyama and Hoshino (1981) and Lee and Ramakrishnan (1982), with results for the observed magnetoconductivity in the forms

$$\Delta\sigma_{\mathrm{WL}} = \Delta\sigma[\tau_{\mathrm{ie}}(T), \tau_{\mathrm{so}}, D, g^*, B] \tag{5.5}$$

$$\Delta\sigma_{\mathrm{EEI}} = \Delta\sigma(F_\sigma, D, g^*, B, T) \ , \tag{5.6}$$

where τ_{so} is the spin-orbit scattering time, D is the diffusion coefficient, g^* is the Landé factor, and F_σ is the Coulomb interaction parameter. In addition, magnetic scattering can break phase coherence and also contribute to the resistivity by direct spin scattering. A corresponding scattering time τ_{s} should then be considered in Eq. (5.5). The EEI contribution in Eq. (5.6) is from the diffusion channel, while the Cooper channel contribution is usually important only in the proximity of a superconducting transition.

Very readable reviews of QIE have been written for the experimentalist by Bergmann (1984) and by Dugdale (1995). A comprehensive theoretical review is that by Lee and Ramakrishnan (1985).

Observations. Early observations of an anomalous magnetoresistance were made by Wong et al. (1987) in i-$Mg_{35}(Al_{52.4}Cu_{12.6})$ and by Baxter et al. (1987) in i-$Mg_{32}(Al_{1-x}Zn_x)_{49}$ for $x = 0.5$ and 0.69. For these samples ρ at 4 K was in the range 60–110 μΩ cm. The magnitude of the magnetoresistance, $|\Delta\rho/\rho|$, was a few times 10^{-4} at 6–8 T, in agreement with the experience from *a* metals. With the discovery of the high-resistivity *i* QCs, the magnetoresistance was observed to be considerably larger, of order of several percent in i-Al-Cu-Fe (Klein et al. 1990a) and in i-Al-Cu-Ru (Biggs et al. 1990), still much larger in subsequently studied QCs, and up to above 100% in i-Al-Pd-Re (Ahlgren et al. 1997a).

As an overview of results in this area, data for the magnetoresistance have been collected in Fig. 5.9 in the form $|\Delta\rho/\rho|^{\max}$ vs ρ(4K). The overall maximum magnetoresistivity for a particular material often occurs at fields larger than 30 T and is seldom experimentally accessible. Therefore $|\Delta\rho/\rho|^{\max}$ was

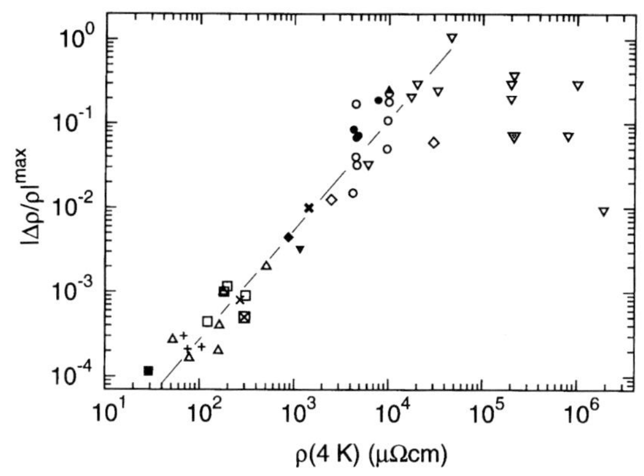

Fig. 5.9. Maximum observed magnetoresistance $|\Delta\rho/\rho|^{\max}$ vs $\rho(4\mathrm{K})$. The straight line $\sim \rho^{1.3}$ emprically summarizes data up to 10^5 μΩ cm. Data: **fcc Cu-Ge**: filled square, Eschner et al. (1984). **Amorphous metals**: open up-traingles, 6 samples, Bieri et al. (1986); open squares, 4 samples, Lindqvist and Rapp (1988), Lindqvist and Fritsch (1989), Lindqvist et al. (1990); cross: *a*-Al-Y, Ström-Olsen et al. (1984). *d*-**Al-Ni-Co**: square with cross, Edagawa et al. (1996a). *i* **samples**: *i*-**Al-Mg-(CuZn)**, plus, 3 samples, Baxter et al. (1987), Wong et al. (1987). *i*-**Al-Cu-Li**: filled rhomboid, Kimura et al. (1989). *i*-**Al-Cu-Cr**: filled down-triangle, Banerjee et al. (1997). *i*-**Al-Fe**: thick cross, Plenet et al. (1992). *i*-**Al-Cu-Fe**: open circles, 8 samples, Klein et al. (1992), Sahnoune et al. (1992), Haberkern et al. (1993a), Ahlgren et al. (1995). *i*-**Al-Pd-Mn**: filled up-triangle, Lindqvist et al. (1995). Filled circles: two *i*-**Al-Cu-Fe** samples, and **approximants** *r* **Al-Cu-Fe** and **cubic Al-Mn-Si**, Lindqvist et al. (1995). *i*-**Al-Cu-Ru**: open rhomboid, 2 samples, Kimura et al. (1991b), Biggs et al. (1990). *i*-**Al-Pd-Re**: open down-triangles, 5 samples with $\rho < 10^5$ μΩ cm, Lindqvist et al. (1995), Bianchi et al. (1997) and 3 samples from Ahlgren et al. (1997a); 6 samples with $\rho > 10^5$ μΩ cm, Poon et al. (1995b) (2 samples), Gignoux et al. (1997), Bianchi et al. (1997), Honda et al. (1994b), Ahlgren et al. (unpublished). The datum of Honda et al. (1994b) at 2×10^6 μΩ cm was obtained only at 2 K and 3 T. Therefore two data at 2×10^5 μΩ cm from Ahlgren et al. (unpublished) are shown, one at 12 T and 1.5 K, and the other, shadowed triangle, at conditions comparable to those of Honda et al. (1994b).

taken to be the maximum value measured under the varying laboratory conditions of each experiment. These conditions are often similar enough that a meaningful comparison can be made. The unusually wide range of data for *i*-Al-Cu-Fe at 4×10^3 μΩ cm in Fig. 5.9 illustrates this point. There is an order of magnitude difference in results for $|\Delta\rho/\rho|^{\max}$ from four different laboratories, with the smallest value observed at 4 K and 5 T (Haberkern et al. 1993a) and the largest value at 1.8 K and 36 T (Klein et al. 1992). Average laboratory conditions are usually in between these facilities, and the trend of the results in Fig. 5.9 is clearly displayed on a logarithmic scale.

The magnetoresistance increases with resistivity up to $\simeq 10^5$ μΩ cm for positive $\Delta\rho/\rho$ as well as negative ones (up to $\simeq 10^3$ μΩ cm), and for a range of materials which includes fcc Cu-Ge, *a* metals, QCs, and approximants. As can be seen from the original publications, QIE describe the observations as a function of B and T at least qualitatively, and in some cases with quantitative precision in this range. In particular, results for $\Delta\rho(B)/\rho$ up to $\simeq$1 are included in this statement (Ahlgren et al. 1997a). For $\rho > 10^5$ μΩ cm, $\Delta\rho(B)/\rho$ is much smaller and of variable sign and this description breaks down. Data in Fig. 5.9 are more scattered in this unexplored region. However, a successively decreasing $\Delta\rho(B)/\rho$ with increasing ρ is clearly suggested. These results will now be discussed from several aspects.

The straight line shown has the form $|\Delta\rho/\rho|^{\max} \sim \rho^{1.3}$, and summarizes data empirically up to $\simeq 1\times10^5$ μΩ cm. Thus the magnetoresistance grows faster than ρ. Furthermore, it is not clear at what temperature an intrinsic resistivity can be observed. The value of ρ was evaluated at 4 K in Fig. 5.9, safely underestimating this slope since ρ(4K) is enhanced due to QIE, and this enhancement increases with increasing resistivity. This result is qualitatively expected within QIE since $\Delta\sigma$ in Eqs. (5.5) and (5.6) depends on D. Both these relations contain a factor $1/\sqrt{D}$ which by Eq. (5.1) leads crudely to $\Delta\rho/\rho = \rho|\Delta\sigma| \simeq \rho^{3/2}$ in qualitative agreement with observations and this discussion.

Figure 5.9 suggests that the quasicrystalline order in itself is not decisive for the anomalous magnetoresistance and that ρ is instead the relevant property. In addition to the wide range of materials displayed, this is supported by the fact that some samples contain second phases without violating this correlation. For *i*-Al-Cu-Cr this was shown already by the authors (Banerjee et al. 1995, 1997). Probably this is also the case for *i*-Al-Fe, which was claimed to be a single *i* phase (Plenet et al. 1992), although other investigators have not found such evidence, as discussed by Stadnik and Müller (1995). The magnetoresistance is apparently unaffected however, as expected when the high-ρ phase is governed by Eqs. (5.5) and (5.6). The main new feature concerning QCs is that the range of $\Delta\rho(B)/\rho$ within QIE has been much expanded. The magnetoresistance of QCs therefore reinforces the problem why the resistivity is so large in these materials. $\Delta\rho(B)/\rho$ itself is rather a tool to help understanding quasicrystalline transport.

The results for *i*-Al-Cu-Ru may not be fully in line with this picture. At 3×10^4 μΩ cm (open rhomboid), $|\Delta\rho/\rho|^{\max}$ is almost an order of magnitude smaller than expected, which is by far the largest discrepancy in the figure. Data were taken to 5 T at 2 K (Biggs et al. 1990) which is limited, but within the range of other data. The resistivity value appears to be consistent, since this sample fits well into the relation in Fig. 5.8 and also in Fig. 5.12 below. Kimura et al. (1991b) on the other hand, obtained a much lower resistivity, but a magnetoresistance of the expected magnitude (open rhomboid at 2×10^3 μΩ cm in Fig. 5.9). Lalla et al. (1995b) reported $\Delta\rho/\rho$ at 8 T and 4 K of only 0.005 for a sample of about 2×10^4 μΩ cm. This sample was excluded from Fig. 5.9 since the R value of 1.4 is much smaller than expected from the resistivity, and sample cracks must be suspected. One might perhaps suggest that the results of Biggs et al. (1990) indicate that QIE break down at smaller ρ values than for i-Al-Pd-Re. However, the authors could account for the observations within WL, albeit excluding EEI. Further measurements of high-resistivity i-Al-Cu-Ru would seem to be necessary to resolve this question.

Analyses. Analyses of the magnetoresistance of the low-resistivity QCs, such as Al-Mg-(Cu,Zn), were made in terms of Eq. (5.5) only (Wong et al. 1987, Baxter et al. 1987), thus neglecting EEI in accordance with results for a metals. For these materials, $\tau_{\text{ie}}(T)$, was found to vary as T^{-p} with p in the range 2–3 above 4 K, and with a tendency towards saturation at lower temperatures for i-Mg-Al-Zn. The value of τ_{so} was of order 5–10 ps, characteristic for weak spin-orbit scattering and a negative magnetoresistance.

For QCs of larger resistivities, detailed analyses of the magnetoresistance by Sahnoune et al. (1992) and by Haberkern et al. (1993a) suggested that EEI were important. F_σ, which reflects the magnitude of the diffusion channel contribution, could reach values close to or above 1, i.e., above the maximum value of 0.93 in a simple Thomas-Fermi description of the screening. This point distinguishes QCs from a metals, where it has been possible to verify a contribution from EEI to the magnetoresistance only in special cases (Lindqvist et al. 1990).

The results for the magnetoresistance show clearly that QIE can describe the observations. A first question concerns the condition $k_{\text{F}}\ell \gg 1$ (with k_{F} the Fermi wave vector and ℓ the electron mean free path) required in the original formulations of QIE. However, theoretical suggestions that corrections from higher order terms in $1/k_{\text{F}}\ell$ were small (Morgan et al. 1985), as well as experimental progress for a metals where $k_{\text{F}}\ell$ is often estimated to be of order unity, have relaxed this condition. For QCs $k_{\text{F}}\ell$ may be larger. For i-Al-Cu-Fe, Burkov et al. (1992) estimated a value of 3–12.

Then one must further ask if such a description of the magnetoresistance of QCs is unique, and to what extent QIE can be used to determine quasicrystalline properties. The number of parameters in QIE is large enough that Eqs. (5.5) and (5.6) can become rather flexible, in spite of independent variation of B and T over sizeable ranges. Different methods have been used to improve convergence. Methods, results, and difficulties encountered will be illustrated by several examples for i-Al-Cu-Fe, one of the best studied QCs. Sahnoune et al. (1992) started with a low-field fit where EEI can be neglected, and then used high-field data to determine the EEI parameters. Haberkern et al. (1993a) determined F_σ from $R_{\text{H}}(T)$, and further assumed a form for the temperature dependence of τ_{ie}. Matsuo et al. (1993) neglected EEI altogether and fitted also g^* within weak localization. Haberkern et al. (1993b) separated WL and EEI by using Hall effect measurements where WL does not enter. Lindqvist et al. (1995) assumed that WL was saturated below a certain low temperature, and separated EEI and WL in this way.

In spite of the differences between these methods, the results for the temperature dependence of τ_{ie} are similar, and thus particularly well supported.

Writing $\tau_{\text{ie}}(T) = \tau_0 T^{-p}$, the results for p in the four papers above where it was derived from the measurements were in the range 1–1.5. For phonon scattering p is 2–4 (Schmid 1973, Keck and Schmid 1976) and results for p in electron-electron scattering in the limit of Eq. (5.4) are in the range 1–2 (Schmid 1974, Isawa 1984). These results therefore consistently suggest that the low-temperature electron scattering processes in *i*-Al-Cu-Fe are dominated by electron-electron scattering.

The parameters τ_{so} and F_σ could be less well determined. Results for τ_{so} in these five papers span a range from 0.02 to 0.8 ps. Also in *a* metals this parameter is the most difficult one to determine from QIE analyses. Although the majority of the papers discussed agreed on the importance of F_σ, the details of the results were quite different. F_σ was found to have an irregular dependence on resistivity, with a minimum at intermediate resistivities (Sahnoune et al. 1992) or a maximum (Haberkern et al. 1993a). As pointed out by Lindqvist et al. (1995), one would expect F_σ to increase with decreasing $N(\epsilon_{\text{F}})$, i.e., with increasing ρ. The only two *i*-Al-Cu-Fe samples analyzed by these authors showed this trend.

In a somewhat different approach to improved fitting procedures, Ahlgren et al. (1995) made detailed analyses of *i*-Al-Cu-Fe over a larger temperature range from 0.08 to 80 K. All data were analyzed simultaneously. $\tau_{\text{ie}}(T)$ and a constant τ_{so} were fitted to all data and in addition F_σ to data up to 10 K. QIE were found to describe observations at all fields and temperatures studied with an unprecedented precision for 3D metals. This appears to open the possibility to use the magnetoresistance in QCs as a tool to study previously inaccessible problems. Three such questions were adressed by Ahlgren et al. (1997c). (i) Up to what temperature can QIE describe the magnetoresistance? (ii) What is the lowest upper bound on the elastic scattering time? (iii) Can the expected decrease and vanishing of F_σ with increasing temperature be experimentally determined? Briefly summarizing limits and success of the answers to these questions: (i) a WL contribution to the magnetoresistance was still observable at 277 K, (ii) τ was found to be smaller than several femtoseconds, and (iii) F_σ was found to decrease with increasing temperature above 10 K and to vanish in the range 80–150 K, an estimate which was found to depend strongly on the details of the analysis method. One result of these analyses is shown in Fig. 5.10, with some details of fitting procedures in the caption. Although the estimate of τ is much smaller than often quoted assumed value of $\sim 10^{-13}$ s (Mayou et al. 1993), it is consistent with the result of 0.5 fs for the dc relaxation time obtained in a model of Burkov et al. (1992) when analyzing the optical conductivity observed by Homes et al. (1991).

For work on the magnetoresistance, the strong temperature dependence of ρ in the most interesting QCs is by itself a severe problem, both for the experiments as well as in the analyses. In experiments, temperature regulation in magnetic field becomes a major concern. Already in *i*-Al-Cu-Fe, R

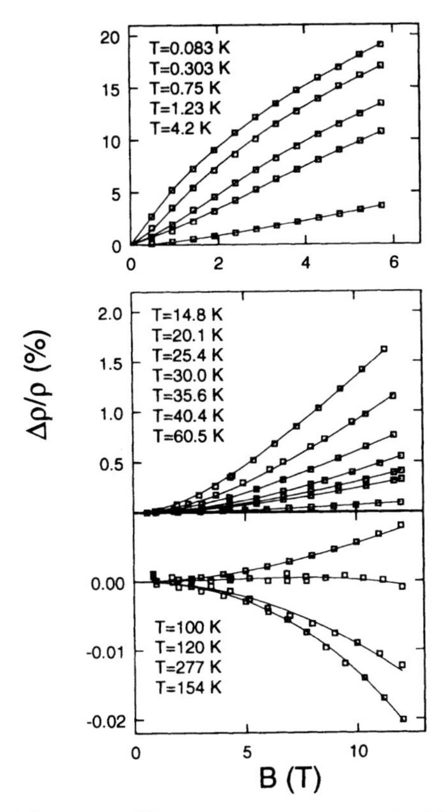

Fig. 5.10. The magnetoresistance of *i*-Al-Cu-Fe from 80 mK to 277 K. The temperature varies from top to bottom in each panel in the order given in that panel. The curves were calculated from QIE theories in the following scheme: ρ was measured, D was obtained from Eq. (5.2) with experimental results from specific heat and $\rho(300\ \mathrm{K})$, corresponding to the assumption that all of the temperature dependence of ρ below 300 K is due to QIE. It was assumed that $g* = 2$. τ_{so} was assumed to be temperature independent and was fitted together with freely varying $\tau_{\mathrm{ie}}(T)$ and $F_\sigma(T)$ to $\Delta\sigma(B,T)$ at each temperature. After Ahlgren et al. (1997c).

is large enough that a temperature stability below 20 mK must be achieved during field sweeps at high temperatures in order to determine the magnetoresistance. This is the reason that there are so few reliable measurements at higher temperatures, including the demanding region of the sign change of $\Delta\rho(B,T)$, and presumably also the reason for frequently occurring implications in the literature that the observations of QIE in QCs would be limited to low temperatures, or even to very low temperatures. As seen in Fig. 5.10, this is not the case.

Furthermore, $\Delta\sigma(B,T)$ should be calculated from a background where there are no QIE contributions, while the measured temperature dependence of ρ, $\Delta\rho(0,T)$, may contain strongly temperature dependent QIE contributions. The problem of determining the temperature range for such contributions in $\Delta\rho(T)$ has not been solved. Ahlgren et al. (1997c) used two extreme assumptions, likely to encompass the real situation, i.e., they assumed that all or nothing of $\Delta\rho(T)$ up to 300 K was due to QIE.

A third problem relevant at low temperatures can also be mentioned. In all papers considering both WL and EEI, it has been stated or understood that these contributions are independent and can be added, with the justification that this would seem reasonable when they are both of order 10^{-3} or smaller in $\Delta\rho/\rho$. For QCs this is no longer the case, but there is no theory treating possible interferences between these effects.

The success of QIE in the description of the magnetoresistance of QCs is impressive. However, this approach remains empirical, and does of course not rigorously prove the relevance of QIE contributions to transport. One of the few papers to look for an alternative is that by Banerjee et al. (1997). However, as also noted by the authors, Kohlers rule is not applicable, and modifications by empirical fits to a B^n behavior give little physical insight. In fact, there is no alternative theory available, which can provide for any quantitative comparison with experiments on the magnetoresistance. An approach to transport properties of QCs based on QIE therefore at present appears to be quite fruitful. However, there are likely exceptions also to this statement. Quasicrystals with $\rho(4\text{ K})$ above 10^5 μΩ cm obviously behave differently. The high-temperature conductivity illustrated in Fig. 5.2 is apparently of different origin.

From QIE descriptions of the magnetoresistance of QCs with $\rho(4K) < 10^5$ μΩ cm it follows that these materials are metals, in the strong scattering limit of Eq. (5.4) (conventionally called the dirty limit), or equivalently, metals with a disordered electron gas in the language of conventional disorder theories. Quasicrystals thus incorporate two concepts which were previously thought to be mutually exclusive: atomic long-range order and electronic disorder. This is yet another conundrum of QCs.

5.3.4 ρ(T) at Low and Intermediate Temperatures

The temperature dependence of the resistivity of QCs is a more difficult problem than the magnetoresistance. $\rho(T)$, albeit unusual, is a smooth function of temperature with few features. Almost any suggestion which employs a few adjustable parameters may appear reasonable. In terms of QIE this is readily appreciated from Eqs. (5.5) and (5.6) with an excessive number of free parameters for $B = 0$. As mentioned above, this problem, except for the low-temperature region, has never been fully solved for *a* metals. Indications are, when no preconceived opinion is allowed, that more than one conductivity mechanism likely contributes (Lindqvist et al. 1988).

To obtain reliable information one can combine information from different measurements. This was made, e.g., by Klein et al. (1990a) who found that the magnitude of the low-temperature $\sqrt{T}$ term in *i*-Al-Cu-Fe was half of the corresponding coefficient of $R_{\mathrm{H}}(\mathrm{T})$, as predicted from EEI theories, and thus presents a strong argument for an EEI contribution in both properties.

Haberkern et al. (1993c) determined the EEI contribution to $\Delta\rho(T)$ from $R_{\mathrm{H}}(\mathrm{T})$ of *i*-Al-Cu-Fe. $\Delta\rho(T)$ could then be described in terms of WL and EEI up to 35 K. From the EEI contribution (Lee and Ramakrishnan 1985)

$$\frac{\Delta\rho}{\rho}(0,T) = -\rho\frac{\mathrm{e}^2}{2\pi^2\hbar}\frac{0.915}{2}[\frac{4}{3} - \frac{3}{2}F_\sigma]\sqrt{\frac{\mathrm{k_B T}}{\hbar \mathrm{D}}} \tag{5.7}$$

and the expected increase of F_σ with increasing ρ discussed above, one can see that $\mathrm{d}\rho(T)/\mathrm{d}T$ changes sign from negative to positive with increasing ρ when F_σ becomes larger than 8/9. The results of Haberkern et al. (1993c) included both a low-resistivity sample with $\mathrm{d}\rho(T)/dT < 0$ and high-resistivity samples with positive derivatives, supporting this interpretation.

Similar attempts using information from the magnetoresistance were less successful until measurements were extended to low enough temperatures that τ_{ie} obtained from analyses of the magnetoresistance was found to saturate at low temperatures (Ahlgren et al. 1995). With a constant τ_{ie}, there is no temperature dependent WL contribution. F_σ determined from $\Delta\rho(B)$ was then found to give a good description of $\Delta\rho(T)$ (Fig. 5.11).

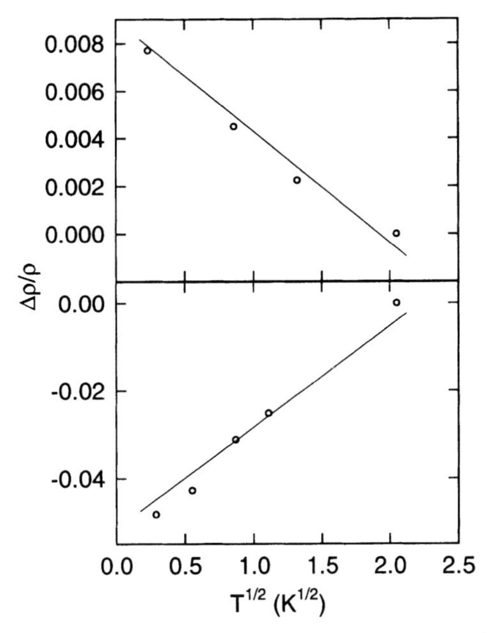

Fig. 5.11. The low-temperature resistance of two *i*-Al-Cu-Fe samples. The slopes of the straight lines shown were calculated with no adjustable parameter from Eq. (5.7) with F_σ obtained from the magnetoresistance. The top panel shows a low-resistivity sample with $F_\sigma = 0.72$ ($< 8/9$) and the lower panel a high-resistivity sample with $F_\sigma = 1.16$ [calculated from results of Ahlgren et al. (1995)].

The problem up to what temperatures QIE contribute to $\Delta\rho(T)$ was studied by a similar approach (Rodmar et al. 1995, Ahlgren et al. 1997c). For example, all information on the T dependence of the parameters was taken from the more stringent analyses of $\Delta\rho(B)$, and only two constants, τ_{so} and $F_\sigma(0\mathrm{K})$, were fitted to $\Delta\rho(T)$. This gave an excellent description of $\Delta\rho(T)$ to 150–200 K. The range over which such descriptions were accurate was strongly dependent on details of the fitting procedures, with an upper limit in the range 150–300 K. These investigations thus indicate that QIE account for the dominating temperature dependence of $\rho(T)$ of *i*-Al-Cu-Fe up to at least 150 K, and probably this description extends to a somewhat higher temperature.

5.3.5 Is There a Metal-Insulator Transition in Icosahedral Al-Pd-Re?

Already the first discovery of high resistivity in QCs raised the question whether these materials were metallic or not. Lanco et al. (1992) displayed data for $\sigma(4\mathrm{K})$ as a function of $\sigma(300\ \mathrm{K})$ for several samples from *i* Al-Cu-Fe, Al-Cu-Ru, and Al-Pd-Mn systems. These results are shown in Fig. 5.12, and may suggest a low-temperature metal-insulator transition at $\sigma(300K) \approx 120$ $(\Omega\,\mathrm{cm})^{-1}$.

For *i*-Al-Pd-Re it soon became clear however that the high-resistivity alloys did not follow the same relation, but instead showed a slower decrease of $\sigma(4\mathrm{K})$ with decreasing $\sigma(300\mathrm{K})$ than expected from the extrapolated straight line in Fig. 5.12 (Honda et al. 1994b). For *i*-Al-Pd-Re it was instead found that $\rho(4.2\ \mathrm{K})$ is almost linear in the resistance ratio R (Ahlgren et al. 1997b). This relation is illustrated in Fig. 5.13.

In *i*-Al-Pd-Re samples studied by Honda et al. (1994b) the relation between $\rho(4.2\ \mathrm{K})$ and R was irregular. However, a similar trend as in Fig. 5.13

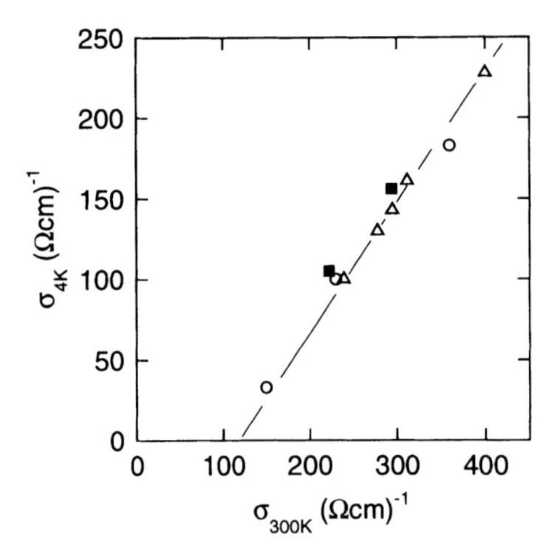

Fig. 5.12. The conductivity at 4 K as a function of conductivity at 300 K for several *i* QCs ***i*-Al-Cu-Fe**: open triangles, ***i*-Al-Pd-Mn**: filled squares, ***i*-Al-Cu-Ru**: open circles. After Lanco et al. (1992).

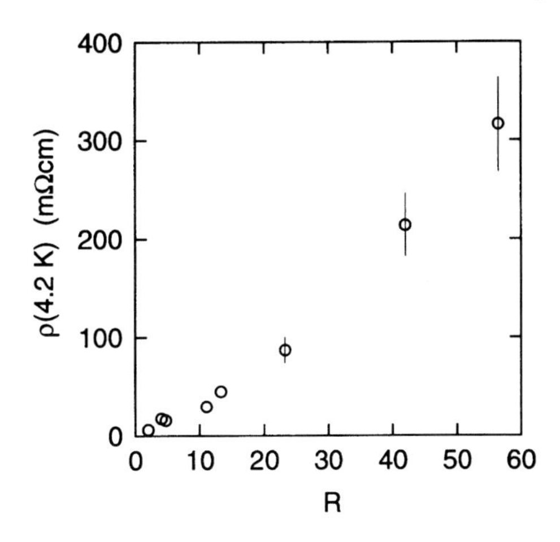

Fig. 5.13. The resistivity at 4 K for *i*-Al-Pd-Re as a function of the resistance ratio R. All samples were of a single i phase in standard x-ray diffraction. Bars corresponding to 15% error in the measurements of ρ have been marked on several samples. After Ahlgren et al. (1997c).

was found by Lin, C. R. et al. (1997), although in this case there was only one data point above $R = 15$. Further studies (Gignoux 1996) indicate that the relation in Fig. 5.13 continues to describe data up to $R \simeq 200$. Since there is a large error in resistivity measurements on small samples, R can thus be used instead as a convenient parameter for sample characterization. This relation is different from that observed for other i samples illustrated in Fig. 5.12, which is not understood.

The possibility of studying a metal-insulator transition in an alloy with only metallic elements is fascinating and has aroused intensive interest. Five papers from 1996–97 studying this question will be discussed below, with an attempt to summarize the strong and weak points in the arguments advanced for and against the observation of a metal-insulator transition.

Guo and Poon (1996) studied i-$Al_{70.5}Pd_{21}Re_{8.5-x}Mn_x$ with Mn concentration up to $x = 5$. If i-Al-Pd-Re is insulating this could provide a method to monitor a transition to the metallic state by increasing x. Extrapolations to $T = 0$ K from temperatures above 0.45 K indicated zero conductivity at a finite temperature for $x < 3$. $\sigma(T)$ for x in the range 2 to 3.5 was found to follow Efros-Shklovskii (1975) VRH, which includes Coulomb repulsion, and predicts $\ln\sigma(\mathrm{T}) \simeq T^{-1/2}$. For $x = 0$, the Mott VRH was considered with $\ln\sigma(T) \simeq T^{-1/4}$. A transition from $T^{-1/2}$ to $T^{-1/4}$ can perhaps be interpreted as a sign of the decreased importance of EEI further into the insulating side of the transition. However, in order to describe data in this way it was necessary to subtract a residual conductivity, $\sigma(0)$ ascribed to impurities. It is not clear why small amounts of impurities in high quality samples should contribute a parallel conductivity channel. Furthermore, although the QIE description of the magnetoresistance broke down below $x = 4$ as expected, it could not be described on the purported insulating side of the transition, for smaller x, by VRH theories.

Lin, C.R. et al. (1996) studied related i-$Al_{70}Pd_{22.5}(Re_{1-x}Mn_x)_{7.5}$ samples. $\sigma(T)$ was analyzed by the expression

$$\sigma(\mathrm{T}) = \sigma(0)(1 + \sqrt{\mathrm{T}/\Delta}) \tag{5.8}$$

found by Grest and Lee (1983) for interacting disordered electrons. Δ is referred to as the correlation gap. Although the slope of $\sigma(T)$ vs $\sqrt{T}$ varied by a factor of three among about 20 samples of i-Al-Pd-Re and Mn-doped i-Al-Pd-Re, Δ varied by a factor of 50 000 and the overall trend of these data was nevertheless described by $\Delta \simeq \rho(0)^{-2}$ over more than two orders of magnitude in $\rho(0)$. This result is interesting. However, the connection with proximity to a metal insulator transition is not fully clarified. $\sigma(0)$ was obtained by linearly extrapolating $\sigma(T)$ vs $\sqrt{T}$ from 4.2 K, and results discussed below (Bianchi et al. 1997, Ahlgren et al. 1997c) show that such a procedure would underestimate $\sigma(0)$ by a σ-dependent factor due to the saturation of $\sigma(T)$ at low T. Furthermore, one must be able to distinguish between Eqs. (5.7) and (5.8), which both predict a $\sqrt{T}$ behavior. Data on i-Al-Cu-Fe were included by Lin, C.R. et al. (1996), where a description based on EEI appears to be well founded.

Gignoux et al. (1997) studied i-Al-Pd-Re with R up to 190 and made qualitative comparisons between several properties and expectations from both sides of a metal-insulator transition. They noted that the electronic specific heat remained metallic-like, and also that even down to 70 mK, $\sigma(T)$ could not be described by a VRH expression. An unusual temperature dependence of the conductivity was pointed out, with a single temperature exponent, $\sigma(T) \simeq T^{1.37}$ from 7 to 700 K. A similar result was found for i-Al-Pd-Mn, which is clearly metallic at low temperatures. Bianchi et al. (1997) observed a linear $\sigma(T)$ in i-Al-Pd-Re, although limited to temperatures below 300 K and to smaller R-values. A suggestive argument by Gignoux et al. is shown in Fig. 5.14. The magnetoresistance was found to have a qualitatively similar behavior to some known systems with a metal-insulator transition. However, in contrast to these insulators, and in qualitative agreement with the results in Fig. 5.9, the magnetoresistance of i-Al-Pd-Re decreases rapidly with increasing R at large R values.

Bianchi et al. (1997) extended measurements of $\sigma(T)$ for two i-Al-Pd-Re samples down to 40 mK. Their high-resistivity sample had ρ(4K) about 200 mΩ cm, and $R \approx 7$, which appears to be rather too small for the large resistivity (Table 5.2). The results clearly indicated that the conductivity extrapolated to $T = 0$ would remain finite. Ahlgren et al. (1997b) used a similar approach and measured $\sigma(T)$ for a series i-Al-Pd-Re samples with R in the range 2 to 57 down to 40 mK. $\sigma(T)$ was found to saturate below a temperature which increased with R (and ρ), and furthermore, the values of $\sigma(0)$ extrapolated from these relations decreased exponentially with R^{-1} in a way which would suggest that $\sigma(0)$ remained finite also for samples with higher R values. Deducing a $\sigma(0)$ larger than zero is of course a strong indication of a metallic ground state. Nevertheless, in spite of the well-characterized

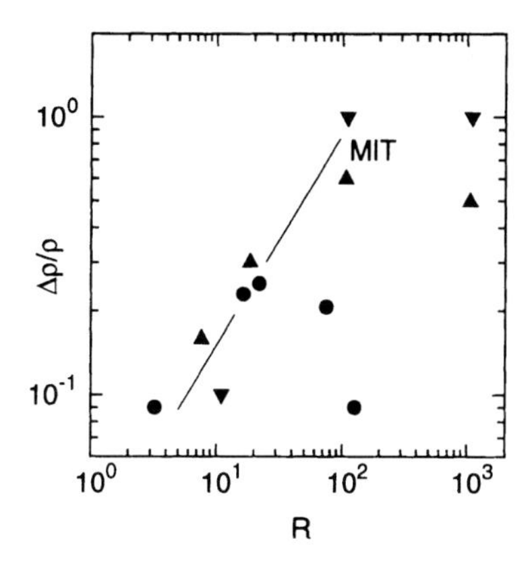

Fig. 5.14. The magnetoresistivity at 9 T and 1.7 K vs the resistance ratio R for *i*-**Al-Pd-Re**, circles, and two systems undergoing a metal-insulator transition: **Mo-Ge**, down traingles, and **Al-Al_2O_3**, up triangles. The straight line is a guide to the eye for data in the region where $\Delta\rho/\rho$ increases with R. For *i*-Al-Pd-Re, the largest $\Delta\rho/\rho$ corresponds from Fig. 5.13 to a sample with ρ somewhat below 0.1 Ω cm, consistent with Fig. 5.9. Replotted from Gignoux et al. (1997).

samples used by Ahlgren et al. (1997c), one must worry about a possible influence of sample imperfections and impurities. What are the reasons for the strongly varying R values in samples which are phase pure in standard x-ray diffraction? Can small amounts of impurities contribute to the conductivity by other mechanisms, e.g., tunneling? A tentative answer to the first question is a small homogeneity range for the i phase and an extreme sensitivity to composition. A small solubility range is in fact indicated by careful structural work (Gignoux 1996, Sawada et al. 1998). Klein and Symko (1994) observed a linear relation between current and voltage in *i*-Al-Cu-Fe over seven orders of magnitude, which appears to rule out some previous propositions of tunneling (Philips and Rabe 1991). Conduction mechanisms in different QCs could be different though. Further improved sample control appears to be required for studying these objections.

The correlation between γ and ρ in Fig. 5.8 can also be mentioned. Clearly $\gamma = 0$ is a sufficient condition for an insulator, but it is not necessary due, e.g., to possible contributions from localized states. However, the three *i*-Al-Pd-Re samples in Fig. 5.8, all above 10^5 μΩ cm in Fig. 5.9, apparently follow the same trend for γ as for the metallic samples, without any clear trend to deviate, e.g., to decrease more strongly on the insulating side as in Si:P (Thomas et al. 1981). This qualitatively supports that also these *i*-Al-Pd-Re samples remain metallic.

5.4 Concluding Remarks

With respect to the field of work to be covered, this review is brief, and omissions are unavoidable. These include new QCs, after *i*-Al-Pd-Re, with so far less spectacular but interesting properties. One example is *i*-Zn-Mg-Y; a QC without Al and with a surprisingly high resistivity at 4 K of $\simeq$ 1 mΩ cm (Kondo et al. 1995). Further work on *i*-Al-Pd-Mn also appears to be fruitful. The interpretation of transport properties, at least $\Delta\rho(T)$, has been complicated by the possibilities of an influence of minor magnetic Mn impurities or a fraction of magnetic Mn sites in the QC. *i*-Al-Pd-Mn can be grown in large single grains, giving an opportunity to investigate if anisotropy can be observed in transport properties. Such work is going on (Yokoyama et al. 1992, Rodmar et al. 1996, 1998), and indications are that anisotropy can be observed in the magnetoresistance. Again, the role, if any, of minor Mn precipitates must be carefully controlled. If such results can be confirmed, this technique would seem to provide a powerful experimental tool for further studies of electronic properties of icosahedral QC.

Data for the electronic specific heat coefficient and the electrical resistivity at room temperature were compiled for a large range of *i* QCs. A simple relation between ρ and γ over more than two orders of magnitude in $\rho(295\text{K})$ suggests that an increasing ρ in *i* QCs is associated both with a decreasing DOS and with a decreasing diffusion constant. Another interesting result is the relation between the maximum magnetoresistance observed in the laboratory and the resistivity at 4 K. The magnetoresistance for samples with $\rho(4\text{K})$ up to $\simeq 10^5 \mu\Omega\,\text{cm}$, with some moderate scatter, is a universal function of ρ, independent of atomic arrangements, whether crystalline, *a*, or quasicrystalline.

The magnetoresistance (up to this limit) has been emphasized. It is, as yet, the only electronic transport property which can be quantitatively described within an existing theoretical framework. This appears to have important consequences for the description of transport at not too elevated temperatures, and also to be a useful testing ground for competing propositions of conduction mechanisms. The success of QIE theories leads to a description of these QCs as electronically disordered metals.

This is but a small fraction of the anomalous results of QCs. The magnetoresistance for samples with $\rho(4\text{K}) > 10^5\ \mu\Omega\,\text{cm}$ is not understood, neither within weak localization in metals, nor from the expected behavior of insulators. The high resistivity itself is an outstanding problem. The strongly increasing conductivity with increasing temperature, often described by a surprisingly simple behavior with a single but sample-dependent exponent, further compounds this problem. The extreme sensitivity to composition and sample conditions is an unsolved problem, as well as the unusual temperature and concentration dependence of the thermoelectric power and the Hall coefficient. Moreover, a metal-insulator transition is as yet neither proved, nor fully disproved. After 13 years with QCs, the subject of electronic transport continues to be a rich research area, which gives us new insights as well as unexpected and unexplained results at a rate with no sign of saturation.

Some New Results Added in Proof (May 1998). A few papers are commented upon to exemplify the most recent work on electronic transport properties.

An overview of current theoretical approaches to understanding electronic transport has been published by Roche et al. (1997). Nonballistic propagation between two collision events, interband transition mechanism and hopping transport, quantum interference and correlated disorder, and Kubo-Greenwood conductivity in quasiperiodis systems with disorder are discussed. This short list illustrates the breadth of the current approach to the problem of electronic transport in QCs. Still this work is in its beginning and quantitative results for detailed comparisons with experiments have not been obtained. The exception, as mentioned above, is the magnetoresistance of quasicrystals with $\rho(4\mathrm{K}) < 10^5\ \mu\Omega\,\mathrm{cm}$.

The resistance and the magnetoresistance of $Al_{65}Cu_{20}Co_{15-x}Fe_x$ was studied by Singh et al. (1997), apparently the first such measurements in this alloy system. For $x = 0.15$ the alloy is icosahedral. The magnetoresistance at 4.2 K and 4 T was about 0.015, which is on the low side of the correlation in Fig. 5.9. However, this may be due to the small measurement range and to the possibly somewhat too large $\rho(4\mathrm{K})$ of 10 $\mathrm{m}\Omega\,\mathrm{cm}$, as suggested by an R value of only 1.3. At $x = 0$ however, in a decagonal phase, the low temperature resistivity was a factor of 30 smaller, but the magnetoresistance was about the same. This result and the anomalous temperature dependence of the magnetoresistance indicate a different origin than quantum corections.

Mizutani(1998) recently extended his previous work on correlations between ρ and γ to include also high resistivity quasicrystals. Part of this work is similar to the result reported in Fig. 5.8. In particular we have independently found the relation $\gamma \simeq 1/\sqrt{\rho}$. However, i-AlPdRe was excepted from this correlation by Mizutani, since there were deviations for these samples in the direction of higher resistivities when data from several alloy systems were plotted in the form $\rho(4.2\ \mathrm{K})$ vs γ. Mizutani interpreted this as an "indication that the AlPdRe quasicrystal is no longer in the metallic regime". From Fig. 5.8 the conclusion would rather be the opposite as mentioned in Sect. 5.3.5.

Krajčí and Hafner (1998) studied anisotropic transport properties of decagonal quasicrystals and approximants. *Ab initio* calculations were performed using the LMTO technique and conventional transport theory. In particular, compounds close to Al_3Mn were included, where experimental results by Volkov and Poon (1995) are illustrated in Fig. 5.4. Krajčí and Hafner found a conductivity gap in a single crystallographic direction in agreement with the experimental results but there was no agreement as to the direction along which this effect was observed. This discrepancy was attributed to the strong sensitivity of calculated properties on structural details, and in particular to the fact that their calculation was performed for the composition $Al_{13}Mn_4$, different from that given by Volkov and Poon (1995).

Acknowledgments

Coworkers over the years, and in particular P. Lindqvist, M. Ahlgren, M. Rodmar in Stockholm, and C. Berger in Grenoble, are warmly thanked for fruitful collaborations and great stimulation. I am also grateful to F. Cyrot-Lackmann for her sustained initiatives in establishing several collaborations. M. Ahlgren, C. Berger, D. Oberschmidt, and M. Rodmar willingly read the manuscipt and gave useful comments, and A. Rydh shared with me of his insights into the LaTeX system. Financial support from the Swedish Natural Science Research Council and the EU network on quasicrystals is gratefully acknowledged.

References

Ahlgren, M., Rodmar, M., Klein, T., Rapp, Ö. (1995): Phys.Rev. B **51**, 7287
Ahlgren, M., Rodmar, M., Sommer, M., Rapp, Ö. (1996a): J. Non-Cryst. Solids **205-207**, 21
Ahlgren, M., Rodmar, Brangefält, J., Berger, C., Gignoux, C., Rapp, Ö. (1996b): Czech. J. Phys. **46**, 1989
Ahlgren, M., Rodmar, M., Gignoux, C., Berger, C., Rapp, Ö. (1997a): Mater. Sci. Eng. A **226-228**, 981
Ahlgren, M., Gignoux, C., Rodmar, M., Berger, C., Rapp, Ö. (1997b): Phys. Rev. B **55**, R11 915
Ahlgren, M., Lindqvist, P., Rodmar, M., Rapp, Ö. (1997c): Phys. Rev. B **55**, 14 847
Akiyama, H., Honda, Y., Hashimoto, T., Edagawa, K., Takeuchi, S. (1993a): Jpn. J. Appl. Phys. B **32**, L1003
Akiyama, H., Hashimoto, T., Shibuya, T., Edagawa, K., Takeuchi, S. (1993b): J. Phys. Soc. Jpn. **62**, 639
Banerjee, S., Goswami, R., Chattopadhyay, K., Raychaudhuri, A.K. (1995): Phys. Rev. B **52**, 3220
Banerjee, S., Raychaudhuri, A.K., Goswami, R., Chattopadhyay, K. (1997): J. Phys. Condens. Matter **9**, 6643
Barnard, R.D. (1972): Thermoelectricity of Metals and Alloys. Taylor and Francis, London
Baxter, D.V., Richter, R., Strom-Olsen, J.O. (1987): Phys. Rev. B **35**, 4819
Belin, E., Miyoshi, Y., Yamada, Y., Ishikawa, T., Matsuda, T., Mizutani, U. (1994): Mater. Sci. Eng. A **181-182**, 730
Berger, C. (1994): in Lectures on Quasicrystals, Hippert, F., Gratias, D. (eds). Les Editions de Physique, Les Ulis, p 463
Berger, C., Lasjaunias, J.C., Tholence, J.L., Pavuna, D., Germi, P. (1988): Phys. Rev. B **37**, 6525
Berger, C., Belin, E., Mayou, D. (1993a): Ann. Chim. Fr. **18**, 485
Berger, C., Grenet, T., Lindqvist, P., Lanco, P., Grieco, J.C., Fourcaudot, G., Cyrot-Lackmann, F. (1993b): Solid State Commun. **87**, 977
Berger, C., Mayou, D., Cyrot.Lackmann, F. (1995a): in Proceedings of the 5th International Conference on Quasicrystals, Janot, C., Mosseri, R. (eds). World Scientific, Singapore, p 423
Berger, C., Gignoux, C., Tjernberg, O., Lindqvist, P., Cyrot-Lackmann, F, Calvayrac, Y. (1995b): Physica B **204**, 44
Bergman, G. (1984): Phys. Rep. **107**, 1
Bessière, M., Quivy, A., Lefebvre, S., Devaud-Rzepski, J., Calvayrac, Y. (1991): J. Phys. (Paris) I **1**, 1823

Bianchi, A.D., Chernikov, M.A., Felder, E., Ritsch, S., Ott, H.R. (1996): Czech. J. Phys. **46**, 2553
Bianchi, A.D., Bommeli, F., Chernikov, M.A., Gubler, U., Degiorgi, L., Ott, H.R. (1997): Phys. Rev. B **55**, 5730
Bieri, J.B., Fert, A., Creuzet, G., Schuhl, A. (1986): J. Phys. F **16**, 2099
Biggs, B.D., Poon, S.J., Munirathnam, N.R. (1990): Phys. Rev. Lett. **65**, 2700
Biggs, B.D., Li, Y., Poon, S.J. (1991): Phys. Rev. B **43**, 8747
Biggs, B.D., Pierce, F.S., Poon, S.J. (1992): Europhys. Lett. **19**, 415
Burkov, S.E., Timusk, T., Ashcroft, N.W. (1992): J. Phys. Condens. Matter **4**, 9447
Burkov, S.E., Varlamov, A.A., Livanov, D.V. (1996): Phys. Rev. B **53**, 11 504
Chernikov, M.A., Bernasconi, A., Beeli, C., Schilling, A., Ott, H.R. (1993a): Phys. Rev. B **48**, 3058
Chernikov, M.A., Bernasconi, A., Beeli, C., Ott, H.R. (1993b): Europhys. Lett. **21**, 767
Chernikov, M.A., Degiorgi, L., Bernasconi, A., Beeli, C., Ott, H.R. (1994): Physica B **194-196**, 405
Chernikov, M.A., Bianchi, A., Ott, H.R. (1995): Phys. Rev. B **51**, 153
Chernikov, M.A., Bianchi, A., Felder, E., Gubler, U., Ott, H.R. (1996): Europhys. Lett. **35**, 431
Dugdale, J.S. (1995): The Electrical Properties of Disordered Metals. Cambridge University Press, Cambridge
Edagawa, K., Naito, N., Takeuchi, S. (1992): Philos. Mag. B **65**, 1011
Edagawa, K., Yamaguchi, S., Suzuki, K., Takeuchi, S. (1996a): in Advances in Physical Metallurgy, Banerjee, S., Ramanujan, R.V. (eds). Gordon and Breach, Amsterdam, p 96
Edagawa, K., Chernikov, M.A., Bianchi, A.D., Felder, E., Gubler, U., Ott, H.R. (1996b): Phys. Rev. Lett. **77**, 1071
Efros, A.L., Shklovskii, B.I. (1975): J. Phys. C **8**, L49
Eschner, W., Gey, W., Warnecke, P. (1984): in Proceedings of the International Conference on Localization, Interaction, and Transport Phenomena in Impure Metals, Schweitzer, L., Kramer, B. (eds). Phys. Techn. Bundesanstalt, Braunschweig, p 327
Friedel, J., Dénoyer, F. (1987): C.R. Acad. Sci. Paris **305**, 171
Fujiwara, T., Mitsui, T., Yamamoto, S. (1996): Phys. Rev. B **53**, R2910
Fukuyama, H., Hoshino, K. (1981): J. Phys. Soc. Jpn. **50**, 2131
Gignoux, C. (1996): Thèse de l'Université Joseph Fourier-Grenoble I
Gignoux, C., Berger, C., Fourcaudot, G., Grieco, J.C., Cyrot-Lackmann, F. (1995): in Proceedings of the 5th International Conference on Quasicrystals, Janot, C., Mosseri, R. (eds). World Scientific, Singapore, p 452
Gignoux, C., Berger, C., Fourcaudot, G., Grieco, J.C., Rakoto, H. (1997): Europhys. Lett. **39**, 171
Giroud, F., Grenet, T., Berger, C., Lindqvist, P., Gignoux, C., Fourcaudot, G. (1996): Czech. J. Phys. **46**, 2709
Graebner, J.E., Chen, H.S. (1987): Phys. Rev. Lett. **58**, 1945
Grenet, T., Lindqvist, P., Berger, C., Gignoux, C., Jansen, A.G.M. (1995): in Proceedings of the 5th International Conference on Quasicrystals, Janot, C., Mosseri, R. (eds). World Scientific, Singapore, p 456
Grest, G.S., Lee, P.A. (1983): Phys. Rev. Lett. **50**, 693
Guo, Q., Poon, S.J. (1996): Phys. Rev. B **54**, 12 793
Haberkern, R., Fritsch, G. (1995): in Proceedings of the 5th International Conference on Quasicrystals, Janot, C., Mosseri, R. (eds). World Scientific, Singapore, p 460
Haberkern, R., Fritsch, G., Schilling, J. (1993a): Z. Phys. B **92**, 383

Haberkern, R., Fritsch, G., Härting, M. (1993b): Appl. Phys. A **57**, 431
Haberkern, R., Lindqvist, P., Fritsch, G. (1993c): J. Non-Cryst. Solids **153-154**, 303
Hashimoto, K., Yamada, Y., Yamauchi T., Tanaka, T., Matsudam T., Mizutani, U. (1994): Mater. Sci. Eng. A **181-182**, 785
Homes, C.C., Timusk, T., Wu, X., Altounian, Z., Sahnoune, A., Ström-Olsen, J.O. (1991): Phys. Rev. Lett. **67**, 2694
Honda, Y., Edagawa, K.,Yoshioka, A., Akyiama, H., Hashimoto, T., Takeuchi, S. (1994a): Mater. Sci. Forum **150-151**, 465
Honda, Y., Edagawa, K.,Yoshioka, A., Hashimoto, T.,Takeuchi, S. (1994b): Jpn. J. Appl. Phys. A **33**, 4929
Honda, Y., Edagawa, K., Takeuchi, S., Tsai, A.-P., Inoue, A. (1995): Jpn. J. Appl. Phys. A **34**, 2415
Howson, M.A., Gallagher, B.L. (1988): Phys. Rep. **170**, 265
Hurd, C.M. (1972): The Hall Effect in Metals and Alloys. Plenum, New York
Isawa, Y. (1984): J. Phys. Soc. Jpn. **53**, 2865
Janot, C. (1996): Phys. Rev. B **53**, 181
Keck, B., Schmid, A. (1976): J. Low Temp. Phys. **24**, 611
Kimura, K., Yamane, H., Hashimoto, T., Takeuchi, S. (1987): Mater. Sci. Forum **22-24**, 471
Kimura, K., Iwahashi, H., Hashimoto, T., Takeuchi, S., Mizutani, U., Ohashi, S., Itoh, G. (1989): J. Phys. Soc. Jpn. **58**, 2472
Kimura, K., Takeuchi, S. (1991a): in Quasicrystals, The State of the Art, DiVincenzo, D.P., Steinhardt, P.J. (eds). World Scientific, Singapore, p 313
Kimura, K., Kishi, K., Hashimoto, T., Takeuchi, S., Shibuya, T. (1991b): Mater. Sci. Eng. A **133**, 94
Kishi, K., Kimura, K., Hashimoto, T., Takeuchi, S. (1990): J. Phys. Soc. Jpn. **59**, 1158
Klein, T., Symko, O.G. (1994): Phys. Rev. Lett. **73**, 2248
Klein, T., Gozlan, A., Berger, C., Cyrot-Lackmann, F., Calvayrac, Y., Quivy, A. (1990a): Europhys. Lett. **13**, 129
Klein, T., Gozlan, A., Berger, C., Cyrot-Lackmann, F., Calvayrac, Y., Quivy, A., Fillion, G. (1990b): Physica B **165-166**, 283
Klein, T., Berger, C., Mayou, D., Cyrot-Lackmann, F. (1991): Phys. Rev. Lett. **66**, 2907
Klein, T., Rakoto, H., Berger, C., Fourcaudot, G., Cyrot-Lackmann, F. (1992): Phys. Rev. B **45**, 2046
Klein, T., Berger, C., Fourcaudot, G., Grieco, J.C., Lasjaunias, J.C. (1993): J. Non-Cryst. Solids **153-154**, 312
Klein, T., Symko, O.G., Paulsen, C. (1995): Phys. Rev. B **51**, 12 805
Kondo, R., Honda, Y., Hashimoto, T., Edagawa, K., Takeuchi, S. (1995): in Proceedings of the 5th International Conference on Quasicrystals, Janot, C., Mosseri, R. (eds). World Scientific, Singapore, p 476
Krajčí, M., Windisch, M., Hafner, J., Kresse, G., Mihalkovič, M. (1995): Phys. Rev. B **51**, 17 355
Krajčí, M., Hafner, J. (1998): unpublished
Lalla, N.P., Tiwari, R.S., Srivastava, O.N. (1995a): J. Phys. Condens. Matter **7**, 2409
Lalla, N.P., Tiwari, R.S., Srivastava, O.N., Schnell, B., Thummes, G. (1995b): Z. Phys. B **99**, 43
Lanco, P., Klein, T., Berger, C., Cyrot-Lackmann, F., Fourcaudot, G., Sulpice, A. (1992): Europhys. Lett. **18**, 227

Lanco, P., Berger, C., Cyrot-Lackmann, F., Sulpice, A. (1993): J. Non-Cryst. Solids **153-154**, 325
Lasjaunias, J.C., Tholence, J.L., Berger, C., Pavuna, D. (1987): Solid State Commun. **64**, 425
Lasjaunias, J.C., Sulpice, A., Keller, N., Préjean, J.J., de Boissieu, M. (1995): Phys. Rev. B **52**, 886
Lee, P.A., Ramakrishnan, T.V. (1982): Phys. Rev. B **26**, 4009
Lee, P.A., Ramakrishnan, T.V. (1985): Rev. Mod. Phys. **57**, 287
Lin, C.R., Chou, S.L., Lin, S.T. (1996): J. Phys. Condens. Matter **8**, L725
Lin, C.R., Lin, S.T., Wang, C.R., Chou, S.L., Horng, H.E., Cheng, J.M., Yao, Y.D., Lai, S.C. (1997): J. Phys. Condens. Matter **9**, 1509
Lin, S.Y, Wang, X.M., Lu, L., Zhang, D.L., He, L.X., Kuo, K.X. (1990): Phys. Rev. B **41**, 9625
Lin, S.Y., Li, G.H., Zhang, D.L. (1996): Phys. Rev. Lett. **77**, 1998
Lindqvist, P., Fritsch, G. (1989): Phys. Rev. B **40**, 5792
Lindqvist, P., Rapp, Ö. (1988): J. Phys. F **18**, 1979
Lindqvist, P., Petrovic, P., Liu, Z.-Y., Rapp, Ö., (1988): Mater. Sci. Eng. **99**, 235
Lindqvist, P., Rapp, Ö., Sahnoune, A., Ström-Olsen, J.O. (1990): B **41**, 3841
Lindqvist, P., Berger, C., Klein, T., Lanco, P., Cyrot-Lackmann, F., Calvayrac, Y. (1993): Phys. Rev. B **48**, 630
Lindqvist, P., Lanco, P., Berger, C., Jansen, A.G.M., Cyrot-Lackmann, F. (1995): Rev. B **51**, 4796
Markert, J.T., Cobb, J.L., Bruton, W.D., Bhatnagar, A.K., Naugle, D.G., Kortan, A.R. (1994): J. Appl. Phys. **76**, 6110
Martin, S., Hebard, A.F., Kortan, A.R., Thiel, F.A. (1991): Phys. Rev. Lett. **67**, 719
Matsuo, S., Nakano, H., Saito, K., Mori, M., Ishimasa, T. (1993): Solid State Commun. **86**, 707
Matsuo, S., Nakano, H., Ishimasa, T., Mori, M. (1994): Solid State Commun. **92**, 811
Matsuo, S., Nakano, H., Saito, K., Ishimasa, T., (1995): in Proceedings of the 5th International Conference on Quasicrystals, Janot, C., Mosseri, R. (eds). World Scientific, Singapore, p 488
Mayou, D., Berger, C., Cyrot-Lackmann, F., Klein, T., Lanco, P. (1993): Phys. Rev. Lett. **70**, 3915
Mizutani, U. (1993): Mater. Sci. Eng. B **19**, 82
Mizutani, U., (1998): J. Phys. Condens. Matter **10**, 4609
Mizutani, U., Yoshida, T. (1982): J. Phys. F **12**, 2331
Mizutani, U., Sakabe, Y., Matsuda, T. (1990a): J. Phys. Condens. Matter **2**, 6153
Mizutani, U., Sakabe, Y., Shibuya, T., Kishi, K., Kimura, K., Takeuchi, S. (1990b): J. Phys. Condens. Matter **2**, 6169
Morgan, G.J., Howson, M.A., Šaub, K. (1985): J. Phys. F **15**, 2157
Mott, N.F. (1990): Metal-Insulator Transitions. Taylor and Francis, London
Nakamura, Y., Mizutani, U. (1994): Mater. Sci. Eng. A **181-182**, 790
Naugle, D.G. (1984): J. Phys. Chem. Solids **45**, 367
Naugle, D.G., Bruton, W.D., Rathnayaka, K.D.D., Kortan, A.R. (1996): J. Non-Cryst. Solids **205-207**, 17
Nguyen-Manh, D., Mayou, D., Morgan, G.J., Pasturel, A. (1987): J. Phys. F **17**, 999
Pavuna, D. (1985): Solid State Commun. **54**, 771
Perrot, A., Dubois, J.M. (1993): Ann. Chim. Fr. **18**, 501
Philips, J.C., Rabe, K.M. (1991): Phys. Rev. Lett. **66**, 923
Pierce, F.S., Bancel, P.A., Biggs, B.D., Guo, Q., Poon, S.J. (1993a): Phys. Rev. B **47**, 5670
Pierce, F.S., Poon, S.J., Biggs, B.D. (1993b): Phys. Rev. Lett. **70**, 3919

Pierce, F.S., Poon, S.J., Guo, Q. (1993c): Science **261**, 737
Pierce, F.S. Guo, Q., Poon, S.J. (1994): Phys. Rev. Lett. **73**, 2220
Plenet, J.C., Perez, A., Rivory, J., Frigerio, J.M., Laborde, O. (1992): Phys. Lett. A **162**, 193
Poon, S.J. (1992): Adv. Phys. **41**, 303
Poon, S.J. (1993a): J. Non-Cryst. Solids **153-154**, 334
Poon, S.J. (1993b): Mater. Sci. Eng. B **19**, 72
Poon, S.J. (1994): in AIP Conference Proceedings 286, Srivastava, V., Bhatnagar, A., Naugle, D.G. (eds). AIP, New York, p 54
Poon, S.J., Drehman, A.J., Lawless, K.R. (1985): Phys. Rev. Lett. **55**, 2324
Poon, S.J., Pierce, F.S., Guo, Q., Volkov, P. (1995a): in Proceedings of the 5th International Conference on Quasicrystals, Janot, C., Mosseri, R. (eds). World Scientific, Singapore, p 408
Poon, S.J., Pierce, F.S., Guo, Q. (1995b): Phys. Rev. B **51**, 2777
Poon, S.J., Guo, Q., Volkov, P., Pierce, F.S. (1996): J. Non-Cryst. Solids **205-207**, 1
Quivy, A., Quiquandon, M., Calvayrac, Y., Faudot, F., Gratias, D., Berger, C., Brand, R.A., Simonet, V., Hippert, F. (1996): J. Phys. Condens. Matter. **8**, 4223
Rapp, Ö., Hedman, L., Klein, T., Fourcaudot, G. (1993): Solid State Commun. **87**, 143
Roche, S., Mayou, D. (1997): Phys. Rev. Lett. **79**, 2518
Roche, S., Trambly de Laissardière, T., Mayou, D. (1997): J. Math. Phys. **38**, 1997
Rodmar, M., Ahlgren, M., Rapp, Ö. (1995): in Proceedings of the 5th International Conference on Quasicrystals, Janot, C., Mosseri, R. (eds). World Scientific, Singapore, p 518
Rodmar, M., Ahlgren, M., Berger, C., Sulpice, A., Beyss, M., Tamura, N., Urban, K., Rapp, Ö. (1996): Czech. J. Phys. **46**, 2703
Rodmar, M., Ahlgren, M., Berger, C., Sulpice, A., Beyss, M., Tamura, N., Urban, K., Rapp, Ö. (1998): in Proceedings of the 6th International Conference on Quasicrystals, Takeuchi, S., Fujiwara, T. (eds). World Scientific, Singapore, p 692
Sahnoune, A., Ström-Olsen, J.O., Zaluska, A. (1992): Phys. Rev. B **46**, 10 629
Sahnoune, A., Ström-Olsen, J.O., Altounian, Z., Homes, C.C., Timusk, T., Wu, X. (1993): J. Non-Cryst. Solids **153-154**, 343
Saito, K., Matsuo, S., Nakano, H., Ishimasa, T., Mori, M. (1994): J. Phys. Soc. Jpn. **63**, 1940
Samwer, K., von Löhneysen, H. (1982): Phys. Rev. B **26**, 107
Sawada, H., Tamura, R., Kimura, K., Ino, H. (1998): in Proceedings of the 6th International Conference on Quasicrystals, Takeuchi, S., Fujiwara, T. (eds). World Scientific, Singapore, p 329
Schmid, A. (1973): Z. Phys. **259**, 421
Schmid, A. (1974): Z. Phys. **271**, 251
Shibuya, T., Hashimoto, T., Takeuchi, S. (1990): J. Phys. Soc. Jpn. **59**, 1917
Singh, K., Bahadur, D., Radha, S., Nigan, A. K., Prasad, S. (1997): J. Alloys Comp. **257**, 57
Smith, A.P., Ashcroft, N.W. (1987): Phys. Rev. Lett. **59**, 1365
Srinivas, V., Dunlap, R.A. (1991): Philos. Mag. B **64**, 475
Stadnik, Z.M., Müller, F. (1995): Philos. Mag. B **71**, 221
Ström-Olsen, J.O., Olivier, M., Altounian, Z., Cochrane, R.W. (1984): in Proceedings of the International Conference on Localization, Interaction, and Transport Phenomena in Impure Metals, Schwitzer, L., Kramer, B. (eds). Phys. Techn. Bundesanstalt, Braunschweig, p 55
Takeuchi, S. (1994): Mater. Sci. Forum **150-151**, 35

Takeuchi, S., Akiyama, H., Naito, N., Shibuya, T., Hashimoto, T., Edagawa, K., Kimura, K. (1993): J. Non-Cryst. Solids **153-154**, 353
Takeuchi, T., Yamada, Y., Mizutani, U., Honda, Y., Edagawa, K., Takeuchi, S. (1995): in Proceedings of the 5th International Conference on Quasicrystals, Janot, C., Mosseri, R. (eds). World Scientific, Singapore, p 534
Tamura, R., Waseda, A., Kimura, K., Ino, H. (1994a): Mater. Sci. Eng. A **181-182**, 794
Tamura, R., Waseda, A., Kimura, K., Ino, H. (1994b): Phys. Rev. B **50**, 9640
Tamura, R., Waseda, A., Kimura, K., Ino, H. (1995a): in Aperiodic'94, Proceedings of the International Conference on Aperiodic Crystals, Chapuis, G., Paciorek, W. (eds). World Scientific, Singapore, p 212
Tamura, R., Kirihara, K., Kimura, K., Ino, H. (1995b): in Proceedings of the 5th International Conference on Quasicrystals, Janot, C., Mosseri, R. (eds). World Scientific, Singapore, p 539
Thomas, G.A., Ootuka, Y., Kobayashi, S., Sasaki, W. (1981): Phys. Rev. B **24**, 4886
Trambly de Laissardière, G., Mayou, D., Nguyen Manh, D. (1993): Europhys. Lett. **21**, 25
Tsai, A.P., Chen, H.S., Inoue, A., Masumoto, T. (1991): Phys. Rev. B **43**, 8782
Volkov, P., Poon, S.J. (1995): Phys. Rev. B **52**, 12 685
Wagner, J.L., Biggs, B.D., Wong, K.M., Poon, S.J. (1988): Phys. Rev. B **38**, 7436
Wagner, J.L., Wong, K.M., Poon, S.J. (1989): Phys. Rev. B **39**, 8091
Wagner, J.L., Biggs, B.D., Poon, S.J. (1990): Phys. Rev. Lett. **65**, 203
Wang, C.R., Lin, S.T., Chen, Y.C. (1994): J. Phys. Condens. Matter **6**, 10 747
Wang, K., Scheidt, C., Garoche, P., Calvayrac, Y. (1993): J. Non-Cryst. Solids **153-154**, 357
Wang, Y.P., Zhang, D.L. (1994): Phys. Rev. B **49**, 13 204
Wong, K.M., Lopdrup, E., Wagner, J.L., Shen, Y., Poon, S.J. (1987): Phys. Rev. B **35**, 2494
Yokoyama, Y., Miura, T., Tsai, A.-P., Inoue, A., Masumoto, T. (1992): Mater. Trans., Jpn. Inst. Met. **33**, 97
Yoshioka, A., Edagawa, K., Kimura, K., Takeuchi, S. (1995): Jpn. J. Appl. Phys. **34**, 1606
Zhang, D.L., Lu, L., Wang, X.M., Lin, S.Y., He, L.X., Kuo, K.H. (1990): Phys. Rev. B **41**, 8557
Zhang, D.L., Cao, S.C., Wang, Y.P., Lu, L, Wang, X.M., Ma, X. L., Kuo, K.H. (1991): Phys. Rev. Lett. **21**, 2778

6. Theory of Electronic Structure in Quasicrystals

Takeo Fujiwara

6.1 Introduction

Shechtman, Blech, Gratias, and Cahn discovered a sharp diffraction pattern of icosahedral (i) symmetry in the Al-Mn alloy rapidly quenched from the melt and reported the i phase of metallic alloys (Shechtman et al. 1984). Levine and Steinhardt (1984) showed that the three-dimensional (3D) Penrose lattice gives a diffraction pattern of densely distributed δ functions with i symmetry. The notion of quasicrystals (QCs) is characterized by the following facts (Steinhart and Ostlund 1987):

1. the scattering intensity is a sum of densely distributed δ functions,
2. δ function spots can be specified with a set of integers whose number exceeds the space dimension, and
3. the pattern of scattering spots has a rotational symmetry forbidden in crystals.

The consequence of this definition is that the quasicrystalline lattice is homogeneous with bond orientational order and any local pattern of a radius d can be found in a region nearby within a distance of order $2d$.

The Al-Mn QC was first found in a metastable phase and now many different kinds of QCs can be grown in thermodynamically stable phases. QCs, both i and decagonal (d), are cluster compounds. Icosahedral QCs can be classified into two different types (Henley 1990); one is the Mackay icosahedron (MI) type and the other is the triacontahedron (TC) type Typical examples of the MI-type QCs are i-Al-Mn-Si and i-Al-Cu-Fe, where a unit of packing is an i cluster called *Mackay icosahedron* and glue atoms are necessary to tie up these clusters. MI-type QCs contain transition metal (TM) elements and also, in many cases, Al atoms. Examples of the TC-type QCs are i-Al-Cu-Li and i-Al-Mg-Zn, where a fundamental unit is a triacontahedral atom cluster sharing several atoms with each other and the packing does not need any glue atoms. TC-type QCs contain simple sp-metal elements instead of TM elements.

The above classification also correlates with the different ratios of the atomic diameter d to the quasilattice constant a_R

$$\frac{d}{a_R} \simeq 0.61 \text{ (MI)} , \qquad \frac{d}{a_R} \simeq 0.57 \text{ (TC)} , \tag{6.1}$$

and different electron-per-atom ratios

$$\frac{e}{a} \simeq 1.6 - 1.8 \text{ (MI)} , \quad \frac{e}{a} \simeq 2.1 - 2.25 \text{ (TC)} . \tag{6.2}$$

Stable QCs seem to be stabilized by an electronic mechanism. This is because stable i QCs can be grown only in a narrow region of the e/a ratio.

Decagonal QCs form a new class of anisotropic materials with a crystalline axis perpendicular to a quasicrystalline plane with fivefold symmetry. The Al-Mn, Al-Fe, and Al-Pd systems are metastable, whereas the Al-Cu-Co, Al-Ni-Co, and Al-Pd-Mn ones are stable. The constituent unit of d QCs is a column cluster, i.e., 1D atom column with fivefold symmetry. Atom columns share common atom layers perpendicular to the periodic direction. The distance between atom layers is about 2 Å. Decagonal Al-Cu-Co and Al-Ni-Co have a two-layered structure with 4 Å periodicity. Other possibilities of d QCs are 4, 6, 8, 12, or 18 layered structures and the periodicity of them are 8 Å (Al-Co, Al-Ni), 12 Å (Al-Mn, Al-Pd-Mn), 16 Å (Al-Pd, Al-Cu-Fe), 24 Å or 36 Å, respectively.

The non-crystalline rotational symmetry or the quasiperiodicity of QCs is specified by a certain irrational number. The d and i QCs are characterized by the golden ratio $\tau = (\sqrt{5}+1)/2 = 1.6180\cdots$, which can be expressed by a continued fraction as

$$\tau = 1 + \cfrac{1}{1 + \cfrac{1}{1 + \cfrac{1}{1 + \cdots}}} .$$

Several physical properties in crystalline compounds of stoichiometry close to that of a QC are very similar to those in the QC. Crystalline compounds are called *crystalline approximants* when the lattice periodicity is specified by an approximated rational number of the irrational number. The approximant of the golden ratio is given as a series of 1/1, 2/1, 3/2, 5/3, $\cdots\cdots$ which is a truncation of the above continued fraction at every finite step. This series is given as F_{n+1}/F_n, where $F_{n+1} = F_n + F_{n-1}$, $F_0 = F_1 = 1$, and F_n is called the Fibonacci number. In some compounds, various crystalline approximants were found. Crystalline approximants are, therefore, compounds of a commensurate structure. Their structures can also be obtained by interchanging atomic sites (phason flip) in the quasicrystalline structure, and local atomic environments in crystalline approximants are hardly distinguishable from those in QCs.

The diffraction pattern of the dense δ functions indicates that electrons are scattered by densely distributed reciprocal lattice points. There is no characteristic wavelength for these scatterings. Bloch's theorem is not applicable due to the lack of translational periodicity. Then electron wave functions may be non-extended states. On the other hand, electrons cannot be exponentially localized because of the following reason. Consider a wave packet of an exponential tail in a region of a diameter d. The identical local environment can be found nearby within a distance of $2d$. Thus the wave packet can transfer to the region of the identical local environment with a help of a tail overlap of wave functions. These two facts are contradictory to each other. Therefore, any wave function may be neither extended nor localized (Tsunetsugu et al. 1986) and naturally propagates along an atomic alignment with similar local environment or an aggregation of atom clusters in the whole lattice.

The electronic structures in 1D and 2D quasilattices have been thoroughly investigated (Kohmoto et al. 1983, Ostlund et al. 1983, Kohmoto et al. 1987, Tsunetsugu et al. 1986) and it is now well known that the energy spectrum and spatial extent of the wave functions are quite anomalous. These investigations are very important in the understanding of new concepts about electronic structures and wave functions in QCs (Fujiwara and Tsunetsugu 1991, Mayou 1994, Sire 1994).

For several years, it was believed that QCs might have physical properties similar to those of crystalline or amorphous metals, with characteristics between those of crystalline and amorphous metals. When stable QCs with an almost perfect structural order were discovered, it was realized that this is not the case (Kimura and Takeuchi 1991, Poon 1992, Berger 1994).

The electronic properties of QCs are quite exotic:

1. they have anomalously large resistivity at low temperatures, e.g., $\sim 1.0\ \Omega\,\mathrm{cm}$ in Al-Pd-Re at 4.2 K,
2. their resistivity decreases with increasing temperature, e.g., $\rho(4\mathrm{K})/\rho(300\mathrm{K}) \simeq 190$ in Al-Pd-Re,
3. the higher structural ordered samples of QCs, the lower the conductivity,
4. the Hall coefficient and thermoelectric power are temperatuere dependent,
5. a very complicated behavior of the low-field magnetoconductivity at low temperatures and the importance of the quantum interference effects are observed, and
6. there is no Drude peak in infrared spectra.

The conductivity can be expressed over a wide temperature range as

$$\sigma(T) = \sigma_0 + \Delta\sigma(T) \ , \tag{6.3}$$

where σ_0 is the residual conductivity at low temperatures and $\Delta\sigma(T)$ is the temperature (T) dependent term. This formula is contradictory to the Matthiessen rule where the T-dependent and T-independent parts of the resistivity are additive. The sensitive effects of the structural disorder appear only in the first term in most QCs. The latter term $\Delta\sigma(T)$ is increasing monotonously with T as

$$\Delta\sigma(T) \sim T^{\beta} \quad (1 < \beta < 1.5) \tag{6.4}$$

over a very wide temperature range. These facts remind us of a certain scaling character of transport properties. Thus the Boltzmann theory with the relaxation-time approximation cannot be a good starting point for an understanding of the transport phenomena in QCs.

The conductivity in d QCs is also very exotic. The absolute value and the temperature dependence are normal metallic along the periodic direction. On the other hand, the conductivity within the quasiperiodic plane is smaller and its T-dependence is anomalous.

In several idealized and realistic approximants of QCs, band structures have been calculated (Fujiwara 1989, Hafner and Krajčí 1992). A pseudogap with a width of $0.5 - 1$ eV is observed in the density of states (DOS) at the Fermi energy, to which the energetic stability is ascribed.

The theoretical studies of electronic structures and transport phenomena in QCs are reviewed in the present article. In Sect. 6.2, characteristics of energy spectra and wave functions are discussed in 1D and 2D models of QCs. The basic concept of an energy spectrum is explained. Section 6.3 is devoted to discussions about calculated electronic states for idealized and realistic structure models of crystalline approximants. In Sect. 6.4, we will discuss the transport properties and the scaling property of wave functions. The scenario for anomalous transport in QCs is presented and the role of elastic scattering by randomness is clarified. The anomalous transport in QCs and the weak localization of the Anderson localization will be discussed.

6.2 Electronic Structure in One- and Two-Dimensional Quasilattices

6.2.1 One-Dimensional Quasilattice: Fibonacci Lattice

6.2.1.1 Fibonacci Lattice and Quasiperiodicity. An infinite sequence consisting of two numbers τ and 1 is called the Fibonacci sequence when the sequence holds a self-generation rule $\tau \to \tau 1$ and $1 \to \tau$. This rule, starting from 1, generates a sequence $\tau 1 \tau \tau 1 \tau 1 \tau \tau 1 \tau \tau 1 \tau 1 \tau \tau 1 \tau 1 \tau \cdots$, which can be defined also as a limit of a recursive sequence

$$S_{n+1} = S_n S_{n-1} \quad \text{with} \quad S_0 = \{1\} \quad \text{and} \quad S_1 = \{\tau\}. \tag{6.5}$$

This sequence is not periodic and the ratio of the total numbers of τ and 1 is, in the limit $n \to \infty$, the golden ratio $\tau = (\sqrt{5} + 1)/2$. When a 1D lattice has a structure equivalent to the Fibonacci sequence in an alignment of hopping integrals or atomic potentials, the system is called the Fibonacci lattice.

The Fibonacci lattice may not be a QC because it does not hold any noncrystalline rotational symmetry. However, the golden ratio can characterize the translational quasiperiodicity not only in the Fibonacci lattice but also that in several quasicrystalline lattices. Thus the Fibonacci lattice is quite important and provides several key concepts of electronic structures in QCs.

6.2.1.2 Singular Continuous Spectra and Critical Wave Functions. In general, there are three types of energy spectra: *absolute continuous*, *point*, and *singular continuous* (Reed and Simon 1972). The absolute continuous spectrum is that with a spectral measure $\mathrm{d}\mu = n(E)\mathrm{d}E$ and a smooth DOS, $n(E)$. This is the case for the DOS in crystals. The point spectrum has a spectral measure of a set of δ functions defined on a countable number of points $\{E_i\}$ and is the case for spectra in 1D and 2D random systems. The third is the singular continuous spectrum, whose integrated number of states below a certain energy is continuous but non-differentiable, e.g., the Cantor function. The DOS cannot be defined in the singular continuous spectrum. All three types of energy spectra appear in the solutions of the Harper equation as the strength of the incommensurate potentials is changed (Fujiwara and Tsunetsugu 1991).

Wave functions are also classified into three types, *extended*, *localized*, and *critical*, corresponding to the energy spectra, absolute continuous, point, and singular continuous, respectively. Extended wave functions are defined with an asymptotic uniform amplitude as

$$\int_{|\mathbf{r}|<L} \left|\psi(\mathbf{r})\right|^2 d\mathbf{r} \sim L^D \, , \tag{6.6}$$

where D is the spatial dimension. Localized wave functions are specified as the square integrability:

$$\int_{|\mathbf{r}|<L} \left|\psi(\mathbf{r})\right|^2 d\mathbf{r} \sim L^0 . \tag{6.7}$$

The critical wave function is neither extended nor localized:

$$\int_{|\mathbf{r}|<L} \left|\psi(\mathbf{r})\right|^2 d\mathbf{r} \sim L^{-2\nu+D} \qquad (0 < 2\nu < D) \, . \tag{6.8}$$

A typical example of the third type may be a power-law type wave function $\psi(\mathbf{r}) \sim |\mathbf{r}|^{-\nu}$ with $\nu < D/2$, or a wave function in the case where the dimension of the support of the wave function is less than the space dimension.

6.2.1.3 Energy Spectrum and Wave Functions in Fibonacci Lattice. The 1D tight-binding model on the Fibonacci lattice is given as

$$t_{m+1}\psi_{m+1} + t_m\psi_{m-1} = E\psi_m \, , \tag{6.9}$$

where t_m $(m = 1, 2, \cdots)$ takes two values t_A and t_B arranged in the Fibonacci sequence. Here, t_A and t_B correspond to τ and 1, respectively. We should notice that the tight-binding Schrödinger equation of the Fibonacci lattice is a discrete analogue of the Harper equation. This equation allows an exact renormalization-group (RG) equation (Kohmoto et al. 1983, 1987)

Wave functions of the Fibonacci lattice [Eq. (6.9)] are written as

$$\Psi_j = M(t_{j+1}, t_j)\Psi_{j-1} \tag{6.10}$$

with

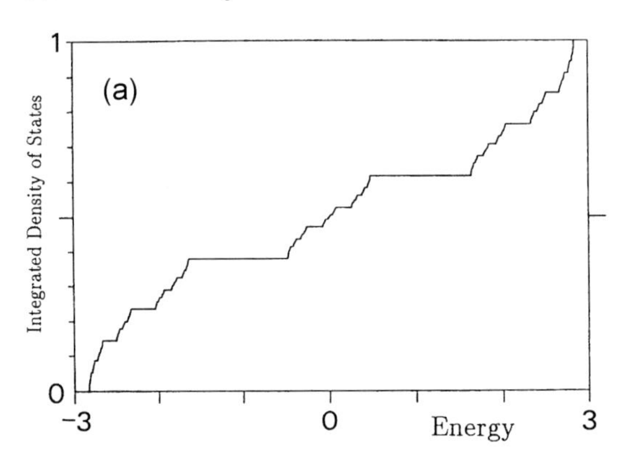

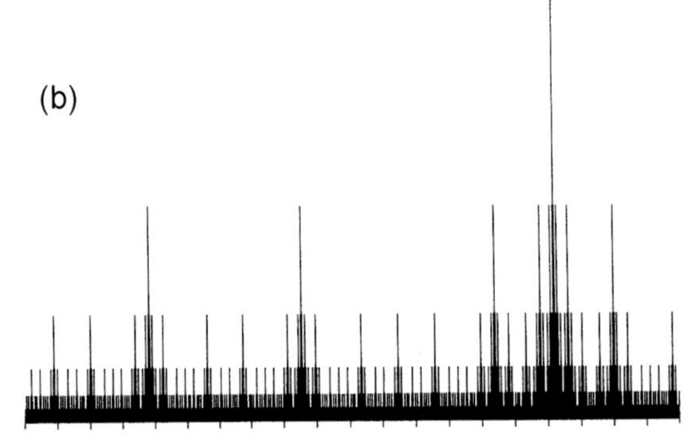

Fig. 6.1. (a) Integrated number of states in 1D Fibonacci lattice. Flat parts correspond to the gaps of the energy spectrum. (b) The wave function at the center of the spectrum in 1D Fibonacci lattice.

$$\Psi_j = \begin{pmatrix} \psi_{j+1} \\ \psi_j \end{pmatrix} \quad \text{and} \quad M(t_{j+1}, t_j) = \begin{pmatrix} E/t_{j+1} & -t_j/t_{j+1} \\ 1 & 0 \end{pmatrix} . \tag{6.11}$$

This equation can be solved with recursive operation of the transfer matrices $M(t_{j+1}, t_j)$ to the boundary function Ψ_0. We obtain a RG map of the transfer matrices $M^{(n)}$:

$$M^{(n+1)} = M^{(n-1)} M^{(n)} , \tag{6.12}$$

with an initial condition

$$M^{(1)} = M(t_A, t_A) \quad \text{and} \quad M^{(2)} = M(t_A, t_B) M(t_B, t_A) . \tag{6.13}$$

The integrated number of states is shown in Fig. 6.1a. The energy spectrum is a self-similar Cantor set with zero Lebesgue measure (i.e., singular continuous spectrum) and is multifractal.

Some wave functions of the Fibonacci lattice, e.g., at the band edge and the band center, follow a periodic cycle of the RG map (Kohmoto et al.

1987). Figure 6.1b shows the self-similar wave function at the center of the energy spectrum. We should notice that the self-similar wave function is critical. Distribution of amplitudes of a wave function ψ_m ($m = 0, 1, 2, \cdots$) can be specified by a continuous distribution of the exponent index α_m of $\psi_m \propto (1/N)^{\alpha_m}$ and is multifractal. A distribution of the multifractal index $f(\alpha)$ at the center of the band can be obtained rigorously (Fujiwara et al. 1989). States almost everywhere in the spectrum correspond to chaotic orbits of the RG map. The corresponding wave functions are also chaotic. This is because the $f(\alpha)$ function of finite approximant chains does not converge monotonously with increasing the length of the chain. All eigenstates in the Fibonacci lattice are critical, irrespective of the values of hopping parameters.

6.2.1.4 Transport in Fibonacci Lattice. The conductance G can be evaluated by the Landauer formula (Landauer 1070, Anderson et al. 1980). The Landauer formula can be rewritten for the Fibonacci lattice as (Kohmoto 1986, Sutherland and Kohmoto 1987, Hiramoto and Kohmoto 1992)

$$\frac{e^2}{h} \cdot \frac{1}{G} = \frac{1}{4} \mathrm{Tr}\{T(n)^\dagger T(n)\} - \frac{1}{2} \tag{6.14}$$

with $T(n) = M(t_{n+1}, t_n) \cdots M(t_2, t_1)$. Once one chooses an energy just on the δ function spectral measure, the conductance decays as a power of the sample length L :

$$G \sim L^{-\alpha}, \quad 0 < \alpha < 1 . \tag{6.15}$$

Compare this with the behavior of the Anderson localization $G \sim \exp(-\gamma L)$ and that of the Ohmic law $G \sim L^{D-2}$. The power-law decaying conductance is directly related to the critical nature of wave functions. Of course, if one chooses an energy in a band gap (almost everywhere in the spectrum), the conductance shows the crossover behavior from the power-law to the exponential decaying, with increase of the system length L.

6.2.2 Two-Dimensional Quasilattice: Penrose Lattice

A typical example of 2D quasilattices is the Penrose lattice (Gardner 1977, Bruijn 1981). Solving a Schrödinger equation on a 2D quasilattice is more difficult compared with that on the Fibonacci lattice, and one should solve it numerically. Another difficulty is a choice of the boundary condition of quasiperiodic systems, but it can be avoided by adopting crystalline approximants of the Penrose lattice (Tsunetsugu et al. 1986, 1991). A lattice corresponding to an approximant $\tau_3 = 3/2$ is shown in Fig. 6.2.

Electronic structures can be analyzed by observing the behavior of the spectra and wave functions with change of the size of the periodic units of the crystalline approximants. Here we discuss a simple tight-binding Hamiltonian

$$\sum_{\langle ij \rangle} t_{ij} \psi_j = E \psi_j \ , \tag{6.16}$$

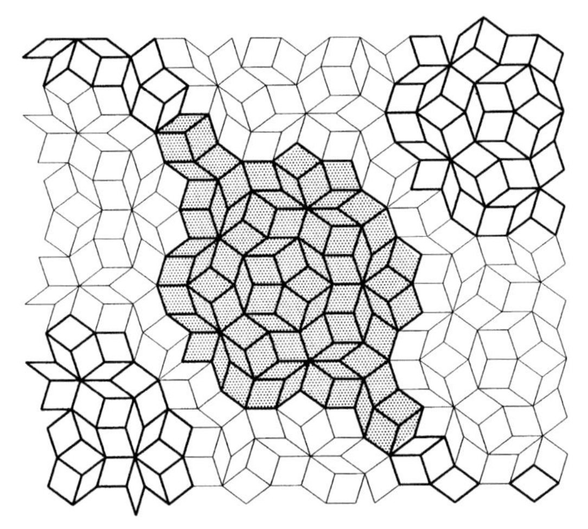

Fig. 6.2. A crystalline approximant of the Penrose lattice corresponding to $\tau_3 = 3/2$. The unit cell is shown by shade.

where one s-type atomic orbital is placed on the center of a rhombus with a constant hopping interaction $t = -1$ between adjacent rhombi (a center model) (Tsunetsugu et al. 1986, 1991).

The DOS in crystalline approximants is shown in Fig. 6.3. The calculated results show two important features. First, the DOS becomes less smooth and consists of sharper spikes as the system size is increased. A distribution of the interval of the energy eigenvalues ΔE in a periodic unit cell with N rhombi is scaled as $\Delta E \sim N^{-\gamma}$, with $\gamma < 1$, except for a few finite energy gaps. Second, the spikiness depends on the energy regions.

The spatial extent of wave functions can be studied by calculation of the $2p$ norm of wave functions defined as

$$\| \psi \|_{2p} = \frac{\sum_i |\psi_i|^{2p}}{\left\{\sum_i |\psi_i|^2\right\}^p} \sim N^{-(p-1)D_p} \; . \tag{6.17}$$

This is the generalization of the participation ratio which is given by $p = 2$, and D_p is the multifractal dimension of the spatial extent of the wave function. We can obtain essentially the same information as that from the multifractal analysis if we vary the parameter p. The behavior of the $2p$ norm shows that most eigenfunctions are power-law decaying in real space; $\psi(\mathbf{r}) \sim |\mathbf{r}|^{-\nu}$, with exponents $3/8 < \nu < 5/8$ (Tsunetsugu et al. 1986, 1991). Since the power-law index ν is less than 1, wave functions cannot be normalized and are critical. An example of typical wave functions is shown in Fig. 6.4 where the amplitudes of the wave function spread nonuniformly, but over regions of characteristic tiling configuration (Yamamoto and Fujiwara 1995). The spatial extent of this type of wave functions may be multifractal.

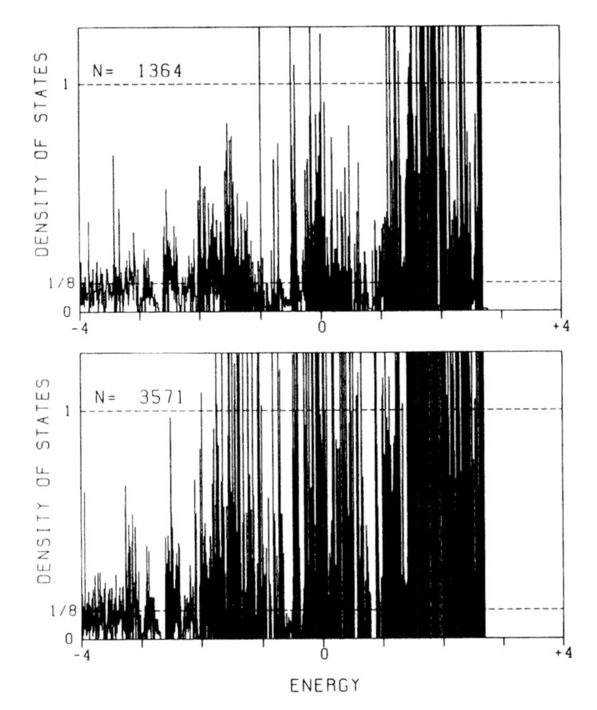

Fig. 6.3. The density of states for a unit cell of $N = 1364$ and $N = 3571$ for crystalline approximants of the 2D Penrose lattice.

It should be mentioned that very recently the energy spectrum in 3D Penrose lattice (Amman-Kramer lattice) was investigated numerically using a similar technique (Rieth and Schreiber 1998). The numerical analysis showed the followings: (1) power-law decaying wave functions in 3D quasilattices cannot be concluded from the analysis of the participation ratio, but (2) the results indicate a tendency towards localization in the energy region of small DOS.

Very special states, called *confined states*, are found also in the Penrose lattice. The states have finite amplitudes only on special tiling alignments and vanish outside (Fujiwara et al. 1988, Arai et al. 1988, Tokihiro et al. 1988, Krajčí and Fujiwara 1988).

The conductance in a 2D Penrose lattice was calculated by using the Landauer formula

$$G = \frac{e^2}{h} \mathrm{Tr}\{\mathbf{t}^\dagger \mathbf{t}\} , \tag{6.18}$$

where $\mathbf{t}$ is the transmission matrix of the whole lattice. The resultant conductance has a power-law dependence

$$G \sim L^{-\nu}, \quad \nu \simeq 0.2 \sim 0.3 \tag{6.19}$$

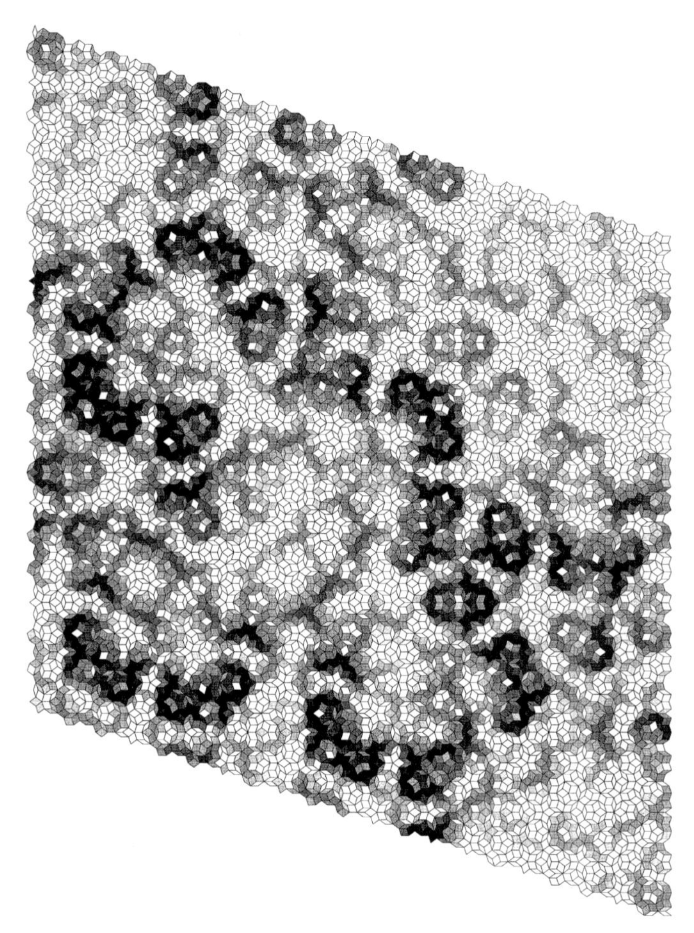

Fig. 6.4. A typical example of wave functions in a unit cell of $N = 9349$ for crystalline approximants of the 2D Penrose lattice. This state is non-degenerate.

with the sample length L (Tsunetsugu and Ueda 1991). The conductance fluctuates very rapidly as a function of the Fermi energy due to the spiky structure of the energy spectrum. The fluctuation ΔG is of the same order as the conductance itself. This fact indicates that the wave functions are not exponentially localized. The power-law dependence of the conductance directly shows the power-law decaying of wave functions.

The effects of static randomness in the DOS and the conductance was also studied (Yamamto and Fujiwara 1995). The randomness smoothes the sharp spikes in the DOS and opens the conductance channels. In other words, the randomness mediates hopping of electrons between certain power-law decaying eigenstates. wave functions become more homogeneous with increasing randomness. This fact is consistent with the observed higher conductivity in samples of the less ordered structures.

6.3 Electronic Structure in Quasicrystals

6.3.1 Method of Calculations: Tight-Binding LMTO and Related Methods

The tight-binding linear-muffin-tin-orbital (TB-LMTO) method is generally very useful for a system with a large unit cell (Andersen et al. 1985). The TB-LMTO method with the atomic-sphere approximation (ASA) and the frozen core approximation are used in the electronic structure calculations in crystalline approximants.

The basis functions of the LMTO are written as

$$|\bar{\chi}_{\mathbf{R}L}\rangle = |\phi_{\mathbf{R}L}\rangle + \sum_{\mathbf{R}',L'} |\dot{\bar{\phi}}_{\mathbf{R}'L'}\rangle h_{\mathbf{R}'L'RL} \tag{6.20}$$

with

$$|\dot{\bar{\phi}}_{\mathbf{R}L}\rangle = |\dot{\phi}_{\mathbf{R}L}\rangle + |\phi_{\mathbf{R}L}\rangle \bar{o}_{\mathbf{R}L} \ , \tag{6.21}$$

where $\mathbf{R}$ specifies atomic sites and the index L denotes angular momenta (l, m). The parameter $\bar{o}_{\mathbf{R}L}$ is determined so that the muffin-tin orbitals $\bar{\chi}_{\mathbf{R}L}(\mathbf{r})$ are properly localized. We usually include one s, three p, and five d orbitals for each atom. $\phi_{\mathbf{R}L}$ is a solution of the Schrödinger equation with a fixed energy E_ν and $\dot{\phi}_{\mathbf{R}L}$ is its energy derivative. The wave functions $\phi_{\mathbf{R}L}$ and $\dot{\phi}_{\mathbf{R}L}$ are constructed within the local-density approximation (LDA) (Kohn and Sham 1965, Jones and Gunnarson 1989). They satisfy equations

$$\begin{aligned} (-\nabla^2 + U - E_\nu)|\phi\rangle &= 0 \\ (-\nabla^2 + U - E_\nu)|\dot{\phi}\rangle &= |\phi\rangle \end{aligned} \tag{6.22}$$

inside an atomic sphere. Here U is the LDA self-consistent potential of the atomic sphere and E_ν should be appropriately chosen in the interesting energy region.

The wave function $\psi_j(\mathbf{r})$ of the eigenstate j may be expanded using the LMTO basis functions $\bar{\chi}_{\mathbf{R}L}$ as

$$|\psi_j\rangle = \sum_{\mathbf{R}L} |\bar{\chi}_{\mathbf{R}L,\infty}\rangle u_{\mathbf{R}L,j} \ . \tag{6.23}$$

The Hamiltonian can be expressed in a tight-binding form. The TB-LMTO method is the first-principle tight-binding method and is powerful for complex systems.

The whole space is treated as a sum of overlapping atomic spheres within the ASA. One can include a term correcting an error by sphere overlap (called the combined correction) but we neglect it in most cases. The approximation of overlapping spheres is reasonable in metallic systems because the system is densely packed and the volume of overlapping region is relatively small.

The DOS is calculated by using of the tetrahedron method with $30-200$ $\mathbf{k}$ points in the irreducible Brillouin zone (BZ) (Rath and Freeman 1975). Therefore, the calculation of the DOS does not introduce any artifact if the number of $\mathbf{k}$-points is large enough. It should be important to analyze wave functions and the structure of the energy bands after obtaining the DOS in QCs in order to discuss the exotic characteristics of electronic structures.

Sometimes the recursion-LMTO method is useful to calculate the DOS (Fujiwara 1984, Novak et al. 1991). We should notice that the recursion-LMTO method could not reproduce fine structures of the DOS because the recursion method uses a small but finite imaginary energy in order to count weights of energy spectra.

The standard matrix diagonalization method, e.g., the band structure calculation method, is only applicable to systems of several hundred atoms of nine orbitals per atom in a unit cell. For calculations of several tens of thousands atoms, one should use the recursion-LMTO method. The detailed analysis of exact wave functions is very desirable in order to understand the spatial extent of eigenstates in QCs. In fact, detailed analysis of exact wave functions in larger systems of several thousands atoms (i.e., several tens of thousands orbitals) may only be possible by using exact diagonalization methods with the TB-LMTO Hamiltonian, e.g., the forced oscillator method (Yamamoto and Fujiwara 1995) or the inverse iteration method (Fujiwara et al. 1996).

6.3.2 Quasi-Brillouin Zone and Modification of DOS of Model Icosahedral Al

The Bragg scattering strongly modifies the energy spectrum at BZ edges $\pm\mathbf{K}_0$ in periodic systems. The periodic potential splits degenerate band states $\mathbf{k} \simeq \mathbf{K}_0$ and $-\mathbf{K}_0$ at the BZ edges and the DOS shows valleys and peaks.

One cannot define a BZ in quasiperiodic systems. However, bisection planes tangential to scattering vectors of the principal diffraction spots play a role of principal scattering planes. These planes form a polyhedron of highly spherical shape called a quasi-BZ.

A model calculation in a hypothetical QC of Al by using pseudopotentials indeed shows a pronounced modification of the DOS from that of free electrons due to the scattering by the quasi-BZ (Fig. 6.5) (Smith and Ashcroft 1987). The singularity and modification of the DOS are enhanced by a large multiplicity of the diffraction peaks associated with the i symmetry. The presence of this large modification of the DOS indicates the Fermi surface-quasi-BZ interaction responsible for the stability of QCs.

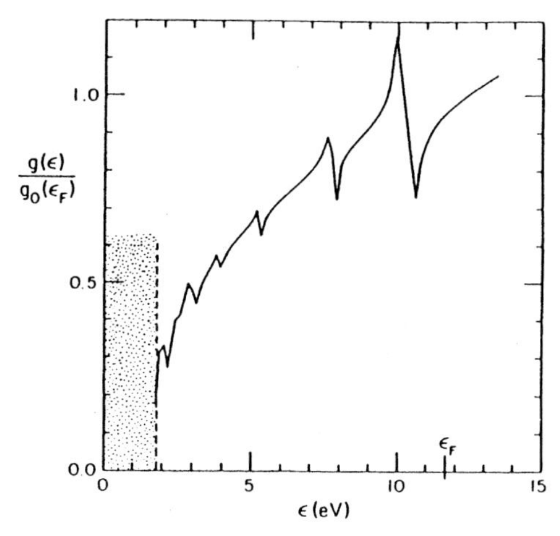

Fig. 6.5. Density of states in a hypothetical quasicrystal of Al.

6.3.3 Electronic Structure in MI-Type Icosahedral Quasicrystals

6.3.3.1 Al-Mn. The electronic structure of an idealized crystalline approximant α-Al-Mn-Si was calculated (Fujiwara 1989, Fujiwara et al. 1994) on the basis of the Elser-Henley model (Elser and Henley 1985, Guyot and Audier 1985). The model structure is the 1/1 approximant, with 114 Al and 24 Mn atomic spheres (138 atoms in a unit cell). Favorable positions of Si atoms are not known experimentally, and all Si atoms are substituted by Al atoms.

The total DOS (Fig. 6.6) reveals a depression or a pseudogap at the Fermi energy (E_F) with a width of 0.5 eV and consists of a dense set of spikes with a width of 10–50 meV or even narrower. The position of the pseudogap measured from the bottom of the Al-*sp* bands corresponds to a momentum $k_F = 1.6$ Å^{-1}. The value of $2k_F$ fairly coincides with a momentum transfer of strong scattering spots (211111) and (221001). Therefore, the formation of the pseudogap is due to the Fermi surface-quasi-BZ interaction and the energy stabilization is ascribed to the pseudogap. The Fermi energy in the calculated DOS is located slightly below the minimum of the pseudogap. In the stable structure, a few Al sites are actually substituted by Si atoms and the Fermi energy would climb towards the minimum of the pseudogap. This would be the stability enhancing mechanism of Si substitution.

The local and projected DOS show a usual resonance shape of the Mn *d* states embedded in a continuous spectrum formed by Al and Mn *s* and *p* states (Fig. 6.7). The pseudogap separates the Mn *d* states into two. The hybridization between Mn *d* states and Al *sp* states is strong. The contribution of Al *d* states is not very small in the local DOS. The Al *d* states push down the Al *p* states because of the orthogonality within the same atomic sphere. The Mn *d* states are also pushed down due to the hybridization with Al *p*

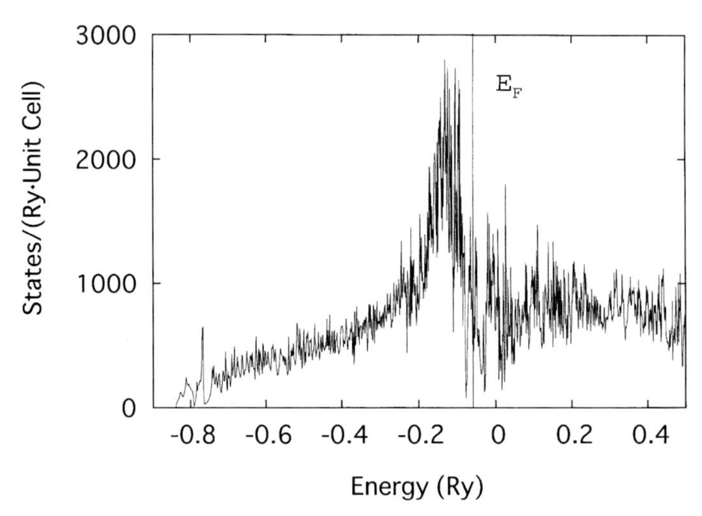

Fig. 6.6. Density of states in α-Al-Mn. The position of the Fermi energy is shown by the vertical line. The main peak originates from the Mn d states, and Mn $4s$, Al $3s$, and $3p$ states spread over the lower and higher energy sides.

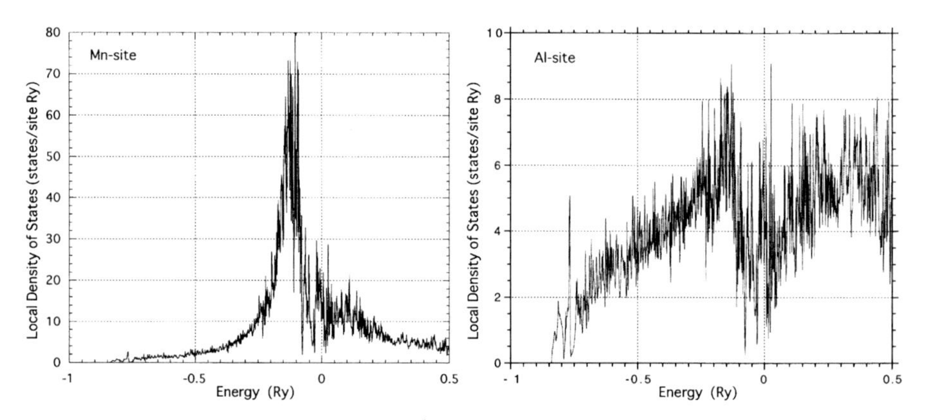

Fig. 6.7. Local density of states in 1/1 approximant of Al-Mn.

states. The resultant Fermi energy shifts downwards by about 2 eV, but the bottom of the conduction bands does not (Fujiwara et al. 1994).

The cluster atoms create many narrow peaks at deeper energies, which may be almost localized within a cluster. The volume of the center of the Mackay icosahedron is not large enough for Al or Mn atoms. Then the center of the Mackay icosahedron should remain vacant.

The spikes in the DOS are clearly seen. One may imagine that it is also common in random systems, such as amorphous metals. The randomness actually achieve the self-averaging of the DOS and one may observe a smooth DOS in amorphous metals. In the model structure of QCs, there is no disorder and the feature of the spikes is real. In a realistic structure, there is certain randomness but it may be not too strong to smear out the structure of spikes. This point is discussed later (Sect. 6.3.6.2).

6.3.3.2 Al-Cu-Fe. Al-Cu-Fe has the highly ordered structure of a face-centered quasilattice. The electronic structure was calculated in several model structures. One calculation (Trambly de Laissardière and Fujiwara 1994a) is based on a hypothetical model of the 1/1 crystalline approximant $Al_{80}Cu_{32}Fe_{15}$ with 127 atoms in a unit cell (Cokayne et al. 1993) The other calculation (Roche and Fujiwara 1998a) is based on a realistic cubic 1/1 approximant structure $(Al,Si)_{84}Cu_{36}Fe_{14}$, with 134 atoms in a unit cell, which was determined experimentally (Yamada et al. 1998).

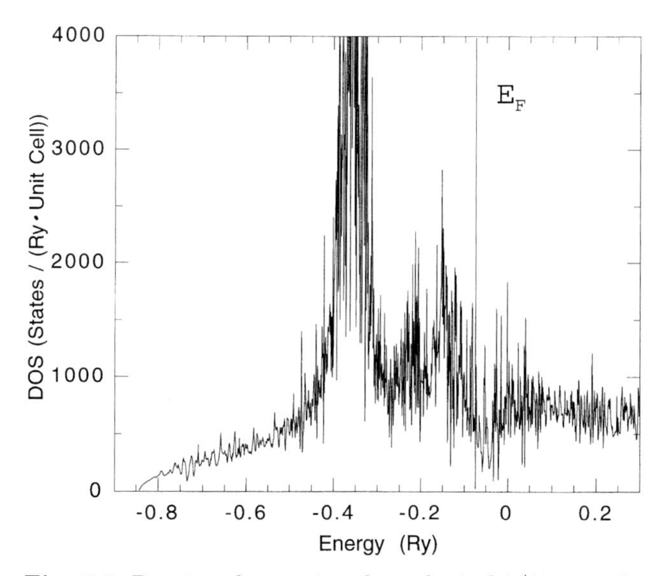

Fig. 6.8. Density of states in a hypothetical 1/1 approximant of Al-Cu-Fe.

The total DOS of $Al_{80}Cu_{32}Fe_{15}$ is shown in Fig. 6.8. The Cu d peak is well separated from Fe d peaks in the DOS. A pseudogap is found in the vicinity of E_F with a width of 0.035 Ry and the value of the DOS at E_F is about 30% of that of pure Al. Sharp spikes can also be observed, especially in the energy region of Fe and Cu d states. The local DOS are quite different from each other for inequivalent sites of the same kind of atoms. In particular, inequivalent Fe sites give very different local DOS This difference is attributed to covalency and the role of glue atoms. The Cu atoms have a d peak at lower energies than that of Fe atoms, and have predominant stabilization effects. Similar resonance effects at the E_F can be seen in crystalline ω phase Al_7Cu_2Fe (Trambly de Laissardière et al. 1995) which is not an approximant but a metallic compound similar in composition to i-Al-Cu-Fe.

The band dispersion of the E–$\mathbf{k}$ relation is very small, typically about 100 meV or even less. The area of the Fermi surface is very small even around highly symmetric points. This is because a band oscillates as a function of $\mathbf{k}$ and crosses the Fermi energy many times. The calculation of the conductivity by the Boltzmann theory shows that the quantity (n/m^*), the ratio of the valence electron density n and the effective mass m^*, is very small and fluctuates rapidly as a function of energy. The amplitude of fluctuations of (n/m^*) is of the same order as its value (see Sect. 6.4.4). Therefore, the small conductivity is attributed to small n and large m^* due to very small Fermi surfaces.

6.3.3.3 Al-Pd-Mn. Icosahedral Al-Pd-Mn is a typical stable system of the face-centered quasilattice which shows anomalous electronic transport properties. The lowest realistic crystalline approximant is the 2/1 phase (Waseda et al. 1992). The model crystalline approximants were constructed on the basis of the Katz-Gratias model and structural relaxation. The electronic structures were calculated from 1/1 to 8/5 approximants using the recursion-LMTO method (Krajčí et al. 1995). The DOS does not show the fine structures (Fig. 7.14 in Chap. 7). The bottom of the pseudogap appears just above the E_{F} in the 1/1 approximant, and the pseudogap shifts downwards with increasing the unit cell size. The pseudogap in the 8/5 approximant locates just at the E_{F}. This may suggest a possible scenario of pronounced stabilization in QCs compared with the crystalline approximants.

6.3.3.4 Al-Pd-Re. Al-Pd-Re is a unique i QC because of its anomalous behavior in the transport properties. It has $\rho \sim 1\Omega$cm at 4.2K and the ratio $\rho(4\mathrm{K})/\rho(300\mathrm{K}) \simeq 190$ (Akiyama et al. 1993, Pierce et al. 1993, Guo and Poon 1996, Bianchi et al. 1997, Gignoux et al. 1997). Such transport properties as the electrical conductivity, the Hall coefficient, and the magnetoconductivity, are very sensitive to the e/a ratio, which correlates strongly with the structural ordering (Sawada et al. 1998). High resistivity samples with $e/a \simeq 1.79$ are reported to show the metal-insulator transition at about 1K. The observed temperature dependence of the conductivity and magnetoconductivity of high-resistive samples at very low temperatures could only be explained by the electron-electron interaction in the almost insulating regime but could not be explained by the weak-localization theory (Tamura et al. 1998). It should be mentioned that other groups reported finite values of the conductivity extrapolated to $T = 0$ K (Ahlgren et al. 1997). Photoconductivity with complex relaxation processes was recently observed (Takeda et al. 1998). The experimentally observed scaling behavior of the electron Coulomb gap with resistivity at 0 K was reported (Lin et al. 1996).

The electronic band structure in a crystalline approximant is expected to be similar to that of Al_2Ru (Manh et al. 1992) which has a real band gap. A recent calculation shows the real gap of a width of 0.15 eV slightly above the E_{F} in the 1/1 approximant $Al_{68.75}Pd_{15.6}Re_{15.6}$ (Krajčí and Hafner 1998) of the Katz-Gratias model of Al-Pd-Mn. However, this real gap disappears in higher approximants.

6.3.4 Electronic Structure in TC-Type Icosahedral Quasicrystals

6.3.4.1 Al-Cu-Li. Icosahedral Al-Cu-Li can be grown through a thermodynamically equilibrate peritectic process but it always contains unavoidable disorder. The 1/1 crystalline approximant of *i*-Al-Cu-Li is called the *R* phase (Al_3CuLi_3).

Several structural models of the hypothetically ordered approximant Al-Cu-Li were considered (Fujiwara and Yokokawa 1991, Windisch et al. 1994). The position of the pseudogap measured from the bottom of the Al-*sp* bands corresponds to a momentum $k_F = 1.64$ Å^{-1}. The value of $2k_F$ coincides with the principal peaks of the diffraction spots (222100) and (311111)/(222110). A clear pseudogap appears even in a hypothetical model without Cu atoms ($Al_{108}Li_{52}$). The position and the width of the pseudogap are only slightly dependent on the atomic composition but the Fermi energy shifts. This assures the stabilization by the pseudogap due to the Fermi surface-quasi-BZ interaction. The Cu atoms do not contribute to formation of the pseudogap but supply additional electrons. The most appropriate content of Cu atoms is determined so that the Fermi energy is located exactly on the bottom of the pseudogap. The height of the DOS at the bottom of the pseudogap is nearly 1/3 of the free-electron parabola and is consistent with the observed reduction of the electronic contribution to the specific heat. This feature of the pseudogap is not essentially altered by disorder.

6.3.4.2 Al-Mg-Zn. Electronic structures of crystalline approximants of the TC-type *i* QCs Al-Mg-Zn were calculated using the Elser-Henley model by the recursion-LMTO method (Hafner and Krajčí 1992, 1993).

Systematic atomic composition dependence of the structure of the 1/1 crystalline approximants was carefully studied by the Rietveld analysis (Mizutani et al. 1997). The electronic structure was calculated for various atomic compositions $Al_xMg_{40}Zn_{60-x}$ ($15 < x < 53$) using the refined realistic structural models (Mizutani et al. 1998). The position and the width of the pseudogap do not change appreciably with change of the atomic composition but the Fermi energy gradually moves across the valley of the pseudogap (Fig. 6.9). These spectra agree with the observed x-ray photoelectron spectroscopy (XPS) valence band spectra and also with the observed concentration dependence of the electronic specific heat coefficient (Takeuchi and Mizutani 1995).

6.3.5 Electronic Structure in Decagonal Quasicrystals

6.3.5.1 Al-Cu-Co. Several structural models of *d*-Al-Cu-Co were proposed (model 1, Burkov 1991, 1992 and model 2, Burkov 1993) and the electronic

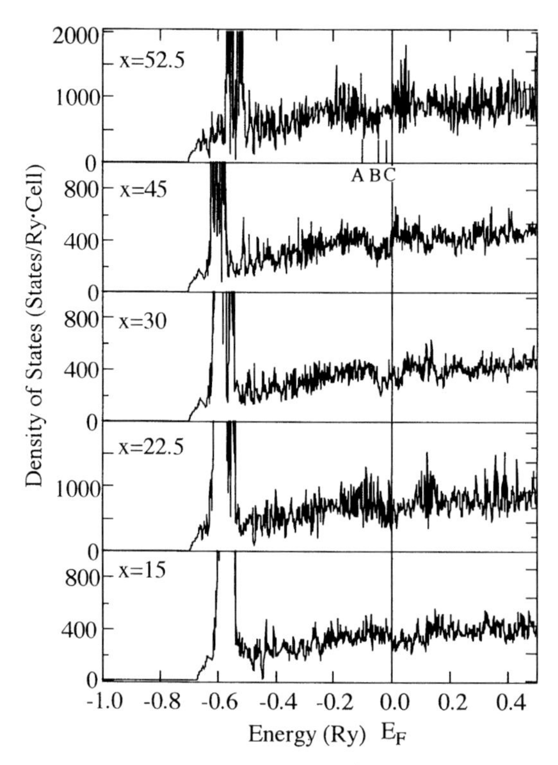

Fig. 6.9. Density of states of 1/1 approximants of $Al_xMg_{40}Zn_{60-x}$ as a function of composition.

structures were calculated based on these models. Because the models do not specify the difference between Cu and Co sites, there remains some ambiguity about the atom positions. Each atom column has the two-layered periodicity.

The Co sites in model 1 are chosen in such a way that a pseudogap appears (Trambly de Laissardière and Fujiwara 1994b). The models contain two types of clusters. A large cluster contains 40 atoms in three shells with a vacancy at the center, 10 Cu in the inner shell, 10 Al atoms in the middle shell, and 20 Al and 10 Co atoms in the outer shell. A small cluster contains one Al atom at the center, and five Al and five Cu atoms in one shell. Two large clusters share with each other two Al and two Co atoms. There are no glue atoms.

The DOS based on model 1 is shown in Fig. 6.10. The pseudogap is sensitive to the structural models, but the width of the $E(\mathbf{k})$ curves is not. A strong hybridization and a large anisotropy of the energy band structure are observed (Fig. 6.11). The $E(\mathbf{k})$ curves are very flat with large effective masses within the quasiperiodic plane, and those along the periodic direction are similar to the free-electron bands (Fig. 6.11c). A bands in Fig. 6.11c correspond to (m_1, m_2) band in Fig. 6.11b, B bands - to (m_2, m_3), and C

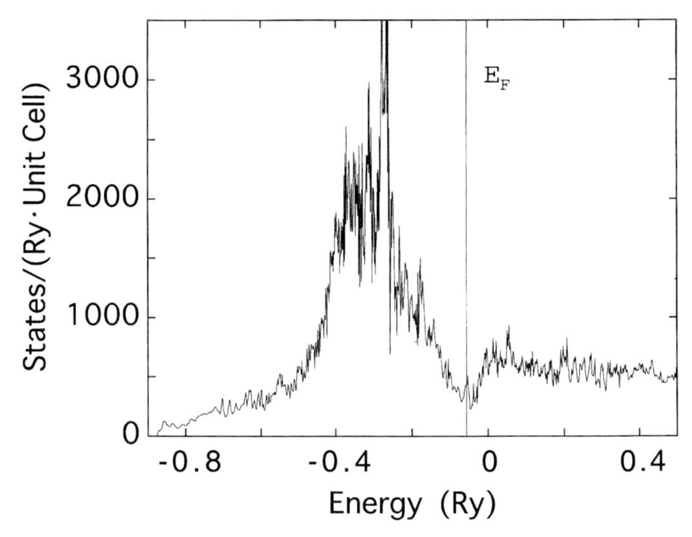

Fig. 6.10. Density of states of *d*-Al-Cu-Co based on the Burkov model 1.

bands - to (m_4, m_5). The value of the effective masses of the (m_1, m_2) band is about half of that of A bands because of the strong hybridization. The bands A are associated with the reciprocal lattice points on the plane perpendicular to the periodic axis and containing the $\mathbf{K} = 0$ point. The bands B and C are associated with those on the perpendicular plane and off the $\mathbf{K} = 0$ point. If the layers periodicity increases, the C bands spread more uniformly along the periodic axis because the number of the C bands is proportional to the

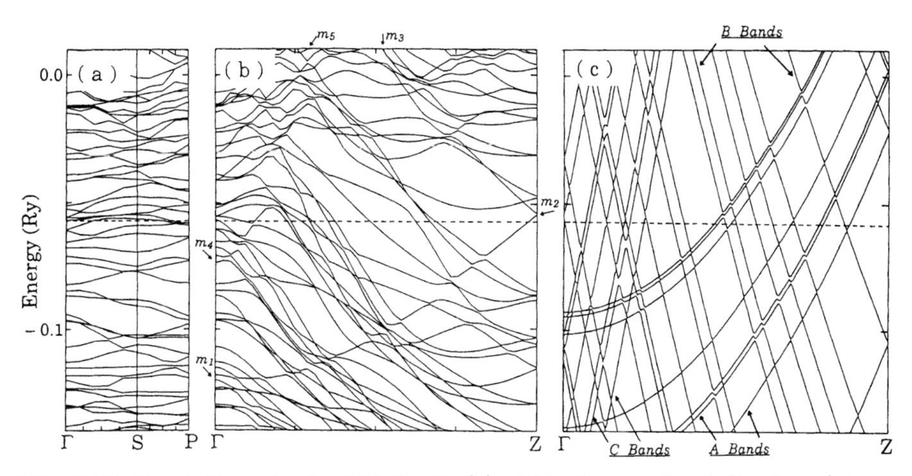

Fig. 6.11. Band dispersion in *d*-Al-Cu-Co (a) within the quasiperiodic plane (along ΓS and SP lines) and (b) along the periodic direction (ΓZ). (c) The free-electron band along the periodic direction.

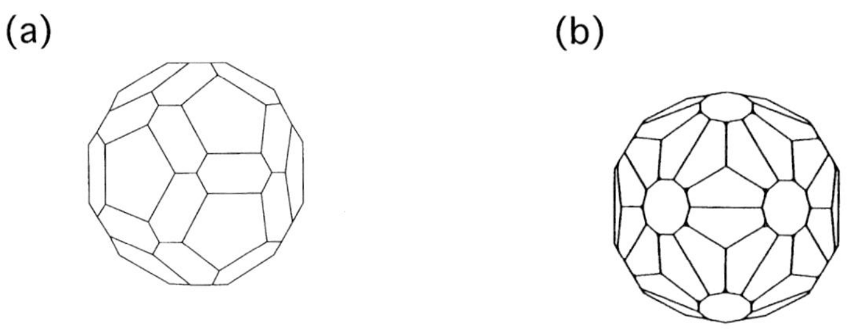

Fig. 6.12. Quasi-BZ's constructed (a) from the [211111] and [221001] planes (MI type), and (b) from the [222100] and [311111]/[222110] planes (TC type).

number of periodic layers. Therefore, the two-layered structure causes the pseudogap to become more pronounced and a periodicity perpendicular to the layers may be realized much easier in the two-layered structures.

Decagonal QCs have the 1D periodicity. Thus the structure of spikes is more smooth in *d* QCs as compared with that in *i* QCs. The distribution of amplitudes of wave functions will be discussed in Sect. 6.3.6.3.

In *d*-Al-Cu-Co and *d*-Al-Ni-Co, the ratio of electric conductivity along periodic and quasiperiodic direction ranges from 4 to 12. The temperature dependence of conductivity is metallic along the periodic direction and similar to that in *i* QCs within the quasiperiodic plane (Kimura and Takeuchi 1991, Martin et al. 1991, Wang et al. 1993, Wang and Zhang 1994)

The calculated DOS based on model 2, which enforces a matching rule on clusters, does not show a pseudogap (Krajčí et al. 1997). A recent careful ultraviolet photoelectron spectroscopy (UPS) experiment shows the existence of the pseudogap, both in *d*-Al-Cu-Co and *d*-Al-Ni-Co (Stadnik et al. 1997).

6.3.5.2 $Al_{13}Fe_4$ and $AlMn_3$. Several electronic structures were calculated in $Al_{13}Fe_4$ (Fujiwara and Yokokawa 1991) and $AlMn_3$ (Fujiwara 1997). The DOS of these systems have the pseudogap in the energy region of the Fermi energy. $AlMn_3$ has a very deep pseudogap, or perhaps a real gap.

6.3.6 General Characteristics of DOS and Wave Functions

6.3.6.1 Pseudogap at Fermi Energy. The DOS modification may be more efficient in QCs than in crystalline approximants because of the pronounced sphericity of the quais-BZ in QCs (Fujiwara and Yokokawa 1991). Several examples of the quais-BZ are shown in Fig. 6.12. The quasi-BZ of the MI type is constructed with the [211111] and [221001] planes, and that of the TC type with the [222100] and [311111]/[222110] planes.

The pseudogap of the DOS does not change the total electron energy if the Fermi energy is outside the pseudogap. If the Fermi energy is nearby or just at the pseudogap, which is always the case in the examples above, one can expect a reduction of the total band energy due to the pseudogap. This is

Table 6.1. Reciprocal lattice vectors and the corresponding valence electron per atom. The m_i denote the six-dimensional components of the reciprocal lattice vectors (Vaks et al. 1988).

G						Multiplicity of **G**	Valence electron per atom Z_c	
m_1	m_2	m_3	m_4	m_5	m_6		TC-type (Al-Cu-Li)	MI-type (Al-Mn)
2	1	1	1	1	1	12	1.28	1.5
2	2	1	0	0	1	30	1.49	1.75
2	2	2	1	0	0	60	2.17	2.55
3	1	1	1	1	1	72	2.42	2.84
3	2	2	1	0	1	60	3.91	4.59

the stabilization mechanism of the Fermi surface-quasi-BZ interaction, which is also called the Hume-Rothery mechanism (Friedel 1988). Stable QCs and crystalline approximants can be prepared only in a narrow region of the e/a ratio. We presume that this fact is evidence of the stabilization mechanism of QCs through the Fermi surface-quasi-BZ interaction. The width of the pseudogap may be about 0.5 eV.

Table 6.1 shows the principal diffraction spots with the 6D indices and corresponding number of valence electrons per atom Z_c (Vaks et al. 1988, Carlsson and Phillips 1991). Evidence of the stabilization by the Fermi surface-quasi-BZ interaction can be seen in Table 6.1. The relation $2k_{\mathrm{F}} \simeq K_p$ between the diameter of the Fermi sphere and the magnitude of the reciprocal-lattice vector of the principal diffraction spots is satisfied in the TC-type i phase for the strong diffraction spots (222100) and (311111)/(222110) with $Z_c \simeq 2.1 \sim 2.5$ when we assume $Z_{Cu} = 1$. It is also satisfied in the MI-type i phase for the scattering spots (211111) and (221001) with $Z_c < 2.0$ if we adopt the concept of the negative valencies of TM ions. The negative valencies of TM ions may be attributed to the effect of strong *sp-d* hybridization (Trambly de Laissardière et al. 1995).

The pseudogaps may be widened and deepened by an aggregation of i clusters or their hierarchical aggregation (Janot 1996, 1997) The pseudogap is also deepened by the d-orbital resonance or the *sp-d* hybridization due to the virtual bound states in an i atom cluster (Trambly de Laissardère and Mayou 1997).

6.3.6.2 Spikes in the Density of States. Another very characteristic structure in the DOS is a dense set of sharp spikes with a width of 10–50 meV. The energy bands are very flat. A typical example of the band dispersion is shown in Fig. 6.13 for $Al_{80}Cu_{32}Fe_{16}$ crystalline approximant. In larger approximants, the width of each spike would be narrower.

Many bands cross in E–$\mathbf{k}$ space, mix with each other, and split. The overlap of bands is quite small and each band is separated in energy. Then bands become very flat and the Van Hove singularities in crystalline approximants form a set of sharp spikes. Therefore, the character of the wave functions

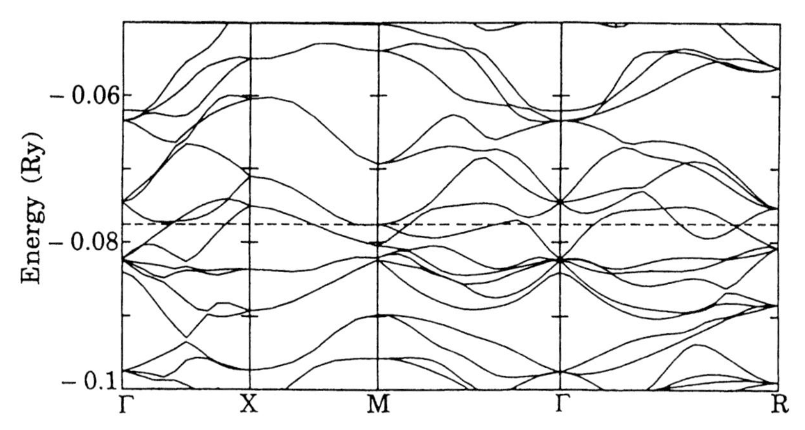

Fig. 6.13. Self-consistent band dispersion in $Al_{80}Cu_{32}Fe_{16}$ crystalline approximant. The horizontal dashed line indicates the position of the Fermi energy.

within one band varies very rapidly with change of the wave vector. Moreover, wave functions in different spikes also have different features in real space.

Once we enlarge the unit cell size of the approximant, new spikes appear and corresponding wave functions are located on new spatial configurations appearing in the enlarged unit cell. wave functions in QCs or crystalline approximants are weakly (e.g., power-law) localized and they are mixed with overlapping tails. Therefore, once new states appear in a larger unit cell, they are mixed with each other and split in energy and in space. There is no possibility of self-averaging or smoothing the spikes of the DOS in the ideal model structures.

The spikiness of the DOS is, in general, not only due to the large number of atoms in a unit cell but is also due to associated effects of the high local symmetry of atom clusters (or aggregation of atom clusters) and the large number of atoms in a unit cell. The width of each flat band is about 10–100 meV even in lower crystalline approximants and is comparable to the width of the spikes. Thus each spike corresponds to one band. The number of bands contributing to the Fermi energy is very small because of the flatness of the bands. The energy difference between the two nearest bands is about 10 meV at the same **k** points. These facts imply an important role of the interband transitions in the electron transport caused by a finite temperature or randomness.

The spikiness of the calculated DOS structure is not changed when one observes the DOS behavior with increasing the unit cell size in 2D systems. In 3D crystalline approximants, the spikes are not changed with the increasing number of **k**-points in the BZ. With increase of the unit cell size, the density of spiky peaks increases. The structure of a dense set of spikes in the DOS is probably not an artifact in the calculation. In real systems we cannot neglect

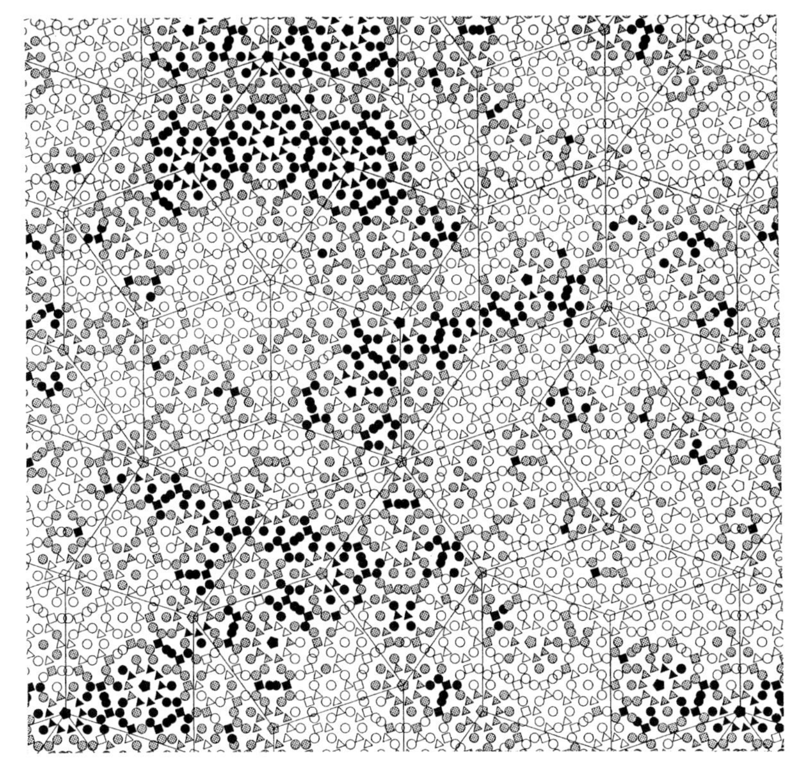

Fig. 6.14. One of the exact eigenstates of *d*-Al-Cu-Co near the Fermi energy in a system of 2728 atoms in a unit cell. The symbols show atomic positions; ○: Al, △: Cu, □: Co. Those atoms represented by the solid symbols are most probable, with the total probability being 60%. Together, the shaded and closed symbols represent those atoms which have a total probability of 90%. The remaining atoms are represented by the open symbols.

the effect of finite lifetime of electrons and thus each spike should have a finite width.

6.3.6.3 Spatial Extent of Wave Functions. The spatial extent of wave functions was analyzed in *d*-Al-Cu-Co (Fujiwara et al. 1996). The quasiperiodic plane of *d*-Al-Cu-Co consists of two kinds of clusters. A large cluster contains 40 atoms in three shells. A small cluster contains 10 atoms in one shell and one atom at the center. A small cluster contains no Co atoms.

The tight-binding parameters were first determined self-consistently in a small crystalline approximant with 110 atoms in a unit cell. Then the exact diagonalization technique was applied with fixing the tight-binding parameters in larger systems of 644, 1686, 2728, and 4414 atoms in a unit cell.

An example of the exact eigenstates near E_{F} is shown in Fig. 6.14. The eigenstates have large amplitudes of wave functions over large clusters. The states at about 0.2 Ry below E_{F} originate from the Co *d* states. However,

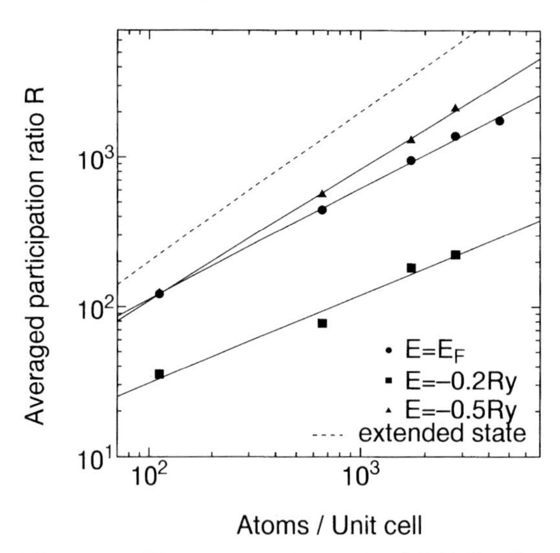

Fig. 6.15. The participation ratio of *d*-Al-Cu-Co averaged over about 50 eigenstates near the Fermi energy E_{F} (circle), 0.2 Ry below E_{F} (square), and 0.5 Ry below E_{F} (triangle) as a function of the number of atoms N in a unit cell.

wave functions at 0.5 Ry below E_{F} spread over both large and small clusters because these states are mainly of Al $3s$, $3p$, Co $4s$, and Cu $4s$ character. This fact implies that the small clusters play the role of a glue and the electronic transport channel is along the alignment of large clusters.

The spatial extent of a wave function can be analyzed quantitatively by the participation ratio defined as

$$P \equiv \frac{(\sum |u_{\mathbf{R}L,j}|^2)^2}{\sum |u_{\mathbf{R}L,j}|^4} \ . \qquad (6.24)$$

Here, $P \simeq 1$ for localized states and $P \simeq N$ for extended states (N is the number of atoms). Figure 6.15 shows $P(\psi(E))$ averaged over about 50 exact eigenstates around E_{F} as a function of N. The results clearly show that the participation ratio can be scaled by the power-law as $P(\psi(E \simeq E_{\mathrm{F}})) \propto N^{0.74}$. Similar behavior can be seen in different energy regions as $P(\psi(E \simeq E_{\mathrm{F}} - 0.2\mathrm{Ry})) \propto N^{0.59}$ and $P(\psi(E \simeq E_{\mathrm{F}} - 0.59\mathrm{Ry})) \propto N^{0.88}$. It should be noted that values of the exponents depend on the energy regions. In the energy region of 0.2 Ry below E_{F}, the states originate mainly from Co $3d$ orbitals. The states are relatively localized on Co sites and the exponent is smaller. In the region of 0.5 Ry below E_{F} the states are of Cu $4s$, Co $4s$, and Al $3s$, $3p$ character. Thus the states are uniformly extended and the exponent approaches 1. The power-law dependence of the participation ratio confirms the fact that wave functions near E_{F} extend spatially on some specific alignment of atom clusters. The wave functions may be multifractal.

6.3.6.4 Fermi Surface of Crystalline Approximants. An attempt to observe the de Haas-van Alphen effect in magnetic susceptibility has been made (Haanappel et al. 1996). Generally, it is very difficult to observe the de Haas-van Alphen effect in systems of high resistivity. Another difficulty in observing a signal of the de Haas-van Alphen effect in QCs comes from the characteristics of the Fermi surface.

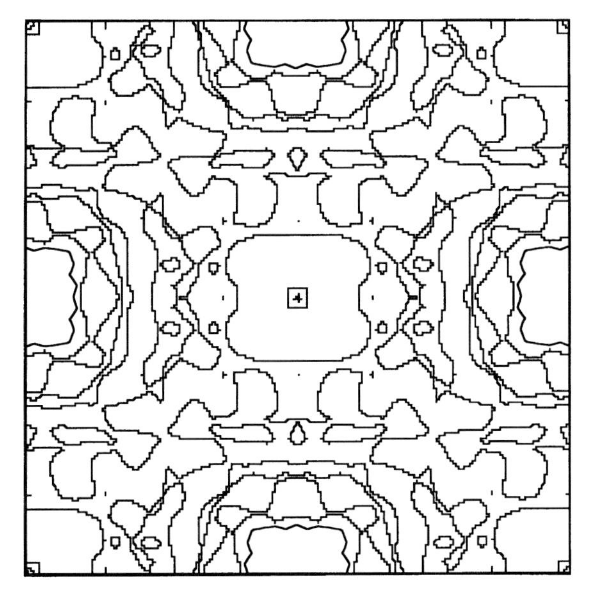

Fig. 6.16. Contour of the Fermi surface on the $k_z = 0$ plane in $Al_{45}Mg_{40}Zn_{15}$.

The structure of the Fermi surface in crystalline approximants Al-Mg-Zn is shown in Fig. 6.16 (Fujiwara 1998). One can see Fermi surfaces formed by eight bands with many small electron and hole pockets. One band forms small pockets but does not appear on the $k_z = 0$ plane. Six bands provide several closed orbits and one band forms an open orbit along the [110] direction together with several closed orbits. An open orbit causes unsaturated magnetoresistivity in high magnetic fields.

The bands are very flat and wavy with changing **k**. Thus each band crosses the Fermi energy many times and creates many electron and hole pockets in crystalline approximants. One should expect many infinitesimally small electron and hole pockets in QCs. In order to discuss the stability, we use an idea based on the quasi-BZ because the stability is essentially a local property. However, a notion of the BZ is not mathematically meaningful in quasiperiodic systems. A cross section of the Fermi surface in crystalline approximants is folded down an infinite number of times. Thus the cross section of each part of the Fermi surface is infinitesimally small. The resonance strength of a magnetic field should be very large because it is inversely proportional to the cross section of the Fermi surface. Even so, open/closed orbits in real space may be meaningful. An electron orbit in real space may be diffusive because of the finite electron scattering time, and one could expect a diffusive orbit in k-space. For this reason resonance frequencies due to the de Haas-van Alphen effects may be still observable, but a large damping of the oscillation signals should be expected. Furthermore, a sufficiently strong magnetic field may cause magnetic breakdown.

6.3.7 Experimental Study of Electronic Structures

Direct measurements of the electronic structures have been done by the XPS, UPS, IPES (inverse photoemission spectroscopy), and SXE (soft x-ray emission) and SXA (soft x-ray absorption) spectroscopy. All these spectroscopic experiments show the existence of the pseudogap (Belin and Traverse 1991, Mori et al. 1991, Matsubara et al. 1991, Belin and Mayou 1993, Belin et al. 1994, Stadnik et al. 1997). Among them the recent experiment by ultra-high resolution photoemission spectroscopy, with a resolution of 5 meV, is conclusive about the pseudogap (Stadnik et al. 1997). The Fermi edge of several i and d QCs (i-Al-Cu-Fe, i-Al-Cu-Ru, i-Al-Cu-Os, i-Al-Pd-Mn, i-Al-Pd-Re, i-Zn-Mg-Y, d-Al-Cu-Co, d-Al-Ni-Co) was analyzed on the basis of the edge-shape function and the temperature dependence. The observed temperature dependence of the Fermi edge at low temperatures in all systems can be explained by the assumed edge-shape function of a pseudogap and the temperature dependence of the Fermi-Dirac distribution function. The electron energy-loss spectroscopy (EELS) is useful to observe the pseudogap and may be promising in the observation of additional fine structures of the DOS (Terauchi et al. 1996).

Tunneling spectroscopy can provide a shape of the pseudogaps (Klein et al. 1994, 1995). NMR results (Hippert et al. 1992) and temperature dependence of magnetic susceptibility, and specific heat also provide some information on pseudogaps. A recent NMR experiment reported the observation of a spike at the Fermi energy with a width of 20 meV (Tang et al. 1997).

6.4 Transport Properties in Quasicrystals

6.4.1 Scenario of Transport in Random Systems

Propagating waves are scattered elastically by static randomness or random impurities. The localization in random systems (the Anderson localization) originates from the interference effect between propagating and backwards scattered waves (Lee and Ramakrishnan 1985, Bergmann 1984). First, the elastic mean free path ℓ should be much larger than the wavelength $1/k_{\mathrm{F}}$ to ensure the coherence between propagating and backwards scattered waves;

$k_F \ell \gg 1$. Second, inelastic scattering plays a crucial role as a dephasing process for wave functions. Inelastic scattering processes are electron-phonon, electron-electron, spin-orbit interactions, magnetic scattering, etc. The dephasing inelastic scattering time τ_i should be larger than the elastic scattering time τ_e, or the dephasing inelastic scattering length $L_{Th} = (D\tau_i)^{1/2}$ should be larger than the elastic mean free path ℓ. Here D is the diffusion constant. At high temperatures, $L_{Th} < \ell$ and one can observe a normal T-dependent conductivity governed by the inelastic scattering process.

L_{Th} increases with decrease of temperature, and below a certain temperature, $L_{Th} > \ell$. The T-dependence of the dephasing length is $L_{Th} \sim T^{-p/2}$. Thus the conductivity decreases as $T^{p/2}$ with decreasing temperature. This is called a *weak localization effect.* The weak localization can be observed only at low temperatures.

At lower temperatures, the cross interaction of the electron-electron scatterings and an impurity scattering plays a crucial role. The Fermi liquid description cannot be applied to electrons in random systems at very low temperatures. This correlation effect causes the $\Delta\sigma \propto T^{1/2}$ dependence in the conductivity in 3D systems (*electron-electron interaction effect*).

The magnetoconductivity behaves in a very complicated way at very low temperatures. Essential length scales are L_{Th}, the Landau orbit size $L_H = (eH/\hbar c)^{-1/2}$, and spin-orbit scattering length $L_{so} = (D\tau_{so})^{1/2}$. The spin-orbit interaction forbids a spin to be a good quantum number and causes another dephasing process. In the absence of the spin-orbit scattering,

$$\begin{aligned} \frac{\Delta\sigma(H,T)}{\sigma(H,T)} &\sim H^2 \quad , \quad L_{Th} < L_H \quad (\text{low } H) \\ &\sim \sqrt{H} \quad , \quad L_H < L_{Th} \quad (\text{high } H). \end{aligned} \tag{6.25}$$

Here the magnetoconductivity is positive. This is because the presence of the magnetic field destroys the time-reversal symmetry, and thus the coherence between propagating and backwards scattered waves. In the case of strong spin-orbit scattering regime ($\tau_{so} < \tau_i$),

$$\begin{aligned} \frac{\Delta\sigma(H,T)}{\sigma(H,T)} &\sim -H^2 \quad , \quad L'_{so} < L_{Th} < L_H \quad (\text{low } H) \\ &\sim -\sqrt{H} \quad , \quad L'_{so} < L_H < L_{Th} \quad (\text{intermediate } H) \\ &\sim \sqrt{H} \quad , \quad L_H < L'_{so} < L_{Th} \quad (\text{high } H), \end{aligned} \tag{6.26}$$

where ${L'_{so}}^{-2} = L_{Th}^{-2} + L_{so}^{-2}$. For low magnetic fields, spin-orbit scattering is essential and the magnetoconductivity is negative. Higher magnetic fields can overcome the spin-orbit scattering and the magnetoconductivity becomes positive. The coefficient for the H-dependent term is the T-dependent. The temperature dependence comes from this coefficient and also from the T-dependence of L_{Th} through the inelastic scattering time τ_i.

In an insulating regime, the conductivity is governed by the hopping process between localized states. The temperature dependence of σ in the 3D case may be represented as

$$\sigma \sim e^{-(T/T_0)^{1/4}} . \tag{6.27}$$

The exponent of T/T_0 depends on the space dimensionality. The electron-electron interaction in the insulating regime creates the Coulomb gap and the resulting conductivity is represented as

$$\sigma \sim e^{-(T/T_0)^{1/2}} . \tag{6.28}$$

The exponent of T/T_0 is independent on the space dimensionality.

6.4.2 Experimental Observations

The conductivity in QCs is usually expressed as $\sigma(T) = \sigma_0 + \Delta\sigma(T)$ and σ_0 is anomalously small. The temperature dependent part $\Delta\sigma$ shows the crossover behavior from $T^{1/2}$ below 30 K to T^{β} $(1.0 < \beta < 1.5)$ above 30 K in Al-Li-Cu (TC-type), Al-Ru-Cu (MI-type), Al-Pd-Mn (MI-type), and Al-Pd-Re (MI-type) (Kimura and Takeuchi 1991).

The magnetoconductivity shows the crossover behavior $\Delta\sigma \sim -H^2$ (low H) to $\Delta\sigma \sim -\sqrt{H}$ (high H) below $T < 30$ K in Al-Li-Cu. In Al-Cu-Fe, the magnetoconductivity behaves as $\Delta\sigma \sim -\sqrt{H}$ and its absolute value is reduced at higher temperatures (Sahnoune et al. 1992, Klein et al. 1992).

The T-dependence of conductivity in QCs is similar to that in the metallic region of the Anderson localization. However, unique characteristics are the lower conductivity for the higher structural ordered samples and anomalously small conductivity at very low temperatures. Furthermore, several observations cannot be explained by the weak localization effects (Gignoux et al. 1997). The behavior $\Delta\sigma \sim T^{\beta}$ was found over a quite wide temperature range (7 K $< T <$ 700 K) in Al-Pd-Mn and Al-Pd-Re. A more complicated temperature behavior of magnetoconductivity was also reported (Tamura et al. 1998).

Basic assumptions of the weak localization theory are a large mean free path $k_F\ell \gg 1$ and a large value of the Drude conductivity. The quantity $1/k_F$ is the characteristic length scale for the plane wave description. If the spatial extent of wave functions is power-law decaying in QCs, there is no characteristic length scale. Consequently there is no correspondence to $1/k_F$. The plane waves are scattered in any length scale in QCs. In other words, the scattering lengths are widely distributed over a length scale ranging from an atomic distance to mesoscopic scale.

6.4.3 Effects of Randomness

QCs containing certain disorder become more conductive (Akahama et al. 1989). The conductance of 2D Penrose lattice or 3D Fibonacci lattice with random phason was calculated as a function of the system size, the random phason density, and the Fermi energy by using either the Landauer formula

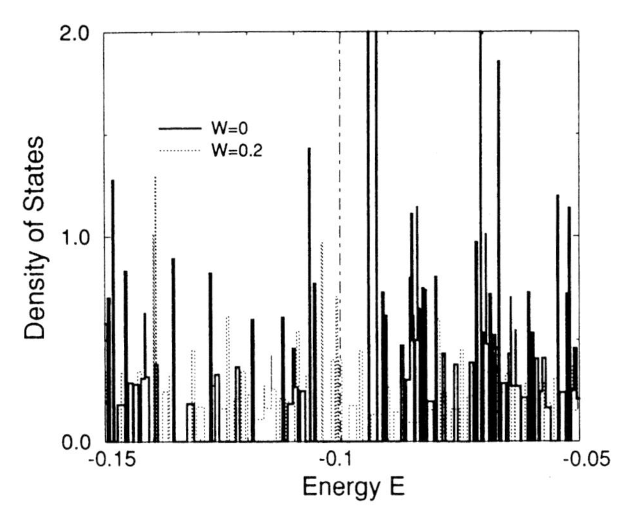

Fig. 6.17. The effect of randomness on the DOS in the 2D Penrose lattice without randomness ($w = 0$) and with randomness ($w = 0.2$). The DOS is smeared out by random phason. The number of rhombi is $N = 3571$.

(Yamamoto and Fujiwara 1995) or the Kubo-Greenwood formula (Roche and Mayou 1997).

Static disorder destroys the quasiperiodicity of the system. The spikes in the DOS are smeared and the channels of the electron transport open. This can be seen in Fig. 6.17 (Yamamoto and Fujiwara 1995). In other words, the interband transition due to elastic scattering by randomness assists hopping of electrons between weakly localized states and the conductivity increases. Wave functions scattered by randomness become smooth and homogeneous. This means that the randomness breaks the coherence essential for the anomalous wave functions.

In a model calculation of the 2D systems one observes an additional crossover behavior from the power-law conduction to the exponential localization with further increase of the system length. This results from the fact that all states should be localized in a large enough 2D random system.

6.4.4 Boltzmann Theory and Relaxation-Time Approximation

The transport properties in crystalline approximants were analyzed using the calculated band structures and the Boltzmann theory. The conductivity (due to the intra-band scattering) is given as

$$\sigma_{\alpha\beta} = e^2\tau(\frac{n}{m^*})_{\alpha\beta} = \frac{e^2\tau}{\Omega_0}\sum_{n,\mathbf{k}} v_{n\alpha}(\mathbf{k})v_{n\beta}(\mathbf{k})(-\frac{\partial f}{\partial E_n(\mathbf{k})}) , \tag{6.29}$$

$$\sigma_{\alpha\beta\gamma} = -\frac{e^3\tau^2}{\hbar\Omega_0}\sum_{n,\mathbf{k}} v_{n\alpha}(\mathbf{k})[\mathbf{v}_n(\mathbf{k}) \times \nabla_{\mathbf{k}}]_\gamma v_{n\beta}(\mathbf{k})(-\frac{\partial f}{\partial E_n(\mathbf{k})}) , \quad (6.30)$$

where $E_n(\mathbf{k})$, Ω_0, τ, T, and f are the energy of the band n at $\mathbf{k}$, the unit cell volume, the relaxation time, the temperature, and the Fermi-Dirac function, respectively, and $\mathbf{v}_n(\mathbf{k}) = \nabla_{\mathbf{k}}E_n(\mathbf{k})/\hbar$. The Hall coefficient and thermoelectric power can be evaluated from these quantities. If we assume that one collision mechanism does not disturb another collision mechanism, we immediately obtain the additive rule of the collision rate $W = \sum_j W_j$, or the relaxation-time approximation $1/\tau = \sum_j 1/\tau_j$. Thus one can obtain the Matthiessen rule of the resistivity.

An example of the calculated dc conductivity is shown in Fig. 6.18 for the Al-Mn 1/1 approximant (Fujiwara et al. 1993). The calculated conductivity is generally very small if one assumes a reasonable value for the relaxation time τ. The conductivity σ is fluctuating as a function of the Fermi energy and the fluctuation amplitude $\Delta\sigma$ is of the order of σ. The small conductivity cannot be explained only by the pseudogap or the small carrier density. In other words, large effective masses are very important (Fujiwara et al. 1994). The spikiness of the DOS originates from the flatness of bands and results in the large values of the effective mass. This analysis has been applied also to *i*-Al-Cu-Fe (Trambly de Laissardière and Fujiwara 1994a) and *d*-Al-Cu-Co (Trambly de Lassardière and Fujiwara 1994b).

The calculated values of the effective masses m^* and the ratio (n/m^*) are $m^* = \hbar^2/\frac{\partial^2 E}{\partial k^2} \simeq (2-10)m$ (m is the free electron mass) and $(n/m^*)_{\alpha\beta} < 0.5$ per unit cell in 1/1 approximants of Al-Mn or Al-Cu-Fe. The unit cell volume of the 1/1 approximant is about 100 times larger than that of a simple metal and, therefore, $(n/m^*)_{\alpha\beta}$ is several hundred times smaller than in normal metals. The Fermi energy measured from the bottom of the electron pocket may be $\sim (0.01 - 0.1)$ eV or less (Fig. 6.13). Using a formula $E_{\mathrm{F}} = (m^*/2)v_{\mathrm{F}}^2$, one can estimate $v_{\mathrm{F}} = \sqrt{2E_{\mathrm{F}}/m^*} \leq \sqrt{2 \times 0.003/10} \simeq 0.024$ (atomic unit) $= 0.5 \times 10^7$ cm/s. This velocity for conduction electrons is smaller by two orders of magnitude than that in simple metals in which $v_{\mathrm{F}} = 4.20/(r_s/a_0) \times 10^8$ cm/s with $E_{\mathrm{F}} = 50.1/(r_s/a_0)^2$ eV. Here r_s/a_0 is the dimensionless sphere radius occupied by one electron.

The energy difference between adjacent bands is about 10 meV. Consequently a temperature effect cannot be neglected in the transport phenomena. For example, the randomness due to the thermal motion of atoms smears the fine structure of the DOS and the conductivity increases. In other words, inelastic scattering mediates the hopping of electrons in a weakly localized eigenstate to other unoccupied eigenstates. This is the T-dependent part of the conductivity caused by the *inter-band* scattering. The resultant conductivity should be

$$\sigma(T) = \sigma_0 + \Delta\sigma_{\mathrm{intra}}(T) + \Delta\sigma_{\mathrm{inter}}(T) , \quad (6.31)$$

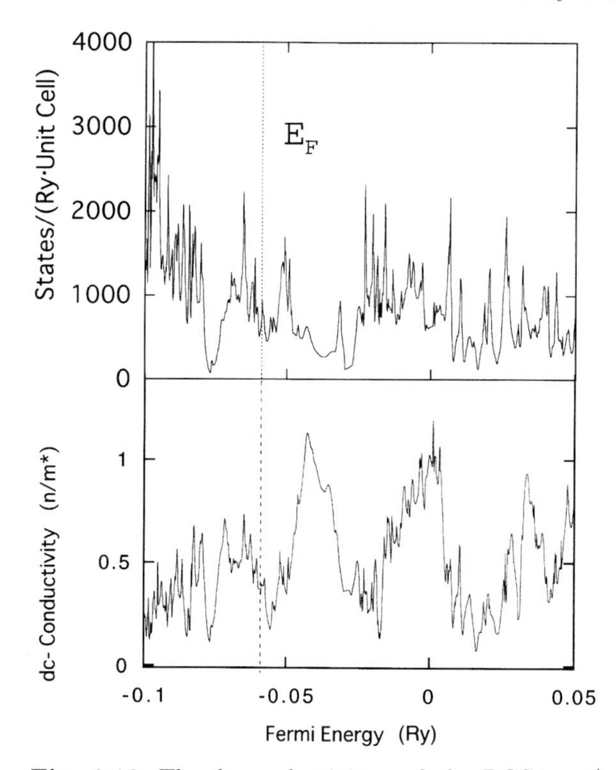

Fig. 6.18. The dc conductivity and the DOS in 1/1 approximant of Al-Mn as a function of the Fermi energy. The vertical dashed line shows a position of the calculated Fermi energy.

where $\Delta\sigma_{\text{intra}}$ and $\Delta\sigma_{\text{inter}}$ are T-dependent contributions due to inelastic processes of intra- and inter-band transitions. Thus the conductivity may follow the additive law [Eq. (6.3)].

It is still difficult to give a unified picture of anomalous transport properties from the viewpoints of the band structure effects and the Boltzmann theory. First, the Boltzmann theory cannot evaluate the magnitude of the electron lifetime. Second, the lifetime can hardly be estimated in QCs by the perturbation theory because wave functions are very sensitive to the local environment and an electron-phonon scattering lifetime depends on the spatial extent of wave functions. Third, the relaxation-time approximation cannot explain the observed dependence $\sigma(T) = \sigma_0 + \Delta\sigma(T)$. In fact, the relaxation-time approximation can hardly be valid in QCs because the perfect structure itself causes the anomalous wave functions or coherent backwards scattering of a wave packet. Elastic scattering due to structural randomness destroys the coherence. A unified explanation of the exotic transport behavior may be only given when one goes beyond the Boltzmann theory.

6.4.5 Anomalous Diffusion

A characteristic length of propagation of wave packets in the strong scattering regime may follow a law

$$L(t) = At^{\beta} , \quad 0 < \beta < 1 \tag{6.32}$$

for a long time t (Mayou et al. 1993, Sire 1994, Schulz-Baldes 1997, Bellissard and Schulz-Baldes 1998). For ballistic and localized cases, $\beta = 1$ and 0, respectively. The region $\beta < 1$ specifies a regime of weak localization of electrons or anomalous diffusion. If we assume L to be the mean free path, then

$$L = v_{\mathrm{F}} \tau , \tag{6.33}$$

where v_{F} is the mean velocity of a wave packet and τ is the collision time. Therefore, the anomalous characteristic length may be related to the anomalous power-law behavior of the mean velocity $v_{\mathrm{F}} \sim L^{1-1/\beta}$. The mean velocity v_{F} should be zero for a large system if $\beta < 1$ because of accumulated backscattering. The diffusivity is related to the characteristic length $L(\tau)$ through

$$D(\tau) = \frac{L^2}{3\tau} \sim \tau^{2\beta-1} . \tag{6.34}$$

Then the conductivity σ is governed by the Einstein relation $\sigma = e^2 N(E_{\mathrm{F}}) D(\tau)$. If the DOS at the Fermi energy $N(E_{\mathrm{F}})$ is smooth and independent on L, the conductivity is scaled as

$$\sigma \sim \tau^{2\beta-1}$$

or

$$\sigma \sim L^{2-1/\beta} \sim \sigma_0 L^{2-2/\beta} ,$$

where σ_0 is the Drude conductivity proportional to τ. A conductance G is defined as $G \sim L^{d-2}\sigma(L)$ with a dimensionality d, and the above argument is consistent with our observation of the power-law decaying conductance in the 2D Penrose lattice (Tsunetsugu and Ueda 1991).

6.4.6 Scaling Behavior

The characteristic spatial extent of wave functions shows the power-law behavior of the participation ratio as a function of the system size. The finite size scaling analysis can be also applied to the conductivity in QCs.

A zero-temperature dc conductivity is expressed by the Kubo-Greenwood formula as

$$\sigma_{\alpha\beta} = \frac{2\pi e^2 \hbar}{V} \int dE \left(-\frac{df}{dE}\right) \sum_i \delta(E - E_i) \tilde{D}_i^{\alpha\beta}(E), \tag{6.35}$$

$$\tilde{D}_i^{\alpha\beta}(E) = -\frac{1}{\pi} \lim_{\gamma \to 0+} \Im \langle i | \hat{v}_\alpha \frac{1}{E + i\gamma - \hat{H}} \hat{v}_\beta | i \rangle \equiv \lim_{\gamma \to 0+} \tilde{D}_i^{\alpha\beta}(E : \gamma) \ , \tag{6.36}$$

where $\hat{H}$ is the TB-LMTO Hamiltonian and $\hat{v}_\alpha = 1/(i\hbar)[\hat{x}_\alpha, \hat{H}]$ is the velocity operator. The function f is the Fermi-Dirac distribution function and V is the volume of the system. This equation defines the "diffusion constant" of an *individual* state $\tilde{D}_i^{\alpha\beta}(E_i)$ of the i–th eigenstate with an energy E_i. The diffusion constant in a finite crystalline system should grow in proportion to the system size, and this is indeed the case [Eq. (6.36)].

The γ in Eq. (6.36) should go to zero after one takes the limit $V \to \infty$. In case of finite temperatures or randomness in the system, γ should remain finite. $\tilde{D}_i^{\alpha\alpha}(E : \gamma)$ is calculated for various values of the parameter γ in a finite system of volume V. When γ is unphysically small (much smaller than the averaged level interval δE proportional to $1/N$), then a behavior of $\tilde{D}_i^{\alpha\alpha}(E : \gamma) \propto \gamma$ should be observed. For larger γ, $\tilde{D}$ may be varying slowly. The crossover value γ_{cr} for two different types of the behavior of $\tilde{D}$ is the smallest limit of γ in a finite system.

This analysis was applied to d-Al-Cu-Co and the scaling properties of the diffusion constant were discussed (Fujiwara et al. 1996). By enlarging the unit cell size, the crossover region is observed to shift gradually to the smaller γ side and the smaller $\tilde{D}$ value. We could express the behavior of the crossover point by a formula $\tilde{D} \sim \gamma_{\mathrm{cr}}^{0.25\cdots}$. The diffusion constant $\tilde{D}$ may be written as $\tilde{D} \sim \langle r^2 \rangle \gamma_{\mathrm{cr}}$, where $\langle r^2 \rangle$ is a spatial extent of a wave function and $\hbar/\gamma_{\mathrm{cr}}$ is the mean free time. Assuming that a wave function is not strongly localized, we obtain $\langle r^2 \rangle \sim L^2$, where L is the effective relaxation length proportional to the linear dimension of the system $\sqrt{N}$. We assume a scaling relation $\gamma_{\mathrm{cr}} \sim L^{-1/\beta}$. By using the observed behavior $\tilde{D} \sim \gamma_{\mathrm{cr}}^{1-2\beta} \simeq \gamma_{\mathrm{cr}}^{0.25\cdots}$, one obtains the value of $\beta \sim 1.33\cdots$ in d-Al-Cu-Co and the diffusion constant of a finite system can be written as $\tilde{D} \sim L^{2-1/\beta} \sim N^{-0.33\cdots}$. The diffusion constant $\tilde{D}$ should vanish in the limit $V \to \infty$ $(N \to \infty)$ at 0 K. The above argument is certainly consistent with the anomalous diffusion.

The value of the power-law index β is not universal and depends upon the energy range and the system itself. The qualitative behavior may be quite universal both in d and i QCs. The diffusion constant $\tilde{D}$ varies monotonously but slowly as a function of the system size and the inelastic scattering time. By increasing the lifetime of electrons by a factor of 10^4, $\tilde{D}$ is reduced only by a factor of 10. This indicates that the diffusion constant does not vanish in a sample of macroscopic size at very low but finite temperatures.

The temperature dependence of the observed conductivity can be discussed in relation to the scaling behavior of the diffusion constant and the temperature dependence of the electron occupations [Eq. (6.35)]. The value of

γ increases with increasing temperatures, and the dephasing inelastic scattering length L_0 decreases. As a result, the whole system becomes equivalent to an array of blocks of length $L_0 \sim \gamma_0^{-\beta}$ of perfect QCs. The bulk conductivity is a sum of small values of the conductivity of individual block of mesoscopic size.

6.4.7 Scenario of Transport in Quasicrystals

A unique topological alignment of atom clusters is essential to ensure the anomalous wave functions. In other words, the anomalous wave functions in QCs are formed by the interference effects due to the quasiperiodic lattice. Presumably the pseudogap is important to make wave functions anomalous (Rieth and Schreiber 1998). The dephasing inelastic scatterings destroy the anomalous wave functions. As a result, the conductivity increases with increasing temperatures as $\Delta\sigma \propto T^{\beta}$ (Roche and Fujiwara 1998b). The coherence of wave functions is ensured in QCs by the quasiperiodic alignment of atom clusters and the elastic scattering by static randomness destroys this coherence. The effect of elastic scattering is opposite to that in the weak localization regime of the Anderson locarization.

The value of the mean velocity v_{F} in the 1/1 approximants of Al-Mn or Al-Cu-Fe is estimated to be $\sim 0.5 \times 10^7$ cm/s. Thus the mean free path ℓ may be much shorter than the interatomic distance if we assume the elastic scattering time $\tau_e \simeq 10^{-15\sim-16}$ s: $\ell = v_{\mathrm{F}}\tau_e \sim 0.5 \times 10^{-8}$ cm . In amorphous metals, we can evaluate the diffusion constant $D \simeq (1-10)$ cm^2/s, the mean velocity $v_{\mathrm{F}} \simeq 10^8$ cm/s, and the elastic scattering time $\tau_e = D/v_{\mathrm{F}}^2 \simeq 10^{-15\sim-16}$ s (Howson and Gallagher 1988). Thus, once one starts from the plane wave description, the coherence between propagating and backwards scattered plane waves cannot be ensured.

We do not have any characteristic length scale for the anomalous wave functions in QCs except the inter-cluster distance. In crystalline approximants of the MI-type QCs, the 1/1 approximant shows quite similar anomalous transport properties as in QCs. In crystalline approximants of the TC-type QC Al-Mg-Ga-Zn, the conductivity of the 1/1 approximant is metallic whereas that of the (2/1) approximant is anomalous (Takeuchi and Mizutani 1995). Therefore, we would expect the traveling path a of at least 15 Å to ensure anomalous wave functions in quasiperiodic lattices. This is a few times larger than the inter-cluster distance.

The dephasing inelastic scattering length becomes shorter with increasing temperature. Consequently the wave functions become less anomalous and the conductivity increases. The band structure viewpoint may explain this feature of the conduction as resulting from the interband transitions between anomalous eigenstates. The anomalous conduction is also explained from a viewpoint of a competition between anomalous wave functions and dephasing inelastic scattering on the basis of the scaling analysis. To preserve the anomaly of wave functions in quasiperiodic systems, a wave packet should travel for a distance of $a = 15$ Å. Again, if we assume $D \simeq (1 - 10)$ cm^2/s and $\tau_i \simeq 10^{-10}/T^2$ s, then $L_{Th} \geq 10^{-5}/T$ cm. So up to a few hundreds K, the dephasing inelastic scattering length may be longer than $a = 15$ Å and one can expect the power-law behavior of $\Delta\sigma \sim T^{\beta}$.

The above analysis explains the characteristic features of conductivity σ in QCs: (i) the very small value of σ at very low temperatures, (ii) the temperature dependence $\Delta\sigma \sim T^{\beta}$ over a wide temperature range, and (iii) the increase of σ with increasing disorder.

6.5 Summary

Recent progress of the theoretical studies of the electronic structures in QCs has been reviewed. The new concepts of energy spectrum and wave functions, i.e., singular continuous spectrum and critical wave functions, were explained. The energy band structures were studied for several crystalline approximants of either idealized or realistic structures. There are two essential characteristics in the DOS, i.e., the pseudogap and the the dense set of spikes. The shape of the Fermi surface was also discussed.

Though some approximants or related compounds have narrow real gaps of the DOS near or just at the Fermi energy, the anomalous transport properties cannot be explained by the real gap of the band structure. It was also emphasized that the relaxation-time approximation and the Boltzmann theory could not explain the anomalous transport phenomena in QCs.

The exact eigenstates in several crystalline approximants of d QCs were analyzed by the finite size scaling and the states near the Fermi energy show the clear power-law behavior of the system size. The amplitudes of wave functions are distributed over specific atom clusters but not uniformly over the whole system. The coherence is essential to ensure the anomalous wave functions and there is no characteristic length scale in QCs. Exotic electronic transport properties and the scaling character of the diffusion constant were also discussed from the viewpoint of these characteristics. The diffusion constant was shown to vanish at absolute zero in an infinite system. However the diffusion constant in finite systems is finite with a slow dependence on the system size. An additional mechanism, such as the electron-electron interaction, may be necessary to cause the metal-insulator transition at finite temperatures. The quantum interference and dephasing inelastic scatterings were discussed. Finally, the essential difference from the weak-localization effect in the Anderson localization was clarified.

Several concrete pictures of the electronic structures and the electronic properties in QCs were presented. QCs are really marginal metals. It should be stressed that further theoretical efforts are necessary in order to understand quantitatively the electronic structures and the transport properties of QCs.

Acknowledgments

This work was supported by a Grant-in-Aid for COE Research and a Grant-in-Aid from the Japan Ministry of Education, Science and Culture.

References

Ahlgren, M., Gignoux, C., Rodmar, M., Berger, C., Rapp, Ö. (1997): Phys. Rev. B **55**, R11 915
Akahama, Y., Mori, Y., Kobayashi, M., Kawamura, H., Kimura, K., Takeuchi, S. (1989): J. Phys. Soc. Jpn. **58**, 2231
Akiyama, H., Honda, Y., Hashimoto, T., Edagawa, K., Takeuchi, S. (1993): Jpn. J. Appl. Phys. B **32**, L1003
Andersen, O. K., Jepsen, O., Grötzel, D. (1985): in Highlight of Condensed Matter Theory, Bassani, F., Fumi, F., Tosi, M. (eds). North-Holland, Amsterdam, p 59
Anderson, P. W., Thouless, D. J., Abrahams, E., Fisher, D. S. (1980): Phys. Rev. B **22**, 3519
Arai, M., Tokihiro, T., Fujiwara, T., Kohmoto, M. (1988): Phys. Rev. B **38**, 1621
Belin, E., Traverse, A. (1991): J. Phys. Condens. Matter **3**, 2157
Belin, E., Mayou, D. (1993): Phys. Scr. T **49**, 356
Belin, E., Dankhazi, Z., Sadoc, A., Dubois, J.M. (1994): J. Phys. Condens. Matter **6**, 8771
Bellissard, H., Schulz-Baldes, H. (1998): unpublished
Berger, C. (1994): in Lectures on Quasicrystals, Hippert, F., Gratias, D. (eds). Les Editions de Physique, Les Ulis, p 463
Bergmann, G. (1984): Phys. Rep. **107**, 1
Bianchi, A.D., Bommeli, F., Chernikov, M.A., Gubler, U., Degiorgi, L., Ott, H.R. (1997): Phys. Rev. B **55**, 5730
Bruijn, N. G. de (1981): Math. Proc. A **84**, 27; ibid. 39; ibid. 53
Burkov, S.E. (1991): Phys. Rev. Lett. **67**, 614
Burkov, S.E. (1992): J. Phys. (Paris) I **2**, 695
Burkov, S.E. (1993): Phys. Rev. B **47**, 12 325
Carlsson, A.E., Phillips, R. (1991): in Quasicrystals, The States of the Art, DiVincenzo, D.P., Steinhardt, P.J. (eds). World Scientific, Singapore, p 361
Cokayne, E., Phillips, R., Kan, X.B., Moss, S.C., Robertson, J.L., Ishimasa, T., Mori, M. (1993): J. Non-Cryst. Solids **153-154**, 140
Elser, V., Henley, C.L. (1985): Phys. Rev. Lett. **55**, 2883
Friedel, J. (1988): Helv. Phys. Acta **61**, 538
Fujiwara, T. (1984): J. Non-Cryst. Solids **61-62**, 1039
Fujiwara, T. (1989): Phys. Rev. B **40**, 942
Fujiwara, T. (1997): unpublished
Fujiwara, T. (1998): in Proceedings of the 6th International Conference on Quasicrystals, Takeuchi, S., Fujiwara, T. (eds). World Scientific, Singapore, p 591
Fujiwara, T., Tsunetsugu, H. (1991): in Quasicrystals, The States of the Art, DiVincenzo, D.P., Steinhardt, P.J. (eds). World Scientific, Singapore, p 343
Fujiwara, T., Yokokawa, T. (1991): Phys. Rev. Lett. **66**, 333
Fujiwara, T., Arai, M., Tokihiro, T., Kohmoto, M. (1988): Phys. Rev. B **37**, 2797
Fujiwara, T., Kohmoto, M., Tokihiro, T. (1989): Phys. Rev. B **40**, 7413
Fujiwara, T., Yamamoto, S., Trambly de Laissardière, G. (1993): Phys. Rev. Lett. **71**, 4166

Fujiwara, T., Trambly de Laissardière, G., Yamamoto, S. (1994): Mater. Sci. Eng. A **179-180**, 118
Fujiwara, T., Mitsui, T., Yamamoto, S. (1996): Phys. Rev. B **53**, R2910
Gardner, M. (1977): Sci. Am. **236**, 110
Gignoux, C., Berger, C., Fourcaudot, G., Grieco, J.C., Rakoto, H. (1997): Europhys. Lett. **39**, 171
Guo Q., Poon, S.J. (1996): Phys. Rev. B **54**, 12 793
Guyot, P., Audier, M. (1985): Philos. Mag. B **52**, L15
Haanappel, E.G., Rabson, D.A., Mueller, F.M. (1996): in Proceedings of Physical Phenomena at High Magnetic Fields - II, Fisk, Z., Gor'kov, L., Meltzer, D., Schrieffer, R. (eds). World Scientific, Singapore, p 362
Hafner, J., Krajčí, M. (1992): Phys. Rev. Lett. **68**, 2321
Hafner, J., Krajčí, M. (1993): Phys. Rev. B **47**, 11795
Henley, C.L. (1990): in Quasicrystals, Fujiwara, T., Ogawa, T. (eds). Springer-Verlag, Berlin, p 38
Hippert, F., Kandel, L., Calvayrac, Y., Dubost, B. (1992): Phys. Rev. Lett. **69**, 2086
Hiramoto, H., Kohmoto, M. (1992): Int. J. Mod. Phys. B **6**, 281
Howson, M.A., Gallagher, B.L. (1988): Phys. Rep. **170**, 265
Janot, C. (1996): Phys. Rev. B **53**, 181
Janot, C. (1997): J. Phys. Condens. Matter **9**, 1493
Jones, R.O., Gunnarsson, O. (1989): Rev. Mod. Phys. **61**, 689
Kimura, K., Takeuchi, S. (1991): in Quasicrystals, The State of the Art, DiVincenzo, D.P., Steinhardt, P.J. (eds). World Scientific, Singapore, p 313
Klein, T., Symko, O.G. (1994): Phys. Rev. Lett. **73**, 2248
Klein, T., Rakoto, H., Berger, C., Fourcaudot, G., Cyrot-Lackmann, F. (1992): Phys. Rev. B **45**, 2046
Klein, T., Symko, O.G., Davydov, D.N., Jansen, A.G.M. (1995) Phys. Rev. Lett. **74**, 3656
Kohmoto, M. (1986): Phys. Rev. B **34**, 5043
Kohmoto, M., Kadanoff, L.P., Tang, C. (1983): Phys. Rev. Lett. **50**, 1870
Kohmoto, M., Sutherland, B., Tang, C. (1987): Phys. Rev. B **35**, 1020
Kohn, W., Sham, L.J. (1965): Phys. Rev. **140**, A1133.
Krajčí, M., Fujiwara, T. (1988): Phys. Rev. B **38**, 12 903
Krajčí, M., Hafner, J. (1998): in Aperiodic'97, Proceedings of the International Conference on Aperiodic Crystals, de Boissieu, M., Currat, R., Verger-Gaugry, J.-L. (eds). World Scientific, Singapore, in press
Krajčí, M., Windisch, M., Hafner, J., Kresse, G., Mihalkovič, M. (1995): Phys. Rev. B **51**, 17 355
Krajčí, M., Hafner, J., Mihalkovič, M. (1997): Phys. Rev. B **55**, 843
Landauer, R. (1970): Philos. Mag. **21**, 863
Lee, P.A., Ramakrishnan, T.V. (1985): Rev. Mod. Phys. **57**, 287
Levine, D., Steinhardt, P.J. (1984): Phys. Rev. Lett. **53**, 2477
Lin, C.R., Chou, S.L., Lin, S.T. (1996): J. Phys. Condens. Matter **8**, L725
Manh, D.N., Trambly de Laissardière, G., Julien, J.P., Mayou, D., Cyrot-Lackmann, F. (1992): Solid State Commun. **82**, 329
Martin, S., Hebard, A.F., Kortan, A.R., Thiel, F.A. (1991): Phys. Rev. Lett. **67**, 719
Matsubara, H., Ogawa, S., Kinoshita, T., Kishi, K., Takeuchi, S., Kimura, K., Suga, S. (1991): Jpn. J. Appl. Phys. A **30**, L389
Mayou, D. (1994): in Lectures on Quasicrystals, Hippert, F., Gratias, D. (eds). Les Editions de Physique, Les Ulis, p 417
Mayou, D., Berger, C., Cyrot-Lackmann, F., Klein, T., Lanco, P. (1993): Phys. Rev. Lett. **70**, 3915

Mizutani, U., Iwakami, W., Takeuchi, T., Sakata, M., Takata, M. (1997): Philos. Mag. Lett. **76**, 349
Mizutani, U., Iwakami, W., Fujiwara, T. (1998): in Proceedings of the 6th International Conference on Quasicrystals, Takeuchi, S., Fujiwara, T. (eds). World Scientific, Singapore, p 579
Mori, M., Matsuo, S., Ishimasa, T., Matsuura, T., Kamiya, K., Inokuchi, H., Matsukawa, T. (1991): J. Phys. Condens. Matter **3**, 767
Nowak, H.J., Andersen, O.K., Fujiwara, T., Jepsen, O., Vargas, P. (1991): Phys. Rev. B **44**, 3577
Ostlund, S., Pandit, R., Rand, D., Schellnhuber, H.J., Siggia, E.D. (1983): Phys. Rev. Lett. **50**, 1873
Pierce, F.S., Poon, S.J., Guo, Q. (1993): Science **261**, 737
Poon, S.J. (1992): Adv. Phys. **41**, 303
Rath, J., Freeman, A.J. (1975): Phys. Rev. B **11**, 2109
Reed, M., Simon, B. (1972): Functional Analysis. Academic Press, New York, p 19
Rieth, T., Schreiber, M. (1998): J. Phys. Condens. Matter **10**, 783
Roche, S., Fujiwara, T. (1998a): in Aperiodic'97, Proceedings of the International Conference on Aperiodic Crystals, de Boissieu, M., Currat, R., Verger-Gaugry, J.-L. (eds). World Scientific, Singapore, in press
Roche, S., Fujiwara, T. (1998b): unpublished
Roche, S., Mayou, D. (1997): Phys. Rev. Lett. **79**, 2518
Sahnoune, A., Ström-Olsen, J.O., Zaluska, A. (1992): Phys. Rev. B **46**, 10 629
Sawada, H., Tamura, R., Kimura, K., Ino, H. (1998): in Proceedings of the 6th International Conference on Quasicrystals, Takeuchi, S., Fujiwara, T. (eds). World Scientific, Singapore, p 329
Schulz-Baldes, H. (1997): Phys. Rev. Lett. **78**, 2176
Shechtman, D., Blech, I., Gratias, D., Cahn, J.W. (1984): Phys. Rev. Lett. **53**, 1951
Sire, C. (1994): in Lectures on Quasicrystals, Hippert, F., Gratias, D. (eds). Les Editions de Physique, Les Ulis, p 505
Smith, A.P., Ashcroft, N.W. (1987): Phys. Rev. Lett. **59**, 1365
Stadnik, Z.M., Purdie, D., Garnier, M., Baer, Y., Tsai, A.-P., Inoue, A., Edagawa, K., Takeuchi, S., Buschow, K.H.J. (1997): Phys. Rev. B **55**, 10 938
Steinhardt, P.J., Ostlund, S. (eds) (1987): The Physics of Quasicrystals. World Scientific, Singapore
Sutherland, B., Kohmoto, M. (1987): Phys. Rev. B **36**, 5877
Takeda, M., Tamura, R., Sakairi, Y., Kimura, K. (1998): in Proceedings of the 6th International Conference on Quasicrystals, Takeuchi, S., Fujiwara, T. (eds). World Scientific, Singapore, p 571
Takeuchi, T., Mizutani, U. (1995): Phys. Rev. B **52**, 9300
Tamura, R., Sawada, H., Kimura, K., Ino, H. (1998): in Proceedings of the 6th International Conference on Quasicrystals, Takeuchi, S., Fujiwara, T. (eds). World Scientific, Singapore, p 631
Tang, X.-P., Hill, E.A., Wonnell, S.K., Poon, S.J., Wu, Y. (1997): Phys. Rev. Lett. **79**, 1070
Terauchi, M., Tanaka, M., Tsai, A.-P., Inoue, A., Masumoto, T. (1996): Philos. Mag. Lett. **74**, 107
Tokihiro, T., Fujiwara, T., Arai, M. (1988): Phys. Rev. B **38**, 5981
Trambly de Laissardière, G., Fujiwara, T. (1994a): Phys. Rev. B **50**, 5999
Trambly de Laissardière, G., Fujiwara, T. (1994b): Phys. Rev. B **50**, 9843
Trambly de Laissardière, G., Mayou, D. (1997): Phys. Rev. B **55**, 2890
Trambly de Laissardière, G., Manh, D.N., Magaud, L., Julien, J.P., Cyrot-Lackmann, F., Mayou, D. (1995): Phys. Rev. B **52**, 7920
Tsunetsugu, H., Ueda, K. (1991): Phys. Rev. B **43**, 8892

Tsunetsugu, H., Fujiwara, T., Ueda, K., Tokihiro, T. (1986): J. Phys. Soc. Jpn. **55**, 1420

Tsunetsugu, H., Fujiwara, T., Ueda, K., Tokihiro, T. (1991): Phys. Rev. B **43**, 8879

Vaks, V.G., Kamyshenko, V.V., Samolyuk, G.D. (1988): Phys. Lett. A **132**, 131

Wang, Y.-P, Lu, L., Zhang, D.-L. (1993): J. Non-Cryst. Solids **153-154**, 361

Wang, Y.-P, Zhang, D.-L. (1994): Phys. Rev. B **49**, 13 204

Waseda, A., Morioka, H., Kimura, K., Ino, H. (1992): Philos. Mag. Lett. **65**, 25

Windisch, M., Krajčí, M., Hafner, J. (1994): J. Phys. Condens. Matter **6**, 6977

Yamada, H., Iwakami, W., Takeuchi, T., Mizutani, U., Takata, M., Yamaguchi, S., Matsuda, T. (1998): in Proceedings of the 6th International Conference on Quasicrystals, Takeuchi, S., Fujiwara, T. (eds). World Scientific, Singapore, p 664

Yamamoto, S., Fujiwara, T. (1995): Phys. Rev. B **51**, 8841

7. Elementary Excitations and Physical Properties

Jürgen Hafner and Marian Krajčí

7.1 Introduction

7.1.1 Quasiperiodic Structure

It has been known for more than twenty years that perfect structural order can exist in systems with broken translational symmetry. In the seventies the scientific interest concentrated on the so-called incommensurably modulated crystalline (IC) phases (see, e.g., Currat and Janssen 1988). The discovery by Shechtman et al. (1984) of icosahedral (i) symmetry (which implies the absence of translational periodicity) in certain alloys introduced the new field of quasicrystals (QCs). Soon after it was found that quasicrystallinity is not restricted to the i phases. Today we know QCs characterized by the classically forbidden fivefold (icosahedral), eightfold (octagonal), tenfold (decagonal), and twelvefold (dodecagonal) symmetry axes (for an introduction and reviews, see Steinhardt and Ostlund 1987, DiVinzenzo and Steinhardt 1991, Janot 1994). Icosahedral QCs are quasiperiodic in all three dimensions, while the other QCs are quasiperiodic only in the plane perpendicular to the non-crystallographic axis and periodic along it. IC phases and QCs are members of a family of quasiperiodic materials. Quasiperiodic structures can be characterized by a diffraction pattern characterized by sharp spots at the positions given by

$$\boldsymbol{k} = \sum_{i=1}^{n} h_i \, \boldsymbol{a}_i^* \,, \quad \text{where} \quad h_i \quad \text{is an integer} \,. \tag{7.1}$$

If $n = 3$ and if the three basis vectors $\boldsymbol{a}_i$ are linearly independent, the vectors $\boldsymbol{k}$ define a reciprocal lattice and the structure shows translational periodicity. If, however, $n > 3$ than there are no translation vectors having an integer scalar product with all the vectors $\boldsymbol{k}$ defined by Eq. (7.1). The set of vectors in Eq. (7.1) is called the Fourier module of the structure, the number n of independent basis vectors is called the rank of the Fourier module. A structure is quasiperiodic if the rank n of its Fourier module is finite and $n > 3$.

The Fourier module can be considered as the projection onto the three-dimensional (3D) physical space of a reciprocal lattice in nD space. To this reciprocal lattice L_3^* belongs a periodic structure in nD space. If the density of a quasiperiodic system is described by

$$\rho_s(\boldsymbol{r}_s) = \sum_{\boldsymbol{k}_s \in L_3^*} \hat{\rho}(\boldsymbol{k}) \exp(\mathrm{i}\boldsymbol{k}\boldsymbol{r}) , \tag{7.2}$$

where the sum is over all vectors belonging to the Fourier module L_3^* (Janssen 1988a), then the function is defined by

$$\rho_s(\boldsymbol{r}_s) = \sum_{\boldsymbol{k}_s \in L_n^*} \hat{\rho}(\boldsymbol{k}) \exp(\mathrm{i}\boldsymbol{k}_s\boldsymbol{r}_s) , \tag{7.3}$$

where $\boldsymbol{k}_s$ is a vector of the nD reciprocal lattice L_n^* and $\boldsymbol{r}_s$ is an arbitrary nD vector, is a lattice-periodic function in n dimensions. It is invariant under all translations forming the nD lattice L_n and whose scalar product with all vectors $\boldsymbol{k}_s \in L_n^*$ is an integer multiple of 2π. This means that the crystallography of quasiperiodic systems may be formulated in terms of periodic lattices in an nD superspace which is the direct product of the 3D physical space (sometimes also called "parallel" space) and the additional space called perpendicular space. Points $\boldsymbol{r}_s$ in superspace are pairs of points $\boldsymbol{r}_{\|}$ in physical (parallel) space $E_{\|}$ and points $\boldsymbol{r}_{\perp}$ in perpendicular space $E_{\perp}$. For a quasiperiodic point lattice the periodic structure embedded in the superspace consists of $(n-3)$D hypersurfaces, which are called atomic surfaces. For an IC structure characterized by a modulation vector $\boldsymbol{Q}$ incommensurate with the underlying periodic 3D lattice (i.e., a displacively modulated lattice), we have n = 4 and the embedded structure is a periodic array of lines. For an i QC the rank of the Fourier module is six, the atomic surfaces are $(n-3) = 3$ dimensional polytopes (triacontahedra for the case of a simple i lattice). The crystallography in higher-dimensional space has been reviewed by Mermin (1992).

7.1.2 Physical Properties

Quasicrystals have attracted the interest of scientists not only because of their interesting structural properties. A question that arises immediately is why nature should prefer quasiperiodic to periodic order. There have been early suggestions, based mainly on electron-counting arguments, that the energetic stability of quasicrystalline phases is promoted by a Hume-Rothery-like mechanism leading to the formation of a structure-induced minimum in the electronic density of states (DOS) at E_{F} (Bancel and Heiney 1986, Smith and Ashcroft 1987, Friedel and Denoyer 1987). Indeed, both experimental (Poon 1992, Stadnik et al. 1997) and theoretical (Krajčí and Hafner 1997a, 1997b) investigations have demonstrated that a pseudogap of a few tenths of an eV is a generic property of i alloys, although it is in many cases not a specific property associated with quasicrystallinity. Whether a similar statement holds also for the QCs with 2D quasiperiodicity is still a subject of debate (Stadnik et al. 1997, Krajčí and Hafner 1997a, Krajčí et al. 1997a, 1997b). In addition, we know today that QCs, and in particular the thermodynamically stable i and decagonal (d) QCs, possess a number of really exotic physical properties. The following are the most outstanding observations. (i) A very low electrical conductivity σ which may be comparable or even lower than Mott's minimum metallic conductivity (Poon 1992, Kimura and Takeuchi 1991). For *i*-Al-Pd-Re extrapolations of the temperature dependence of σ even suggest a vanishing σ at finite T, placing this alloy on the insulating side of a metal-insulator transition (Pierce et al. 1993). (ii) A strong temperature-dependence of the electrical resistivity ρ, characterized by the values of the resistivity ratio $R = \rho(4.2\ K/\rho(300\ K)$ which may be as high as ~ 50. (iii) An extreme sensitivity of the electric transport properties to minimal changes in the composition and to the variations in the thermal treatment of the samples. (iv) Variations of the magnetoresistance with R, spanning the interval from giant positive values at small R (Ahlgren et al. 1997) to small positive or even negative values comparable to those observed in systems undergoing a metal-insulator transition (Honda et al. 1994, Guo and Poon 1996). (v) A strong anisotropy of the electrical conductivity in *d* QCs, with high values and metallic behavior in the periodic directions and low values and almost semiconducting behavior in the quasiperiodic planes (Martin et al. 1991, Yun-ping and Dian-lin 1994). (vi) A very unusual optical conductivity $\sigma(\omega)$, with a very small Drude contribution (due to the low dc-conductivity) and a frequency dependence dominated by a strong interband adsorption term (Homes et al. 1991, Burkov et al. 1993). (vii) Low values of the electronic specific heat, indicating a reduced electronic DOS at E_{F} (Mizutani 1994), and confirmed by nuclear magnetic resonance (NMR) measurements (Hippert et al. 1992). (viii) Low thermal conductivities, resulting from low electronic and lattice contributions (Legault et al. 1995) with a temperature-dependence similar to that observed in amorphous metals. (ix) Highly exotic magnetic properties, ranging from a paramagnetic susceptibility with a Curie-Weiss behavior and a low-temperature spin-glass-like cusp in *i*-Al-Mn phases (Hauser et al. 1986, Gozlan et al. 1991) to diamagnetism in certain *i* (Klein et al. 1991, Lanco et al. 1992) and *d* alloys (Lück and Kek 1993).

There is now quite general agreement that the unusual physical properties of QCs cannot be understood in terms of peculiarities of the electronic DOS alone. at E_{F} Even a very deep pseudogap at the Fermi energy E_{F} or the "spiky" structure of the DOS (with pseudogaps of the order of several 0.01 eV) are insufficient to explain the experimental observations. More has to be learnt about the character of the eigenstates in QCs in order to explain their unusual physical properties.

7.1.3 Spectral Properties of Quasiperiodic Hamiltonians

The eigenvalue spectrum of a Hamiltonian can be classified as absolutely continuous, singular continuous, or discrete, corresponding respectively to extended, critical, and localized eigenstates. Absolutely continuous and point spectra are characteristic for periodic and random systems, respectively. The existence of extended eigenstates and of a continuous spectrum in crystals are the consequence of Bloch's theorem which states that eigenstates are plane-waves, modulated by a lattice-periodic function. In random systems strong disorder can lead to the formation of localized eigenstates with an exponential decay of the amplitudes.

For 1D quasiperiodic chains renormalization-group studies have demonstrated that the spectrum is singular continuous and that the eigenstates are critical, i.e., they have a complex structure characterized by a power-law decay of the amplitudes (Kohmoto et al. 1987, Fujiwara et al. 1989). The existence of critical eigenstates is intimately related to the competition between the broken translational invariance (leading to localized wave functions) and the self-similarity of quasiperiodic structures causing resonances between equivalent configurations. According to Conway's theorem (Gardner 1977), a given atomic pattern of diameter D repeats within a distance less than a few D's from the original cluster. Hence, there is always a finite probability that an eigenstate on one of the clusters will tunnel to the next cluster where it assumes an identical form, except for a finite damping factor (Sire 1994). Note that the argument for the criticality of eigenstates is independent of the dimension of the system. This would suggest that critical eigenstates can exist in real 3D QCs as well.

For quasiperiodic systems, a version of Bloch's theorem can be expressed as (Hofstadter 1976)

$$\psi_n = \mathrm{e}^{\mathrm{i}nk} g(\sigma n) \ , \tag{7.4}$$

where k is the wave number and $g(x) = g(x+1)$ is a periodic function with period 1. If σ is a rational number, $\sigma = N/M$ (N, M are integers), this is Bloch's theorem. In this case $g(x)$ is defined only on a discrete set of points $x \in \{i/M\}, 0 \leq i < M$, which determine the wave function on the infinite number of sites of the crystal lattice. If σ is irrational, $g(x)$ is defined on a countably infinite dense set of points, and the transition from ψ_n to g corresponds merely to a reordering of the data; in this case Eq. (7.4) has no predictive value: the wavefunction must still be calculated on an infinite number of points. It has been demonstrated, however, that for certain classes of irrational numbers and if the quasiperiodic potential is not too strong, $g(x)$ will be continuous and this implies that the eigenstates are extended. Indeed, for 2D QCs (essentially Penrose tilings with a nearest-neighbour coupling between the vertices) similar results have been obtained numerically. In addition, some mathematically rigorous results concern certain exact degenerate eigenstates at particular energies, called confined states (Tsunetsugu et al. 1986). The confined states are strictly localized and their degeneracy is due to Conway's theorem. However, they are not a characteristic property of quasiperiodicity since they are a consequence of a special local topology (Rieth and Schreiber 1995a). Similar confined states have also been identified on 3D quasilattices by Krajčí and Fujiwara (1988). For the overwhelming majority of states the analysis of the dependence of the participation number P

(which is essentially a measure for the number of sites on which the amplitude of the eigenstate is significantly non-zero) on the system size shows a scaling relation $P \propto N^{\beta}$, with $\beta \sim 0.9$. This corresponds to a decay of the envelope of the eigenstates, $\psi(r) \propto r^{-\alpha}$ with $\alpha \lesssim 0.5$, i.e., to the power-law localization (Rieth and Schreiber 1995a, 1998) that is more distinct near the band center (Fig. 7.1a).

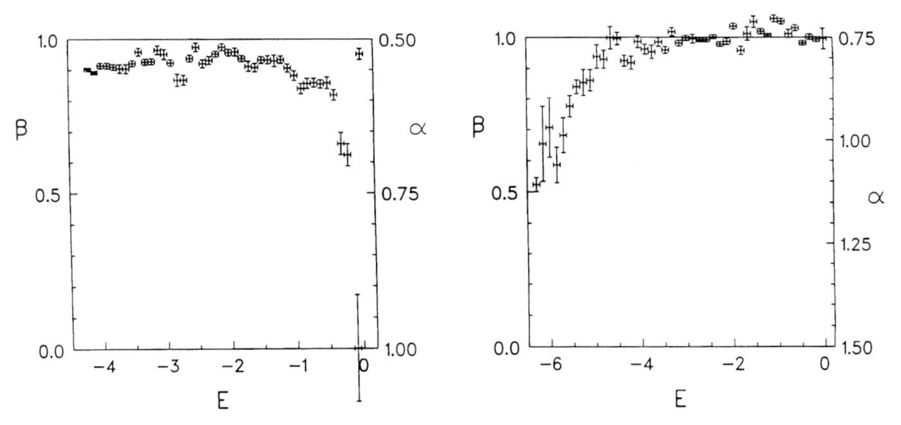

Fig. 7.1. Exponent β for the scaling of the participation number P of eigenstates with energy E (measured relative to the band-center) and exponent α for the power-law decay of the eigenstates. Left-panel: 2D Penrose tiling; right panel: 3D Penrose tiling, vertex model with nearest neighbour coupling. After Rieth and Schreiber (1998).

We note that this is characteristically different from the predictions of the Anderson model for localization on random structures where localization is expected to occur at the band edges. A more detailed analysis shows that the power-law decay parallels a self-similar character of the eigenstates. On the other hand, for the 3D Penrose lattice states at the band edges are more strongly localized than those at the band center (which have $\beta \approx 1$ as expected for extended states) (Fig. 7.1b). Critical behavior and self-similarity can be expected (if at all) only in the tails of the DOS, this time in correspondence to the Anderson model. Degenerate confined states have been found both in two and three dimensions. However, in the 3D case confined states occur in much smaller numbers and are not separated from the rest of the spectrum by a finite gap.

A further important property of quasiperiodic structures is their self-similarity. Self-similarity implies that deflation/inflation procedures map the structure onto itself. It has been argued that self-similarity will lead to a multi-fractal character of the eigenstates confined in specific regions of the QC. Although it has been demonstrated (Sutherland 1986, Tokihiro et al. 1988) that, by choosing a tight-binding Hamiltonian according to a complex rule, a self-similar groundstate with critical and multifractal behavior can be obtained, it is not clear whether such a result has general relevance. Again, numerical studies by Rieth and Schreiber (1998) confirm the multifractal character of the eigenstates for the 2D Penrose tiling, but not for its 3D analog.

For realistic models of QCs self-similarity leads to a hierarchical arrangement of "noncrystallographic", mostly i clusters: the elemental clusters are arranged in the form of inflated clusters of the same topology (for i QCs the scaling factor would be τ^3, where τ is the golden mean), e.g., in the form of an "icosahedron of icosahedra". Janot and de Boissieu (1994) have argued that many of the anomalous electronic properties of QCs, such as the spikiness of the DOS, could be explained in terms of eigenstates confined to the elemental clusters.

7.2 Quasiperiodicity, Symmetry, and Elementary Excitations

The concept of elementary excitations or quasiparticles (crystal electrons, phonons, magnons, ...) has proved to be very helpful for exploring the relationship between symmetry and physical properties. The description of elementary excitations in crystals is based on the concept of Bloch waves and Brillouin zones (BZs) in reciprocal space. For quasiperiodic systems BZs and hence dispersion relations for elementary excitations in the usual sense do not exist. However, it has been shown that the characterization of QCs in terms of their diffraction patterns, which are the Fourier spectra of their density, allows it to be possible to describe their symmetry in terms of the operations of their point- and space-groups (Mermin 1992, Bienenstock and Ewald 1962), but leads also to a very useful generalization of the concept of BZs.

The diffraction pattern of a periodic or quasiperiodic lattice is described by Eq. (7.1). The symmetry of the lattice may be described as the set of proper or improper rotations which leave the diffraction pattern invariant. For a periodic crystal, this has the consequence that the rotations leave the density in real space invariant within a lattice translation. No such symmetry exists for QCs. The densities of the rotated and unrotated QCs agree only in a statistical sense, i.e., any bounded substructure of the rotated QC can be found in the unrotated QC, albeit at distances that increase proportionally to the diameter of this region (cf. Conway's theorem). The precise mathematical definition of two densities being indistinguishable is that they have the same positionally averaged m-point correlation functions for all m. As shown by Mermin (1992) and Rokhsar et al. (1988), this generalized notion of invariance is the key to the definition of the symmetry group of a QC.

Even if for a QC the lattice L_3^* is dense in wave-number space, the amplitude of the Fourier components $\rho(\boldsymbol{k})$ of the density [$\simeq$ the intensity of the diffraction peaks, see Eq. (7.2)] is modulated. This is due to symmetry; there may even be lattice vectors where $\hat{\rho}(\boldsymbol{k})$ is required to vanish because of symmetry (extinction rules). The set of lattice vectors $\boldsymbol{k}$ for which $\hat{\rho}(\boldsymbol{k})$ exceeds experimentally measured values is essentially discrete. In this sense QCs meet the International Union of Crystallography's new definition of a crystal as a solid "with an essentially discrete diffraction pattern" (The International Union of Crystallography 1992).

As has already been emphasized, the Fourier module (or the reciprocal lattice) L_3^* of the quasiperiodic structure may be viewed as the projection of a regular hypercubic reciprocal lattice L_n^* in n dimensions onto 3D space. This forms the basis for the generalization of the concept of BZs. Each Bragg peak may be considered as the projection of the center of the BZ of the nD reciprocal lattice. Similarly, the projections of high-symmetry points corresponding to the corners and midpoints of faces and edges of the nD BZ defined on L_n^* determine a set of high-symmetry points in 3D wave-number space (Niizeki and Akamatsu 1990). For an i lattice, for example, there exist eight types of special points denoted as $\Gamma, R, X_5, M_5, X_3, M_3, X_2, and M_2$. They correspond respectively to 6D wave vectors (000000), (hhhhhh), (h00000), (0hhhhh), (hhh000), (000hhh), (hh0000), and (00hhhh). The point group symmetry of the special points is the i group I_h for Γ and R, and D_{2d}, D_{3d} and D_{5d} for X and M, depending on whether the subscript is 2, 3 or 5. The projections of the 6D special points onto 3D reciprocal space define sets of quasiperiodically distributed special points. The special points are dense everywhere, but with intensities modulated by interference effects and described by a generalized structure factor (Niizeki 1989, Niizeki and Akamatsu 1990) accounting for the symmetry of the quasiperiodic structure. R points, for example, have intensities on the five- and threefold symmetry axes, but vanish on the twofold axes. X_2 and M_2 points have intensities on the twofold axes, but vanish on the other symmetry axes. Quite generally, the distribution of the special points can be described by a generalized structure factor. Figure 7.2 shows the distribution of the Γ and X_2, M_2 points of an i lattice in a plane containing the two-, three-, and fivefold symmetry axes.

The most intense special points define a quasi-discrete set of quasi-BZ centers and quasi-BZ boundaries, which is the quasiperiodic analogue to the extended-zone scheme for a crystal. One of the important results of the last few years is the realization that the spectrum of elementary excitations is determined largely by this quasi-BZ structure of wave-number space. Free-particle dispersion relations are found around the most intense Γ points, whereas the most intense quasi-BZ boundaries lead to stationary states and further to a hierarchical structure in the DOS by a mechanism that is closely related to the formation of Van Hove singularities in the spectrum of crystalline materials.

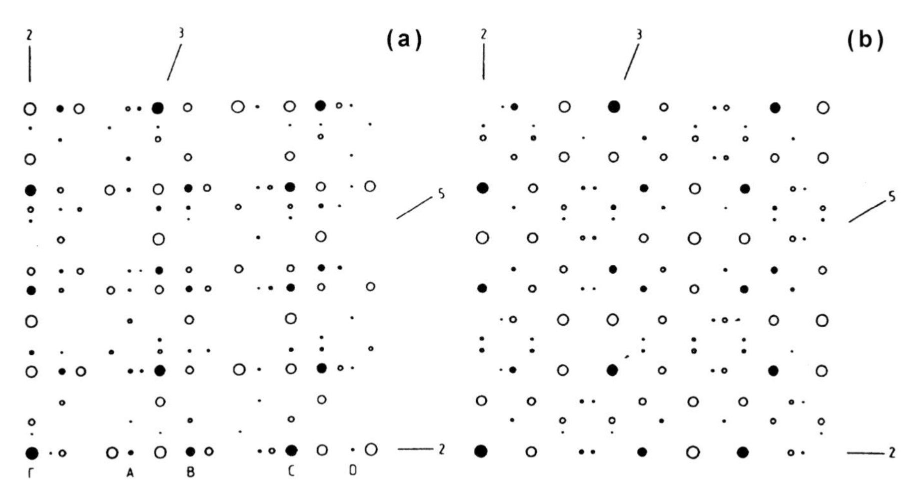

Fig. 7.2. Distribution of the X_2 (a) and M_2 (b) special points of an i lattice in a plane containing the two-, three-, and fivefold symmetry axes. The special points are represented by open circles, the full circles show the Γ points. In each case the intensity is proportional to the generalized structure factor. After Niizeki and Akamatsu (1990).

7.3 Modelling Quasicrystalline Structures and Approximant Phases

The first step in a numerical investigation of elementary excitation has to be the construction of a structural model of the QC by specifying the positions of all atoms (which are not uniquely determined by diffraction experiments). There are two different strategies for creating quasicrystalline model structures.

7.3.1 Icosahedral Quasicrystals

The first approach consists in the "decoration" of quasiperiodically arranged structural units. For example, models for the simple i (SI) QCs of the Al-Cu(Zn)-Mg(Li) class are based on the decoration of the prolate and oblate rhombohedra and of a composite structural unit (a rhombic dodecahedron) forming a 3D Penrose tiling with atomic motifs derived from the crystal structure of the Frank-Kasper phase $(Al,Zn)_{49}Mg_{32}$, as proposed by Henley and Elser (1986). The underlying Penrose tiling is constructed by the projection method of Bak (1986). Alternatively, the canonical-cell-tiling (CCT) method proposed by Henley (1991), Newman and Henley (1993), and Mihalkovič and Mrafko (1993) may be used to create the i lattice. A face-centered-icosahedral (FCI) lattice, such as that formed by Al-Mg-Li QCs (Niikura et al. 1993), may be created by introducing a chemical order on the SI lattice and decorating the odd and even vertices with (Al-Mg) and

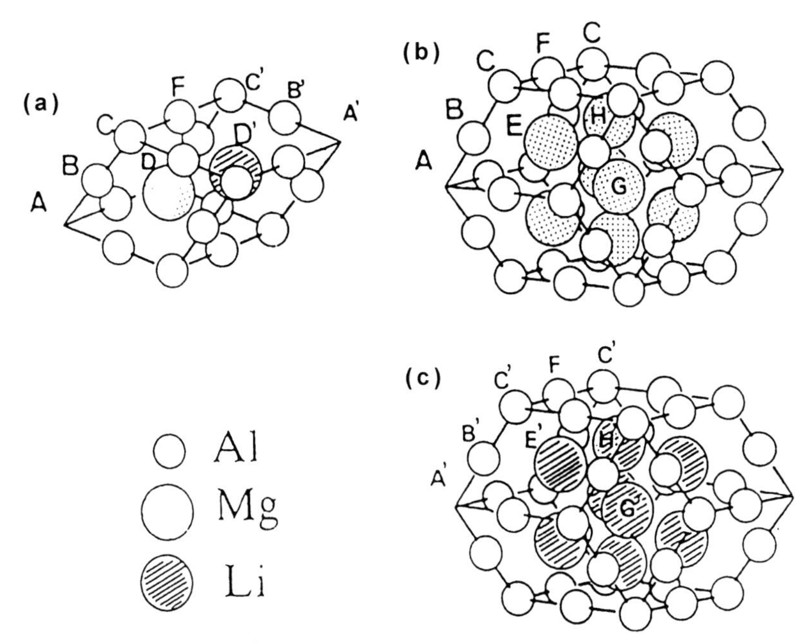

Fig. 7.3. Decoration of the prolate rhombohedron (a) and of the rhombic dodecahedron (b,c) in face-centered-icosahedral Al-Mg-Li. The long axis of the rhombohedron always connects an even (A) with an odd (A') vertex, whereas the long twofold axis of the rhombic dodecahedron connects vertices of equal parity. The parity of these vertices determines whether the sites in the interior are occupied by Mg or Li atoms, respectively.

(Al-Li) clusters, respectively (Tsai et al. 1997, Dall'Acqua et al. 1997). The decoration proposed for the FCI lattice is shown in Fig. 7.3.

In the second approach, the structural model is generated directly by projection from 6D hyperspace. The atomic positions and their occupation by different chemical species are determined by the position, shape, and chemical character of 3D atomic surfaces in perpendicular space. The model for FCI Al-Pd-Mn or Al-Cu-Fe was proposed by Katz and Gratias (1993) and Cockayne et al. (1993) and is based on the analysis of the diffraction data, including contrast-variation techniques to determine the chemical order (Boudard et al. 1992, de Boissieu et al. 1994). It consists of a triacontahedron with a fivefold radius τ^2 (in units of the edge of the rhombohedra of the underlying Penrose lattice) at the even vertices of L_6^*, an identical triacontahedron at the odd vertices (except truncated by its intersections with its 12 images displaced by τ^3 along the fivefold axes), and a small triacontahedron with radius 1 at the body-centered positions. The division into subshells for Al, Mn, and Pd determines the chemical order on the FCI lattice (Fig. 7.4). For i QCs it is, in general, possible to achieve excellent agreement be-

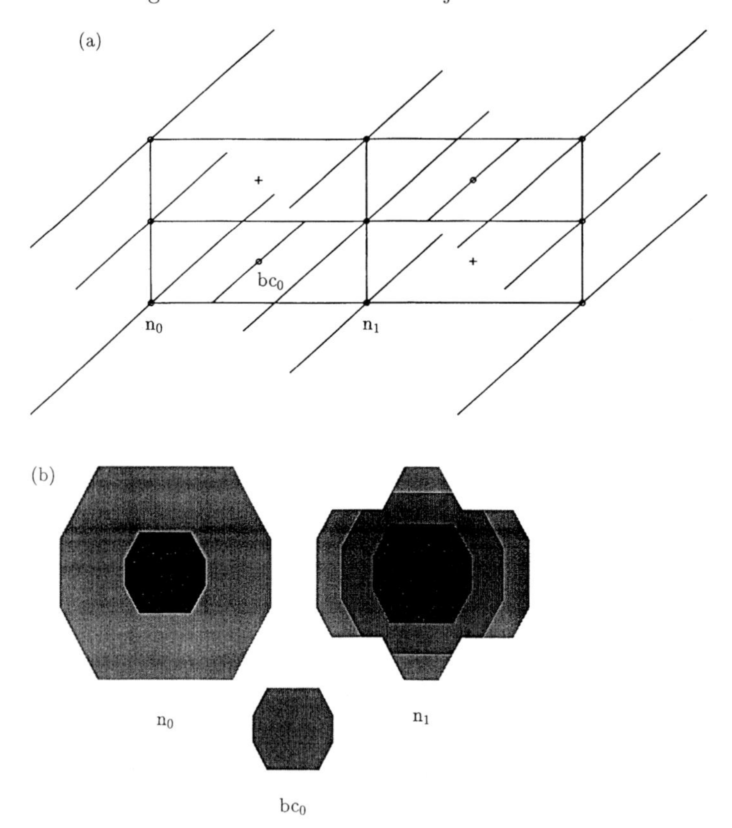

Fig. 7.4. Illustration of the atomic surfaces for face-centered *i*-Al-Pd-Mn in 6D wave-number space. (a) Section through 6D space containing the five fold axes in physical and in perpendicular space. Atomic surfaces are centred on the even and odd vertices (n_0, n_1) and on the body-centered positions (bc_0). (b) Cross sections through the atomic surfaces. The Mn cores of the triacontahedra at n_0 and n_1 are represented by the darkest shading. The lighter shading represents the outer Pd and Al shells at these nodes and the Pd surface at the bc_0 position.

tween a well-constructed model and the available structural information from diffraction and other eperiments.

7.3.2 Decagonal Quasicrystals

For *d* QCs the general strategy remains the same. For *d*-Al-Pd-Mn, Hiraga and Sun (1993) proposed a tiling-decoration model which is based on the tiling of the quasiperiodic plane perpendicular to the tenfold axis with decagons (D), pentagonal stars (P), and hexagons (H), and the decoration of all vertices with pentagonal clusters (Krajčí et al. 1997a). Mihalkovič (1995) has shown that this DPH tiling has only a nontrivial 5D representation in hyperspace with fractal atomic surfaces, that is related to the closest packing of discs with d symmetry. For d-Al-Cu-Co, Burkov (1991) proposed a model that can be viewed as a cluster-decoration model of a binary tiling. A second model proposed by Burkov (1993), which is based on the Klotz-triangle tiling of Baake et al. (1990), can also be defined in terms of atomic surfaces in 5D hyperspace (Krajčí et al. 1997b). However, in this case the available diffraction data are not sufficient to specify the chemical decoration. For d QCs electronic structure calculations can play an important role in refining structural models, particularly in determining the local chemical order (Krajčí et al. 1997b).

7.3.3 Approximant Structures

Numerical studies of spectral properties must necessarily be performed on finite models. To all QCs there exist periodic "approximant" structures such that the basic building blocks and their atomic decoration are the same as in the quasiperiodic phase. The calculation of the spectrum of elementary excitations for a series of approximant structures allows to deduce its properties in the quasiperiodic limit.

7.3.3.1 Cubic Approximants to Icosahedral Phases. Within the cut-and-projection method, there is a simple way to construct an infinite hierarchy of cubic approximants to an i lattice. The basic principle is a rotation of the physical space $E_{\|}$ within the 6D hyperspace such that its "slope" becomes rational. Lubensky et al. (1986) and Qiu and Jarić (1990) have shown that this is equivalent to the introduction of a linear phason strain. Mathematically, the prescription is very simple. In the matrix which relates the 6D-periodic space with perpendicular space, the golden mean $\tau = \frac{1+\sqrt{5}}{2}$ is replaced by one of its rational approximants $\tau_n, \tau_n = F_{n+1}/F_n$, where F_n is a Fibonacci number, $F_{n+1} = F_n + F_{n-1}, F_0 = 0, F_1 = 1)$. The periodic approximants to the quasicrystalline structure are characterized by the rational approximants to τ, i.e., by the sequence (1/1), (2/2), (3/2), (5/3), (8/5), ...

Such a definition gives a well-defined relation between the rhombohedral cell parameter of the i phase a_R and the lattice constant a_n of the n-th order approximant, $a_n = \frac{2}{\sqrt{\tau+2}}\,(F_{n+1}\tau + F_n) \times a_R$. In many cases, the low-order approximants are well-known crystalline structures. For example, the 1/1 approximants to i-Al-Mg-Zn and i-Al-Cu-Li are the Frank-Kasper phases $(Al,Zn)_{49}Mg_{32}$ and the R-Al-Cu-Li, respectively. The (2/1) and (3/2) approximant phases have been reported for Al-Mg-Zn by Edagawa et al. (1992) and Mukhopadhyay et al. (1991). This relation between crystalline and quasicrystalline phases has been used in the modelling of i QCs (Henley and Elser 1986, Krajčí and Hafner 1992, Windisch et al. 1994a). As an example, Table 7.1 lists the lattice constants, the number of atoms per cell (note that the cell volume and the number of atoms increase at each step by a factor $\tau^3 \approx 4.24$), the chemical composition, and the space-group symmetry of the approximants to i-Al-Pd-Mn. For the FCI QCs all approximants are simple cubic, for SI

QCs the approximants are body-centered cubic for mod(n,3) = 1 and simple cubic otherwise (Krajčí and Hafner 1992) For Al-Pd-Mn, the as-quenched $Al_{70}Pd_{26}Mn_4$ QC transforms on heating into a 2/1-crystalline approximant (Waseda et a. 1992).

Table 7.1. Lattice constant a_n, number of atoms N_{at}, chemical composition (in at.%) and space-group symmetry of the rational approximants to i-Al-Pd-Mn.

n	τ_n	a_n (Å)	N_{at}	c_{Al}	c_{Pd}	c_{Mn}	space group
1	1/1	12.45	128	68.8	15.6	15.6	$P2_13$
2	2/1	20.15	544	68.4	22.8	8.8	$P2_13$
3	3/2	32.60	2292	70.3	20.6	9.1	$P2_13$
4	5/3	52.75	9700	70.6	20.7	8.7	$P2_13$
5	8/5	85.36	41068	70.7	20.6	8.6	$P2_13$
6	13/8	138.11	173936	70.8	20.6	8.6	$P2_13$

If different levels of approximations are chosen for the three Cartesian axes, the approximant has orthorhombic symmetry. An example is the orthorhombic (3/2, 2/1, 2/1) approximant phase in the Al-Zn-Ga-Mg system described by Edagawa et al. (1991). The discussion given above refers to the 3D Penrose-tiling approach to QCs. Techniques for constructing rational approximants within the canonical cell-tiling scheme have been given by Newman and Henley (1993) and Mihalkovič and Mrafko (1993). Note that the CCT tiling can be decorated with Penrose rhombohedra, providing a set of packing rules for the rhombohedral tiles.

7.3.3.2 Quasicrystalline and Approximant Phases in the Al-Pd-Mn System. Stable i and d phases have been identified in the ternary Al-Pd-Mn phase diagram. At certain compositions a transformation from the d to the i phase has been observed. There is a unique orientation relationship with a fivefold axis of the i phase parallel to the tenfold axis of the d phase (Audier et al. 1993). In addition, orthorhombic approximants related either to the d or the i phase have been described. Within linear phason strain field theory, a crystalline phase can be considered as an approximant to the quasicrystalline structure if the basis vectors of the approximant phase can be obtained by projecting a linearly sheared lattice in hyperspace which has at least three nD lattice vectors lying in physical space (Duneau and Audier 1994). The Mn-rich orthorhombic T-Al_3Mn or $Al_{11}Mn_4$ (space group Pnma) and R phases $Al_{60}Mn_{11}Ni_4$ (space group Cmcm) are approximants to the d phase, with the b axis of the orthorhombic cell coinciding with the tenfold axis of the QC. In addition, a phase with τ^2-inflated a and c lattice parameters ($\tau^2 - R$ phase) has been identified. A corresponding $\tau^2 - T$ phase can be constructed, but has not yet been experimentally confirmed. A discussion of the d approximants within phason strain field theory can be found in Duneau and Audier (1994) and a detailed description in terms of a pentagonal DPH tiling has been

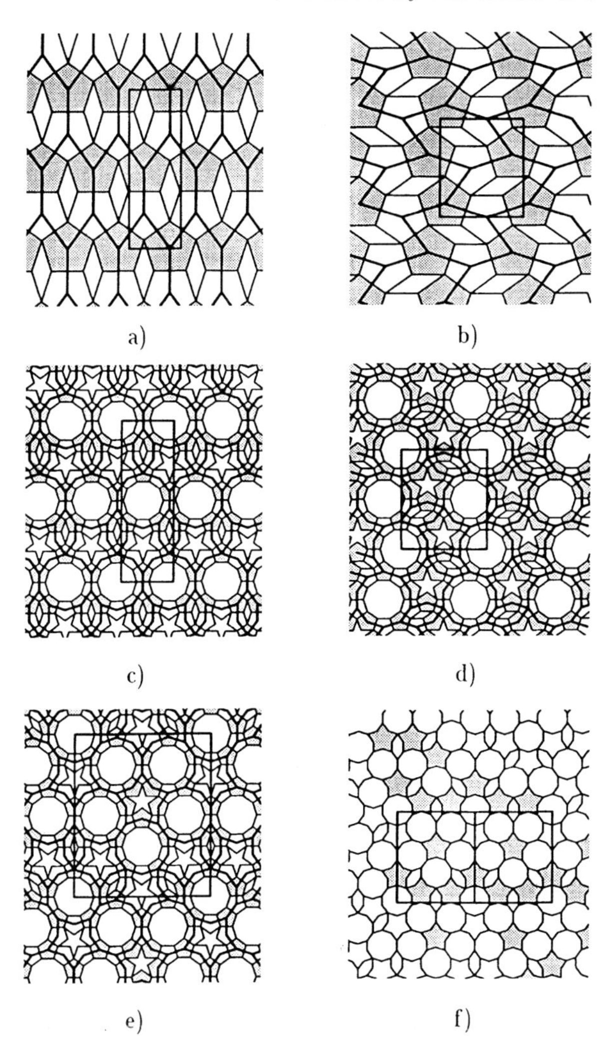

Fig. 7.5. Structural models for approximants to *d*-Al-Pd-Mn defined in terms of the DPH tiling: (a) R phase, (b) T phase described by different arrangements of H tiles, (c,d) $\tau^2 - R$ and $\tau^2 - T$ phases, (e) $\tau^3 - T$ phase described by D, P, and H tiles. (f) DPH tiling derived from high-resolution electron microscopy, showing two unit cells of the τ^3 approximant. The thin lines show the decomposition of the DPH tiling into pentagonal clusters. After Krajčí et al. (1997a).

given by Krajčí et al. (1997a). Figure 7.5 shows schematic representations of the tiling models to the d approximants.

On the Pd-rich side, the i phase is in equilibrium with the orthorhombic ξ phase $Al_{74}Pd_{21.6}Mn_{4.4}$, which is closely related (but not identical) to the cubic 2/1 approximant $Al_{70}Pd_{22}Mn_8$. The i phase also has a remarkable structural relationship to the crystalline Al_6Mn compound (Nicol 1953) which is, however, not an approximant that conforms with the definition given above (for details, see the review by Duneau and Audier 1994). The construction of rational approximants to *d*-Al-Cu-Co and their relation to crystalline phases has been discussed by Krajčí et al. (1997b).

7.4 Numerical Characterization of Elementary Excitations

7.4.1 Direct Diagonalization

The physically relevant information about phonons and electrons is contained in the eigenvalue spectrum and in the eigenvectors of the dynamical and Hamiltonian matrices, respectively. For periodic approximants Bloch's theorem is valid and the complete spectrum of elementary excitations can be obtained by taking the sum over the eigenstates with fixed wave number $\boldsymbol{k}$ within the BZ of the periodic approximant.

Given the atom coordinates from the model structure and the interatomic force-law, the dynamical matrix is readily constructed in the harmonic approximation, using standard techniques (Maradudin et al. 1971). For Al-based alloys (which constitute the majority of QC formers) interatomic forces are well described by effective pair-interactions. For *sp*-bonded alloys the calculation may be based on pseudopotential perturbation theory (Hafner 1987). For alloys containing transition-metals (TMs) the covalent *d*-*d* interactions and *sp*-*d* hybridization must be considered (see, e.g., Carlsson 1993 and reference given therein).

The description of electrons in condensed matter is generally based on the local-density approximation (LDA) (Jones and Gunnarsson 1989), with the Kohn-Sham Hamiltonian represented in an appropriate basis. Self-consistent potentials, charge-densities, and eigenstates are obtained by an iterative diagonalization of the LDA Hamiltonian. Given the large number of atoms in quasicrystalline approximants, the choice of a basis is of crucial importance. A very appropriate choice are linear-muffin-tin-orbitals (LMTO's) constructed in an atomic-sphere approximation (ASA), as proposed by Andersen (1975) and Andersen et al. (1987), leading to a minimal basis set of nine *s*,*p*,*d* orbitals per atom. However, even with this minimal basis set, to store the complex Hamiltonian matrix for a 2/1 approximant to *i*-Al-Pd-Mn requires $\sim$ 400 MB core memory (and a memory of equal size is needed to store the matrix of the eigenvectors) for each $\boldsymbol{k}$ point. For each successive level of approximants, the memory requirement scales with a factor $\tau^6 \approx 18$. The space group symmetry of the approximants (Table 7.1) allows one to block-diagonalize the Hamiltonian. At the Γ point the point-group symmetry is T and this allows to reduce the maximal size of eigenvalue problem by a factor of 4. Exploiting symmetry reduction, approximants up to 3/2 [$\widehat{=}$ 2292 atoms for FCI Al-Pd-Mn(Re)] have been treated via exact diagonalization by Krajčí and Hafner (1998b).

For such large approximants, charge self-consistency becomes a serious problem. In this context the self-similarity of quasiperiodic structures, together with the representation of the LMTO Hamiltonian in terms of structure constants and structure-independent potential parameters, are very helpful. It has been shown that a canonical transformation from the standard

LMTO basis to the most localized tight-binding-LMTO basis leads to a very effective screening of the structure constants, reducing their range essentially to second neighbours (Andersen et al. 1987, Bose et al. 1988, Nowak et al. 1991). Within such short distances, only a finite number of local environments exists in QCs, repeated in space by virtue of Conway's theorem. Hence, within the ASA only the self-consistent calculation of the charge density within each topologically inequivalent muffin-tin sphere is required. After each diagonalization of the Hamiltonian, the new charge density is calculated inside the spheres and new potential parameters are determined by the solution of the Schrödinger (or scalar relativistic Dirac) equation. The procedure is repeated until self-consistency is achieved. A good approximation to the self-consistent solution may be obtained by assigning to each topological inequivalent site in a high-order approximant the potential parameters calculated self-consistently for the sites of the same topological class in a low-order approximant (Hafner and Krajčí 1993b, Windisch et al. 1994b).

The exact diagonalization provides information not only on the eigenvalue spectrum, but also on the exact eigenstates. Hence, it allows the characterization of the localization of the eigenstates as well as the calculation of the transport properties via Boltzmann or Kubo-Greenwood theories (for details, see below). Using the local spin-density approximation of Hedin and Lundqvist (1971), the electronic structure calculations have been extended to treat magnetic properties by Hafner and Krajčí (1998). However, the calculation of the electronic properties via self-consistent iterative diagonalization is strictly limited to 3/2 and smaller approximants.

7.4.2 Real-Space Recursion

For very large ensembles, a very efficient approach to the spectral properties of elementary excitations involves the calculation of diagonal elements of the Green's function $G_0(z) = \langle\psi_0|\,(z-H)^{-1}\,|\psi_0\rangle$ (where H is the Hamiltonian or a dynamical matrix) using real-space recursion techniques (Haydock 1980). Starting from an arbitrary initial state ψ_0, a completely orthogonal basis in which the Hamiltonian has a tridiagonal representation is constructed via a two-step recursion relation. The diagonal element of the resolvent $(z-H)^{-1}$ can then be expressed in the form of a continued fraction. The formally infinite continued fraction must be truncated by introducing an appropriate terminator after a finite number L of recursion steps (Luchini and Nex 1987). The DOS projected on the initial state ψ_0 is proportional to the imaginary part of $G_0(z)$.

In contrast to the direct diagonalization, the recursion approach does not yield eigenvalues and eigenstates, but only spectral information of finite resolution. The resolution ΔE is determined by the number of exact eigenvalues ($\widehat{=}$ number of recursion steps) in relation to the band width W, $\Delta E \sim W/L$. To avoid finite size effects, the size of the model must increase proportionally to L.

The choice of the initial state determined the physical information that can be drawn from the projected DOS. (a) If ψ_0 is a local orbital on site i with the angular momentum l (or a local atomic displacement with a polarization vector $\boldsymbol{e}_i$), the resulting DOS is a local angular-momentum (polarization) decomposed DOS. (b) If $\psi_{0,i}$ is a plane-wave with wave number $\boldsymbol{k}$, $\psi_{0,i} \propto \mathrm{e}^{\mathrm{i}\boldsymbol{k}\boldsymbol{R}_i}$, a Bloch spectral function $f(\boldsymbol{k}, E)$ for electrons or phonons is obtained. For phonons one may in addition differentiate between propagating states with longitudinal or transverse polarization. (c) If the initial state is a linear combination of local orbitals (local atomic displacements) with random phases, an average over a number of such starting states converges towards the total DOS of the system (Hafner 1983). (d) Alternatively, the total DOS may be calculated by averaging the local partial DOS over sites and orbitals (polarizations) or by integrating the spectral functions over a large volume in $\boldsymbol{k}$-space.

Using recursion techniques, approximants with up to $\sim 50\,000$ atoms have been investigated. All technical details are described in Hafner (1983), Hafner and Krajčí (1993a–1993c), Windisch et al. (1994a, 1994b), and Krajčí and Hafner (1998a).

7.4.3 Comparison With Experiment

The calculations provide very detailed information on eigenvalues, eigenstates, and spectral properties of elementary excitations. Experimentally the total and partial DOS are accessible via a variety of spectroscopic techniques (photoelectron and soft x-ray spectroscopy for electrons, inelastic incoherent neutron- or x-ray scattering for phonons). Electronic DOS at E_F may be estimated from specific heat, NMR Knight-shift, and other measurements. For crystals, angle-resolved photoelectron spectroscopy allows for the determination of the dispersion relations of electronic eigenstates and inelastic coherent neutron scattering allows a measurement of phonon dispersion relations. Inelastic neutron scattering has also successfully been applied to QCs by Goldman et al. (1992) and de Boissieu et al. (1993b), yielding important information on the character of eigenstates. The angle-resolved photoelectron spectroscopy has been attempted only very recently by Wu et al. (1995). At least for a selected valence-band state in i-Al-Pd-Mn, a quasiperiodic dispersion could be confirmed.

7.5 Phonons in Quasicrystals

7.5.1 Interactomic Force Law and Quasiperiodicity – Modulated Quasicrystals

Before we use the harmonic approximation to construct the dynamical matrix of a solid, we must make sure that its structure is in equilibrium under the action of the interatomic forces (for the harmonic theory of lattice dynamics, see, e.g., Maradudin et al. 1971). Hence the problem of the existence of propagating collective excitations (phonons) in QCs is intimately related to the question of the mechanism determining quasicrystalline stability. In metallic systems with dominant *sp* bonding the form of interatomic potentials is determined by the interplay of a short-range screened Coulomb interaction and a long-range oscillatory potential whose wavelength is given by $\lambda_F = 2\pi/2k_F$, where k_F is the Fermi momentum (for a detailed discussion, see Hafner and Heine 1983,1986). A given crystal structure is stable if the periodicity of the lattice matches the Friedel wavelength λ_F – this is precisely a reformulation of the Hume-Rothery criterion. Hafner and Heine (1983) have demonstrated that the trends in the crystal structures of the elements may be explained in terms of the variation of the interatomic potential with valence atomic volume and electron-ion interaction. Hafner and Kahl (1984) extended the argument to the liquid structures. In this parameter space, there is a net separation between close-packed structures in a region where short-range packing and long-range electronic bonding are "in phase", and open structures in a region where electronic effects dominate. Smith (1991) was the first to extend these arguments to QCs. He demonstrated that in a narrow region where short- and long-range effects compete, a quasicrystalline structure may be lower in energy than any crystalline lattice. Recently Denton and Hafner (1997) have demonstrated, by using classical density functional techniques, that in the region where the crystal structures lose mechanical stability, QCs become thermodynamically stable.

Whereas in crystals there exists only a small number of topologically inequivalent sites, in QCs all sites are topologically inequivalent if we consider the infinite environment. Even if we limit ourselves to distances corresponding to the range of the interatomic forces (which is large in metallic systems), the number of inequivalent sites will be very large. Hence, there is an unavoidable conflict between the location of the atoms in an ideal quasicrystalline lattice (which is determined solely by the geometrical nature of its basic building blocks) and the forces acting on the atoms which depend on the global environment, and hence are different in different local units. This results in static displacements of the atoms, which, however, do not break the symmetry of the QC (Hafner and Krajčí 1992, Windisch et al. 1994a). The result is what Janssen (1988b) has called a displacively modulated QC. These displacive modulations are an intrinsic property of quasicrystalline structures. The extent of the modulation will depend on the relation between the characteristic diameter of the quasiperiodically arranged tiles and the range of the effective interatomic potentials (Krajčí and Hafner 1992, Windisch et al. 1994a, Hafner et al. 1995, Krajčí et al. 1995a).

7.5.2 Phonons in Icosahedral Quasicrystals

7.5.2.1 Phonon Dispersion Relations. A Bloch spectral function $f_{\boldsymbol{e}}(\boldsymbol{k},\omega)$ for propagating collective excitations with a wave vector $\boldsymbol{k}$ and polarization $\boldsymbol{e}$ is given by the projection of the atomic displacements (described by the eigenvectors $\boldsymbol{u}_\nu(i)$ of an eigenmode with frequency ω_ν) onto a plane wave $\boldsymbol{e}\cdot\mathrm{e}^{\mathrm{i}\boldsymbol{k}\boldsymbol{R}}$

$$f_{\boldsymbol{e}}\,(\boldsymbol{k},\omega) = 2\omega\sum_{\nu}\sum_{i,j}\,\boldsymbol{e}\cdot\boldsymbol{u}_\nu(i)\,\mathrm{e}^{-\mathrm{i}\boldsymbol{k}\boldsymbol{R}_i}\,\boldsymbol{e}\cdot\boldsymbol{u}_\nu(j)\,\mathrm{e}^{\mathrm{i}\boldsymbol{k}\boldsymbol{R}_j}\,\delta(\omega^2-\omega_\nu^2)\,. \quad (7.5)$$

In crystals, $f_{\boldsymbol{e}}(\boldsymbol{k},\omega)$ is given by a series of delta functions whose location in $(\boldsymbol{k},\omega)$ space defines the dispersion relations of phonons. In QCs, the wave vector is not a conserved quantity and hence the Bloch spectral functions are continuous functions of $\boldsymbol{k}$ and ω. The FWHM at fixed $\boldsymbol{k}$ is inversely proportional to the lifetime of a "phonon" in a QC.

Figure 7.6a shows the spectral function $f_{T2}^{NN}\,(\boldsymbol{k},\omega)$ for transverse density fluctuations in *i*-Al-Cu-Li propagating along a twofold symmetry direction, and the dispersion relations for transverse phonons defined in terms of the maxima of the spectral functions are shown in Fig. 7.6b. The calculation has been performed for a 5/3 approximant (12 244 atoms) using a real-space recursion by Windisch et al. (1994a) and Hafner et al. (1995a, 1995b). There are several characteristic features. (a) A series of sharp peaks with linear dispersion relations issuing from the most intense Bragg peaks (Γ points) on the twofold axis. These peaks represent long-wavelength acoustic modes with frequencies $\omega = c_T|\boldsymbol{k}|$, where c_T is the transverse sound velocity. (b) Close to the $\boldsymbol{k}$ points corresponding to quasi-BZ boundaries (X_2 and M_2 points for the given symmetry direction, cf. Sect. 7.2) the dispersion relations become stationary. (c) At higher frequencies one finds a series of broad peaks, again with very little dispersion. (d) The stationary peaks also produce peaks in the vibrational DOS, similar to the Van Hove singularities in a crystal. Using direct diagonalization, very similar results have been obtained by Los and Janssen (1990) and Los et al. (1993) on simple 3D Penrose lattices with short-range interactions.

The picture of propagating collective excitations in a QC corresponds to a straightforward generalization of an extended-BZ representation of phonons in a crystal, with the exception that the periodically repeated BZ is replaced by a hierarchy of dense, quasiperiodically spaced BZ-centers and boundaries (here only the most intense features are shown).

Experimentally, one measures a dynamical structure factor $S(\boldsymbol{k},\omega)$. The dynamical structure factor may be expressed in terms of the vibrational spectral functions and Debye-Waller factors, appropriately weighted with the concentrations and the coherent neutron-scattering lengths of the different atomic species, as described by Hafner (1983) and Krajčí and Hafner (1998a). At this level a comparison between theory and experiment becomes possible. Figure 7.7 shows a comparison of the low-energy part of the disper-

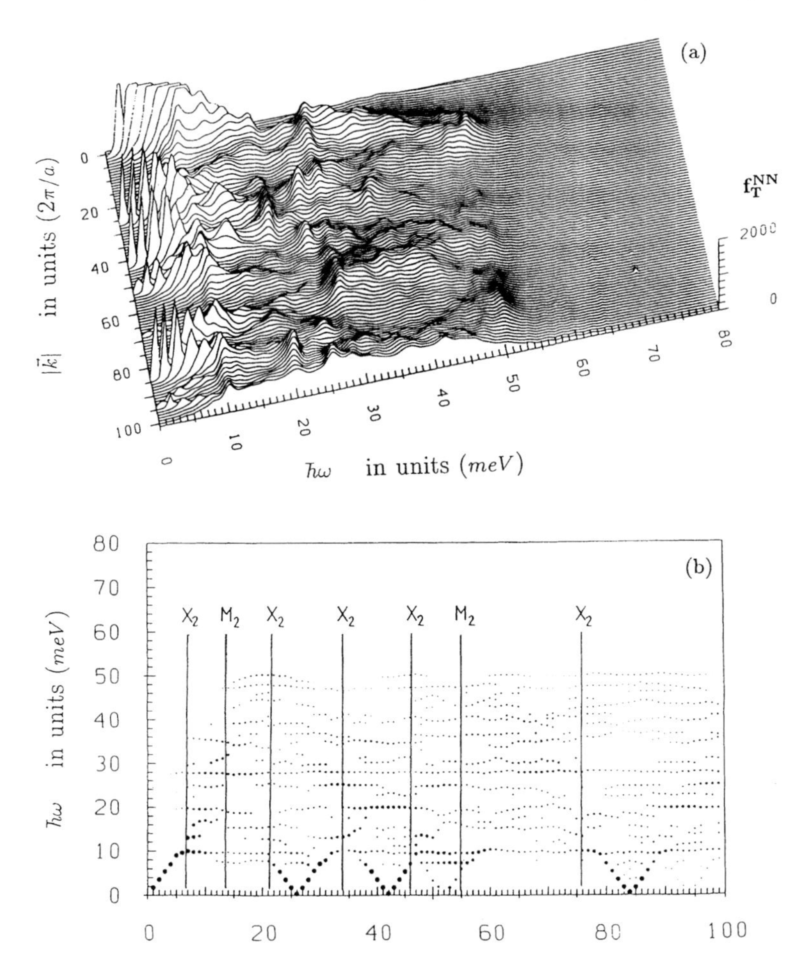

Fig. 7.6. (a) Vibrational spectral function f_{T2}^{NN} $(\boldsymbol{k}, \omega)$ for transverse density fluctuations in SI Al-Cu-Li propagating along a twofold symmetry direction. (b) Dispersion relations defined in terms of the positions of the peaks in the spectral functions. The positions of the most intense "quasi-BZ boundaries" X_2 and M_2 are marked. After Windisch et al. (1994a) and Hafner et al. (1995b).

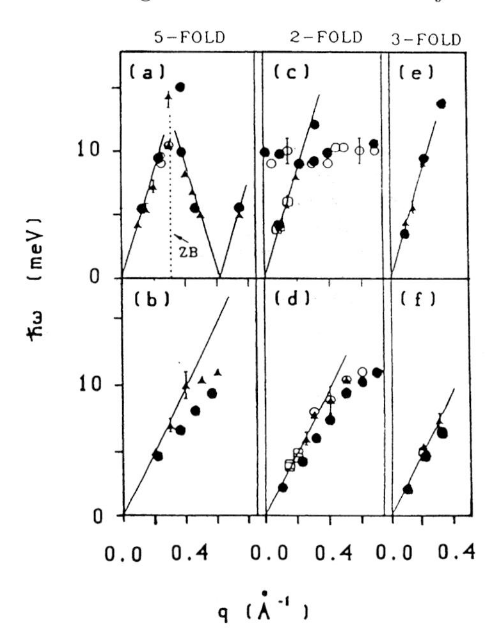

Fig. 7.7. Comparison of measured (open symbols) and calculated (full dots) dispersion relations for longitudinal (top row) and transverse (bottom row) phonons propagating along the two-, three- and fivefold symmetry directions in SI Al-Cu-Li. $\boldsymbol{q} = |\boldsymbol{k} - \boldsymbol{G}|$ measures the distances from the closest intense Γ-point. The straight lines mark the slope of the long-wavelength dispersion as determined by the experimental velocity of sound. After Windisch et al. (1994a).

sion relations calculated for SI Al-Cu-Li and the neutron-scattering data of Goldman et al. (1992), demonstrating the excellent agreement between theory and experiment. In particular, the existence of low-energy stationary modes is confirmed. Stationary modes at a higher energy ("optic" modes) have been confirmed by recent work on FCI Al-Pd-Mn by de Boissieu et al. (1993b).

7.5.2.2 Localized or Confined Modes. The existence of a large number of dispersionless modes raises the question of the character of the stationary eigenstates. If we diagonalize the dynamical matrix, each eigenstates ν can be characterized by the participation ratio $p(\omega_\nu)$

$$p(\omega_\nu) = \left(\sum_j \frac{|\boldsymbol{e}_\nu(\boldsymbol{R}_j|^2}{m_j} \right)^2 \left(N \sum_j \frac{|\boldsymbol{e}_\nu(\boldsymbol{R}_j|^4}{m_j} \right)^{-1} , \tag{7.6}$$

where $\boldsymbol{e}_\nu$ is the eigenvector of mode ν at the site $\boldsymbol{R}_j$ and m_j is the mass of the atom occupying that site. For extended states we have $p \sim 1$, whereas for strictly localized states $p \sim \mathcal{O}(N^{-1})$.

Figure 7.8 shows the total and partial vibrationalDOS DOS and the participation ratio of vibrational eigenmodes of a 3/2-approximant to i-Al-Zn-Mg (2920 atoms), calculated by direct diagonalization of the symmetry-reduced dynamical matrix. The symmetry is Im3 for the 1/1 approximant (Frank-Kasper phase), Pa$\bar{3}$ for the 2/1, 3/2, 8/5 approximants, and I$2_1$3 for the 5/3 approximant (Krajčí and Hafner 1992). The inset shows the participation

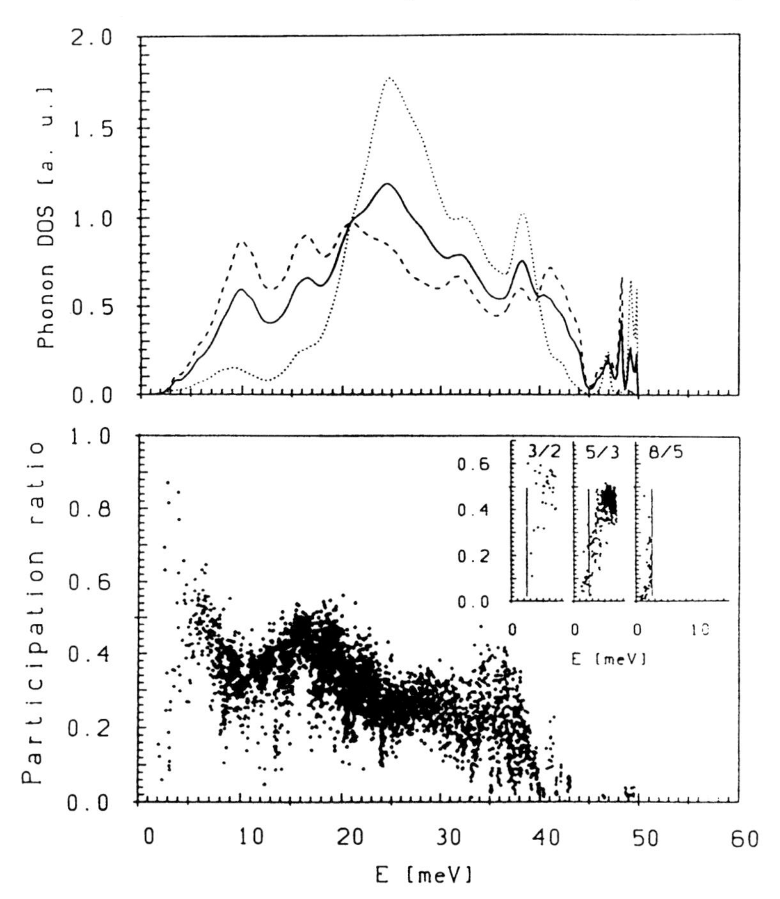

Fig. 7.8. Vibrational DOS (full lines – total DOS, broken lines – Al(Zn), dotted lines – Mg partial DOS) and participation ratio of eigenmodes in *i*-Al-Zn-Mg (3/2 approximant). The inset shows the participation ratio of low-frequency modes with F_{1u} symmetry for 3/2, 5/3, and 8/5 approximants. After Krajčí and Hafner (1995).

ratio of the low energy modes of symmetry F_{1u} for the 3/2, 5/3, and 8/5 approximants (up to 52 440 atoms), which was calculated (Krajčí and Hafner 1995) by using an iterative diagonalization approach.

The result is quite different from that found on simple 3D Penrose lattices with vertex decoration and short-range interaction (see, e.g., Rieth and Schreiber 1998). The important results are as follows. (a) The participation ratios are much more widely scattered. (b) At the edges of the spectrum and in several narrow frequency intervals (close to $\hbar\omega \sim 9, 12 - 16, 24$ meV) low participation ratios are calculated for certain groups of modes. (c) These groups of modes are characterized by a distinct topology of the sites supporting these modes. Low-energy modes are strictly confined to the 13-fold coordinated sites (Hafner and Krajčí 1993a, 1993c), modes in the range 9

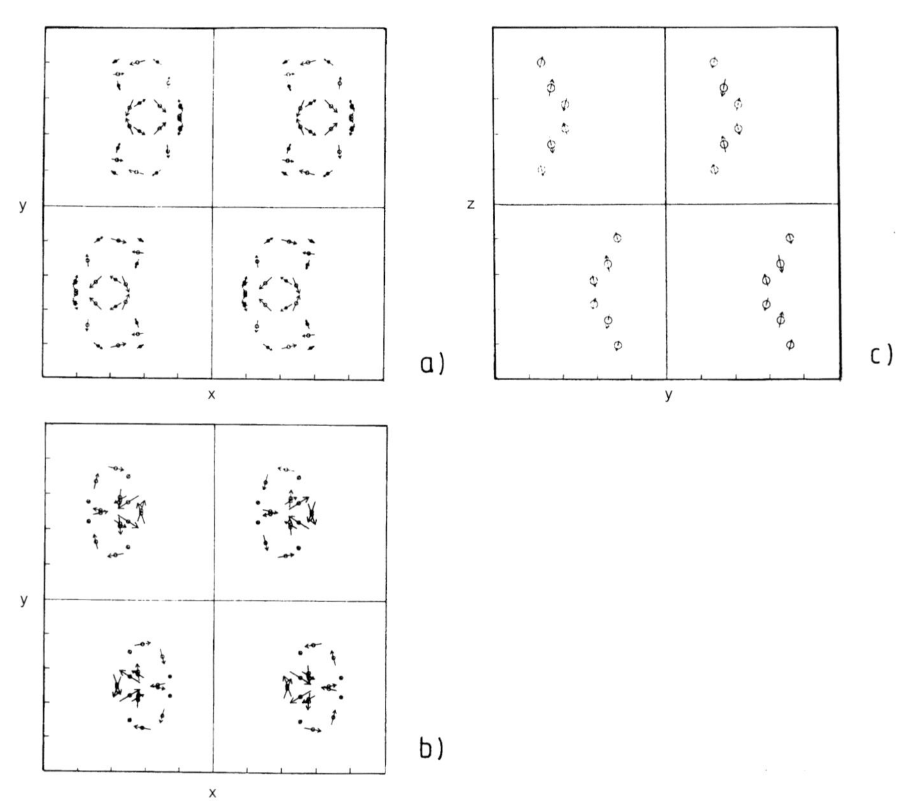

Fig. 7.9. Projection of the displacement pattern of three types of confined phonon eigenstates into a plane. (a) Low-energy mode located at the 13-fold coordinated Al(Zn) sites ($\hbar\omega_\nu = 3.18\,\text{meV}$, $p_\nu = 0.09$). (b) Eigenstate at $\hbar\omega_\nu = 12.44$ meV, $p_\nu = 0.05$ localized at 12-fold coordinated Al(Zn) sites. (c) Localized mode centred at some of the 16-fold coordinated Mg sites ($\hbar\omega_\nu = 24.13$ meV, $p_\nu = 0.09$). The arrows represent projections of the eigenvectors.

meV $< \hbar\omega < 16\,\hbar\omega$ - to the 12-fold coordinated sites (all occupied by Al or Zn), and the modes in the narrow range 23.5 meV $< \hbar\omega < 24.5\,\hbar\omega$ - to the 16-fold coordinated sites occupied by Mg (Fig. 7.9). (d) The frequencies at which the confined modes are detected coincide with the frequencies of the stationary modes at the quasi-BZ boundaries (but not all stationary modes are confined). (e) Although it is very difficult to extend these studies to high-order approximants, calculations of the eigenstates with the lowest energies based on an iterative diagonalization of the smaller blocks of the dynamical matrix using an iterative approach [for a conjugate-gradient-based matrix-diagonalization, see, e.g., Kresse and Hafner (1994)] demonstrate that the lowest value of the participation ratio scales inversely proportional to τ^3, as expected. However, the results are not sufficient to deduce a scaling exponent for the confined states.

These results are remarkable in several respects. (i) Confined eigenstates are more frequent in realistic *i* QCs than in idealized models such as the 3D Penrose vertex model of Rieth and Schreiber (1995a, 1998). (ii) Although the structure of *i*-Al-Mg-Zn is based on $Al(Zn)_{24}Mg_{20}$ Bergman clusters forming the nuclei of larger *i* units (see Sect. 7.3.1), the support of the confined mode is in no way related to these clusters. The atoms participating in the vibrations of a confined mode are not localized on one of these *i* units. The atoms with the largest displacement amplitudes are rather distributed over several *i* clusters and all these sites share a certain topological characteristic. A detailed analysis by Hafner and Krajčí (1993c) has demonstrated that the confined modes are best characterized in terms of disclination lines characterizing local deviations from *i* symmetry (Nelson 1983, Nelson and Spaepen 1989). The 13-fold coordinated sites, for example, are links of positive and negative declination lines, wherease the 16-fold sites are nodes of four negative desclination lines. This result contradicts claims that the spectrum of elementary excitations in *i* QCs contains eigenstates localized on *i* clusters and is largely dominated by the self-similar scaling transformations of the *i* units (Janot and de Boissieu 1994).

7.5.3 Phonons in Decagonal Quasicrystals

In *d* QCs quasiperiodicity is restricted to the plane perpendicular to the tenfold axis. As a consequence of the reduced dimensionality of the quasiperiodicity, one would expect that its influence on the vibrational spectrum to be more pronounced than for *i* case. This is indeed the case, as confirmed by the vibrational DOS and the distribution of the participation ratio of the eigenstates of *d*-Al-Mn shown in Fig. 7.10 (Hafner et al. 1995a–1995c, 1996). The calculations are based on a variant of the Burkov model by Henley (1993) and pair-forces constructed using the method of Phillips et al. (1994) (see also Mihalkovič et al. 1995). We find that the vibrational spectrum of the *d* phase differs from that of *i*-Al-Mn alloys of similar composition by (i) eigenvalues that are as high as $\hbar\omega \sim 70$ meV, compared with the highest eigenvalues of $\hbar\omega \sim 50$ meV in the *i* phase, (ii) many sharp peaks in the DOS, in contrast to a smooth DOS with a single broad and flat peak for the *i* alloy reported by Poussigue et al. (1994), (iii) lower average values of the participation ratio and very low values of p_ν at both edges of the spectrum and at several intermediate frequencies coincident with some of the most pronounced features of the DOS ($p_\nu \leq 0.01$ to 0.04, indicating that the atomic displacements are confined to groups of 10 to 40 atoms).

These features are intimately connected to the *d* symmetry. Modes propagating along the periodic axis show a behavior analogous to that known for complex crystalline structures (Figs. 7.11a and 7.11b). Longitudinal and transverse modes propagating in the quasiperiodic plane and polarized parallel to that plane are characterized by acoustic dispersion relations centred at the most intense Γ points and stationary modes around the most intense

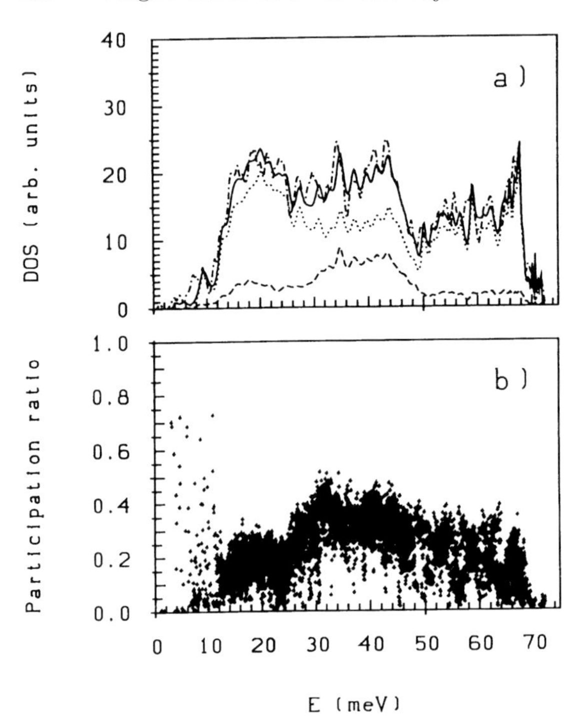

Fig. 7.10. Vibrational density of states (a) and participation ratio of the phonon eigenstates (b) in *d*-Al-Mn. Full line – total DOS; dotted line – partial Al DOS; dashed line – partial Mn DOS calculated via real-space recursion; dash-dotted line – total DOS calculated via exact diagonalization. After Hafner et al. (1996).

quasi-BZ boundaries (Figs. 7.11c and 7.11d). Transverse modes with wave vectors in the quasiperiodic plane but polarized perpendicular to it are stationary almost everywhere in $\boldsymbol{k}$-space, except in the immediate environment of the most intense Γ points (Figs. 7.11e and 7.11f). The frequencies of the stationary modes coincide with the most pronounced peaks in the DOS and with the eigenvalues of the states with the lowest participation ratio, as shown by Hafner et al. (1996).

One also notes a coincidence of stationary transverse modes with polarization parallel to the periodic axis and longitudinal modes propagating along this axis. This indicates a strong interaction between these two types of eigenstates. This coincidence has been confirmed by the investigation of the spatial distribution of the amplitude of these states.

A further characteristic feature is the low-energy peak in the spectrum at about 6 meV. These modes exist also in approximant phases, as shown for the Al-Mn T phase by Krajčí and Hafner (1997b). There is evidence that these modes are localized in the less dense region of the d structure. Note that the

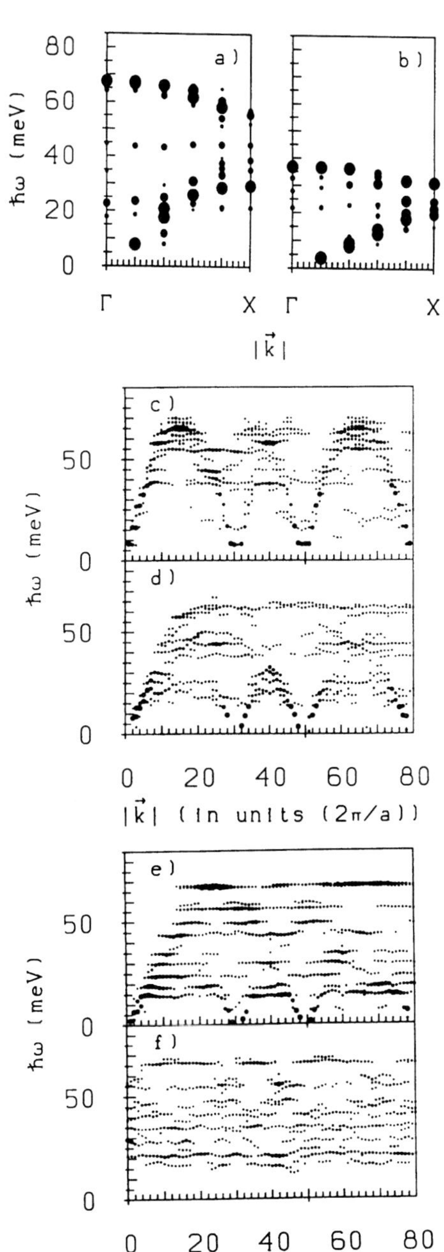

Fig. 7.11. Dispersion relations for propagating collective excitations in *d*-Al-Mn. (a,b) Longitudinal and transverse modes propagating along the tenfold (periodic axis). (c,d) Longitudinal and transverse modes propagating along a twofold axis in the quasiperiodic plane. The transverse modes (d) have their polarization vector in this plane. (e,f) Transverse modes with $\boldsymbol{k} \parallel$ twofold axis and $\boldsymbol{e} \parallel$ tenfold axis, with $k_z = 0$ (e) and $k_z = \pi/c$ (f), respectively. After Hafner et al. (1996).

energy of these modes lies considerably below that of the lowest stationary modes in *i* phases (Hafner and Krajčí 1993b, Windisch et al. 1994a).

7.5.4 Phonons – Summary

The results discussed in the preceding sections have demonstrated that the vibrational eigenstates of both *i* and *d* QCs are largely determined by the hierarchical BZ structure of quasiperiodic lattices. For *i* QCs both the existence of propagating acoustic modes located near the most intense Bragg reflections and of stationary "optic" modes near the quasi-BZ boundaries has been confirmed by experiment. In addition to the results on SI Al-Cu-Li, we refer, in particular, to the extensive experimental work on FCI Al-Pd-Mn (Goldman et al. 1992, de Boissieu et al. 1993a, 1993b, 1995). For *d* QCs such a confirmation is still lacking. Part of the stationary modes is localized (or rather confined) to a rather small number of atomic sites, but these sites are not concentrated in compact clusters. Rather, the sites supporting a confined mode are loosely distributed in space and their identifying feature is their local topology.

So far, little is known about the influence of the low-lying optical modes on the physical properties of QCs. A point that certainly deserves further attention is the lattice contribution to the thermal conductivity λ_{ph}. Because of the characteristic variation with temperature (T^2 dependence at very low T, plateau at intermediate temperatures), λ_{ph} of QCs is often characterized as "glass-like"(Legault et al. 1995, Chernikov et al. 1995). The properties of phonons characterized above (long-wavelength acoustic modes, a broad distribution of stationary low- and high-energy modes) evidently fit quite well to this "glass-like" scenario. A further confirmation of the existence of low-lying stationary modes is provided by a recent careful analysis of deviations of the lattice specific heat from the T^3 law (Lasjaunias et al. 1997).

For incommensurately modulated materials, one finds in the harmonic approximation also atomic oscillations of a character different from those in periodic crystals, i.e., phasons. Phasons correspond to motions of the atomic surfaces. However, this is very different in QCs. Because atomic surfaces are bounded, their motion leads to jumps of atoms between different positions. These phason jumps cannot be considered as harmonic since the displaced atom has to overcome a potential barrier. However, phason jumps may play an important role in diffusion dynamics, as argued by Bellissent et al. (1994).

7.6 Electrons in Quasicrystals

From the brief overview given in the Introduction it should be clear that many of the outstanding physical properties of QCs are intimately related to their electronic spectrum. The preceding discussion of results obtained for model Hamiltonians and for phonons have demonstrated a high sensitivity of the spectrum to details of the quasiperiodic structure and to the coupling parameters. Hence, any discussion of the electronic spectrum of real QCs must be based on very accurate self-consistent calculations performed on a series of approximants.

We expect that the electronic structure of QCs will be characterized by a hierarchical structure of larger and smaller pseudogaps which, depending on the point of view adopted, arise either from the hierarchy of quasi-BZs or from the hierarchical, self-similar cluster structure. The wider pseudogaps in the electronic spectrum (on a scale of 0.5 to 1 eV) are expected to dominate a possible Hume-Rothery-like electronic mechanism for the stabilization of QCs, whereas the smaller pseudogaps (on a scale of a few hundredths of an eV) are generally thought to be related to the electronic transport properties of QCs and to the extreme composition dependence of their physical properties.

7.6.1 s, p-Bonded Icosahedral Alloys as Hume-Rothery Phases

The s, p-bonded simple i alloys of the Al(Ga)-Zn(Cu)-Mg(Li) class can be considered as the ideal testing ground for Hume-Rothery-like mechanism for the stabilization of quasicrystalline structures. Bancel and Heiney (1986), Smith and Ashcroft (1987), and Friedel and Denoyer (1987) have argued on the basis of electron-counting arguments that the electronic structure of QCs should by characterized by a structure-induced DOS minimum at E_{F}.

Fujiwara and Yokokawa (1991), Hafner and Krajčí (1992, 1993), and Windisch et al. (1994a, 1994b) have investigated a series of quite large approximants based on decorated 3D Penrose and canonical-cell tilings with the Henley-Elser decoration. The results may be summarized as follows. (a) The calculation of the Bloch spectral functions $f(\boldsymbol{k}, E)$ demonstrates the existence of well-defined nearly-free-electron like bands around the most intense Γ points, leading to parabolic dispersion relations and a free-electron-like DOS near the bottom of the band (Fig. 7.12). (b) Close to the quasi-BZ boundaries the intersection of the free-electron parabolas issuing from the Γ points leads to the formation of a quasi-periodically distributed series of degenerate states close to E_{F}. Lifting the degeneracy due to the electron-lattice interaction produces the Hume-Rothery minimum in the electronic DOS. Note that the existence of the "pseudogap" is attributed not to a single dominant BZ (in straightforward, but incorrect, generalization of an argumentation appropriate to crystals only), but to the most intense quasi-BZ boundaries (see Sect. 7.2 and Fig. 7.1) distributed all over $\boldsymbol{k}$ space. (c) The depth of the pseudogap varies only insignificantly in the hierarchy of the approximants (Fig. 7.13), in agreement with similar conclusions drawn from NMR data by Hippert et al. (1992) and electronic specific heat measurements summarized by Mizutani (1994). The calculated electronic spectrum is also well confirmed by photoemission and soft x-ray spectroscopies [see Windisch et al. (1994b)

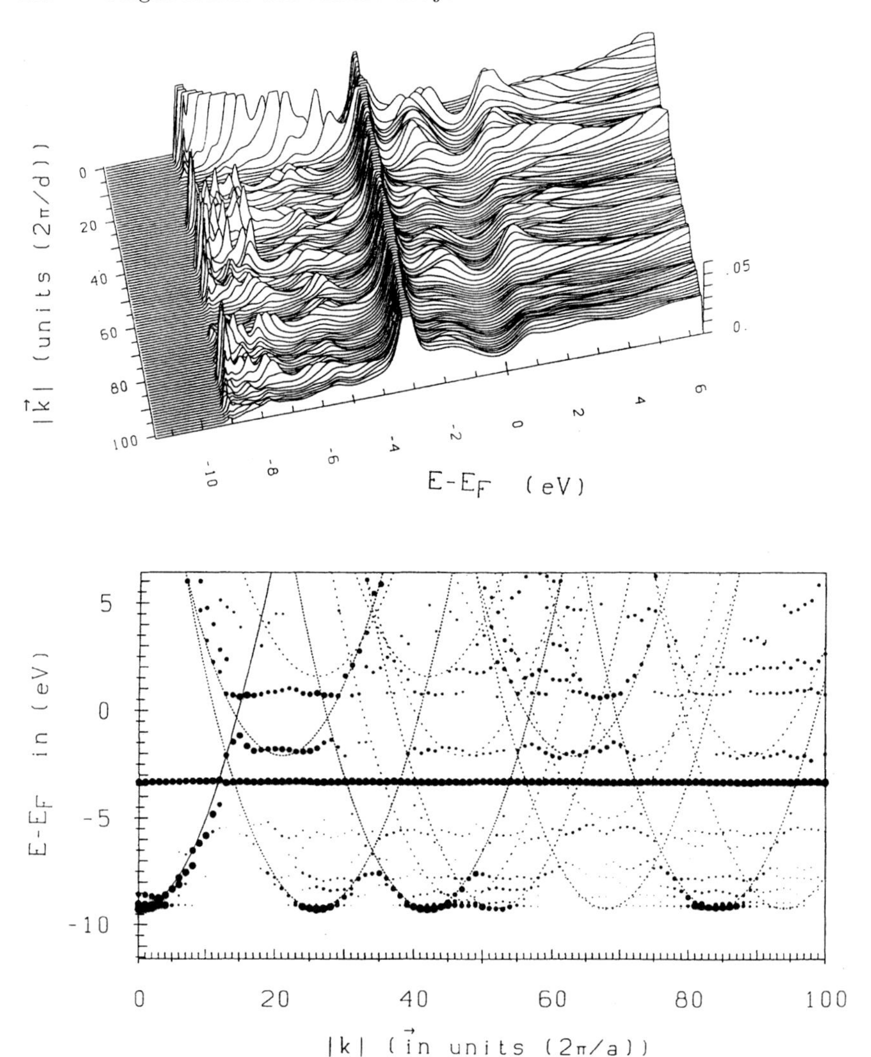

Fig. 7.12. Top: Bloch spectral function $f(\boldsymbol{k}, E)$ for electrons in the 5/3 approximant to *i*-Al-Cu-Li for the $\boldsymbol{k}$ vector oriented along a twofold symmetry direction. Bottom: Dispersion relations for electrons derived from the peak positions in the spectral function. The broken lines show the free-electron parabolas originating from the most intense Γ points on and in the vicinity of the twofold axis. After Windisch et al. (1994b).

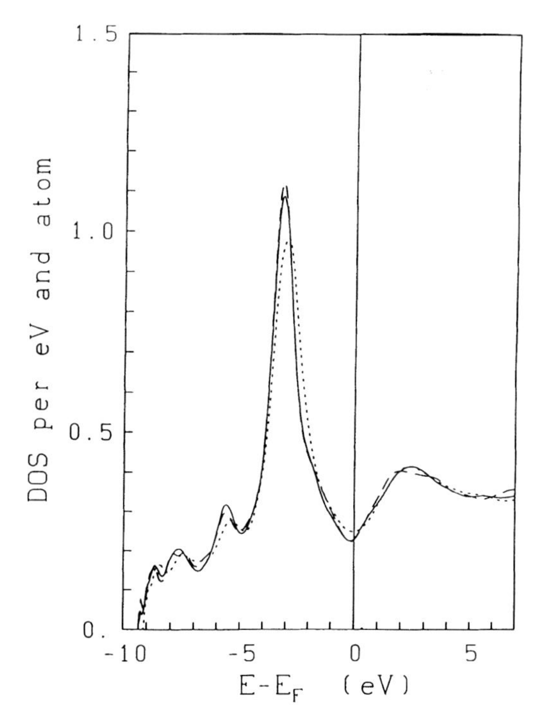

Fig. 7.13. Electronic density of states in a series of approximants to *i*-Al-Cu-Li: dotted line – 3/2, dashed line – 5/3, full line – 8/5 approximant (51752 atoms) based on a CCT tiling. After Windisch et al. (1994b).

for a detailed confrontation of theory and experiment]. (d) Although the existence of a pseudogap is a generic property of this class of QCs, it is not a specific property. Pseudogaps of various intensities exist also in crystalline, amorphous, and even liquid alloys of similar composition, as demonstrated by Hafner and Krajčí (1993b).

For Al-Mg-Zn and Ga-Zn-Mg QCs the almost equal stability of crystalline and quasicrystalline phases is also reflected in the coexistence of a whole series of cubic and even orthorhombic approximants with the quasicrystalline phase (Edagawa et al. 1991, Mukhopadhyay et al. 1991, Edagawa et al. 1992, Takeuchi and Mizutani 1995). A detailed analysis of this class of QCs leads to the conclusion that the main stabilizing mechanism is based, as proposed for the crystalline phases (the lowest-order approximants) decades ago by Frank and Kasper (1958), on the close-packing of spheres of different size. The electronic Hume-Rothery mechanism plays an important role at compositions where the close-packing requirements can no longer be perfectly satisfied but where the criteria for a Hume-Rothery mechanism are met. Very recently Dall'Acqua et al. (1997) have extended the investigations of the electronic structure to the FCI Al-Mg-Li alloys. It has been shown that the chemical order leading to the FCI superstructure (cf. Sect. 7.3.1) induces structures in the DOS at higher binding energy whose origin can be described by a similar mechanism.

7.6.2 Icosahedral and Decagonal Aluminum-Transition Metal Alloys

For quasiperiodic alloys formed by Al and one or two TMs, the situation is much more complex. The TM-d band overlaps with E_F and will eventually cover a structure-induced minimum in the Al-s, p band. On the other hand, the hybridization of sp and d states can induce a "hybridization pseudogap" at the upper edge of the d band. Hence the interplay between structure-induced and hybridization-induced band-structure effects can reduce or reinforce an electronic stabilization mechanism. The role of hybridization effects in crystalline Al-TM Hume-Rothery alloys has been extensively discussed by Trambly de Laissardière et al. (1995). It was demonstrated that in certain alloys, such as Al_6Mn, $Al_{12}Mn$, Al_5Co_2, and Al_3V, sp-d hybridization strongly increases the depth and width of the pseudogap. According to Nguyen-Manh et al. (1992), Al_2Ru and Ga_2Ru compounds with the Al_2Cu structure even form a real gap and show semiconducting behavior. The question arises whether a constructive interference of both effects also contributes to the stabilization of quasiperiodic phases.

7.6.2.1 Icosahedral Al-Pd-Mn Alloys. For the thermodynamically stable FCI $Al_{0.808}Pd_{0.206}Mn_{0.086}$ alloy a detailed investigation of the electronic structure of the 1/1 to 8/5 approximants (see Sect. 7.3 for the characterization of the structure model) has demonstrated (Krajčí et al. 1995b) that a structure-induced DOS minimum exists in the metastable 2/1 phase [identified by Waseda et al. (1992)] slightly above E_F, but exactly at E_F for all higher order approximants. Such a minimum does not occur in the 1/1 approximant which is destabilized by conflicting requirements from local topology and local chemical order (Fig. 7.14). For the higher-order approximants the calculated electronic structure is confirmed in its details by the soft x-ray absorption and emission spectra of Belin et al. (1994) and the resonant photoelectron spectroscopy of Zhang et al. (1994). The important point is that although the structure-induced minimum in the Al-sp band (the mechanism for a gap formation is entirely analogous to that discussed in the previous section) is even reinforced by sp-d hybridization, it is not the decisive factor for the stabilization of the i phase. The following points should be stressed. (a) The pseudogap at E_F is much more pronounced in the crystalline Al_6Mn compound, which is not an approximant phase. (b) In addition to sp-d hybridization, one has to consider the effect of the d-d hybridization arising from the formation of covalent bonds between the two TM constituents. Due to the d-d hybridization the center of gravity of the d band of the TM with a higher band-filling (Pd) is shifted to higher binding energies (3.8 ± 0.3 eV)

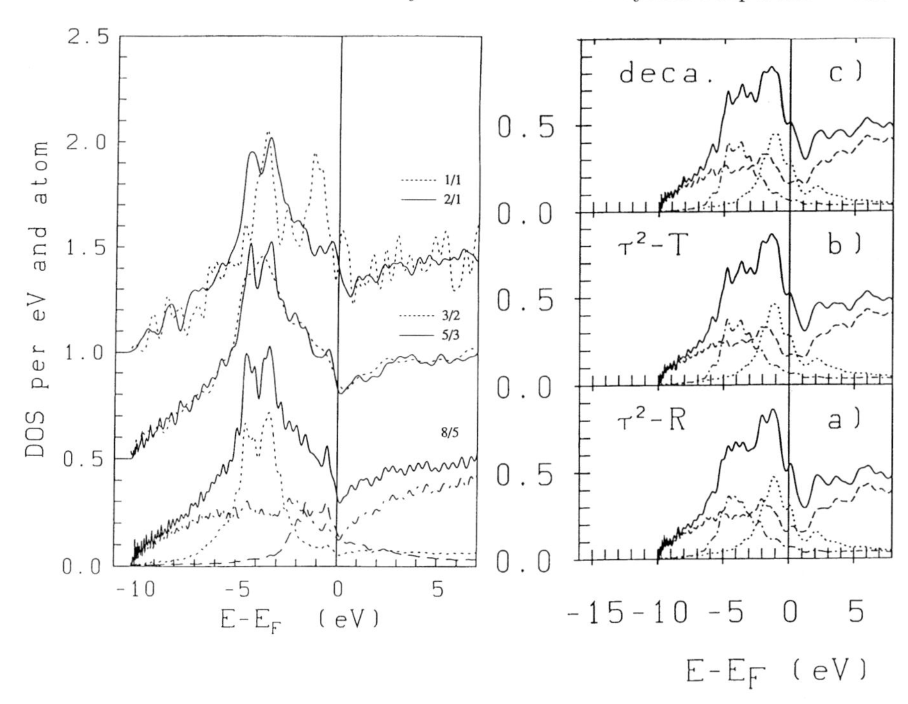

Fig. 7.14. Total and partial electronic DOS in approximants to *i*- (left panel, for the 8/5 approximant the partial Al-, Pd-, and Mn-DOS are shown by the dot-dashed dotted, and dashed lines, respectively) and *d*-Al-Pd-Mn alloys (right panel, partial Al-, Pd-, and Mn-DOS are represented by dashed, dot-dashed, and dotted lines, respectively).

compared to its position in the pure metal (~ 0.9 eV). The shift of the *d*-band lowers the total energy and is also instrumental in reducing the overlap of the *d* band with E_{F} so that the pseudogap is exposed. Again it is important to note that the *d*-band shift is not specific to quasicrystalline alloys, but exists in crystalline and amorphous alloys as well (Jank et al. 1991, Hausleitner et al. 1992). The displacement of the band scales with the heat of formation, i.e., with the strength of the chemical bonds between unlike TM atoms, as pointed out by Oelhafen (1983).

A detailed investigation has been devoted to the relation between the atomic-shell structure of the *i* phase and its electronic structure (Krajčí et al. 1995b). The result is that very large clusters (≥ 200 atoms) have to be used to achieve a reasonable approximation to the quasicrystalline DOS at the central site and that there is nothing in the electronic structure that would indicate a "confinement" of electronic eigenstates to clusters as proposed by Janot and de Boissieu (1994) and Trambly de Laissardière et al. (1997). This conclusion is also corroborated by an analysis of the electronic eigenstates determined by direct diagonalization of the Hamiltonian (cf. below).

For the lowest-order approximants, very similar results have been obtained by Fujiwara et al. (1989,1993) and Trambly de Laissardière and Fujiwara(1994a) for FCI Al-Cu-Fe and SI Al-Mn, confirming the interplay of a structure-induced gap and the hybridization effects.

7.6.2.2 Decagonal Al-Pd-Mn and Al-Cu-Co Alloys. The electronic DOS of a series of approximants to the d-$Al_{0.70}Pd_{0.13}Mn_{0.17}$ phase constructed on the basis of the DPH tiling of the crystalline R and T phases (cf. Sect. 7.3.3.2) is shown in Fig. 7.14b. As the chemical decoration of the tiles cannot be determined on the basis of the available structural information, it has been optimized by Krajčí et al. (1995a, 1996, 1997a) to produce the maximum electronic stabilization, i.e., a deep pseudogap lowering of the total energy. The result demonstrates that a weak structure-induced pseudogap exists also in the d-Al-Pd-Mn and its approximant phases. There are important differences between the i and d phases. (a) Due to the higher Mn content, the broadened Mn band overlaps with E_F, inducing a relatively high DOS at E_F. (b) The influence of *sp-d* hybridization on the width and shape of the pseudogap is even more evident. The DOS minimum in the Al band is clearly split into two subminima coincident with similar features in the Mn band. This structure can be attributed to local atomic arrangements characteristic for the R phase (Krajčí et al. 1997a). (c) The E_F is pinned in the shallower subminimum of the total DOS, in contrast to the i phase where it falls into the deeper main minimum. Altogether, the importance of a Hume-Rothery-like band-gap stabilization decreases in the sequence crystalline $Al_6Mn \rightarrow i$-$Al_{0.70}Pd_{0.22}Mn_{0.08} \rightarrow d$-$Al_{0.70}Pd_{0.13}Mn_{0.17}$.

The electronic structure of d-Al-Cu-Co has been calculated by Krajčí et al. (1997b) for models of Burkov (1993) based on Klotz-triangle tilings which have a simple cut-projection representations in 5D space at the "magic composition" $Al_{\tau-1}Cu_{\tau'}Co_{\tau'}$, $\tau' = (2-\tau)/2$, where τ is the golden mean. Again, the problem is that the available x-ray diffraction data do not allow a determination of the distribution of Cu and Co atoms. In five dimensions, the atomic surfaces are a pentagonal star and a decagon arranged along the commensurate direction in hyperspace (Fig. 7.15). Al atoms occupy the pentagonal star and the outer part of the decagon, whereas the TM atoms are distributed over a pentagonal star inscribed inside the decagon. Figures 7.15b–7.15d define three different possibilities to distribute Cu and Co atoms over the atomic surface proposed by Krajčí et al. (1997b), leading to a different local chemical order. The calculation of the electronic structure for these three different structural models and the comparison with the available photoemission data of Stadnik et al. (1995a) leads to the conclusion that the model placing the Cu atoms into the central region of the atomic surface (and thus supressing direct Cu-Cu contacts while increasing the number of Co-Co neighbours) leads to the best agreement with experiment (Fig. 7.15). Calculations based on the older Burkov model (Burkov 1991) produce a deep pseudogap at E_F, as shown by Krajčí et al. (1997b) and Trambly de Laissardière and Fujiwara (1994b), and hence disagree with the spectroscopic data.

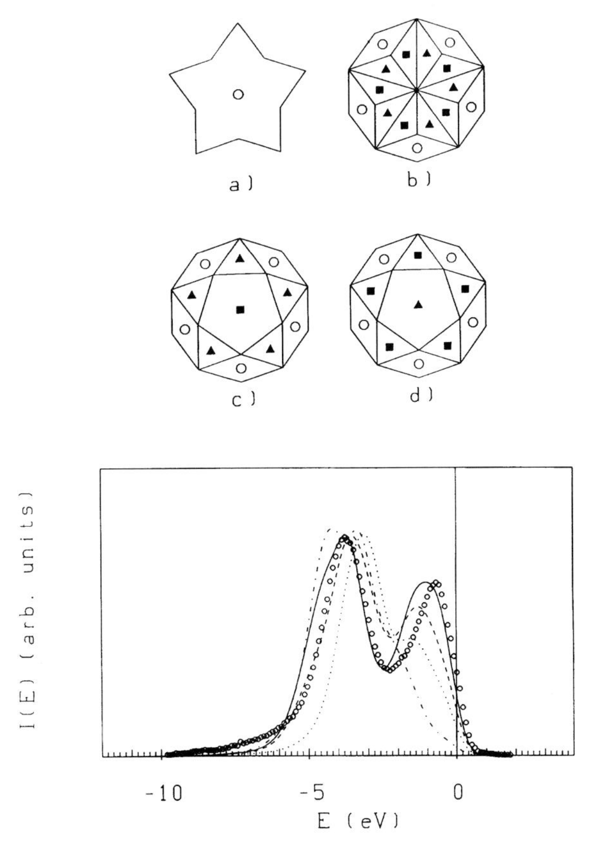

Fig. 7.15. Top: Atomic surfaces of d-Al-Cu-Co in 5D hyperspace. (a) Al atoms in the pentagonal star, (b-d) Al atoms (open circles), Co (squares) and Cu (triangles) distributed in different ways over the decagon. Bottom: Photoemission spectra calculated for different structural models of d-Al-Cu-Co: Dot-dashed line – older Burkov model (Burkov 1991), dashed, full, and dotted lines – models (b), (c), and (d) shown above. The open circles represent the experimental spectra. After Krajčí et al. (1997b).

The important point is again that the structure producing the best agreement with the measured photoemission intensity shows the most pronounced splitting of the Cu and Co d bands and hence optimizes the electronic contribution to the binding energy. The increased binding energy is due to two effects: (a) The larger overlap of the Co band with E_{F} reduces the occupation of antibonding Co-d states. (b) The shift of the Cu-d band increases the Cu-Co bond order – this favours heterocoordination of the TM atoms. We note that this analysis is well confirmed by soft x-ray spectroscopy (Belin-Ferré

1996). On the other hand, x-ray diffraction is insufficient to discriminate between models (b)–(d) in Fig. 7.15. A direct structural confirmation would require neutron-diffraction experiments sensitive to the Cu-Co substitutions. Calculations of the electronic DOS for the orthorhombic $Al_{13}Co_4$ (with a close structural relationship to the *d* phase) and tetragonal Al_7Cu_2Co (showing no structural similarities) lead to a position of E_F in a DOS minimum. Altogether, the analysis of the *d*-Al-Cu-Co phases confirms the conclusions that the importance of Hume-Rothery-like effects in Al-TM alloys decreases in the sequence crystalline $\rightarrow i \rightarrow d$ alloys (although the Al band always shows a structure-induced pseudogap), while the importance of covalent *d-d* bonding increases.

7.6.3 Titanium-Based Quasicrystals

Although to date most work has concentrated in Al-based QCs, a metastable *i* phase in the Ti-Ni-V system was discovered by Zhang et al. (1985), shortly after the pioneering work on *i*-Al-Mn alloys. Most QCs based only on TMs are highly disordered (Libbert et al. 1993). Libbert and Kelton (1995) have argued that the addition of Si or O oxygen is crucial for the stabilization of the quasiperiodic phase. From the point of view of a possible electronic stabilizing mechanism, the investigation of the QCs formed by Ti with a second TM is of great interest. The first investigation based on the structure model of Libbert et al. (1993) was performed by Krajčí and Hafner (1994). Recent work of Hennig and Teichler (1997) is based on the improved structure model of Libbert and Kelton (1995) and considers also the influence of oxygen. The main result of these studies is that the stabilization of the *i* phase over the disordered bcc high-temperature phases is promoted by the formation of a DOS minimum close to E_F. The pseudogap is caused by the interaction of the *d* bands of the two TMs. Quite generally one observes a transition from a common-band regime (no gap) in the case of small differences in the position of the atomic *d* levels (i.e., small differences in *d*-band filling) to a split-band regime when the differences in the *d* levels increases (Hausleitner et al. 1992). The pseudogap is reinforced by chemical ordering. This explains the increasing stability of the *i* phase in the series Ti-V – Ti-Cr – Ti-Mn – Ti-Fe – Ti-Ni.

7.6.4 Fine Structure of the Electronic Spectrum, Pseudogaps, and Real Gaps

The results discussed so far have been based mostly on real-space recursion calculations. Hence the spectrum is obtained with a finite resolution determined essentially by the width of the band (W $\sim 10-15$ eV), divided by the number of recursion levels ($L \sim 80-100$). A resolution of $\Delta E \lesssim 0.2$ eV is sufficient for a quantitative comparison of theory with most spectroscopic experiments, but not for answering the question about the character of the spectrum (continuous or singular continuous). For the description of the fine structure of the electronic DOS and for the characterization of the eigenstates (localized, critical, or extended), eigenvalues and eigenstates must be calculated by exact diagonalization of the Hamiltonian. For the lower-order approximants we have to remember that the eigenstates show dispersion through the BZ that may be larger than the separation of neighbouring eigenvalues. Hence the eigenstates must be calculated on a fine mesh in $\boldsymbol{k}$ space and convergence of the BZ integration must be carefully checked.

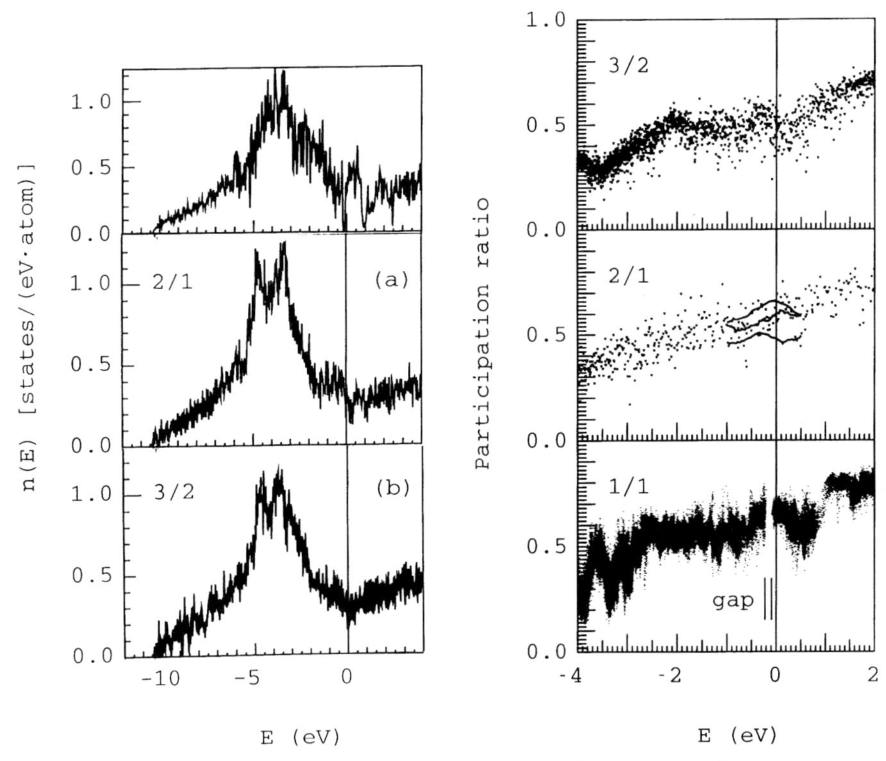

Fig. 7.16. Left column: Total densities of states for 1/1, 2/1, and 3/2 approximants to i-Al-Pd-Re. Right column: Participation ratio p_ν of the eigenstates close to E_F. Cf. text. After Krajčí and Hafner (1998b).

Figure 7.16 shows the DOS for the 1/1, 2/1, and 3/2 approximants to i-Al-Pd-Re calculated on the basis of the 6D model of the homologuous QC Al-Pd-Mn (cf. Sect. 7.3 for details). The result for the 1/1 approximant (128 atoms/cell) is based on a fine mesh of 176 $\boldsymbol{k}$ points in the irreducible BZ; for the 2/1 approximant (544 atoms/cell) four $\boldsymbol{k}$ points have been used, and for the 3/2 approximant (2292 atoms/cell) the calculation has been performed at the Γ point only so that the full point-group symmetry could be exploited to block-diagonalize the Hamiltonian. All calculations have been performed self-consistently by Krajčí and Hafner (1998b). The important result is that the electronic DOS is characterized at all energies by spiky peaks of a width of a few 0.01 eV. This fine structure is superposed to the coarser structures produced by most intense Bragg reflextions and by hybridization effects. From the point of view of the band structure of the periodic approximants, the fine structure of the DOS is the result of folding back the bands into a smaller and smaller BZ as the order of the approximant increases. From the point of view of the diffraction properties of quasicrystalline structures, the spikiness of the spectrum arises from the hierarchical quasi-BZ. It is presumed that the spiky structure of the spectrum survives in the quasiperiodic limit, although there will be a lower limit set by the inelastic scattering of electrons.

The strong structure in the DOS seems to provide a natural explanation for the extreme sensitivity of the Hall effect and thermoelectric power to changes in chemical composition and temperature (Poon 1992). On the other hand, recent attempts of Stadnik et al. (1995b, 1997) and Zhang et al. (1995) to observe the fine structure of the electronic DOS by high-resolution photoemission have failed although with a nominal resolution of 0.02 eV it should be possible to observe the spikiness of the DOS at least within a range of 2 eV from E_{F} (at higher binding energies the fine structure is smeared out due to lifetime broadening effects). The failure to observe the spikiness of the DOS has been attributed to electron-electron scattering and to scattering by defects. Studies of Ebert et al. (1996) based on scanning tunneling microscopy have demonstrated that as-cleaved quasicrystalline surfaces are rough. Their structure has been found to be determined by cluster aggregates with diameters close to 10 Å, which corresponds to the size of the building blocks of the i structure and is nearly as large as the estimated escape depth of the photoelectrons. Hence the photoemission intensity will be dominated to a large extent by the surface electronic structure.

7.6.4.1 Metal-Insulator Transition in Icosahedral Al-Pd-Re. Perhaps the most striking result shown in Fig. 7.16 is the appearance of a real gap of 0.15 eV in the spectrum of the 1/1 approximant, very close to E_{F}. The existence of the gap depends in a very sensitive way on the details of the atomic structure and chemical order – the gap closes even if only a pair of Pd and Re atoms is interchanged. In the higher-order approximants no real gaps, but only a series of very deep pseudogaps, are observed. Evidently this demonstrates a strong interaction between all local orbitals and a strong spatial coherence of the eigenstates, as discussed by Krajčí and Hafner (1998b). This result seems to provide the first explanation of the striking transport properties of Al-Pd-Re and their extreme sensitivity to changes in the chemical composition and annealing conditions. Changes in the chemical decoration and in the thermal treatment will lead to a varying concentration of phason defects. The result for the 1/1 approximant demonstrates that a linear phason strain modifies the atomic structure in such a way that a gap opens at

E_F. Other types of phason defects are likely to introduce similar gaps. Hence the picture of a phason-induced metal-insulator transition emerges. At the moment, however, the conditions leading to the opening of a gap remain incompletely understood.

7.6.4.2 Localization or Confinement of Electrons. In connection with a possible metal-insulator transition, the character of the eigenstates is of particular interest. Figure 7.16b shows the participation ratio p_ν calculated by Krajčí and Hafner (1998b) for the eigenstates in approximants to *i*-Al-Pd-Re. Similar results have been obtained by Krajčí et al. (1995b) for *i*-Al-Pd-Mn and by Krajčí et al. (1997b) for *d*-Al-Cu-Co. The participation ratio is given in terms of the tight-binding eigenvectors $\boldsymbol{e}_{i\alpha}(E_\nu)$ of the states with energy E_ν as

$$p_\nu = \frac{\left[\sum_i \sum_\alpha |e_{i\alpha}(E_\nu)|^2\right]^2}{N \sum_i \left[\sum_\alpha |e_{i\alpha}(E_\nu)|^2\right]^2} . \tag{7.7}$$

The results can be summarized as follows. (a) Most eigenstates are extended ($p_\nu \sim 0.5$). Only in the energy range where the 3*d* or 4*d* bands of the TMs dominate the spectrum, the participation ratio drops to significantly lower values ($p_\nu \sim 0.2 - 0.3$). States belonging to a 5*d* band (Re) remain extended; in this respect there is a distinct difference between Al-Pd-Re and Al-Pd-Mn alloys. This shows clearly that the decrease of the participation ratio is related to the more localized character of 3*d* or 4*d* orbitals than to a structure-induced onset of localization. (b) In *i*-Al-Pd-Re, the participation ratio averaged over a small energy interval decreases systematically with increasing order of the approximant for binding energies between -2 and 1 eV. Such a decrease is compatible with a scaling behavior of the participation number $P_\nu = Np_\nu$, with an exponent $\beta \sim 0.9$ ($P_\nu \propto N^\beta$). A scaling of the particpation number with an exponent $\beta < 1$ corresponds to a decay of the envelope of the eigenstates according to $\psi(r) \propto r^{-\alpha}$, with $\alpha \sim 0.8$, i.e., a very weak power-law localization of the eigenstates is observed. Here we used the approximate relation for the scaling exponents $\beta = 2 - 4\frac{\alpha}{D}$ derived by Rieth and Schreiber (1998), where D is the dimensionality of the system. For *d*-Al-Cu-Co QCs a similar scaling behavior has been reported by Fujiwara et al. (1996). However, we have to emphasize that the size of the model is still too small to allow for any firm conclusions. (c) A few states with a very low participation ratio have been detected in *i*-Al-Pd-Mn (Fig. 7.17). The distribution of the amplitudes of one of these more localized states is given in Fig. 7.17b. One finds that the largest part of the intensity is concentrated on four Mn sites situated at large distances from each other. These four sites are just the images of the projections of the sites with the short distance from the center of the Mn-acceptance domain (Fig. 7.2) at the odd nodes of the hypercubic lattices. Sites projected from the center of the atomic surface have the full point-group symmetry of the node on which the acceptance domain is centered, whereas sites

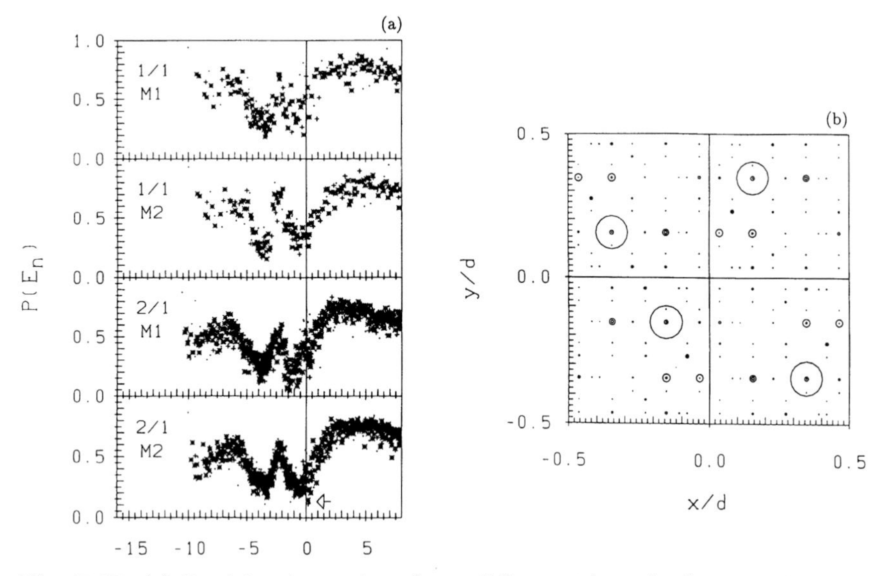

Fig. 7.17. (a) Participation ratios of two different 1/1 and 2/1 approximants to *i*-Al-Pd-Mn. (b) Spatial distribution of the amplitudes of one of the most localized eigenstates [marked by the arrow in (a)]. The size of the symbols scales with the amplitude. After Krajčí et al. (1995b).

with a large distance in perpendicular space have only a reduced symmetry. Thus, there is a correlation between local *i* order and incipient localization (Krajčí et al. 1995). However, we must again emphasize that no hint of electronic eigenstates confined to *i* clusters could be detected. From the point of view of electronic structure calculations there is definitely no indication for the existence of cluster-supported confined eigenstates, as postulated by Janot and de Boissieu (1994) and Trambly de Laissardière et al. (1997).

7.6.5 Band-Structure Effects in Electronic Transport

Attempts to correlate the electronic structure of QCs with the anomalous low conductivity of the stable *i* phases and its large anisotropy in *d* alloys can be made on the basis of Bloch-Boltzmann theory which gives the zero-temperature conductivity as

$$\sigma_{\alpha\beta} = \frac{\mathrm{e}^2}{\Omega_0} D_{\alpha\beta}(E_\mathrm{F}) n(E_\mathrm{F}) , \tag{7.8}$$

where $n(E_\mathrm{F})$ is the DOS at the Fermi level. The electronic diffusivity $D_{\alpha\beta}(E)$ can be calculated in terms of the band velocities $\boldsymbol{v}_n(\boldsymbol{k}) = \hbar^{-1}\boldsymbol{\nabla}_{\boldsymbol{k}} E_n(\boldsymbol{k})$ via

$$D_{\alpha\beta}(E) = \tau \sum_{\boldsymbol{k}} v_{n\alpha}(\boldsymbol{k}) v_{n\beta}(\boldsymbol{k})|_{E_n = E} , \tag{7.9}$$

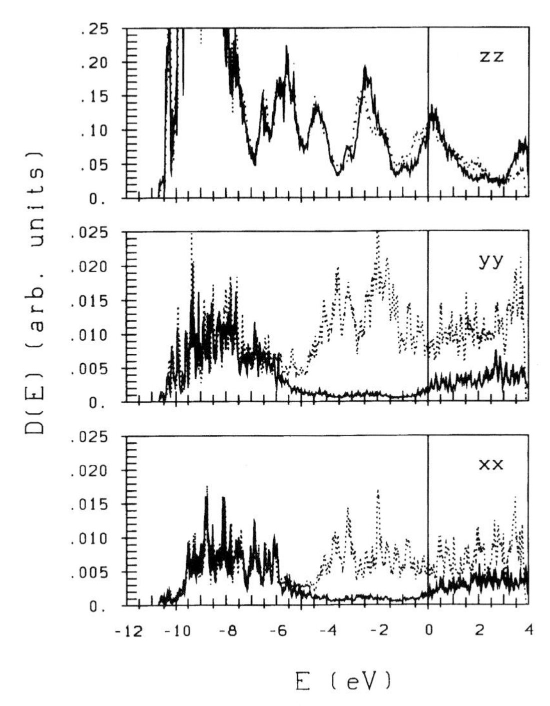

Fig. 7.18. Electronic diffusivity (in arbitrary units) along the three Cartesian axes in *d*-Al-Cu-Co (z is the periodic axis). The full lines represent the result for the equilibrium structure, the dashed lines illustrate the effect of superposed random atomic displacements ($\widehat{=}$ thermal disorder or quenched-in structural disorder). After Krajčí et al. (1997b).

where τ is the relaxation time. Figure 7.18 shows the energy-dependent diffusivity calculated along the three Cartesian axes of the orthorhombic approximant (352 atoms/cell) to *d*-Al-Cu-Co. The most striking result is a very deep and broad depression of the conductivity function in the region from -5 eV up to E_{F} for both orthogonal directions in the quasiperiodic plane, i.e., in a range of high DOS. This contrasts with the strongly modulated energy dependence of the conductivity calculated for the periodic z direction. The anisotropy ratio σ_{zz}/σ_{xx} varies between 20 and 150, depending on the details of the structural model adopted (Krajčí et al. 1997b). The origin of the anisotropy is the almost dispersionless character of the bands in the quasiperiodic plane and the considerable dispersion calculated for the periodic direction.

Band-structure effects can also explain the anomalous dependence of the transport coefficients on the degree of quasicrystalline order: disorder leads to a decrease of resistivity. This effect has been estimated by imposing random displacements with a standard deviation of $\Delta r = 0.15$ Å to all atoms. The effect of disorder on the conductivity (Fig. 7.18) is rather dramatic: the electronic diffusivity in the quasiperiodic plane increases by a factor of five, while the diffusivity along the periodic direction is hardly affected. The reason is that disorder leads to an increased mixing of the bands and hence to an increased dispersion (on a very local scale in $\boldsymbol{k}$ space) of the eigenstates. Important conclusions may also be derived from the calculation of the diffusivity in the crystalline $Al_{13}Co_4$ and Al_7Cu_2Co phases (cf. Sect. 7.6.2.2 for the electronic structure). Low diffusivity and strong anisotropy are found in $Al_{13}Co_4$, which is similar to the *d* phase, but not in Al_7Cu_2Co with no structural relationship. This suggests that the transport properties are determined by a correlation within a range of $\sim$ 20 to 30 Å. Similar results have been obtained for *d*-Al-Pd-Mn and approximant phases.

Although the Bloch-Boltzmann picture establishes a strong link between the electronic band structure and the transport properties, its application to quasicrystalline alloys has been criticized because it does not account for the character of the electronic eigenstates and does not allow for interband transitions (Fujiwara et al. 1996). The natural step would involve the use of Kubo-Greenwood rather than Bloch-Boltzmann theory. Calculations on model systems predict that the scaling properties of the wave functions are reflected in a similar scaling law for the conductance, but for realistic quasicrystalline models the Kubo-Greenwood approach still represents a tremendous challenge. For very low temperatures, quantum interference effects become important for understanding the temperature and magnetic-field dependence of the conductivity [see, e.g., Kimura and Takeuchi (1991), Ahlgren et al. (1997), and Chap. 6].

7.6.6 Magnetic Properties of Quasicrystals

Among the many striking properties of QCs, their magnetic behavior is perhaps the least well understood. Diamagnetism, paramagnetism, ferromagnetism, and spin-glass behavior have been reported for quasicrystalline alloys [for detailed references see, e.g., Hafner and Krajčí (1998) and Chap. 9]. Al-Pd-Mn alloys have been studied most thoroughly. In the binary Al-Mn system, the metastable *i*, *d*, and amorphous phases display a similar general behavior: only a few Mn atoms are magnetic and carry a large effective magnetic moment (Goto et al. 1988, Berger and Prejean 1990, Godinho et al. 1990, Gozlan et al. 1991). A similar conclusion has recently been reached for the stable *i*-Al-Pd-Mn phase by Lasjaunias et al. (1995).

The formation of magnetic moments in crystalline, quasicrystalline, and liquid ($\approx$ amorphous) Al-Pd-Mn phases has been investigated using ab-initio local-spin-density calculations by Hafner and Krajčí (1998). It was shown that in all cases the formation of a local magnetic moment is determined by a local Stoner criterion, i.e., a site will be magnetic if $n_i(E_\mathrm{F})I > 1$ and nonmagnetic if $n_i(E_\mathrm{F})I \leq 1$, where $n_i(E_\mathrm{F})$ is the local DOS at E_F and I is the Stoner intra-atomic exchange integral which, for an itinerant magnet, is determined primarily by Hund's rule exchange. The decisive factor is then the form of the local electronic DOS at the Mn sites: (a) In the crystalline Al_6Mn compound the strong Al-*sp*–Mn-*d* hybridization reinforces the structure-induced pseudogap at E_F and hence the compound is nonmagnetic. (b) Hybridization effects are still quite strong in quasicrystalline alloys (both *i* and *d* alloys). However, quasicrystallinity allows for a wide variety of local environments and on certain sites the criterion for the formation of a magnetic moment can be satisfied. This has been explicitly demonstrated for some lower-order *d*-Al-Mn and *i*-Al-Pd-Mn approximants. Magnetic moments of up to 2.5 to 3 μ_B can be formed on a small fraction of Mn sites ($\lesssim 5$ %). These local sites are characterized by an enhanced Mn-Mn coordination and a looser than average Mn-Al coordination. However, these sites belong to the ideal quasicrystalline lattice. No defects are necessary to form a Mn moment. (c) Melting reduces, but does not completely destroy the local order. Compared to the QC, the sites on which a magnetic moment can be formed are more numerous in a liquid or amorphous Al-Mn alloy. This explains the observed increase of the magnetic susceptibility on melting reported by Hippert et al. (1996). (d) In a substitutionally disordered fcc Al_6Mn alloy the confinement of the atoms to a fixed Al-Mn distance very effectively reduces the Al-*sp*–Mn-*d* hybridization and for the resulting impurity-like DOS the local Stoner criterion is satisfied on all sites. Hence fcc Al-Mn is a concentrated spin glass. For all systems a Ruderman-Kittel-Kasuya-Yoshida-like exchange interaction has been found, confirming the universal character of the magnetism in all these systems.

7.6.7 Electrons – A Summary

The electronic structure of QCs has now been explored for a fairly wide range of quasicrystalline alloys. In analogy to the results obtained for phonons, we can conclude that the electronic spectrum of QCs is determined by the hierarchical quasi-BZ structure: free-electron-like states exist at the bottom of the valence band near the most intense Γ points and stationary states and pseudogaps of various widths are induced by the quasi-BZ boundaries.

The larger of these pseudogaps close to E_F have been associated with a Hume-Rothery-like mechanism for the stabilization of the quasicrystalline structure. However, it is now clear that structure-induced pseudogaps exist also in many crystalline phases and that small variations in the stoichiometry can decide whether the Hume-Rothery mechanism dips the energetic balance in favor of the quasicrystalline or crystalline structure. In addition, it has turned out that the situation may be in reality more complex: the structure-induced DOS minimum in the Al-*sp* band may be covered more or less completely by the *d*-band of a TM component with an incompletely filled *d*-band. In this case, both electronic structure calculations and spectroscopic experiments agree on the important role of covalent *d*-*d* interactions.

The existence of a fine-structure in the electronic DOS has been disputed on the ground of missing experimental evidence. We consider the close agreement between calculations for so many different systems and different techniques as solid evidence for the spikiness of the electronic eigenvalue spectrum. The failure to observe the predicted spikiness in high-resolution photoelectron and tunneling spectroscopies is probably to be attributed to the structure of the surfaces (interfaces) on which the experiments have been performed.

The determination of the character of the electronic eigenstates remains problematic. Only for a very few systems have exact eigenstates been calculated for approximants that are large enough to allow for a rough estimate of the scaling behavior of the participation ratio. The scaling exponents $\beta < 1$ derived for *d*-Al-Cu-Co by Fujiwara et al. (1996) ($\beta \sim 0.6 - 0.9$, depending on the binding energy) and for *i*-Al-Pd-Re by Krajčí and Hafner (1998b) ($\beta \sim 0.9$ for small binding energies) would indicate a power-law decay of the wave functions. The lower scaling exponent for the *d* phase tends to agree with the results obtained on model systems by Rieth and Schreiber (1998). The result for *i*-Al-Pd-Re should not be generalized to other *i* phases because of the exotic properties of this system. The investigation of the spatial distribution of the amplitudes of the eigenstates seems to exclude any "confinement to clusters", although some indications for correlations between localization and local symmetry have been established.

Band-structure effects have been shown to have a decisive influence on the electronic transport properties. Within the Bloch-Boltzmann picture, the electronic diffusivity can be low even in a region of high DOS if the majority of the eigenstates shows only little dispersion. The weak dispersion of the eigenstates is, of course, intimately related to the spikiness of the spectrum, arising from the dense, hierarchical quasi-BZ structure. It has been shown that even a small disorder leads to an increased mixing of neighbouring bands, increased dispersion on a small $\boldsymbol{k}$ scale, and decreasing resistivity with increasing disorder – in contrast to repeated claims that Bloch-Boltzmann theory cannot cope with such a situation.

Symmetry is also broken in an external magnetic field and this could be an important step towards the anomalous magnetorestistance behavior. The next step must consist in a quantum-mechanical calculation of the transport coefficients based on the Kubo-Greenwood formula. The work of Fujiwara et al. (1996) is a first step towards an application of Kubo-Greenwood theory to realistic models of QCs. The very recent work of Roche and Mayou (1997) on real-space recursion techniques applied to the calculation of the Kubo-Greenwood conductivity opens a new access to this difficult problem. However, in view of the difficulties in establishing a scaling behavior of the eigenstates, the way towards an ab-initio confirmation of scaling theories for electric transport promises to be a long one.

Scaling behavior is intimately related to the spatial coherence of the eigenstates. It has been shown for the *i*-Al-Pd-Re alloy that interchanging the occupation of only two sites can lead to the opening/closing of a real gap in the

DOS. Evidently this is possible only if the eigenstates show strong long-range coherence.

The formation of local magnetic moments is another demonstration of the strong dependence of the physical properties on the local order and the surprising effects that can arise from the fluctuations in the local environment in a quasiperiodic structure.

7.7 Final Remarks

The analysis of the spectrum of vibrational excitations of QCs has demonstrated that the characteristic aspects related to quasicrystallinity are now well understood on the basis of the hierarchical quasi-BZ structure. At the energy resolution of most spectroscopic experiments, this applies also to the electronic spectrum of QCs. The situation is different, however, if we probe the spectrum at much higher resolution or if we analyze the transport and magnetic response properties of QCs. These problems remain a challenge to both theory and experiment.

Acknowledgments

The support of our work by the Austrian Science Foundation, the Austrian Ministery for Science and Transport through the Center for Computational Materials Science, and the Slovak Academy of Sciences is gratefully acknowledged.

References

Ahlgren, M., Rodmar, M., Gignoux, C., Berger, C., Rapp, Ö. (1997): Mater. Sci. Eng. A **226-228**, 981

Andersen, O.K. (1975): Phys. Rev. B **12**, 3060

Andersen, O.K., Jepsen, D., Šob, M. (1987) in Electronic Band Structure and its Applications, Youssouff, M. (ed). Springer-Verlag, Berlin, p 1

Audier, M., Durand-Charre, M., de Boissieu, M. (1993): Philos. Mag. B **68**, 607

Baake, M., Kramer, P., Schlottmann, M., Zeidler, D. (1990): Int. J. Mod. Phys. B **4**, 2217

Bak, P. (1986): Phys. Rev. Lett. **56**, 861

Bancel, P.A., Heiney, P.A. (1986): Phys. Rev. B **33**, 7917

Belin, E., Dankházi, Z., Sadoc, A., Dubois, J.M. (1994): J. Phys. Condens. Matter **6**, 8771

Belin-Ferré, E., Dankházi, Z., Fournée, V., Sadoc, A., Berger, C., Müller, H., Kirchmayr, H. (1996): J. Phys. Condens. Matter **8**, 6213

Bellissent, R., de Boissieu, M., Coddens, G. (1994): in Lectures on Quasicrystals, Hippert, F., Gratias, D. (eds). Les Editions de Physique, Les Ulis, p 385

Berger, C., Prejean, J.J. (1990): Phys. Rev. Lett. **64**, 1769
Bienenstock, A., Ewald, P.P. (1962): Acta Crystallogr. **15**, 1253
Bose, S.K., Jaswal, S.S., Andersen, O.K., Hafner, J. (1988): Phys. Rev. B **37**, 9955
Boudard, M., de Boissieu, M., Janot, C., Heger, G., Beeli, C., Nissen, H.-U., Vincent, H., Ibberson, R., Audier, M., Dubois J.M. (1992): J. Phys. Condens. Matter **4**, 10 149
Burkov, S.E. (1991): Phys. Rev. Lett. **67**, 614
Burkov, S.E. (1993): Phys. Rev. B **47**, 12 325
Burkov, S.E., Timusk, T., Ashcroft, N.W. (1993): J. Phys. Condens. Matter **4**, 9447
Carlsson, A.E. (1993): Phys. Rev. B **47**, 2515
Chernikov, M.A., Bianchi, A., Müller, H., Ott, H.R. (1995): in Proceedings of the 5th International Conference on Quasicrystals, Janot, C., Mosseri, R. (eds). World Scientifc, Singapore, p 569
Cockayne, E., Phillips, R., Kan, X.B., Moss, S.C., Robertson, J.L., Ishimasa, T., Mori, M. (1993): J. Non-Cryst. Solids **153-154**, 140
Currat, R., Janssen, T. (1988): in Solid State Physics, Vol. 41, Ehrenreich, H., Turnbull, D. (eds). Academic Press, San Diego, p 210
Dall'Acqua, G., Krajčí, M., Hafner, J. (1997): J. Phys. Condens. Matter **9**, 10 725
de Boissieu, M., Boudard, M., Bellissent, R., Quilichini, M., Hennion, B., Currat, R., Goldman, A.I., Janot, C. (1993a): J. Phys. Condens. Matt. **5**, 4945
de Boissieu, M., Boudard, M., Moudden, H., Quilichini, M., Bellissent, R., Hennion, B., Currat, R., Goldman, A., Janot, C. (1993b): J. Non-Cryst. Solids **153-154**, 552
de Boissieu, M., Stephens, P., Boudard, M., Janot, C. (1994): J. Phys. Condens. Matter **6**, 363
de Boissieu, M., Boudard, M., Kycia, S., Goldman, A.I., Hennion, B., Bellissent, R., Quilichini, M., Currat, R., Janot, C. (1995): in Proceedings of the 5th International Conference on Quasicrystals, Janot, C., Mosseri, R. (eds). World Scientific, Singapore, p 577
Denton, A.R., Hafner, J. (1997): Europhys. Lett. **38**, 189; Phys. Rev. B **56**, 2469
DiVincenzo, D.P., Steinhardt, P.J. (eds) (1991): Quasicrystals, The State of the Art. World Scientific, Singapore
Duneau, M., Audier, M. (1994): in Lectures on Quasicrystals, Hippert, F., Gratias, D. (eds). Les Editions de Physique, Les Ulis, p. 283
Ebert, Ph., Feuerbacher, M., Tamura, N., Wollgarten, M., Urban, K. (1996): Phys. Rev. Lett. **77**, 3827
Edagawa, K., Suzuki, K., Ichihara, M., Takeuchi, S., Kamiya, A., Mizutani, U. (1991): Philos. Mag. Lett. **64**, 95
Edagawa, K., Naito, N., Takeuchi, S. (1992): Philos. Mag. B **65**, 1011
Friedel, J., Dénoyer, F. (1987): C.R. Acad. Sci. Paris **305**, 171
Frank, F.C., Kasper, J.S. (1958): Acta Crystallogr. **11**, 184; ibid. **12**, 483
Fujiwara, T. (1989): Phys. Rev. B **40**, 942
Fujiwara, T., Yokokawa, T. (1991): Phys. Rev. Lett. **66**, 333
Fujiwara, T., Kohmoto M., Tokihiro, T. (1989): Phys. Rev. B **40**, 7413
Fujiwara, T., Yamamoto, S., Trambly de Laissardière, G. (1993): Phys. Rev. Lett. **71**, 4166
Fujiwara, T., Mitsui, T., Yamamoto, S. (1996): Phys. Rev. B **53**, R2910
Gardner, M. (1977): Sci. Am. **236**, 110
Godinho, M., Berger, C., Lasjaunias, J.C., Hasselbach, K., Bethoux, O. (1990): J. Non-Cryst. Solids **117-118**, 808
Goldman, A.I., Stassis, C., de Boissieu, M., Currat, R., Janot, C., Bellissent, R., Moudden, H., Gayle, F.W. (1992): Phys. Rev. B **45**, 10 280
Goto, T., Sakakibara, T., Fukamichi, K. (1988): J. Phys. Soc. Jpn. **57**, 1751

Gozlan, A., Berger, C., Fourcaudot, G., Omari, R., Lasjaunias, J.C., Préjean, J.J. (1991): Phys. Rev. B **44**, 575

Guo, Q., Poon, S.J. (1996): Phys. Rev. B **54**, 12 793

Hafner, J. (1983): Phys. Rev. B **27**, 678

Hafner, J. (1987): From Hamiltonians to Phase Diagrams. Springer-Verlag, Berlin

Hafner, J., Heine, V. (1983): J. Phys. F **13**, 2479

Hafner, J., Heine, V. (1986): J. Phys. F **16**, 1429

Hafner, J., Kahl, G. (1984): J. Phys. F **14**, 2259

Hafner, J., Krajčí, M. (1992): Phys. Rev. Lett. **68**, 2321

Hafner, J., Krajčí, M. (1993a): Phys. Rev. B **47**, 1084

Hafner, J., Krajčí, M. (1993b) : Phys. Rev. B **47**, 11 795

Hafner, J., Krajčí, M. (1993c): J. Phys. Condens. Matter **5**, 2489

Hafner, J., Krajčí, M. (1998): Phys. Rev. B **57**, 2849

Hafner, J., Krajčí, M., Mihalkovič, M. (1995a): in Proceedings of the 5th International Conference on Quasicrystals, Janot, C., Mosseri, R. (eds). World Scientific, Singapore, p 600

Hafner, J., Krajčí, M., Mihalkovič, M. (1996): Phys. Rev. Lett. **76**, 2738

Hafner, J., Krajčí, M., Windisch, M. (1995b): J. Non-Cryst. Solids **192-193**, 212

Hauser, J.J., Chen, H.S., Waszczak, J.V. (1986): Phys. Rev. B **33**, 3577

Hausleitner, C., Tegze, M., Hafner, J. (1992): J. Phys. Condens. Matter **4**, 9557; Phys. Rev. B **45**, 115

Haydock, R. (1980): in Solid State Physics, Vol. 35, Ehrenreich, H., Seitz, F., Turnbull, D. (eds). Academic Press, New York, p 215

Hedin, L., Lundqvist, B.I. (1971): J. Phys. C **4**, 2064

Henley, C.L. (1991): Phys. Rev. B **43**, 993

Henley, C.L. (1993): J. Non-Cryst. Solids **153-154**, 172

Henley, C.L., Elser, V. (1986): Philos. Mag. B **53**, L59

Hennig, R.G., Teichler, H. (1997): Philos. Mag. A **76**, 1053

Hippert, F., Kandel, L., Calvayrac, Y., Dubost, B. (1992): Phys. Rev. Lett. **69**, 2086

Hippert, F., Audier, M., Klein, H., Bellissent, R., Boursier, D. (1996): Phys. Rev. Lett. **76**, 54

Hiraga, K., Sun, W. (1993): Philos. Mag. Lett. **67**, 117

Hofstadter, D.R. (1976): Phys. Rev. B **14**, 2239

Homes, C.C., Timusk, T., Wu, X., Altounian, Z., Sahnoune, A., Ström-Olsen, J.O. (1991): Phys. Rev. Lett. **67**, 2694

Honda, Y., Edagawa, K., Yoshioka, A., Hashimoto, T., Takeuchi, S. (1994): Jpn. J. Appl. Phys. A **33**, 4929

Jank, W., Hausleitner, C., Hafner, J. (1991): Europhys. Lett. **16**, 473

Janot, C. (1994): Quasicrystals, A Primer, 2nd ed. Oxford University Press, New York

Janot, C., de Boissieu, M. (1994): Phys. Rev. Lett. **72**, 1674

Janssen, T. (1988a): Phys. Rep. **168**, 55

Janssen, T. (1988b): in Proceedings of the I.L.L./CODEST Workshop, Janot, C., Dubois, J.M. (eds). World Scientific, Singapore, p 327

Jones, R.O., Gunnarsson, O. (1989): Rev. Mod. Phys. **61**, 689

Katz, A., Gratias, D. (1993): J. Non-Cryst. Solids **153-154**, 187; for a more general discussion, see also Katz, A., Gratias, D. (1994): in Lectures on Quasicrystals, Hippert, F., Gratias, D. (eds). Les Editions de Physique, Les Ulis, p 187

Kimura, K., Takeuchi, S. (1991): in Quasicrystals, The State of the Art, DiVincenzo, D.P., Steinhardt, P.J. (eds). World Scientific, Singapore, p 313

Klein, T., Berger, C., Mayou, D., Cyrot-Lackmann, F. (1991): Phys. Rev. Lett. **66**, 2907

Kohmoto, M., Sutherland, B., Tang, C. (1987): Phys. Rev. B **35**, 1020
Krajčí, M., Fujiwara, T. (1988): Phys. Rev. B **38**, 12 903
Krajčí, M., Hafner, J. (1992): Phys. Rev. B **46**, 10 669
Krajčí, M., Hafner, J. (1994): Europhys. Lett. **27**, 147
Krajčí, M., Hafner, J. (1995): J. Non-Cryst. Solids **192-193**, 338
Krajčí, M., Hafner, J. (1997a): Mater. Sci. Eng. A **226-228**, 950
Krajčí, M. Hafner, J. (1997b): in Aperiodic'97, Proceedings of the International Conference on Aperiodic Crystals, de Boissieu, M., Currat, R., Verger-Gaugry, J.-L. (eds). World Scientific, Singapore, in press
Krajčí, M. Hafner, J. (1998a): in An Introduction to Structure, Physical Properties and Application of Quasicrystalline Alloys, Suck, J.-B., Schreiber, M., Häussler, P., (eds). Springer-Verlag, Berlin, in press
Krajčí, M., Hafner, J. (1998b): unpublished
Krajčí, M., Hafner, J., Windisch, M., Mihalkovič, M. (1995a): in Proceedings of the 5th International Conference on Quasicrystals, Janot, C., Mosseri, R. (eds). World Scientific, Singapore, p 617
Krajčí, M., Windisch, M., Hafner, J., Kresse, G., Mihalkovič, M. (1995b): Phys. Rev. B **51**, 17 355
Krajčí, M., Windisch, M., Hafner, J. (1995c): J. Non-Cryst. Solids **192-193**, 321
Krajčí, M., Hafner, J., Mihalkovič, M. (1996): Europhys. Lett. **34**, 207
Krajčí, M., Hafner, J., Mihalkovič, M. (1997a): Phys. Rev. B **55**, 843
Krajčí, M., Hafner, J., Mihalkovič, M. (1997b): Phys. Rev. B **56**, 3072
Kresse, G., Hafner, J. (1994): Phys. Rev. B **49**, 14 251
Lanco, P., Klein, T., Berger, C., Cyrot-Lackmann, F., Fourcaudot, G., Sulpice, A. (1992): Europhys. Lett. **18**, 227
Lasjaunias, J.C., Sulpice, A., Keller, N., Préjean, J.J., de Boissieu, M. (1995): Phys. Rev. B **52**, 886
Lasjaunias, J.C., Calvayrac, Y., Yang, H. (1997): J. Phys. I (Paris) **7**, 959
Legault, S., Ellman, B., Ström-Olsen, J., Taillefer, L., Kycia, S. (1995): in Proceedings of the 5th International Conference on Quasicrystals, Janot, C., Mosseri, R. (eds). World Scientific, Singapore, p 592; see also Perrot, A., Dubois, J.M., Cassart, M., Issi, J.P., ibid., p 588
Libbert, J.L., Kelton, K.F. (1995): Philos. Mag. Lett. **71**, 153
Libbert, J.L., Kelton, K.F., Gibbons, P.C., Goldman, A.I. (1993): J. Non-Cryst. Solids **153-154**, 53
Los, J., Janssen, T. (1990): J. Phys. Condens. Matter **2**, 9553
Los, J., Janssen, T., Gähler, F. (1993): J. Phys. I (Paris) **3**, 107
Lubensky, T.C., Socolar, J.E.S., Steinhardt, P.J., Bancel, P.A., Heiney, P.A. (1986): Phys. Rev. Lett. **57**, 1440
Luchini, M.U., Nex, C.M.M. (1987): J. Phys. C **20**, 3125
Lück, R.L., Kek, S. (1993): J. Non-Cryst. Solids **153-154**, 329
Maradudin, A.A., Montroll, E.W., Weiss, G.H., Ipatova, I.P. (1971): Theory of Lattice Dynamics in the Harmonic Approximation. Academic Press, New York
Martin, S., Hebard, A.F., Kortan, A.R., Thiel, F.A. (1991): Phys. Rev. Lett. **67**, 719
Mermin, N.D. (1992): Rev. Mod. Phys. **64**, 3
Mihalkovič, M. (1995): in Aperiodic'94, Proceedings of the International Conference on Aperiodic Crystals, Chapuis, G., Paciorek, W. (eds). World Scientific, Singapore, p 552
Mihalkovič, M., Mrafko, P. (1993): J. Non-Cryst. Solids **156-158**, 936

Mihalkovič, M., Zhu, W.-J., Henley, C.L., Newman, M.E.J., Oxborrow, M., Phillips, R.B. (1995): in Aperiodic'94, Proceedings of the International Conference on Aperiodic Crystals, Chapuis, G., Paciorek, W. (eds). World Scientific, Singapore, p 169
Mizutani, U. (1994): in Materials Science and Technology, Vol. 3B, Cahn, R.W., Haasen, P., Kramer, E.J. (eds). VCH, Weinheim, p 97
Mukhopadhyay, N.K., Isihara, K.N., Ranganathan, S., Chattopadhyay, K. (1991): Acta Metall. Mater. **39**, 1151
Nelson, D.R. (1983): Phys. Rev. Lett. **50**, 982
Nelson, D.R., Spaepen, F. (1989): in Solid State Physics, Vol. 42, Ehrenreich, H., Turnbull, D. (eds). Academic Press, Boston, p 1
Newman, M.E.J., Henley, C.L. (1993): J. Non-Cryst. Solids **153-154**, 205
Nguyen Manh, D., Trambly de Laissardière, G., Julien, J.P., Mayou, D., Cyrot-Lackmann, F. (1992): Solid State Comm. **82**, 329
Nicol, A.D.I. (1953): Acta Crystallogr. **6**, 285
Niikura, A., Tsai, A.P., Inoue, A., Masumoto, T., Yamamoto A. (1993): Jpn. J. Appl. Phys. B **32**, L1160
Niizeki, K. (1989): J. Phys. A **22**, 4295
Niizeki, K., Akamatsu, T. (1990): J. Phys. Condens. Matter **2**, 2759
Nowak, H.J., Andersen, O.K., Fujiwara, T., Jepsen, O., Vargas, P. (1991): Phys. Rev. B **44**, 3577
Oelhafen, P. (1983): in Glassy Metals II, Beck, H., Güntherodt, H.J. (eds). Springer-Verlag, Berlin, p 283
Phillips, R., Zhou, J., Carlsson, A.E., Widom, M. (1994): Phys. Rev. B **49**, 9322
Pierce, F.S., Poon, S.J., Guo, Q. (1993): Science **261**, 737; Phys. Rev. Lett. **73**, 2220
Poon, S.J. (1992): Adv. Phys. **41**, 303
Poussigue, G., Benoit, C., de Boissieu, M., Currat, R. (1994): J. Phys. Condens. Matter **6**, 659
Qiu, S.-Y., Jarić, M.V. (1990): in Proceedings of the Anniversary Adriatico Research Conference on Quasicrystals, Jarić, M.V., Lundqvist, S. (eds). World Scientific, Singapore, p 19
Rieth, T., Schreiber, M. (1995a): Phys. Rev. B **51**, 15 827
Rieth, T., Schreiber, M. (1995b): in Proceedings of the 5th International Conference on Quasicrystals, Janot, C., Mosseri, M. (eds). World Scientific, Singapore, p 514
Rieth, T., Schreiber, M. (1998): J. Phys. Condens. Matter **10**, 783
Roche, S., Mayou, D. (1997): Phys. Rev. Lett. **79**, 2518
Rokhsar, D.S., Wright, D.C., Mermin, N.D. (1988): Phys. Rev. B **37**, 8145
Shechtman, D., Blech, I., Gratias, D., Cahn, J.W. (1984): Phys. Rev. Lett. **53**, 1951
Sire, C. (1994) : in Lectures on Quasicrystals, Hippert, F., Gratias, D. (eds). Les Editions de Physique, Les Ulis, p 505
Smith, A.P. (1991): Phys. Rev. B **43**, 11 635
Smith, A.P., Ashcroft, N.W. (1987): Phys. Rev. Lett. **59**, 1365
Stadnik, Z.M., Zhang, G.W., Tsai, A.-P., Inoue, A. (1995a): Phys. Lett. A **198**, 237; Phys. Rev. B **51**, 11 358
Stadnik, Z.M., Zhang, G.W., Tsai, A.-P., Inoue, A. (1995b): Phys. Rev. B **51**, 4023
Stadnik, Z.M., Purdie, D., Garnier, M., Baer, Y., Tsai, A.-P., Inoue, A., Edagawa, K., Takeuchi, S., Buschow, K.H.J. (1997): Phys. Rev. B **55**, 10 938
Steinhardt, P.J., Ostlund, S. (eds) (1987): The Physics of Quasicrystals. World Scientific, Singapore
Sutherland, B. (1986): Phys. Rev. B **34**, 3904; Phys. Rev. B **35**, 9529
Takeuchi, T., Mizutani, U. (1995): Phys. Rev. B **52**, 9300

The International Union of Crystallography (1992): Acta Crystallogr. A **48**, 928
Tokihiro, T., Fujiwara, T., Arai, M. (1988): Phys. Rev. B **38**, 5981
Trambly de Laissardière, G., Fujiwara, T. (1994a): Phys. Rev. B **50**, 5999
Trambly de Laissardière, G., Fujiwara, T. (1994b): Phys. Rev. B **50**, 9843
Trambly de Laissardière, G., Nguyen Manh, D., Magaud, L., Julien, J.P., Cyrot-Lackmann, F., Mayou, D. (1995): Phys. Rev. B **52**, 7920
Trambly de Laissardière, G., Roche, S., Mayou, D. (1997): Mater. Sci. Eng. A **226-228**, 986
Tsai, A.P., Yamamoto, A., Niikura, A., Inoue, A., Masumoto, T. (1994): Philos. Mag. Lett. **69**, 343
Tsunetsugu, H., Fujiwara, T., Ueda, K., Tokohiro, T. (1986): J. Phys. Soc. Jpn. **55**, 1420; Phys. Rev. B **43**, 8879
Waseda, A., Morioka, H., Kimura, K., Ino, H. (1992): Philos. Mag. Lett. **65**, 25
Windisch, M., Hafner, J., Krajčí, M., Mihalkovič, M. (1994a): Phys. Rev. B **49**, 8701
Windisch, M., Krajčí, M., Hafner, J. (1994b): J. Phys. Condens. Matter **6**, 6977
Wu, X., Kycia, S.W., Olson, C.G., Benning, P.J., Goldman, A.I., Lynch, D.W. (1995): Phys. Rev. Lett. **75**, 4540
Yun-ping, W., Dian-lin, Z. (1994): Phys. Rev. B **49**, 13 204
Zhang, G.W., Stadnik, Z.M., Tsai, A.-P., Inoue, A. (1994): Phys. Lett. A **186**, 345; Phys. Rev. B **50**, 6696
Zhang, G.W., Stadnik, Z.M., Tsai, A.-P., Inoue, A., Miyazaki, T. (1995): Z. Phys B **97**, 439
Zhang, Z., Ye, H.Q., Kuo, K.H. (1985): Philos. Mag. A **52**, L49

8. Spectroscopic Studies of the Electronic Structure

Zbigniew M. Stadnik

8.1 Introduction

The discovery of an icosahedral Al-Mn alloy by Shechtman et al. (1984) extended the dichotomous division of solids into either crystalline or amorphous by introducing the notion of quasiperiodic crystals, or quasicrystals (QCs). This new form of matter has a long-range *quasiperiodic* order and long-range orientational order associated with the classically forbidden fivefold (icosahedral), eightfold (octagonal), tenfold (decagonal), and 12-fold (dodecagonal) symmetry axes. The majority of known QCs are either icosahedral (i) or decagonal (d) alloys. A few known octagonal and dodecagonal alloys cannot be produced yet in sufficient quantity to allow studies of their physical properties. A central problem in condensed matter physics is to determine whether quasiperiodicity leads to physical properties significantly different from those of crystalline and amorphous materials.

The electronic structure of solids determines many of their fundamental physical properties. Therefore, experimental techniques which probe this structure directly are among the most important tools for understanding the microscopic properties of solids. Among these techniques, photoemission spectroscopy (PES) (Cardona and Ley 1978, Ley and Cardona 1979, Mahlhorn 1982, Hüfner 1996), which probes the density of states (DOS) below the Fermi level E_F, and inverse photoemission spectroscopy (IPES) and electron energy-loss spectroscopy (EELS) (Fuggle and Inglesfield 1992), which probe the DOS above E_F, are the most powerful in studies of the electronic structure of solids. These electronic states can be also probed, respectively, with soft-x-ray-emission (SXE) and soft-x-ray-absorption (SXA) spectroscopies (Agarwal 1991, Fuggle and Inglesfield 1992). Both the occupied and unoccupied DOS can also be investigated with tunneling spectroscopy (TS) (Wolf 1985, Stroscio and Kaiser 1993). The spectroscopies mentioned above were mainly used in studies of the DOS in QCs. In this chapter, the results based on these spectroscopies are critically reviewed and compared with the theoretical predicitions.

This chapter is organized as follows. The main theoretical predictions of the electronic structure of QCs are briefly reviewed in Sect. 8.2. These predictions are discussed in detail in Chaps. 6 and 7. Spectroscopic results on the electronic structure of i and d QCs are presented in Sect. 8.3. Possible reasons for some discrepancies between theory and spectroscopic data are also discussed. Possible unique features of the electronic structure of QCs are discussed in Sect. 8.4. Section 8.5 attempts to assess the influence of quasiperiodicity on the physical properties of QCs. Concluding remarks and some suggestions for future work are made in Sect. 8.6.

8.2 Theoretical Predictions

8.2.1 Pseudogap in the Density of States

One of the first questions posed after the discovery of QCs was the following: why nature would prefer quasiperiodic to periodic order in some alloys? It was soon pointed out, based on the observation that i alloys form at compositions with a specific electron-per-atom ratio, e/a, for which the relation $Q = 2k_{\mathrm{F}}$ (Q is the magnitude of the reciprocal lattice vector and k_{F} is the radius of the Fermi sphere) is satisfied, that these alloys are the Hume-Rothery phases (Bancel and Heiney 1986, Tsai et al. 1989a, 1989b, Inoue et al. 1990). This was supported by the results of calculations within the nearly-free-electron approximation for models of i alloys (Friedel and Dénoyer 1987, Smith and Ashcroft 1987, Vaks et al. 1988, Friedel 1988, 1992). For the compositions at which the i alloys form, the Fermi surface–effective Brillouin zone interaction results in a minimum of the DOS(E_{F}) (a "pseudogap"). Modern band-structure calculations carried out for approximants (complex crystalline alloys close in composition and with a local structural similarity to QCs) to i alloys do indeed predict the existence of a structure-induced pseudogap at E_{F} (see Figs. 6.6–6.11 in Chap. 6 and Figs. 7.13, 7.14, and 7.16 in Chap. 7) in a number of i alloys (Fujiwara 1989, Fujiwara and Yokokawa 1991, Hafner and Krajčí 1992, 1993, Windisch et al. 1994, Trambly de Laissardière and Fujiwara 1994a, Krajčí et al. 1995, Dell'Acqua et al. 1997, Hennig and Teichler 1997, Krajčí and Hafner 1998). Theory thus predicts that a Hume-Rothery-type electronic mechanism is mainly responsible for the stabilization of i alloys.

Fewer approximants to d alloys are known. Therefore electronic structure calculations are available only for Al-Co-Cu, Al-Pd-Mn, Al-Fe, and Al-Mn d alloys. The first such band-structure calculations (Trambly de Laissardière and Fujiwara 1994b) made for the d approximant $Al_{60}Co_{14}Cu_{30}$, which was based on the Burkov model (Burkov 1991), predicted a broad Hume-Rothery-like pseudogap at E_{F} (see Fig. 6.11 in Chap. 6 and Fig. 7.14 in Chap. 7). However, calculations for several variants of the Burkov model showed (Sabiryanov and Bose 1994, Sabiryanov et al. 1995) that there is no pseudogap, and consequently that the d-Al-Co-Cu alloys are not stabilized by the Hume-Rothery mechanism. In another theoretical study Haerle and Kramer (1998) showed that the location of the transition metal (TM) atoms is crucial for the presence or absence of the pseudogap in the d-Al-Co-Cu system. The presence of a pseudogap was predicted in detailed band-structure calculations by Krajčí et al. (1997a) who argued that the Co and Cu d-band shifts are important for the stabilization of the *d*-Al-Co-Cu phase. These contradictory theoretical predictions with regard to the presence of a pseudogap in the *d*-Al-Co-Cu system probably result from many possible variants of the Burkov model.

Band-structure calculations for approximants to the *d*-Al-Pd-Mn system predict (Krajčí et al. 1996, Krajčí et al. 1997b) that a structure-induced pseudogap exists only in the Al band and stress the importance of the *sp-d* hybridization in its formation. The DOS calculated for approximants to the Al-Fe (Fujiwara and Yokokawa 1991) and Al-Mn (Fujiwara 1997) *d* phases predict, respectively, the presence of a pseudogap and a very deep pseudogap.

8.2.2 Fine Strucure of the Density of States

The DOS of *i* alloys is very unusual in that it consists of many very fine spiked peaks with a width of about 10 meV. This DOS spikiness results from the multiplicity of dispersionless bands. Based on band-structure calculations performed for approximants to *i* alloys, such a fine structure has been predicted for Al-Cu-Li (Fujiwara and Yokokawa 1991), Al-Mg-Li (Dell'Acqua et al. 1997), Al-Zn-Mg (Hafner and Krajčí 1992, Windish et al. 1994), Al-Mn (Fujiwara 1989), Al-Cu-Fe (Trambly de Laissardière and Fujiwara 1994a), Al-Pd-Mn (Krajčí et al. 1995), Al-Pd-Re (Krajčí and Hafner 1998), and Ti-TM (Krajčí and Hafner 1994, Hennig and Teichler 1997) *i* alloys. It has been argued (Trambly de Laissardière and Fujiwara 1994a) that the DOS spikiness is augmented by the presence of the TM atoms in the *i* alloys. The DOS spikiness is predicted even for simple models in which quasiperiodic order is introduced (Roche et al. 1997).

The presence of the DOS spikiness may explain the unusual sensitivity of the electrical conductivity σ of *i* alloys to slight changes in their composition. Such changes shift the position of E_{F}, which results in a dramatic change of the DOS(E_{F}), and consequently of the value of σ (Fujiwara et al. 1993). The unusual composition and temperature dependences of other transport properties, such as the Hall conductivity and the thermoelectric power, can be also qualitatively explained (Fujiwara et al. 1993) by the presence of the DOS fine structure.

The DOS spikiness is predicted to be also present in *d* alloys (Fujiwara and Yokokawa 1991, Trambly de Laissardière and Fujiwara 1994b, Sabiryanov and Bose 1994, Krajčí et al. 1997a, Haerle and Kramer 1998). However, it is attenuated, in comparison to that in *i* alloys, by the effect of the periodic direction (Trambly de Laissardière and Fujiwara 1994b). It thus appears that the predicted fine structure of the DOS is important for explaining unusual dependences on the composition and temperature of some electronic transport properties of QCs.

The fine structure of the DOS has been calculated with different techniques for various approximant systems and is therefore regarded as an intrinsic characterisic of the electronic structure of QCs. However, a concern has been expressed that it may be an artifact resulting from the insufficient computational precision used in the calculations (Haerle and Kramer 1998).

8.3 Experimental Results

In this section an overview of the spectroscopic data on the DOS below and above E_F is presented. An experimental evidence for the presence of the predicted pseudogap and of the DOS spikiness in i and d QCs is discussed.

8.3.1 s, p-Bonded Icosahedral Alloys

The first stable i alloy was discovered in the Al-Cu-Li system (Dubost et al. 1986). There is an intermetallic compund in this system, called the R phase, which is the 1/1 approximant to the i-Al-Cu-Li alloy. The electronic structure calculations carried out for this approximant (Fujiwara and Yokokawa 1991) revealed the presence of a pseudogap in the DOS at E_F. This was confirmed later by Windisch et al. (1994) who found that the DOS minimum occurs just 30 meV below E_F. Calculations for higher-order approximants, up to a 8/5 approximant (Windisch et al. 1994), confirmed that a deep, structure-induced pseudogap slightly below E_F (Fig. 7.13 in Chap. 7) is a generic property of the i-Al-Cu-Li alloys.

PES spectra of the i-Al-Cu-Li alloy and the R phase measured at photon energies $h\nu = 100$ and 40 eV, with an energy resolution respectively of 0.7 and 0.3 eV, are shown in Fig. 8.1a. The measured PES intensity is the weighted average of the local partial DOS, with the weight proportional to the $h\nu$-dependent photoionization cross section, σ_ph (Stadnik et al. 1995a). A strong feature at the binding energy (BE) of about -4.1 eV (the minus sign indicates the energies below E_F which is assigned the value of 0.0 eV) is predominantly due to the Cu $3d$-derived states. This conclusion is based on the fact that σ_ph of Cu $3d$ orbitals is about two orders of magnitude larger than σ_ph of Al and Li sp orbitals and on its weak dependence on $h\nu$ in the $h\nu$ range 40–100 eV (Yeh 1993). Since the values of σ_ph of Al and Li sp orbitals decrease by a factor of about three in the the $h\nu$ range 40–100 eV (Yeh 1993), an enhancement of the broad feature in the BE region between 0 and -3 eV in the PES spectrum measured at $h\nu = 40$ eV (Fig. 8.1a) as compared to the feature in the $h\nu = 100$ eV spectrum shows that this feature is mainly of the Al and Li sp character.

A comparison between the PES spectra calculated (Windisch et al. 1994) on the basis of the DOS for the 5/3 (Fig. 7.13 in Chap. 7) and 1/1 approximants, and the experimental PES spectra from Fig. 8.1a, is presented in Fig. 8.1b. The position of the DOS peak due to the Cu $3d$-like states at $BE \approx -3.2$ eV (Fig. 7.13 in Chap. 7) is different than the corresponding position in the experimental spectrum (Fig. 8.1a). This is due to the

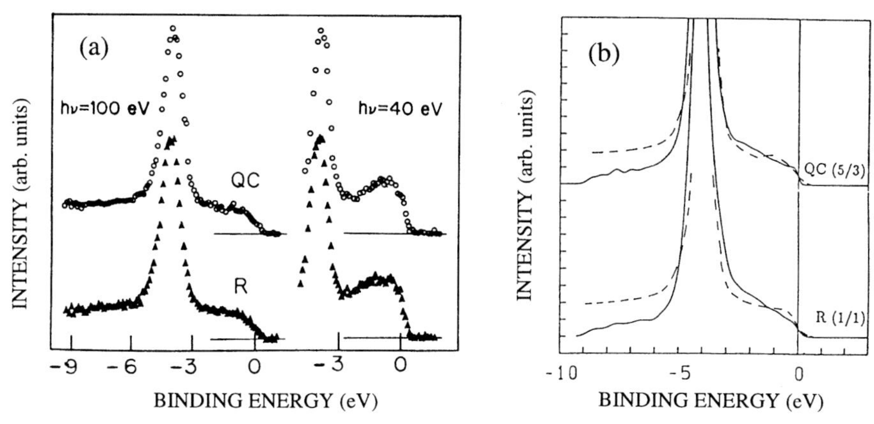

Fig. 8.1. (a) Valence bands of *i* QC $Al_{55}Li_{35.8}Cu_{9.2}$ and *R* phase $Al_{54}Li_{36.8}Cu_{9.2}$ measured at $h\nu = 100$ and 40 eV. After Matsubara et al. (1991). (b) Comparison between the $h\nu = 100$ eV valence bands of *i* QC $Al_{55}Li_{35.8}Cu_{9.2}$ and *R* phase $Al_{54}Li_{36.8}Cu_{9.2}$ (broken lines) from (a) with the calculated PES spectra for a 5/3 approximant and *R* phase (solid lines). After Windisch et al. (1994).

fact that PES spectra involve excited-state eigenvalues, whereas calculated DOS involves ground-state energies (Knapp et al. 1979, Fuggle and Inglesfield 1992). For narrow *d*-band systems, a comparison between the theoretical and experimental PES spectra requires an inclusion of the self-energy correction in theoretical calculations. This correction was simulated (Windisch et al. 1994) by shifting the calculated Cu *d* DOS so as to match the position of the experimental Cu-3*d* peak (Fig. 8.1b). There is reasonable agreement between theory and experiment with respect to the positions and intensities of the main features of the total DOS. The discrepancies at the *BE*'s lower (more negative) than about −5 eV (Fig. 8.1b) are due to the fact that the experimental PES spectra were not corrected for the secondary-electron contribution (Stadnik et al. 1995a).

The experimental PES spectra of the *i*-Al-Cu-Li and its 1/1 approximant (Fig. 8.1a) are virtually the same, although after normalization of the Cu 3*d* intensity Matsubara et al. (1991) found that the PES intensity at E_F of the *i* phase is slightly smaller than that of the *R* phase. This agrees with the observation that the electronic specific heat coefficient γ, which is proportional to DOS(E_F), is slightly smaller for the *i*-Al-Cu-Li phase than for the *R* phase (Mizutani et al. 1991).

The decrease of the experimental PES intensity towards E_F (Fig. 8.1a) is often interpreted as evidence for the existence of a pseudogap in the DOS at E_F. This is an overinterpretation of the experimental data. The PES technique probes the electronic states below E_F, but the pseudogap involves both occupied and unoccupied states. Thus, conclusive evidence for the existence of a pseudogap requires probing the electronic states both below and above

E_F. The only relaible conclusion that can be drawn from the PES intensity decrease towards E_F (Fig. 8.1a) is that it is indicative of the presence of a pseudogap around E_F. Similarly, interpreting the small value of γ found in many QCs as evidence for the existence of a DOS pseudogap at E_F is not justified as γ is a measure of the DOS at only one particular energy (E_F). A small value of γ for a given QC as compared, e.g., to that of Al metal is merely suggestive of the presence of a pseudogap at E_F in that QC, but does not prove its existence.

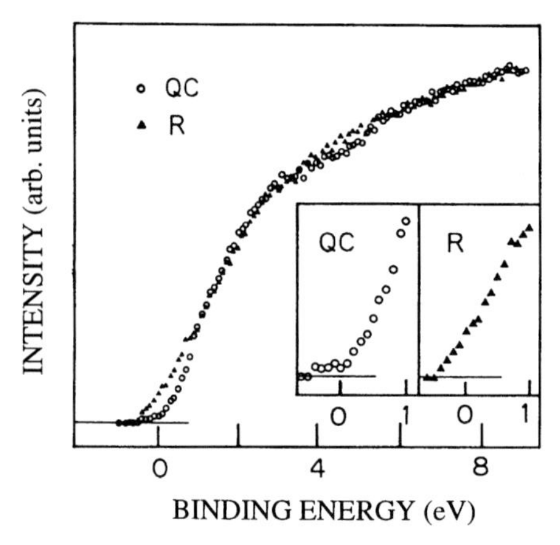

Fig. 8.2. IPES spectra of *i* QC $Al_{55}Li_{35.8}Cu_{9.2}$ and *R* phase $Al_{54}Li_{36.8}Cu_{9.2}$. The relative intensities have been normalized at the high-*BE* region. The insert shows the spectra close to E_F. After Matsubara et al. (1991).

A comparison between the IPES spectra of the *i*-Al-Cu-Li and *R* phases (Fig. 8.2) shows that the pseudogap is opening slightly above, rather than below E_F, and that it is shallower in the latter than in the former. A combination of PES and IPES spectra thus establishes that, as predicted by theory, there is a pseudogap in the DOS in the *i*-Al-Cu-Li alloy and the *R* phase, but slightly above, rather than below E_F. The difference between theory and experiment as to location of the minimum of the DOS is most probably due to the difference between the composition of the samples used in the experiment and that used in the calculations. This is based on the observation (Fujiwara and Yokokawa 1991) that the position of E_F in the Al-Cu-Li alloys shifts with the change of the Cu concentration.

Electronic structure calculations predict the presence of a pseudogap in the DOS at E_F for the stable Frank-Kasper phase $Al_{13}Zn_{36}Mg_{32}$ (the 1/1 approximant to the *i* phase) and for higher-order approximants to the *i* phase in the Al-Zn-Mg system (Hafner and Krajčí 1992, 1993). Such a pseudgap is also predicted (Fig. 6.9 in Chap. 6) for the 1/1 approximants in the $Al_xZn_{60-x}Mg_{40}$ (x = 15–22.5) series (Mizutani et al. 1998). A slight apparent decrease of the intensity in the x-ray photoelectron spectroscopy (XPS) valence bands of the 1/1 approximants $Al_xZn_{60.5-x}Mg_{39.5}$ ($x = 20.5$–45.5) and the *i* QCs $Al_xZn_{56-x}Mg_{44}$ ($x = 13$–25) as compared to the intensity of the band of pure Al (Takeuchi and Mizutani 1995) was taken as evidence for the presence of the pseudogap in these alloys. As mentioned earlier, a conclusive evidence requires probing the states above E_F. XPS valence bands of the *i*-$(Al_xGa_{1-x})_{20.5}Zn_{40}Mg_{39.5}$ ($x = 0.0$–0.8) alloys (Mizutani et al. 1993) show a decrease of the XPS intensity towards E_F which is indicative of the presence of a pseudogap in this system.

It can be concluded that there is a strong spectroscopic evidence for the presence of a pseudogap in the *i*-Al-Cu-Li system and an indication of its presence in the Al-Zn-Mg and Al-Ga-Zn-Mg *i* QCs.

8.3.2 Al-Cu-Transition Metal Icosahedral Alloys

Three stable *i* alloys of high structural quality exist in the Al-Cu-TM (TM=Fe,Ru,Os) system. Electronic structure calculations were carried out for the 1/1 approximant to the *i*-Al-Cu-Fe phase by Trambly de Laissardière and Fujiwara (1994a). They predict the presence of a pseudogap with a width of ~ 0.5 eV whose center is located at ~ 0.3 eV above E_F (Fig. 6.8 in Chap. 6).

In order to establish the origin of the main features in the measured valence bands of *i*-$Al_{65}Cu_{20}TM_{15}$ alloys, the resonant-PES and PES-near-the-Cooper-minimum methods were employed (Mori et al. 1991, 1992, Stadnik and Stroink 1993, Stadnik et al. 1995a–1995c, Zhang et al. 1995). As an example, these methods are illustrated for *i*-$Al_{65}Cu_{20}Os_{15}$. For TM elements the resonance occurs at excitation energies near the np and nf thresholds. For Os one would expect the stronger $5p \rightarrow 5d$ transition at about 44.5 eV ($5p_{3/2} \rightarrow 5d$) and the weaker $4f \rightarrow 5d$ transition to take place at about 50.7 eV ($4f_{7/2} \rightarrow 5d$) (Zhang et al. 1995). The Os $5d$-derived features should be enhanced or suppressed as $h\nu$ is swept through the $5p$-$5d$ and $4f$-$5d$ thresholds. It can be seen in Fig. 8.3a that, as $h\nu$ increases, the relative intensity of the peak at $BE \approx -1.5$ eV with respect to the peak at $BE \approx -3.7$ eV decreases first and reaches its minimum at $h\nu = 44.5$ eV. Then it increases slightly between $h\nu = 50$ and 60 eV. This corresponds to the stronger $5p \rightarrow 5d$ resonance (the $4f \rightarrow 5d$ resonance was too weak to be observed in the measured valence band) and indicates that the feature at $BE \approx -1.5$ eV is predominantly due to the Os $5d$-derived states (Zhang et al. 1995).

The features in the valence bands originating from the $4d$ and/or $5d$ elements can be also identified by making use of the Cooper-minimum effect, which is the minimum in the σ_{ph} values for particular orbitals at a specific value of $h\nu$ (Stadnik et al. 1995a, 1995b, Zhang et al. 1995). Since the decrease in the σ_{ph}(Os $5d$) in the $h\nu$ range 50–200 eV is expected to be two orders of magnitue (Yeh 1993), a strong suppresion of the contribution of the Os $5d$-like states to the valence band of *i*-$Al_{65}Cu_{20}Os_{15}$ is anticipated. Figure 8.3b shows that as $h\nu$ increases from 80 eV, the relative contribution of the feature

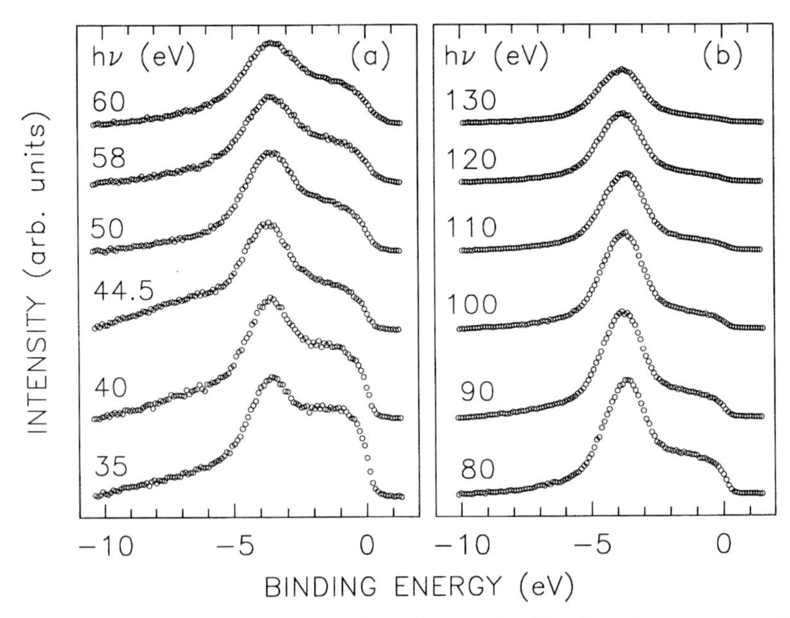

Fig. 8.3. Room-temperature valence bands of *i*-$Al_{65}Cu_{20}Os_{15}$ measured for different photon energies (a) around the Os $5p \rightarrow 5d$ transition and (b) around the Os $5d$ Cooper minimum. After Stadnik et al. 1995b and Zhang et al. (1995).

at $BE \approx -1.5$ eV with respect to the other feature at $BE \approx -3.7$ eV, which originates from the Cu $3d$-derived states, decreases steadily and at $h\nu = 130$ eV is almost completely suppressed. In other words, since $\sigma_{\mathrm{ph}}(\mathrm{Cu}\ 3d)$ decreases much slower with the increase of $h\nu$ than $\sigma_{\mathrm{ph}}(\mathrm{Os}\ 5d)$, and since $\sigma_{\mathrm{ph}}(\mathrm{Cu}\ 3d)$ for $h\nu = 130$ eV is expected to be almost two orders of magnitude larger than $\sigma_{\mathrm{ph}}(\mathrm{Os}\ 5d)$ (Yeh 1993), the $h\nu = 130$ eV valence band in Fig. 8.3b must be almost entirely due to the Cu $3d$ emission.

As the bands for the $h\nu$ values around the Os $5d$ Cooper minimum exhibit much more dramatic changes in the intensity then those for the $h\nu$ values around the Os $5p \rightarrow 5d$ resonance (Fig. 8.3), the former can be used to derive the partial DOS due to Cu $3d$-like and Os $5d$-like states either by a direct subtraction method or by a method described below (Zhang et al. 1995). Assuming a negligible influence of the matrix-element effects, the intensity $I(h\nu, BE)$ of the valence band of *i*-$Al_{65}Cu_{20}Os_{15}$ can be represented by

$$I(E_{\mathrm{B}}) = C(h\nu)[\sigma_{\mathrm{Cu}}(h\nu)D_{\mathrm{Cu}}(BE)/Z_{\mathrm{Cu}} + \sigma_{\mathrm{Os}}(h\nu)D_{\mathrm{Os}}(BE)/Z_{\mathrm{Os}}], \quad (8.1)$$

where D_i ($i =$ Cu,Os) is the partial DOS of the ith element. The instrumental factor is represented by $C(h\nu)$. The partial DOS is assumed to fulfill the normalization condition,

$$\sum_i D_i \Delta BE = N_i Z_i, \quad (8.2)$$

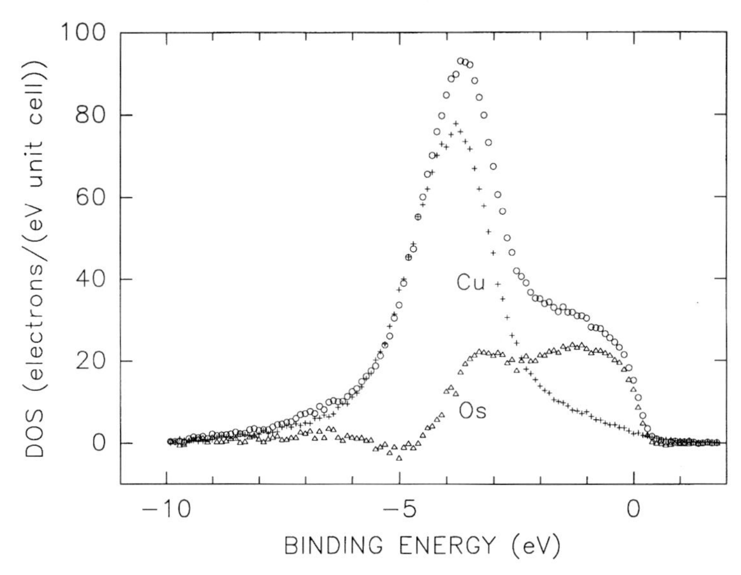

Fig. 8.4. Partial DOS of the Os $5d$ (trangles) and the Cu $3d$ (open circles) character obtained from the $h\nu = 80$ and 130 eV valence bands in Fig. 8.3b and by using the linear relation between the measured intensity and the parial DOS and σ_{ph} (Zhang et al. 1995). Their sum (full circles) represents the total DOS of the d character in i-$Al_{65}Cu_{20}Os_{15}$. After Stadnik et al. (1995b) and Zhang et al. (1995).

where N_i and Z_i are respectively the concentration and the number of d electrons of the ith element. For i-$Al_{65}Cu_{20}Os_{15}$, $N_{\mathrm{Cu}} = 20$, $N_{\mathrm{Os}} = 15$, and $Z_{\mathrm{Cu}} = 10$, $Z_{\mathrm{Os}} = 6$ (Zhang et al. 1995). By using Eq. (8.1) for two different $h\nu$ values, the partial DOS of the Cu $3d$-derived and Os $5d$-derived states can be obtained (Fig. 8.4).

Using Eq. (8.1), one should be able to reproduce the experimental valence bands measured for all $h\nu$ values provided that the Cu $3d$ and Os $5d$ partial DOS in Fig. 8.4 are a good measure of the true Cu $3d$ and Os $5d$ DOS in i-$Al_{65}Cu_{20}Os_{15}$. A good agreement (Fig. 8.5) between the valence bands calculated for all $h\nu$ values with the corresponding experimental valence bands confirms the reliability of the derived partial Cu $3d$ and Os $5d$ DOS (Fig. 8.4).

The results based on the resonant-PES and PES-near-the-Cooper-minimum studies of the i-$Al_{65}Cu_{20}TM_{15}$ QCs established (Mori et al. 1991, 1992, Stadnik and Stroink 1993, Stadnik et al. 1995a–1995c, Zhang et al. 1995) the origin of the two main features in the valence bands of these QCs. Also the partial d-like DOS for the QCs with TM=Ru,Os were derived. For these two QCs, the experimental total and partial DOS could not be compared with theoretical DOS as these have not been calculated yet. A small measured intensity at E_{F} (Figs. 8.3–8.5) is indicative of a small values of DOS(E_{F}) in these QCs. However, due to the medium energy resolution used (~ 0.4 eV) and the fact that the PES experiments were carried out at room tempera-

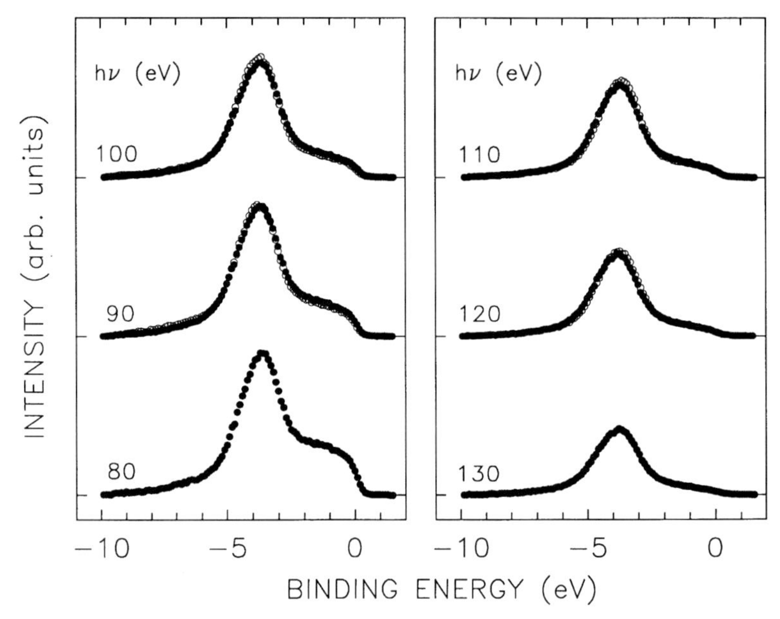

Fig. 8.5. Comparison of the valence bands generated from Eq. (8.1) using the empirical partial Cu $3d$ and Os $5d$ DOS of Fig. 8.4 (full circles) with the experimental valence bands from Fig. 8.3b (open circles). After Stadnik et al. (1995b) and Zhang et al. (1995).

ture, it was not possible to make a meaningful conclusion about the possible pseudogap as the intensity decrease toward E_F could not be separated from the Fermi edge cutoff.

The low-temperature ultraviolet photoelectron spectroscopy (UPS) He II ($h\nu$ = 40.8 eV) valence bands of the i series Al-Cu-TM (TM=Fe,Ru,Os) have a similar two-peak structure (Fig. 8.6a). The feature at $BE \approx -4.1$ eV is mainly due to the Cu $3d$-derived states. The resonant and the Cooper-minimum PES experiments, together with the PES experiments carried out in the constant-initial-state mode, have established (Mori et al. 1991, 1992, Stadnik and Stroink 1993, Stadnik et al. 1995a–1995c, Zhang et al. 1995) that the broad feature close to E_F is predominantly due to states of Fe $3d$, Ru $4d$, and Os $5d$ character, as appropriate. The positions and intensities of the Cu and Fe $3d$ features in the valence band of i-Al-Cu-Fe (Fig. 8.6a) agree relatively well with those obtained from the theoretical DOS (Fig. 6.8 in Chap. 6) appropriately broadened to account for the lifetime broadening effects inherent to the PES technique and the finite energy resolution of a PES experiment (Stadnik et al. 1995c).

The decrease of the spectral intensity toward E_F, which is clearly distinguishable from the Fermi-edge cutoff in the high-energy-resolution, low-temperature UPS spectra (Figs. 8.6a–8.6c), can be shown to be compatible

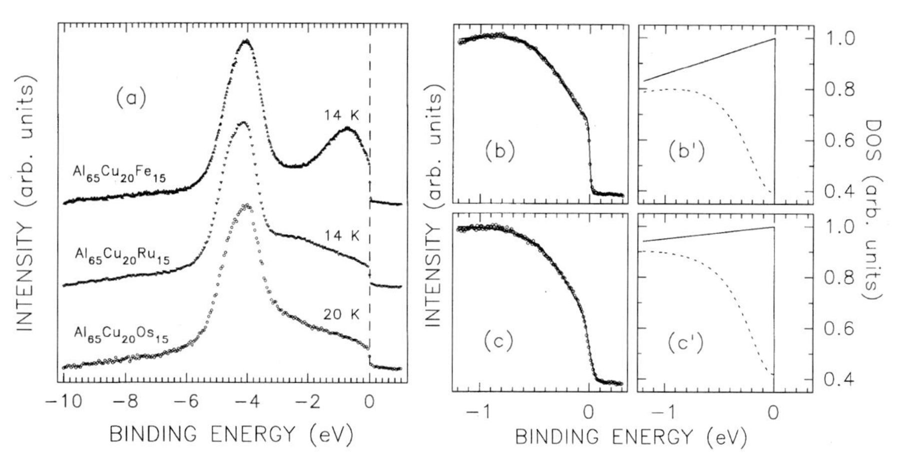

Fig. 8.6. (a) Low-temperature He II valence bands of *i* QCs $Al_{65}Cu_{20}TM_{15}$ (TM=Fe,Ru,Os). The energy resolution is $\sim$ 30 meV. He II valence band regions close to E_F (open circles) of *i*-$Al_{65}Cu_{20}Fe_{15}$ measured at (b) 14 K and (c) 283 K with an energy resolution of 31.6(1.8) meV fitted (solid lines) with Eq. (8.1) to the corresponding model DOS in (b′) and (c′). The solid lines in (b′) and (c′) represent the normal DOS at 0 K, and the broken lines show the Lorentzian dip which must be subtracted from the normal DOS in order to fit the valence-band regions in (b) and (c). After Stadnik et al. (1997).

with the presence of a pseudogap using the model proposed by Mori et al. (1991). As conventional alloys of QC-forming elements do not generally display a DOS minimum close to E_F, the model assumes that a simple linear extrapolation of the spectra for the *BE* range directly before the peak of the valence band feature close to E_F represents the DOS of an alloy without a pseudogap (the normal DOS). The presence of the pseudogap would result in an intensity dip which is assumed to be of Lorentzian shape centered at E_F, characterized by the half-width Γ_L and the dip depth C relative to the normal DOS. Thus, the observed intensity $I(E_B)$ is the convolution of the normal DOS multiplied by the pseudogap Lorentzian function and by the Fermi-Dirac function $f(E_B, T)$, and the experimental resolution Gaussian function

$$I(E_B) = \int N(ax+b)\left(1 - \frac{C\Gamma_L^2}{x^2+\Gamma_L^2}\right) f(x,T)\exp\left[-\frac{(x-E_B)^2}{2s^2}\right] dx, \quad (8.3)$$

where N is a normalization factor, the experimental Gaussian full width at half maximum (FWHM) is related to s through FWHM $= 2\sqrt{2\ln 2}s$, and the constants a and b are determined from a linear fit of the spectra for the *BE* range directly before the peak of the valence band feature close to E_F. The C values of 0 and 100% correspond respectively to the normal DOS (no pseudogap) and no DOS(E_F) (full gap). This model (Figs. 8.6b′ and 8.6c′) fits well the near-E_F region of the valence bands of *i*-$Al_{65}Cu_{20}Fe_{15}$ at 14 and 283 K (Figs. 8.6b and 8.6c) for values of C and Γ_L equal to 60.5(3)%, 0.36(2) eV and 58.0(6)%, 0.33(1) eV, respectively. The width of the pseudogap is comparable to that predicted by theory. The near-$E_{\rm F}$ region of valence bands of other *i*-Al-Cu-TM alloys can be successfully fitted with this model (Stadnik et al. 1996, 1997, Kirihara et al. 1998), thus providing support for the existence of a pseudogap in these *i* alloys. It should be stressed, however, that the model does not prove the existence of the pseudogap since the PES technique probes only the occupied states.

The TS technique has the unique capability of probing the electronic states below and above $E_{\rm F}$ (Wolf 1985, Stroscio and Kaiser 1993). It is thus especially sensitive to any gap in the quasiparticle excitation spectrum at $E_{\rm F}$ which shows up directly as a characteristic feature near zero bias in the raw data conductance curves. The conductance $G = dI/dV$, where I and V are the tunnel current and the dc bias voltage [at negative (positive) sample bias, occupied (unoccupied) electron states are probed]. The G vs V dependencies are proportional to the DOS (Wolf 1985, Stroscio and Kaiser 1993). Compared to other techniques, TS has a very high energy resolution ($\sim k_{\rm B}T$), which is less than 1 meV for spectra measured at liquid helium temperatures. The DOS of the *i*-$Al_{62.5}Cu_{25}Fe_{12.5}$ alloy in the form of a film 3000 Å thick, which was obtained from the tunneling spectrum after the subtraction of the $E^{1/2}$ ($E = eV$) background, exhibits a pseudogap 60 meV wide which is centered at $E_{\rm F}$ (Fig. 8.7a). A similar narrow pseudogap was observed in the tunneling spectra of *i*-$Al_{63}Cu_{25}Fe_{12}$ in the ribbon form (Fig. 8.7b).

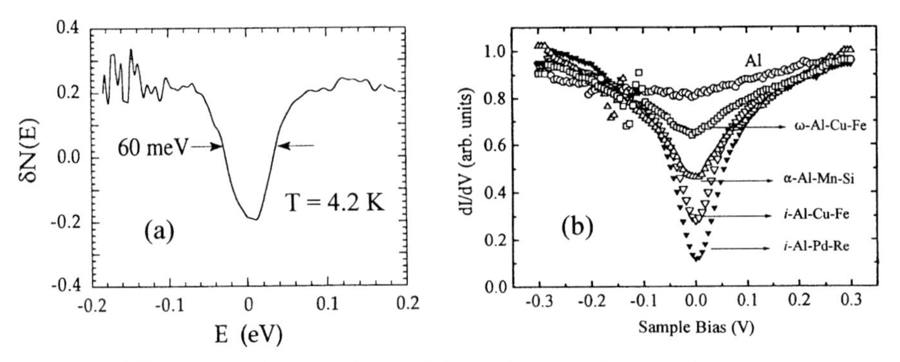

Fig. 8.7. (a) Density of states obtained from the tunneling conductance spectrum of the *i*-$Al_{62.5}Cu_{25}Fe_{12.5}$ 3000 Å-thick film. After Klein et al. (1995). (b) Tunneling spectra of Al single crystal, crystalline ω-Al_7Cu_2Fe and α-$Al_{72.4}Mn_{17.5}Si_{10.1}$ phases, and *i*-$Al_{63}Cu_{25}Fe_{12}$ and *i*-$Al_{70.5}Pd_{21}Re_{8.5}$ at 4.2 K. After Davydov et al. (1996).

The observed width of a pseudogap in the TS spectra of *i*-Al-Cu-Fe is an order of magnitude smaller than that predicted by theory. Also its location at $E_{\rm F}$ is at variance with the predicted location at $\sim$ 0.3 eV above $E_{\rm F}$ (Fig. 6.8 in Chap. 6). It was suggested (Klein et al. 1995, Davydov et al. 1996) that the observed narrow pseudogap corresponds to one of the spiky structures predicted by theory rather than to a wide Hume-Rothery-like pseudogap. If this is true, then one would expect to observe at least several such structures in the tunneling spectra, which is clearly not the case (Fig. 8.7). The possible reasons for this disagreement are discussed in Sect. 8.3.7.

A crystalline alloy Al_7Cu_2Fe (ω phase), which has a composition very close to that of *i*-Al-Cu-Fe alloys but which is not an approximant to the *i*-Al-Cu-Fe phase, is often used as a reference material in studies of the physical properties of *i*-Al-Cu-Fe QCs. It is a metallic alloy with an electrical resistivity ρ of 22 $\mu\Omega$ cm at 4.2 K and a positive temperature coefficient of ρ (Davydov et al. 1996). Electronic structure calculations predict the presence of a pseudogap in this crystalline alloy (Belin and Mayou 1993, Trambly de Laissardière et al. 1995a, 1995b). However, the $V^{1/2}$ dependence of dI/dV observed in this alloy (Fig. 8.7b) indicates the lack of a pseudogap around E_F (Davydov et al. 1996). On the other hand, the SXE and SXA spectra were interpreted as evidence for the presence of a pseudogap in this alloy (Sadoc et al. 1993, Belin and Mayou 1993, Belin et al. 1993, Trambly de Laissardière et al. 1995a). The problem with this interpretation, which has been also used in numerous papers on the SXE and SXA studies of QCs, is related to the method of combining the SXE and SXA spectra at E_F which secures the occurence of a pseudogap in any alloy. The measured intensity in the SXE/SXA or PES/IPES depends not only on the respective DOS but also on other factors intrinsic to the technique which are notoriously difficult to evaluate (Agrawal 1991, Fuggle and Inglesfield 1992, Kortboyer et al. 1989). This, together with a large uncertainty in the determination of the position of E_F in typical SXE/SXA measurements, makes it challenging to have a reliable combination of the SXE/SXA spectra on the same energy and intensity scale. The problem of whether there is a minimum in the DOS(E_F) in crystalline Al_7Cu_2Fe remains an open question.

The Al *s*- and *d*-like unoccupied DOS in *i*-$Al_{65}Cu_{20}Ru_{15}$ was probed with the EELS *L*-edge measurements with an energy resolution of 0.12 eV (Terauchi et al. 1996). As in the UPS spectrum of this alloy (Fig. 8.6a), a clear Fermi edge was observed (Fig. 8.8) which proves that this alloy, in spite of its very high value of ρ, is metallic. As compared to the Al *L*-edge spectrum of Al metal measured with a resolution of 0.2 eV, the L_3 edge is shifted by 0.2 eV to 73.1 eV (Fig. 8.8). The key features of the Al *L*-edge spectrum of *i*-$Al_{65}Cu_{20}Ru_{15}$ is a small rise of intensity at the L_3 edge as compared to that of Al metal and "scooping out" of the DOS near the edge onset. This indicates the occurence of the pseudogap of the estimated half width at half maximum of ~ 0.9 eV in the Al *s* and *d* states at E_F (Terauchi 1996, 1998a).

In summary, spectroscopic data reviewed above (Davydov et al. 1996, Kirihara et al. 1998, Klein et al. 1995, Mori et al. 1991, Stadnik et al. 1996, 1997, Terauchi 1996, 1998a) are compatible with the presence of the pseudogap in the DOS around E_F in the *i*-Al-Cu-TM (TM=Fe,Ru,Os) alloys.

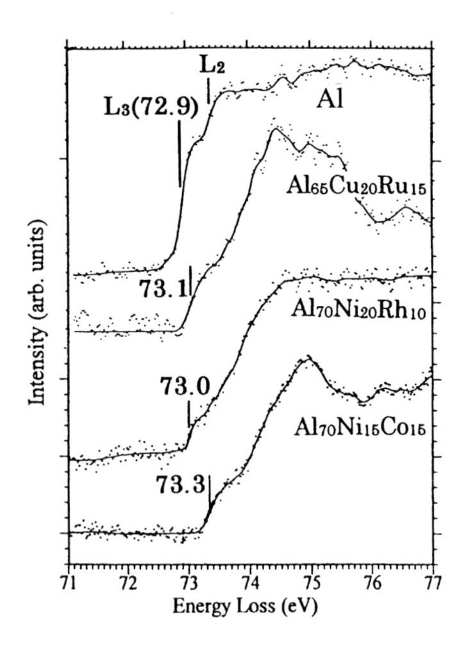

Fig. 8.8. EELS Al L-edge spectra of Al, i QC $Al_{65}Cu_{20}Ru_{15}$, and d QCs $Al_{70}Ni_{20}Rh_{10}$ and $Al_{70}Ni_{15}Co_{15}$. After Terauchi et al. (1998a).

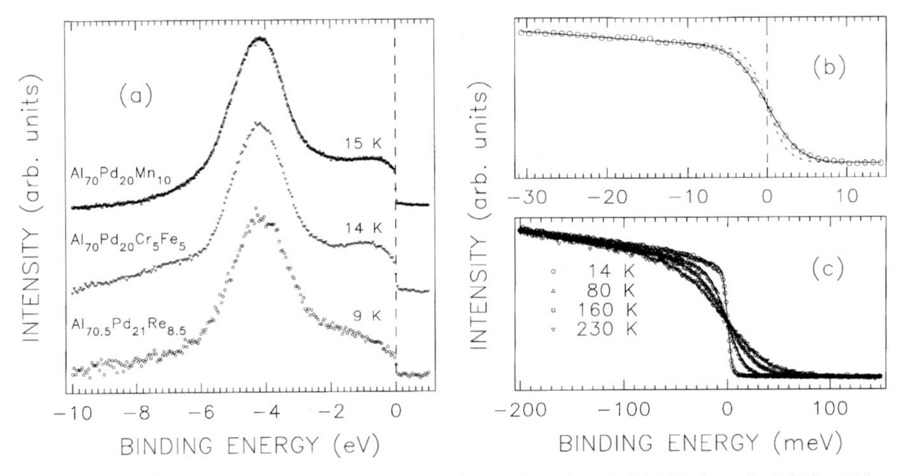

Fig. 8.9. (a) Low-temperature He II valence bands of Al-Pd-based i QCs. The energy resolution is $\sim$ 30 meV. (b) Near-E_F He I valence band of i-$Al_{70}Pd_{20}Mn_{10}$ at 14 K. The solid line is the fit to a linearly decreasing intensity multiplied by the Fermi-Dirac function at 14 K (broken curve) and convoluted with a Gaussian whose FWHM = 5.8(2) meV. The step between the data points is 1 meV. (c) Near-E_F He I valence band of i-$Al_{70}Pd_{20}Mn_{10}$ measured at different temperatures. The solid lines are the fits as described in (b). Note the different BE scales in (b) and (c). After Stadnik et al. (1996, 1997) and Stadnik and Purdie (1997).

8.3.3 Al-Pd-Mn Icosahedral Alloys

The electronic structure calculations carried out for approximants to the *i* phase (Krajčí et al. 1995) predict a structure-induced minimum slightly above E_F in the DOS (Fig. 7.14 in Chap. 7). This minimum is strongly pronounced in the Al band and rather weakly formed in the Pd and Mn bands.

The low-temperature UPS He II valence band of i-$Al_{70}Pd_{20}Mn_{10}$ (Fig. 8.9a) has a two-peak structure. As is shown below, the feature at $BE \approx -4.2$ eV is mainly due to the Pd 4*d*-like states, and the broad feature close to E_F is predominantly due to the states of the Mn 3*d* character. A clear presence of the Fermi edge in the near-E_F UPS He I ($h\nu = 21.2$ eV) valence band (Fig. 8.9b) and its temperature evolution following the Fermi-Dirac function (Fig. 8.9c) show that i-$Al_{70}Pd_{20}Mn_{10}$, in spite of its large ρ, is metallic.

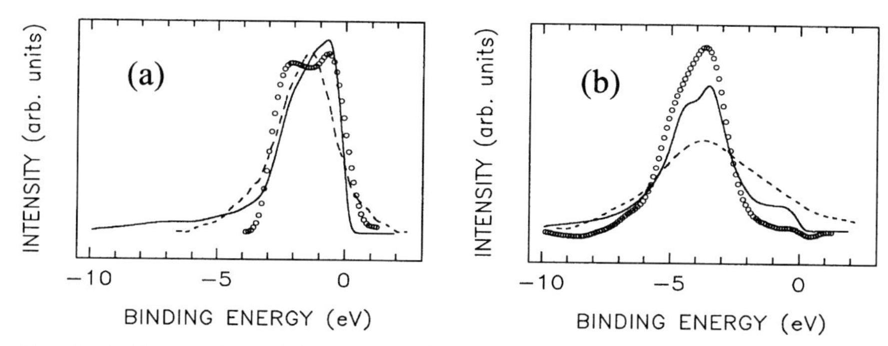

Fig. 8.10. Comparison of the partial DOS of the Mn 3*d* (a) and Pd 4*d* (b) character in *i*-Al-Pd-Mn obtained from the PES spectra (open circles, Zhang et al. 1994) and from the SXE measurements (dashed line, Belin et al. 1994a) with the corresponding broadened theoretical partial DOS (solid line, Krajčí et al. 1995). After Zhang and Stadnik (1995).

The occupied partial *d*-like and total DOS predicted by theory (Krajčí et al. 1995) for *i*-Al-Pd-Mn compare well with those determined from the room-temperature PES measurements on i-$Al_{70}Pd_{20}Mn_{10}$ (Zhang et al. 1994, Zhang and Stadnik 1995). This is illustrated in Fig. 8.10 where the partial DOS of the Mn-3*d* and Pd-4*d* character, which were derived from the PES spectra measured with an energy resolution of ~ 0.4 eV for $h\nu$ values close to the Pd 4*d* Cooper minimum (Zhang et al. 1994), are compared with the corresponding calculated DOS of the 8/5 approximant to the *i* phase (Fig. 7.14 in Chap. 7) which were appropriately broadened (Zhang and Stadnik 1995) to account for the lifetime broadening effects inherent to the PES technique and for the finite resolution of a PES experiment. There is a notable agreement with respect to the overall structure of the partial Mn 3*d* and Pd 4*d* DOS This structure is not present in the SXE spectra (Fig. 8.10) due to severe lifetime broadening effects inherent to the SXE transitions probing the Mn and Pd d states (Zhang and Stadnik 1995).

In order to compare the theoretical total DOS calculated for the 8/5 approximant to the i phase (Fig. 7.14 in Chap. 7) with the valence band of i-$Al_{70}Pd_{20}Mn_{10}$ measured at a given $h\nu$, the theoretical Al, Pd, and Mn partial DOS have to be first multipled by the corresponding σ_{ph} at that $h\nu$. The sum of the products of the theoretical partial DOS and the corresponding σ_{ph}, which was broadened in the same way as described above, compares well with the experimental valence band measured at $h\nu = 100$ eV (Fig. 8.11). It is thus concluded that the theoretical occupied DOS (Krajčí et al. 1995) agrees well with the measured PES spectra (Stadnik et al. 1996, 1997, Zhang et al. 1994, Zhang and Stadnik 1995).

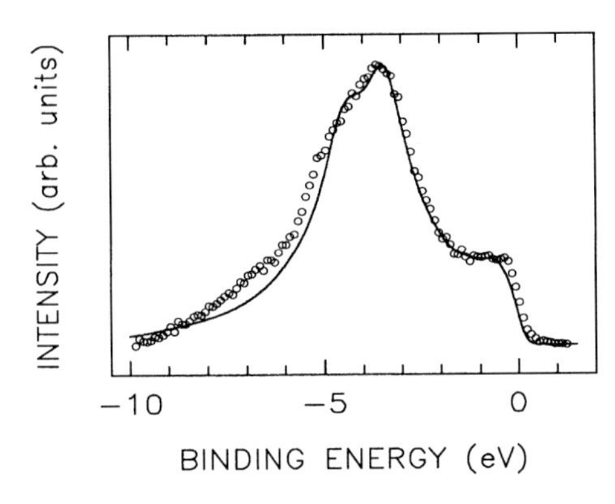

Fig. 8.11. Comparison of the valence band of i-$Al_{70}Pd_{20}Mn_{10}$ measured at $h\nu = 100$ eV (open circles, Zhang et al. 1994) with the broadened theoretical DOS for the 8/5 approximant to the i phase (solid line, Krajčí et al. 1995). After Zhang and Stadnik (1995).

The existence of a pseudogap in the DOS at E_F cannot be determined unambiguously from the PES spectra measured at room temperature with a medium resolution of 0.4 eV (Fig. 8.11) in which the decrease of the spectral intensity due to the possible presence of the pseudogap cannot be distinguished from the Fermi cutoff. The near-E_F region of the low-temperature UPS spectrum of i-$Al_{70}Pd_{20}Mn_{10}$ (Fig. 8.9a) can be fitted well to Eq. (8.3) and this can be interpreted as being compatible with the presence of the pseudogap at E_F with $\Gamma_L = 0.22(2)$ eV (Stadnik et al. 1996, 1997).

The lack of the Fermi edge in the angle-resolved, high-energy-resolution (50 meV), room-temperature PES valence band of a single-grain i alloy $Al_{70}Pd_{21.5}Mn_{8.5}$ was interpreted (Wu et al. 1995) as direct evidence of a pseudogap in the DOS near E_F. However, this is in contradiction to the presence of the Fermi edge in the high-energy-resolution low-temperature spectra of both polyquasicrystalline (Fig. 8.9) and single-grain (Purdie et al. 1998) i-Al-Pd-Mn alloys.

An unambigous experimental evidence for the presence of the pseudogap was provided by a PES experiment with an energy resolution of 70 meV (Neuhold et al. 1998) on a single grain i-$Al_{70.5}Pd_{21}Mn_{8.5}$ at 570 K. In this

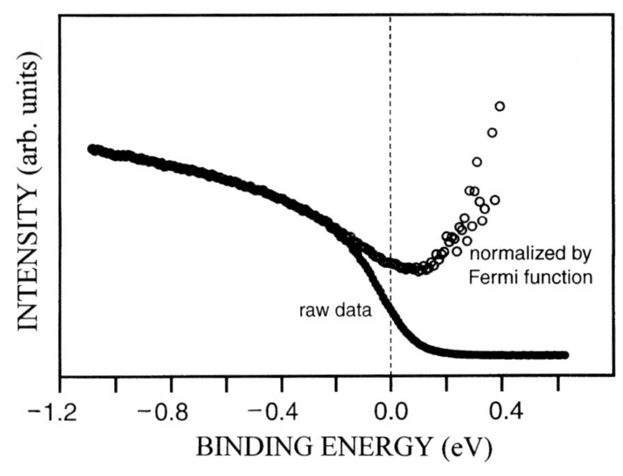

Fig. 8.12. Valence band of i-$Al_{70.5}Pd_{21}Mn_{8.5}$ at 570 K measured at $h\nu = 32.3$ eV (full circles) and the reconstructed spectral function (open circles) near E_F. After Neuhold et al. (1998).

experiment, the states both below and above E_F were probed. The PES technique can probe the states also above E_F is they are populated. According to Fermi-Dirac statistics, in metallic systems these states become populated at nonzero temperatures (Fig. 8.9c). At a temperature of 570 K, a region of several hundred meV above E_F becomes accessible (Fig. 8.12). A spectral function, which is proportional to the DOS, can be reconstructed from the 570 K valence band (Neuhold et al. 1998). It clearly shows a minimum of DOS located at 90 meV above E_F (Fig. 8.12). For BE's higher than ~ 0.3 eV ($6.1 k_B T$), a considerable scatter of the reconstructed spectral function, which is due to the small probability that the states are populated, prevents a meaningful evaluation of the DOS. It should be noted (Neuhold et al. 1998) that accessing states slightly above E_F through the method described above is superior to IPES or SXA because of the much higher energy resolution.

The optical conductivity studies in the energy range 0.001–12 eV of i-$Al_{70}Pd_{21}Mn_9$ showed the presence of a large absorption in the conductivity spectrum at ~ 1.2 eV (Bianchi et al. 1993, 1998). This absorption was interpreted to result from excitations across a pseudogap in the DOS.

The PES core-line asymmetry, whose measure is the asymmetry parameter α in the Doniach–Šunjić formula describing the shape of the PES core-level lines (Cardona and Ley 1978, Mehlhorn 1982), is proportional to the local DOS(E_F): the larger α indicates the larger local DOS(E_F) (Mehlhorn 1982). An increase of α determined from the fits of the Al $2p$ lines measured with the resolution of 180 meV for different values of $h\nu$ (Fig. 8.13a), and thus different electron kinetic energies, E_{kin}, as a function of E_{kin} was observed for the single-grain i-$Al_{70.5}Pd_{21}Mn_{8.5}$ alloy (Fig. 8.13b). No such increase could be observed for a fcc Al metal (Fig. 8.13b). The value of E_{kin} is a measure of

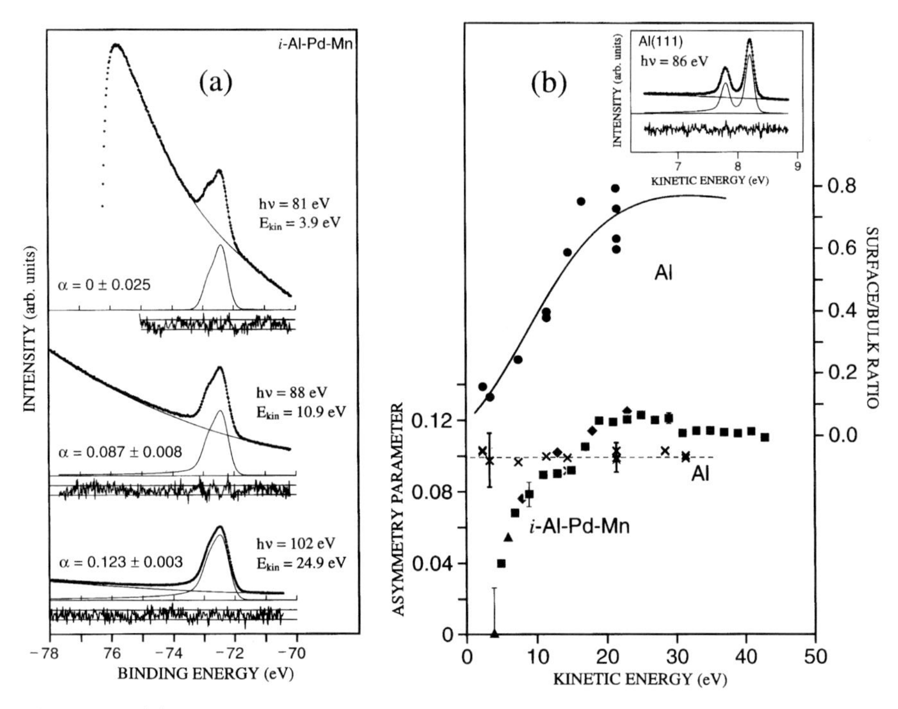

Fig. 8.13. (a) Al $2p$ core-level spectra of i-$Al_{70.5}Pd_{21}Mn_{8.5}$ measured at different photon energies. The solid lines for each spectrum represent the fits of the background and the Doniach–Šunjić doublet. The residuals to the fits are also shown with the upper and lower bounds at $\pm\sqrt{N}$ (N is the number of counts) determined from the Poisson statistics. (b) Lower part: Asymmetry parameter of the Al $2p$ core-level lines of i-$Al_{70.5}Pd_{21}Mn_{8.5}$ (full symbols) and fcc Al (crosses) as a function of the electron kinetic energy. Upper part: Surface to bulk intensity ratio determined from the Al $2p$ core-level lines of a single-crystal Al(100). The solid line is calculated from theory. Inset: Al $2p$ core-level lines of a single-crystal Al(111). After Neuhold et al. (1998).

the sampling depth of a PES experiment by virtue of inelastic electron scattering in a sample: the larger E_{kin}, the smaller the sampling depth, i.e., the more surface contribution to the measured PES intensity (Cardona and Ley 1978, Mehlhorn 1982, Hüfner 1996). The sampling depth dependence upon E_{kin} was measured through the relative intensity of the surface to bulk Al $2p$ core-level lines (Fig. 8.13b) for a single crystal Al(100) for which surface and bulk signals can be separated. A similar dependence upon E_{kin} of α and of the surface to bulk ratio (Fig. 8.13b) was interpreted as indicating that α for atoms near the surface of i-$Al_{70.5}Pd_{21}Mn_{8.5}$ is larger than α for those in the bulk. Since α is proportional to DOS(E_{F}), this observation was claimed to provide evidence for the higher DOS(E_{F}) at the surface than in the bulk in i-$Al_{70.5}Pd_{21}Mn_{8.5}$ (Neuhold et al. 1998).

This interpretation is valid provided that the values of α derived from the fits is meaningful. As it can be seen from the Al $2p$ PES spectrum of Al(111) (inset in Fig. 8.13b), the Al $2p_{1/2}$ and $2p_{3/2}$ components are clearly separated and consequently the value of α can be precisely determined. However, this separation is absent in the corresponding spectra of the i-Al-Pd-Mn alloy (Fig. 8.13a). This is due the multiplicity of Al sites, and thus the multiplicity of the local chemical surroundings (see Sect. 8.4) around Al atoms in a quasicrystalline structure, as demonstrated experimentally by various local probes (Stadnik 1996). One would thus expect that the measured Al $2p$ spectrum results from a superposition of the Doniach–Šunjić lines corresponding to these sites, which naturally leads to the distribution of the parameters characterizing a Doniach–Šunjić profile. It is therefore not obvious whether fitting the measured Al $2p$ line with one Doniach–Šunjić line profile provides a meaningful average value of α corresponding to a given E_{kin}. Since different core levels should have the same α (Hüfner 1996), it would be important to carry out similar measurements for other TM core-level lines in order to establish the validity of the observation that α is smaller at the surface than in the bulk of i alloys.

The availability of high-quality i-Al-Pd-Mn samples made it possible to observe the de Haas–van Alphen effect (Shoenberg 1984) in a single grain i-$Al_{70}Pd_{21.5}Mn_{8.5}$ (Haanappel et al. 1996). Two well-defined frequencies were observed for the magnetic field along the fivefold direction and two other frequencies were detected for the field parallel to the two fold direction. These observations are indicative of the reality of the Fermi surface in QCs, which has probably a very complex shape (Fujiwara 1998). The first attempts to calculate the shape of the Fermi surface in QCs indicate that it has a complex shape. It would be an experimental challenge to determine its size and shape using the the de Haas–van Alphen effect and other spectroscopic techniques.

The spectroscopic results presented above confirm the presence of the predicted minimum of the DOS slighly above E_{F} and are in good agreement with the predicted overall structure of the occupied DOS in i-Al-Pd-Mn (Krajčí et al. 1995).

8.3.4 Al-Pd-Re Icosahedral Alloys

Among all QCs, the i-Al-Pd-Re alloys distinguish themselves in exhibiting the highest ρ and the extreme sensitivity of their physical properties to minute changes in their composition (see Chap. 5). It was even suggested that the i-Al-Pd-Re alloys are insulators at low temperatures (Pierce et al. 1993a, 1994), but this claim was recently disputed (Ahlgren et al. 1997, Bianchi et al. 1997). The recent electronic structure calculations (Krajčí and Hafner 1998) predict that a minute change of stoichiometry of the 1/1 approximant to the i-Al-Pd-Re phase may change it from being a metal with a pseudogap to a narrow-gap semiconductor with a real gap of the width ~ 0.15 eV located just below E_F. For higher-order approximants, no real gaps, but a series of pseudogaps around E_F is predicted (Fig. 7.16 in Chap. 7).

The low-temperature UPS He II valence band of i-$Al_{70.5}Pd_{21}Re_{8.5}$ (see Fig. 8.9a) exhibits a feature at $BE \approx -4.1$ eV mainly due to the Pd $4d$-like states and a very broad feature close to E_F predominantly due to the states of the Re $5d$ character (Stadnik et al. 1996, 1997). The positions and intensities of these features are in good agreement with the calculated DOS (Fig. 7.16 in Chap. 7). There is a noticeable smaller spectral intensity at E_F in the valence band of i-$Al_{70.5}Pd_{21}Re_{8.5}$ in comparison to that of other Al-Pd-based i alloys (Fig. 8.9a), which is indicative of a smaller DOS(E_F) in this alloy than in other i alloys. The near-E_F region of the valence band of i-$Al_{70.5}Pd_{21}Re_{8.5}$ can be well fitted to Eq. (8.3), which can be interpreted as indicative of the presence of a pseudogap at E_F of the FWHM ($2\Gamma_L$) of $\sim$ 200–400 meV (Purdie and Stadnik 1997, Stadnik et al. 1996, 1997). This width is several times wider than that of $\sim$ 60 meV found in the TS spectrum (Fig. 8.7b).

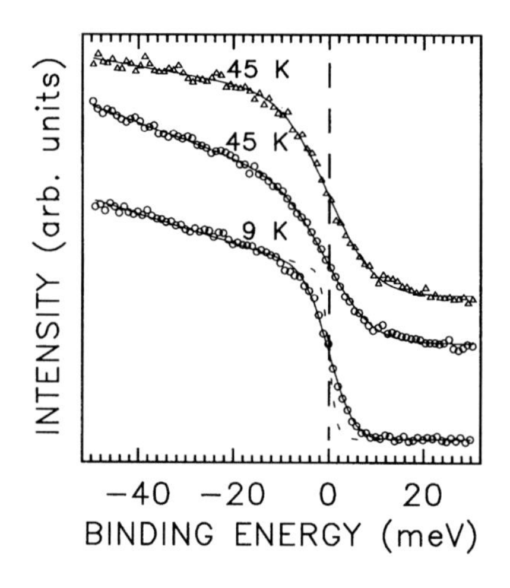

Fig. 8.14. Near-E_F He I valence bands of i-$Al_{70.5}Pd_{21}Re_{8.5}$ (circles) and Ag (triangles) evaporated on it. The solid lines are the fits to a linearly decreasing intensity multiplied by the Fermi-Dirac function at 45 and 9 K (broken curve) and convoluted with a Gaussian whose FWHM is respectively 6.0(5) and 9.2(3) meV for the spectra measured at 45 and 9 K. After Stadnik et al. (1998).

The optical conductivity studies carried out in the energy range 0.001–12 eV on two i samples, $Al_{70}Pd_{20}Re_{10}$ and $Al_{70}Pd_{21.4}Re_{8.6}$, showed a large absorption in the conductivity spectra at $\sim$ 2.6 eV (Bianchi et al. 1997). This absorption was interpreted as resulting from excitations across a pseudogap in the DOS.

The low-temperature near-E_F UPS He I spectra of i-$Al_{70.5}Pd_{21}Re_{8.5}$ measured with the highest energy resolution presently available (Fig. 8.14) clearly show the presence of a Fermi edge. This can be seen by comparing a near-E_F spectrum at 45 K of i-$Al_{70.5}Pd_{21}Re_{8.5}$ with that of Ag evaporated onto the alloy. Additional evidence comes from the perfect fits of the spectra of this alloy measured at 45 and 9 K (Fig. 8.14). This constitutes a direct proof that the i-$Al_{70.5}Pd_{21}Re_{8.5}$ alloy, in spite of its very high ρ, is metallic, supporting the same conclusion based on low-temperature ρ measurements by Ahlgren et al. (1997) and Bianchi et al. (1997). The finite value of G for the zero sample bias (Fig. 8.7b) indicates a finite DOS(E_F), and thus a metallic character of the *i*-Al-Pd-Re QC.

The above analysis of the spectroscopic data on the occupied and unoccupied DOS provides conclusive evidence for the presence of the theoretically predicted pseudogap in the Al-Cu-Li and Al-Pd-Mn *i* alloys. For other investigated *i* QCs, the data are compatible with the presence of the pseudogap. A generally good agreement between the calculated and measured DOS is observed, which indicates that the essential ingredients of the electronic structure of QCs can be obtained from the calculations carried out for their approximants.

8.3.5 Al-Co-Cu Decagonal Alloys

The low-temperature He II valence band of *d*-$Al_{65}Co_{15}Cu_{20}$ exhibits two features (Fig. 8.15). As shown earlier with the resonance PES technique (Stadnik et al. 1995d), the feature close to E_F is predominantly due to states of Co 3*d* character and the feature at $BE \approx -4.2$ eV is mainly due to the Cu 3*d*-derived states. The resonance PES data enabled the derivation of the partial DOS of the Co and Cu 3*d* character and a significant discrepancy between the theoretical (Trambly de Laissardière and Fujiwara 1994b, 1994c) and experimental Co 3*d* DOS was found (Stadnik et al. 1995d). This is illustrated in Fig. 8.16 where a much larger experimental intensity at E_F due to the Co 3*d* states as compared to that expected from the presence of a pseudogap can be seen. This was taken as evidence for the absence of a pseudogap in the *d*-$Al_{65}Co_{15}Cu_{20}$ alloy (Stadnik et al. 1995d).

The electronic structure calculations (Krajčí et al. 1997a) based upon several variants of the original Burkov model for the atomic structure of *d*-Al-Co-Cu show the best agreement with the PES data from Fig. 8.16 (Fig. 7.15 in Chap. 7) for the variant B2Cu. There is no pseudogap in the DOS(E_F) for this variant of the atomic structure of *d*-Al-Co-Cu.

The successful fit of the near-E_F UPS He II valence band of the *d*-$Al_{65}Co_{15}Cu_{20}$ alloy with Eq. (8.3) led to the conclusion (Stadnik et al. 1997) of the presence of the pseudogap in this alloy, which is in contradiction to the conclusion based on the analysis of the PES data (Stadnik et al. 1995d, Krajčí et al. 1997a). There is some arbitrariness in the model expressed by Eq. (8.3) regarding the choice of the BE range for which the data are fitted to a line $aBE + b$ representing a normal DOS. This range was chosen much farther away from E_F for *d* alloys (leading to the normal DOS with a positive slope) than for *i* alloys (Stadnik et al. 1997). In view of the DOS calculated for the structural variant B2Cu [Fig. 8 in Krajčí et al. (1997a)], the assumption of the normal DOS with the positive slope may be unwarranted; the choice of the BE range closer to E_F would lead to no pseudogap in the fit of the near-E_F UPS He II valence band of *d*-$Al_{65}Co_{15}Cu_{20}$. Clearly, further spectroscopic data, especially those probing the unoccupied DOS, are needed to unambiguously determine the presence or lack of the pseudogap in the *d*-Al-Co-Cu system.

8.3.6 Al-Ni-Co and Al-Ni-Rh Decagonal Alloys

No electronic structure calculations have been carried out yet for the two *d* systems, Al-Ni-Co and Al-Ni-Rh, which would allow a comparison to be made with the spectroscopic data on the occupied and unoccupied DOS. The valence band of *d*-$Al_{70}Co_{15}Ni_{15}$ (Depero and Parmigiani 1993, Stadnik et al. 1995d, 1997) consists of a broad feature (Fig. 8.15) resulting from the overlap of the Co and Ni 3*d* states (Stadnik et al. 1995d).

The Al Kα XPS valence band of *d*-$Al_{70}Co_{15}Ni_{15}$, which was measured at room temperature with an energy resolution of 0.3 eV, was interpreted as showing no presence of the pseudogap (Depero and Parmigiani 1993); no justification of this interpretation was given. The presence of significant intensity at the E_F due to Co and Ni 3*d*-like states in the PES bands, which were measured at room temperature for different values of $h\nu$ and with an energy resolution of $\sim$ 0.4 eV, was taken as an indication for the lack of a pseudogap (Stadnik et al. 1995d). On the other hand, the successful fit of the near-E_F UPS He II valence band of *d*-$Al_{70}Co_{15}Ni_{15}$ with Eq. (8.3) led to the conclusion (Stadnik et al. 1997) that there is a pseudogap of the width $2\Gamma_L \approx 2.2$ eV in this alloy. As indicated above, the assumption of the normal DOS with the positive slope, which leads to the pseudogap, may be not justified for *d* alloys. However, a combination of the PES and IPES

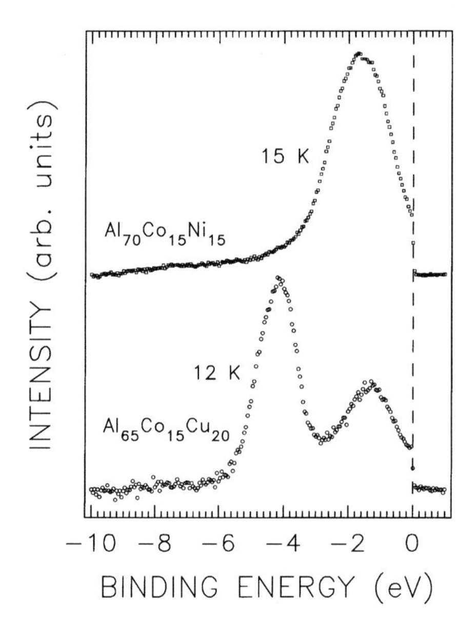

Fig. 8.15. Low-temperature He II valence bands of *d* QCs $Al_{65}Co_{15}Cu_{20}$ and $Al_{70}Co_{15}Ni_{15}$. The energy resolution is $\sim$ 30 meV. After Stadnik et al. (1998).

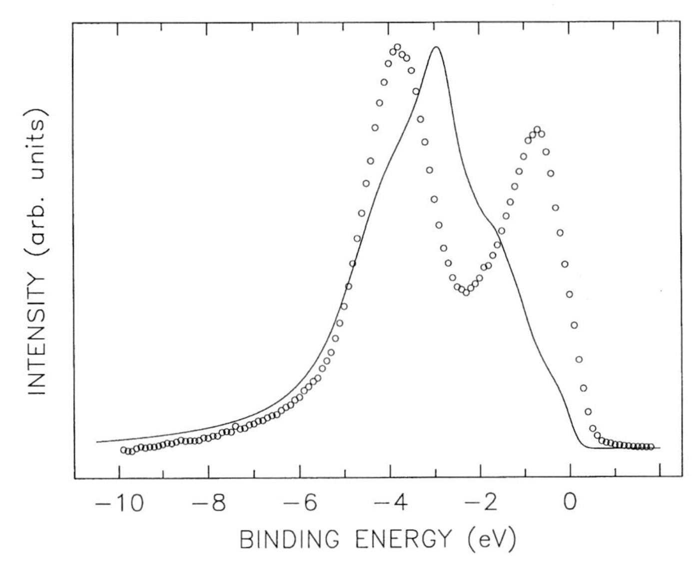

Fig. 8.16. Comparison between the room-temperature valence band of d-$Al_{65}Co_{15}Cu_{20}$ measured at $h\nu = 100$ eV with an energy resolution of ~ 0.4 eV (open circles) and the appropriately broadened theoretical DOS (solid line) calculated by Trambly de Laissardière and Fujiwara (1994b, 1994c). After Stadnik et al. (1995d).

techniques shows the presence of the pseudogap of the width of ~ 1.7 eV in the total DOS of d-$Al_{72}Co_{16}Ni_{12}$ (Fig. 8.17).

The presence of a step structure above a Fermi edge in the EELS Al L-edge spectrum of d-$Al_{65}Co_{15}Ni_{15}$ (Fig. 8.8), which was measured with an energy resolution of 0.15 eV, was interpreted as evidence for the presence of a pseudogap in the DOS of the Al s- and d-like states (Terauchi et al. 1998a, 1998b). A similar conclusion was reached by Soda et al. (1998a) based on the Al $L_{2,3}$ SXE and SXA spectra of d-$Al_{72}Co_{16}Ni_{12}$ (Fig. 8.18).

The optical conductivity studies in the energy range 0.001–12 eV of a single grain d-$Al_{71}Co_{16}Ni_{13}$ (Bianchi et al. 1998) revealed absorptions for the periodic direction at ~ 1 eV, and for a direction in the quasiperiodic plane at ~ 2 eV. These absorptions were interpreted as evidence for the electronic excitations across a pseudogap.

The EELS Al L-edge spectrum of d-$Al_{70}Ni_{20}Rh_{10}$ (Fig. 8.8), which was measured with an energy resolution of 0.12 eV (Terauchi et al. 1998c), exhibits a step structure above the Fermi edge. This was interpreted as evidence for the presence of a pseudogap in the DOS of the Al s- and d-like states (Terauchi et al. 1998c). No other spectroscopic data are available for this alloy system.

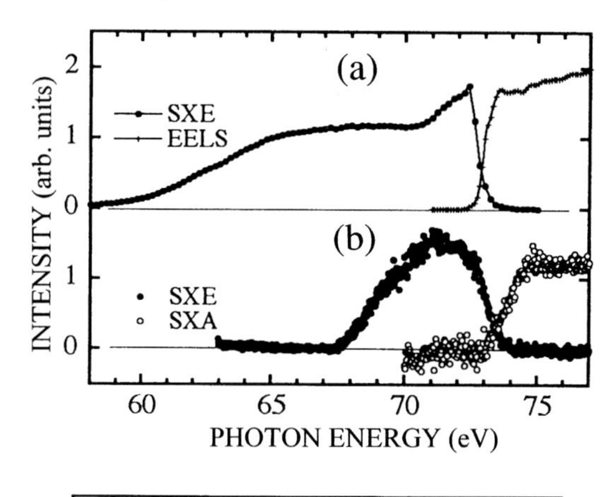

Fig. 8.17. (a) IPES spectrum of Au. (b) PES and IPES spectra of d-$Al_{72}Co_{16}Ni_{12}$. The solid lines are the smoothed experimental data. After Soda et al. (1998a).

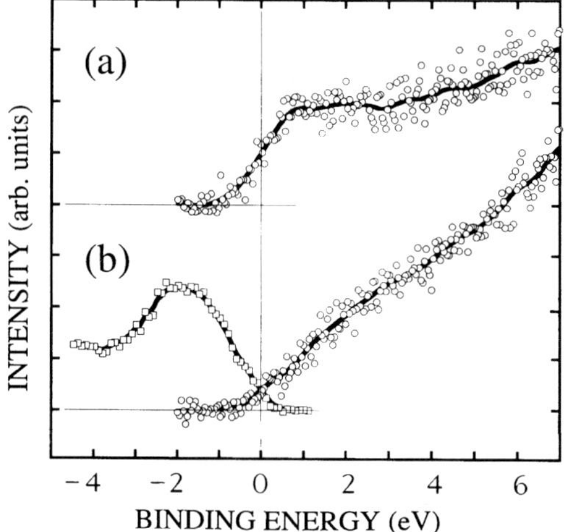

Fig. 8.18. Comparison between (a) SXE Al $L_{2,3}$ and EELS Al L-edge spectra of Al and (b) Al $L_{2,3}$ SXE and SXA spectra of d-$Al_{72}Co_{16}Ni_{12}$. After Soda et al. (1998b).

Thus it can be concluded that there is strong spectroscopic evidence for the existence of the pseudogap in the DOS in the d-Al-Co-Ni system and some indication for its existence in the d-Al-Ni-Rh system.

8.3.7 Fine Structure of the Density of States

In order to assess the possible existence of the predicted DOS spikiness, a meaningful comparison between the measured DOS spectra and the calculated DOS has to be made. This involves modifying the theoretical DOS to account for the finite energy resolution of an experiment, the lifetime broadening effects inherent to a given spectroscopic technique used to measure the DOS, and the sample temperature (Stadnik et al. 1995a). The instrumental broadening is usually represented by a Gaussian with the FWHM, Γ_G, equal to the energy resolution of an experiment. The lifetime broadening effects are usually described by a Lorentzian whose FWHM is in the form $\Gamma_L^0 BE^2$,

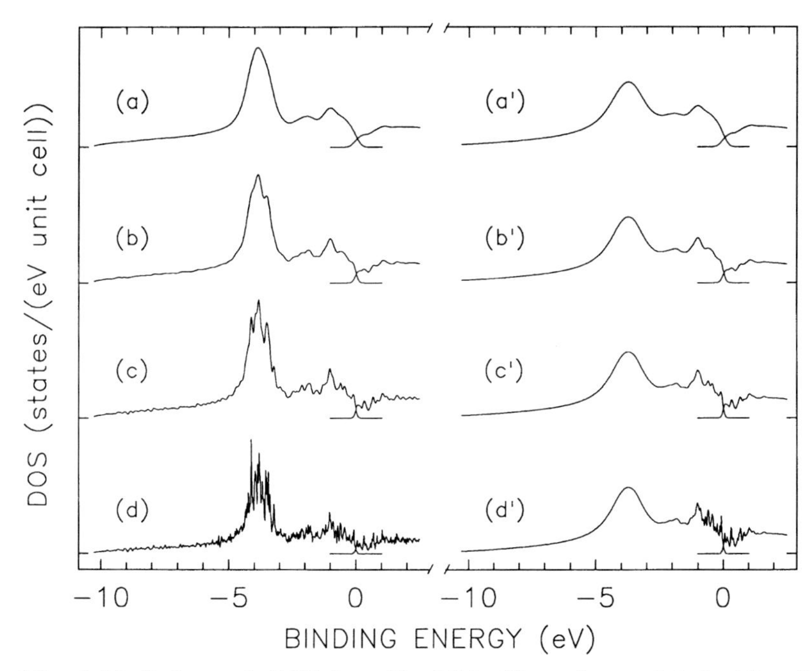

Fig. 8.19. Left panel: DOS from Fig. 6.8 in Chap. 6 convoluted with a Gaussian with Γ_G equal to (a) 0.4 eV, (b) 0.2 eV, (c) 0.1 eV, and (d) 0.025 eV, and multiplied by the Fermi-Dirac function at room temperature. Right panel: DOS from Fig. 6.8 in Chap. 6 convoluted with a Lorentzian with $\Gamma_L^0 = 0.05$ eV^{-1} and a Gaussian with Γ_G equal to (a′) 0.4 eV, (b′) 0.2 eV, (c′) 0.1 eV, and (d′) 0.025 eV, and multiplied by the Fermi-Dirac function at room temperature. The separation between the two neighboring ticks on the ordinate scale corresponds to 380 states/(eV unit cell) and the DOS values in plot (d) were divided by the factor 1.47. After Stadnik et al. (1995a).

where the Γ_L^0 parameter fixes the scale of broadening. Assuming a hypothetical case of a PES experiment on a QC at room temperature with no lifetime broadening effects (left panel in Fig. 8.19), an energy resolution better than ~ 0.2 eV would be required to detect the spikes. The inclusion of the lifetime braodening effects (right panel in Fig. 8.19) smears the sharp features of the DOS in proportion to the square of their distance from E_F. It is clear that the possible DOS spikiness can be observed in a QC sample at room temperature with the spectroscopic techniques with an energy resolution better than $\sim$ 100 meV, and only in the vicinity of E_F.

In view of the analysis presented above, it is not surprising that no spikiness was observed in PES/IPES experiments carried out with a resolution ≥ 0.1 eV (Matsubara et al. 1991, Mori et al. 1992, 1993, Stadnik and Stroink 1993, Stadnik et al. 1995, Takeuchi and Mizutani 1995, Zhang and Stadnik 1995, Zhang et al. 1994, 1995). The claim (Belin et al. 1994b) of experimental evidence of the presence of DOS spikiness in the Al p-like DOS in *i*-$Al_{62.5}Cu_{26.5}Fe_{11}$, which was based on the observation of a low-intensity structurless SXA Al *p* conduction band, is unfounded in view of the poor energy resolution and severe lifetime broadening effects inherent to the SXA technique. However, no DOS spikiness could be observed also in high-energy-resolution PES spectra of QCs at room temperature. Wu et al. (1995) observed no fine structure in the near-E_F valence band of a single grain *i*-$Al_{70}Pd_{21.5}Mn_{8.5}$ measured at $h\nu = 13$ eV with an energy resolution of 50 meV. No DOS spikiness was observed in the UPS He II near-E_F valence band of *i*-$Al_{65}Cu_{20}Fe_{15}$ at 283 K measured with an energy resolution of 32 meV (Fig. 8.6c).

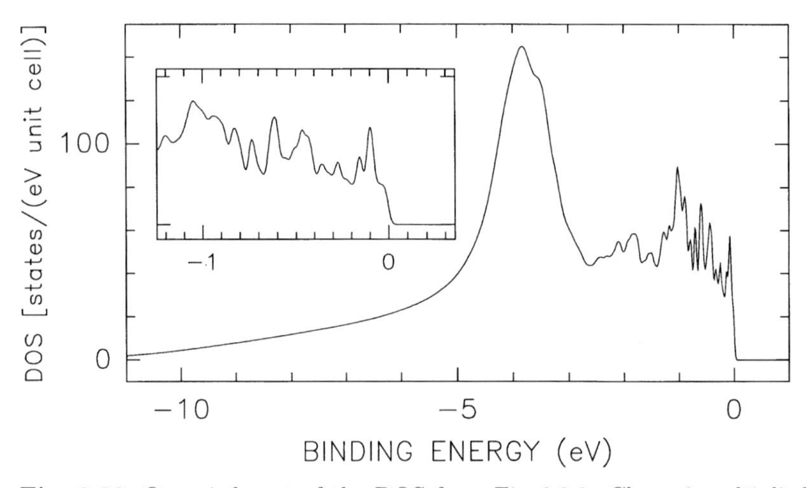

Fig. 8.20. Occupied part of the DOS from Fig. 6.8 in Chap. 6 multiplied by the Fermi-Dirac function at 14 K and convoluted with a Lorentzian with $\Gamma_L^0 = 0.02$ eV^{-1} and a Gaussian with $\Gamma_G = 31.6$ meV. The values of temperature and Γ_G correspond the the experimental spectrum in Figs. 8.6(a) and 8.6(b). The inset shows the DOS in the vicinity of E_F. The two ticks on on the ordinate axis of the inset correspond, respectively, to 0 and 100 states/(eV unit cell). After Stadnik et al. (1995a).

Figure 8.20 shows the expected low-temperature He II valence band of *i*-Al-Cu-Fe. It was calculated based on the DOS from Fig. 6.8 in Chap. 6 which was broadened using the values of temperature and Γ_G corresponding to the experimental spectra of *i*-$Al_{65}Cu_{20}Fe_{15}$ in Figs. 8.6a and 8.6b. The Γ_L^0 value used is a typical value employed for metallic systems (Stadnik et al. 1997). It is clear that the predicted DOS spikiness should be observed for *BE*'s up to a few eV below E_F in low-temperature UPS valence bands of QCs measured with the high energy resolution of a few tens of meV. Obviously, the DOS spikes should be observed even more readily for the ultrahigh energy resolution (< 10 meV). However, no spikes could be found in the low-temperature valence bands of *i* alloys measured with high (Figs. 8.6

and 8.9a) or ultrahigh (Figs. 8.9b, 8.9c, and 8.14) energy resolution (Stadnik et al. 1996, 1997, Kirihara et al. 1998, Neuhold et al. 1998). Such spikes were also not detected in the high- and ultrahigh-enery-resolution UPS valence bands of *d* QCs (Stadnik et al. 1997).

TS experiments, which have an energy resolution perhaps better than 1 meV, showed no evidence of spiky features in the occupied and unoccupied DOS for energies up to 300 meV around E_F (Klein et al. 1995, Davydov et al. 1996). Contradictory conclusions concerning DOS spikiness were reported based on the indirect studies of the DOS with a NMR technique. In a NMR pressure study of *i*-$Al_{65}Cu_{20}Ru_{15}$, DOS fine structure was expected to be probed on the 10 meV energy scale through the pressure dependence of the Knight shift K_s and the spin-lattice relaxation rate R_1 (Shastri et al. 1995). Since no such dependence of K_s and R_1 upon pressure up to $\sim$ 2 kbar was observed, it was concluded that no fine structure in the DOS of *s* character exist in the studied *i* alloy. In another NMR study of Al-Cu-Fe, Al-Cu-Ru, and Al-Pd-Re *i* QCs (Tang et al. 1997), the T^2 dependence of R_1 between 93 and 400 K, which was established on the basis of a fit of fewer than 10 experimental points, was interpreted as a signature of a sharp feature in the DOS at E_F whose width was estimated to be $\sim$ 20 meV. However, another NMR study on *i*-Al-Pd-Re (Gavilano et al. 1997) in the temperature range 18–300 K found a T^3 dependence of R_1.

The coefficient β of the T^3 term of the electronic contribution to the specific heat (Wälti et al. 1998) and the coefficient A of the T^2 contribution to the magnetic susceptibility (Kobayashi et al. 1997) are a function of the first and the second derivatives of DOS(E_F). Large values of these derivatives are expected if the DOS has a spiky structure, which would lead to large values of β and A. This has not been observed experimentally (Kobayashi et al. 1997, Wälti et al. 1998).

A universal linear relation between $1/\rho(300\ \mathrm{K})$ and γ^2 for many *i* QCs has been found independently by Rapp (Fig. 5.10 in Chap. 5) and Mizutani (1998). This relation indicates that $1/\rho$ is proportional to the square of the DOS(E_F). The fact that the experimental data on the ρ-γ diagram lie on a straight line with a slope of -2 was interpreted as evidence that the DOS around E_F must be smooth. If this was not the case, i.e., if the DOS exhibited a fine spiky structure, the data on this diagram would be scattered and no universal relation could be observed (Mizutani 1998).

It can be concluded that there is no experimental evidence based on the the high-enery-resolution PES and TS experiments which probe the DOS directly, as well as on the experiments which probe the DOS indirectly [except the claim by Tang et al. (1997)], for the presence of the predicted DOS spikiness in QCs. The possible reasons for this are discussed in Sect. 8.4.

8.4 Uniqueness of the Electronic Structure of Quasicrystals

The major effort in the theoretical and experimental investigations of the physical properties of QCs has been directed to finding the properties arising from the quasiperiodicity itself. They were expected to be different from the corresponding properties of crystalline and amorphous solids, and thus unique to QCs. Have such properties been found?

A pseudogap in the DOS around E_F predicted by theory for many nonmagnetic QCs was found experimentally in some of them. It thus seems to be a generic property of the electronic structure of nonmagnetic QCs. In magnetically ordered QCs (see Chap. 9), such a pseudogap probably does not occur as a local Stoner criterion, which seems to be valid in QCs (Hafner and Krajčí 1998), requires a high DOS(E_F) for magnetic moment formation. The minimum of DOS in the vicinity of E_F is not, however, a unique characteristics which distinguishes the electronic structure of QCs from that of crystalline and amorphous alloys since the structure-induced pseudogap is believed to exist in both crystalline (Carlsson and Meschter 1995, Trambly de Laissardière et al. 1995b) and amorphous (Häussler 1994) systems.

The predicted DOS spikiness would seem to be such a unique property distinguishing QCs from other materials. The only crystalline material for which such a fine DOS structure was predicted is the cubic $Al_{10}V$ alloy (Trambly de Laissardière et al. 1995b), but no experimental attempts have been reported yet to verify this prediction. In the discussion below several possible reasons are suggested to explain the failure to detect the DOS spikiness in QCs in various experiments probing the DOS directly and indirectly.

It has been argued (Tang et al. 1998, see also Chap. 7) that PES and TS, as surface-sensitive techniques, probe the surface electronic states of QCs rather than those of the bulk, and therefore may be not representative of the bulk DOS. Since TS probes the top-most atomic layer of a studied material, the conductance spectra reflect the surface DOS (Wolf 1985, Lang 1986, Stroscio and Kaiser 1993) which, depending on the material, may substantially differ from the bulk DOS. It is thus conceivable that the narrow ($\sim$ 60 meV wide) pseudogap observed in the TS spectra of *i* alloys (Fig. 8.7) is the gap of the surface DOS. This would be consistent with the recent theoretical results indicating that the pseudogap persists also on the surface of a QC (Janssen and Fasolino 1998). The failure to detect the predicted bulk DOS spikes $\sim$ 10 meV wide in the vicinity of E_F in TS experiments could indicate that the DOS spikiness does not extend to the surface of a QC.

PES, on the other hand, probes deeper into the surface. The so-called "universal curve" of the electron mean-free path, λ, as a function of electron E_{kin}, which defines the surface sensitivity of the technique, has a minimum at a E_{kin} of $\sim$ 100 eV corresponding to a λ of $\sim$ 5–10 Å ; the value of λ increases to $\sim$ 20 Å for lower ($\sim$ 10 eV) and higher ($\sim$ 1000 eV) electron E_{kin}'s (Cardona and Ley 1978, Mehlhorn 1982, Hüfner 1996). This should be compared with the diameter $\sim$ 8–10 Å of the elementary clusters observed in a single-grain *i*-Al-Pd-Mn alloy by scanning tunneling microscopy (Ebert et al. 1996). By changing the value of $h\nu$, one can thus vary the weight of the surface and bulk DOS contributions to the measured PES spectrum. The small λ makes thus PES a surface-sensitive technique, but not to the exclusion of the bulk response. A pseudogap an order of magnitude wider than that found in TS experiments was observed in PES experiments carried out for different values of $h\nu$, as well as by IPES, SXE, SXA, and EELS experiments (see Sect. 8.3). It thus seems that these PES experiments probe deeply enough so that they reflect mainly the bulk DOS. Additional experimental support for this statement comes from good agreement between the shape and position of different features of the PES DOS and those of the DOS predicted by theory. It should be mentioned that differences between more surface-like DOS obtained from TS experiments and more bulk-like DOS determined from PES experiments were observed for other materials (Shih et al. 1989, Takahashi 1989).

Another possible reason to explain the failure of detecting DOS spikiness in high- and ultrahigh-energy-resolution PES experiments is related to the quality of the surfaces of QCs studied. In the preparation of surfaces for these experiments by using a standard procedure of scraping or fracturing (carried out at low temperatures to prevent any structural reorganization of the sample) the polyquasicrystalline samples could, in principle, lead to a destruction of their quasicrystalline order, and thus the disappearance of the fine strucure of the DOS. However, the valence bands of well-characterized single-grain *i*-Al-Pd-Mn alloys (Wu et al. 1995, Neuhold et al. 1998, Purdie et al. 1998) are virtually indistinguishable from those of polyquasicrystalline alloys (Stadnik et al. 1996, 1997), which indicates that the employed surface preparation procedures maintained the sample surface quasicrystallinity.

It may not be possible to detect the predicted DOS spikiness even with the PES experiments of the highest energy resolution because of the existence of chemical and topological disorder in QCs of high structural quality. Such disorder, which is not taken into account in the electronic structure calculations, may wash out the DOS spikiness induced by quasiperiodicity. Furthermore, the concept of quasiperiodicity implies that no two crystallographic positions of a given atom are exactly the same, which can be viewed as a sort of topological disorder. There is growing experimental evidence which shows that the chemical and topological disorder are present in these structurally "perfect" QCs. Diffuse scattering is often observed in x-ray-, electron-, and neutron-diffraction patterns of high-quality QCs (de Boissieu et al. 1995, Frey 1997, Frey et al. 1998, Mori et al. 1998). Its presence indicates that some disorder must exist in the diffracting structure. Local probes, such as Mössaber spectroscopy (Stadnik 1996), NMR (Shastri et al. 1994b, Gavilano et al. 1997), and nuclear quadrupole resonance (Shastri et al. 1994a, 1994b), clearly detect the distribution of the electric quadrupole splittings, P(Δ),

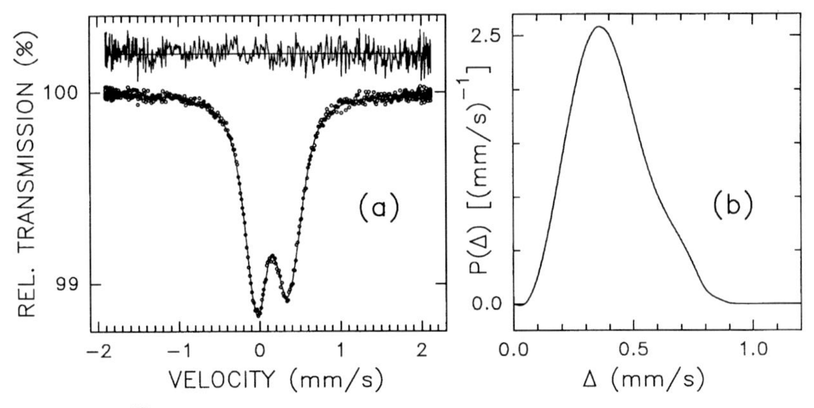

Fig. 8.21. ^{57}Fe room-temperature Mössbauer spectrum of *i*-$Al_{70.5}Pd_{21}Re_{8.45}Fe_{0.05}$ (a). The solid line is the fit with a distribution of quadrupole splittings $P(\Delta)$ (b). The residuals, multiplied by a factor of three, are shown above the spectrum. After Stadnik (1996).

in high-quality stable QCs. This is illustrated in Fig. 8.21a which shows a Mössbauer spectrum of the *i*-$Al_{70.5}Pd_{21}Re_{8.45}Fe_{0.05}$ alloy of high structural quality which consists of two broad lines. These broad lines result from the presence of the distribution $P(\Delta)$ (Fig. 8.21b). Such a distribution can be detected only if there is a chemical and/or topological order in the investigated samples (Stadnik 1996). The apparent success of quantum interference theories (the electron-electron interaction and weak-localization effects), which were originally developed for highly disordered conductors, in accounting for the temperature and field dependencies of σ and magnetoresistance of high-quality QCs (see Chap. 5) also indicates the importance of chemical disorder. The experiments mentioned above demonstrate that chemical and/or topological disorder is present in the high-quality phason-free stable QCs and therefore may smear out the DOS spikiness predicted for disorder-free QCs.

And finally, it is conceivable that the predicted DOS spikiness is an artifact of the calculations. It has been recently demonstrated (Haerle and Kramer 1998) that unphysical spikes do appear in the calculated DOS of a toy model with an assumed *a priori* smooth DOS as a result of insufficient computational precision necessary for analyzing structures with large unit cells. It was concluded that the predicted fine structure at the level of $\sim$ 100 meV may be an artifact. This is rather an alarming possibility and it is important to investigate it in more detail. The prediction of the DOS spikiness for a cubic $Al_{10}V$ (Trambly de Laissardière et al. 1995b) with a large lattice constant of 14.492 Å, but not for other Al-based crystalline alloys of smaller lattice constants, would seem to be in line with the conclusion that DOS spikiness is an artifact (Haerle and Kramer 1998).

The third novel electronic property intrinsic to quasiperiodicity is the concept of critical eigenstates. Whereas in periodic and random systems the eigenstates are respectively extended and localized, they are critical in the quasiperiodic systems (see Chaps. 6 and 7), i.e., neither extended nor localized. It is not clear how these novel eigenstates reflect upon various measured physical properties. In an ^{27}Al NMR study of *i*-$Al_{70}Pd_{21.4}Re_{8.6}$ (Gavilano et al. 1997), a gradual localization of the itinerant electrons, which was deduced from an unusual increase of R_1/T with decreasing T below 20 K, was suggested to give, for the first time, direct evidence for the presence of critical electronic states in a QC. Futher experimental studies are needed to conclusively establish the reality of the critical states in QCs.

8.5 Quasiperiodicity and Unusual Physical Properties

The electronic structure results discussed above show no unusual features that could be the consequence of the quasiperiodic order. Many unusual physical properties of QCs were also found in the approximants to QCs. For example, very high values of ρ, and its increase with temperature, were found in the approximants to the *i*-Mg-Ga-Al-Zn, *i*-Al-Cu-Fe, and *i*-Al-Mn-Si phases (Edagawa et al. 1992, Poon 1992, Mayou et al. 1993, Pierce et al. 1993b, Takeuchi 1994, Berger et al. 1995a, 1995b) and in several *d* approximants along the pseudoquasiperiodic planes (Volkov and Poon 1995). Similarities were also observed between the values and/or temperature and/or magnetic field dependencies of the Hall coefficient (Berger et al. 1995b, Pierce et al. 1995b), thermoelectric power (Pierce et al. 1995b), magnetoresistance (Berger et al. 1995b), optical conductivity (Wu et al. 1993), γ (Mizutani et al. 1991, Pierce et al. 1993b, Berger et al. 1995b), and local hyperfine parameters (Hippert et al. 1992, 1994, Stadnik 1996).

The unexpectedly high value of ρ of many *i* alloys, which consist of normal metallic elements (see Chap. 5), was thought to be indicative of their unique properties induced by their quasiperiodic structure. But as mentioned above, similarly high values of ρ were observed in approximants to QCs. Furthermore, high ρ and its negative temperature coefficient were also observed in some crystalline alloys consisting of normal metallic elements whose structure is unrelated to the structure of QCs. For example, the Heusler-type Fe_2VAl alloy exhibits a semiconductor-like behavior of ρ (Fig. 8.22a), yet its valence band (Fig. 8.22b) clearly shows the presence of the Fermi edge. Electronic structure calculations predict a narrow pseudogap in the DOS around E_F (Guo et al. 1998, Singh and Mazin 1998).

It has to be concluded that, as yet, no conclusive experimental evidence has been found to support the claim of the unusual physical properties of QCs which are due to their quasiperiodic nature and which therefore do not occur in crystalline or amorphous systems. This is perhaps not surprising since there is no solid physical basis to expect that quasiperiodicity should lead to physical properties distinctly different from those found in periodic systems.

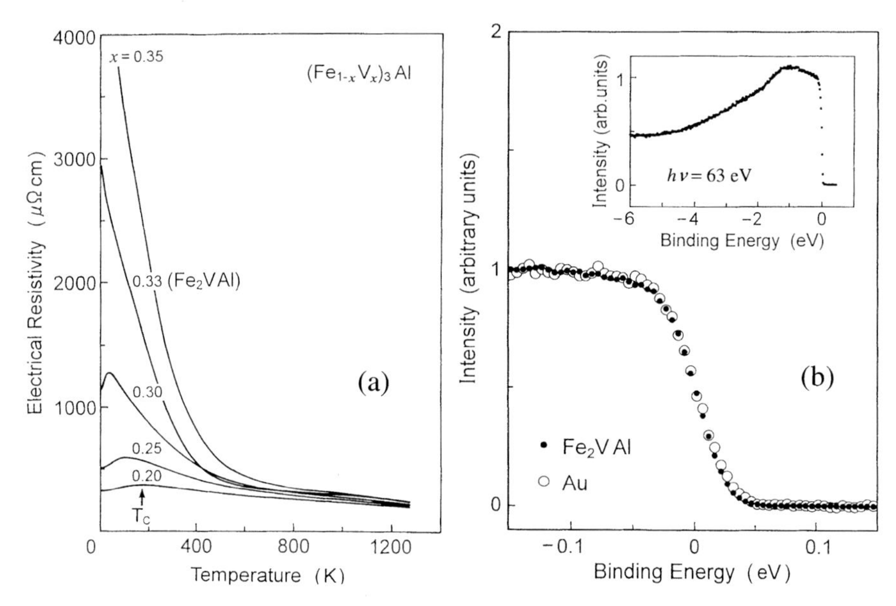

Fig. 8.22. (a) Temperature dependence of electrical resistivity in $(Fe_{1-x}V_x)_3Al$ with $0.20 \leq x \leq 0.35$. The arrow indicates the Curie temperature. Note a semiconductor-like behavior of Fe_2VAl ($x = 0.33$). (b) Near-E_F PES spectra of Fe_2VAl (full circles) and Au (open circles). The inset shows the valence band of Fe_2VAl. The spectra were measured at 40 K and with $h\nu = 63$ eV. After Nishino et al. (1997).

8.6 Conclusions and Outlook

Experimental results obtained with main spectroscopic methods on the electronic structure of QCs were reviewed. Generally good agreement was observed between the overall structure of the calculated DOS for approximants to QCs and that determined experimentally. The presence of a pseudogap in the DOS was conclusively shown to exist in a few *i* alloys. Some evidence for its presence exists for *d* alloys. The theoretically predicted DOS spikiness was not observed experimentally. Various possible reasons for this were considered, including the possibility that the fine structure of DOS is an artifact of calculations. No physical properties of QCs could be identified as resulting directly from their quasiperiodic nature; similar properties are found in crystalline alloys.

The issue of DOS spikiness in QCs is an outstanding problem. On a theoretical side, an effort should be made to establish that DOS spikiness is not an artifact. On the experimental side, near-E_F spectroscopic studies with high energy resolution probing bulk electronic states are necessary to resolve the issue of the DOS spikiness. As more QCs in a single-grain form become available, their surface DOS becomes of interest and theoretical calculations are needed to establish how it differs from the bulk DOS. With single-grain

QCs of high purity, the de Haas–van Alphen effect experiments become feasible; they could conclusively verify the existence and the details of the Fermi surface in QCs.

Acknowledgment

This work was supported by the Natural Sciences and Engineering Research Council of Canada.

References

Agarwal, B.K. (1991): X-Ray Spectroscopy, 2nd ed. Springer-Verlag, Berlin

Ahlgren, M., Gignoux, C., Rodmar, M., Berger, C., Rapp, Ö (1997): Phys. Rev. B **55**, R11 915

Bancel, P.A., Heiney, P.A. (1986): Phys. Rev. B **33**, 7917

Belin, E., Mayou, D. (1993): Phys. Scr. T **49**, 356

Belin, E., Dankhazi, Z., Sadoc, A. (1993): J. Non-Cryst. Solids **156-158**, 896

Belin, E., Dankházi, Z., Sadoc, A., Dubois, J.M. (1994a): J. Phys. Condens. Matter **6**, 8771

Belin, E., Dankhazi, Z., Sadoc, A., Dubois, J.M., Calvayrac, Y. (1994b): Europhys. Lett. **26**, 677

Berger, C., Mayou, D., Cyrot-Lackmann, F. (1995a): in Proceedings of the 5th International Conference on Quasicrystals, Janot, C., Mosseri, R. (eds). World Scientific, Singapore, p 423

Berger, C., Gignoux, C., Tjernberg, O., Lindqvist, P., Cyrot-Lackmann, F., Calvayrac, Y. (1995b): Physica B **204**, 44

Bianchi, A.D., Chernikov, M.A., Beeli, C., Ott, H.R. (1993): Solid State Commun. **87**, 721

Bianchi, A.D., Bommeli, F., Chernikov, M.A., Gubler, U., Degiorgi, L., Ott, H.R. (1997): Phys. Rev. B **55**, 5730

Bianchi, A.D., Bommeli, F., Felder, E., Kenzelmann, M., Chernikov, M.A., Degiorgi, L., Ott, H.R. (1998): Phys. Rev. B **58**, 3046

Burkov, S.E. (1991): Phys. Rev. Lett. **67**, 614

Cardona, M., Ley, L. (eds) (1978): Photoemission in Solids I. Springer-Verlag, Berlin

Carlsson, A.E., Meschter, P.J. (1995): in Intermetallic Compounds, Vol. 1, Westbrook, J.H., Fleischer, R.L. (eds). John Wiley and Sons, Chichester, p 55

Davydov, D.N., Mayou, D., Berger, C., Gignoux, C., Neumann, A., Jansen, A.G.M., Wyder, P. (1996): Phys. Rev. Lett. **77**, 3173

de Boissieu, M., Boudard, M., Hennion, B., Bellissent, R., Kycia, S., Goldman, A., Janot, C., Audier, M. (1995): Phys. Rev. Lett. **75**, 89

Dell'Acqua, G., Krajčí, M., Hafner, J. (1997): J. Phys. Condens. Matter **9**, 10 725

Depero, L.E., Parmigiani, F. (1993): Chem. Phys. Lett. **214**, 208

Dubost, B., Lang, J.-M., Tanaka, M., Sainfort, P., Audier, M. (1997): Nature **324**, 48

Ebert, Ph., Feuerbacher, M., Tamura, N., Wollgarten, M., Urban, K. (1996): Phys. Rev. Lett. **77**, 3827

Edagawa, K., Naiti, N., Takeuchi, S. (1992): Philos. Mag. B **65**, 1011

Frey, F. (1997): Z. Kristallogr. **212**, 257
Frey, F., Hradil, K., Grushko, B., McIntyre, G.J. (1998): in Proceedings of the 6th International Conference on Quasicrystals, Takeuchi, S., Fujiwara, T. (eds). World Scientific, Singapore, p 37
Friedel, J., Dénoyer, F. (1987): C.R. Acad. Sci. Paris **305**, 171
Friedel, J. (1988): Helv. Phys. Acta **61**, 538
Friedel, J. (1992): Philos. Mag. B **65**, 1125
Fuggle, J.C., Inglesfield, J.E. (eds) (1992): Unoccupied Electronic States. Springer-Verlag, Berlin
Fujiwara, T. (1989): Phys. Rev. B **40**, 942
Fujiwara, T. (1997): unpublished
Fujiwara, T. (1998): in Proceedings of the 6th International Conference on Quasicrystals, Takeuchi, S., Fujiwara, T. (eds). World Scientific, Singapore, p 591
Fujiwara, T., Yokokawa, T. (1991): Phys. Rev. Lett. **66**, 333
Fujiwara, T., Yamamoto, S., Trambly de Laissardière, G. (1993): Phys. Rev. Lett. **71**, 4166
Gavilano, J.L., Ambrosini, B., Vonlanthen, P., Chernikov, M.A., Ott, H.R. (1997): Phys. Rev. Lett. **79**, 3058
Guo, G.Y., Botton, G.A., Nishino, Y. (1998): J. Phys. Condens. Matter **10**, L119
Haerle, R., Kramer, P. (1998): Phys. Rev. B **58**, 716
Hafner, J., Krajčí, M. (1992): Phys. Rev. Lett. **68**, 2321
Hafner, J., Krajčí, M. (1993): Phys. Rev. B **47**, 11 795
Hafner, J., Krajčí, M. (1998): Phys. Rev. B **57**, 2849
Haanappel, E.G., Rabson, D.A., Mueller, F.M. (1996): in Proceedings of Physical Phenomena at High Magnetic Fields - II, Fisk, Z., Gor'kov, L., Meltzer, D., Schrieffer, R. (eds). World Scientific, Singapore, p 362
Häussler, P. (1994): in Glassy Metals III, Beck, H., Güntherodt, H.-J. (eds). Springer-Verlag, Berlin, p 163
Hennig, R.G., Teichler, H. (1997): Philos. Mag. A **76**, 1053
Hippert, F., Kandel, L., Calvayrac, Y., Dubost, B. (1992): Phys. Rev. Lett. **69**, 2086
Hippert, F., Brand, R.A., Pelloth, J., Calvayrac, Y. (1994): J. Phys. Condens. Matter **6**, 11 189
Hüfner, S. (1996): Photoelectron Spectroscopy, 2nd ed. Springer-Verlag, Berlin
Inoue, A.,Tsai, A.-P., Masumoto, T. (1990): in Quasicrystals, Fujiwara, T., Ogawa, T. (eds). Springer-Verlag, Berlin, p 80
Janssen, T., Fasolino, A. (1998): in Proceedings of the 6th International Conference on Quasicrystals, Takeuchi, S., Fujiwara, T. (eds). World Scientific, Singapore, p 757
Kirihara, K., Kimura, K., Ino, H., Kumigashira, H., Ashihara, A., Takahashi, T. (1998): in Proceedings of the 6th International Conference on Quasicrystals, Takeuchi, S., Fujiwara, T. (eds). World Scientific, Singapore, p 651
Klein, T., Symko, O.G., Davydov, D.N., Jansen, A.G.M. (1995): Phys. Rev. Lett. **74**, 3656
Knapp, J.A., Himpsel, F.J., Eastman, D.E. (1979): Phys. Rev. B **19**, 4952
Kobayshai, A., Matsuo, S., Ishimasa, T., Nakano, H. (1997): J. Phys. Condens. Matter **9**, 3205
Kortboyer, S.W., Grioni, M., Speier, W., Zeller, R., Watson, L.M., Gibson, M.T., Schäfers, F., Fuggle, J.C. (1989): J. Phys. Condens. Matter **1**, 5981
Krajčí, M., Hafner, J. (1994): Europhys. Lett. **27**, 147
Krajčí, M., Hafner, J. (1998): unpublished

Krajčí, M., Windisch, M., Hafner, J., Kresse, G., Mihalkovič, M. (1995): Phys. Rev. B **51**, 17355
Krajčí, M., Hafner, J., Mihalkovič, M. (1996): Europhys. Lett. **34**, 207
Krajčí, M., Hafner, J., Mihalkovič, M. (1997a): Phys. Rev. B **56**, 3072
Krajčí, M., Hafner, J., Mihalkovič, M. (1997b): Phys. Rev. B **55**, 843
Lang, N.D. (1986): Phys. Rev. B **34**, 5947
Ley, L., Cardona M., (eds) (1979): Photoemission in Solids II. Springer-Verlag, Berlin
Matsubara, H., Ogawa, S., Kinoshita, T., Kishi, K., Takeuchi, S., Kimura, K., Suga, S. (1991): Jpn. J. Appl. Phys. A **30**, L389
Mayou, D., Berger, C., Cyrot-Lackmann, F., Klein, T., Lanco, P. (1993): Phys. Rev. Lett. **70**, 3915
Mehlhorn, W. (ed) (1982): Corpuscules and Radiation in Matter I. Springer-Verlag, Berlin
Mizutani, U. (1998): J. Phys. Condens. Matter **10**, 4609
Mizutani, U., Kamiya, A., Matsuda, T., Kishi, K., Takeuchi, S. (1991): J. Phys. Condens. Matter **3**, 3711
Mizutani, U., Matsuda, T., Itoh, Y., Tanaka, K., Domae, H., Mizuno, T., Murasaki, S., Miyoshi, Y., Hashimoto, K., Yamada, Y. (1993): J. Non-Cryst. Solids **156-158**, 882
Mizutani, U., Iwakami, W., Fujiwara, T. (1998): in Proceedings of the 6th International Conference on Quasicrystals, Takeuchi, S., Fujiwara, T. (eds). World Scientific, Singapore, p 579
Mori, M., Matsuo, S., Ishimasa, T., Matsuura, T., Kamiya, K., Inokuchi, H., Matsukawa, T. (1991): J. Phys. Condens. Matter **3**, 767
Mori, M., Kamiya, K., Matsuo, S., Ishimasa, T., Nakano, H., Fujimoto, H., Inokuchi, H. (1992): J. Phys. Condens. Matter **4**, L157
Mori, M., Ogawa, T., Ishimasa, T., Tanaka, M., Sasaki, S. (1998): in Proceedings of the 6th International Conference on Quasicrystals, Takeuchi, S., Fujiwara, T. (eds). World Scientific, Singapore, p 387
Neuhold, G., Barman, S.R., Horn, K., Theis, W., Ebert, P., Urban, K. (1998): Phys. Rev. B **58**, 734
Nishino, Y., Kato, M., Asano, S., Soda, K., Hayasaki, M., Mizutani, U. (1997): Phys. Rev. Lett. **79**, 1909
Pierce, F.S., Poon, S.J., Guo, Q. (1993a): Science **261**, 737
Pierce, F.S., Bancel, P.A., Biggs, B.D., Guo, Q., Poon, S.J. (1993b): Phys. Rev. B **47**, 5670
Pierce, F.S., Guo, Q., Poon, S.J. (1994): Phys. Rev. Lett. **73**, 2220
Poon, S.J. (1992): Adv. Phys. **41**, 303
Purdie, D., Garnier, M., Hengsberger, M., Baer, Y., Stadnik, Z.M., Lograsso, T.A., Delaney, D.W. (1998): unpublished
Roche, S., Trambly de Laissardière, G., Mayou, D. (1997): J. Math. Phys. **38**, 1794
Sabiryanov, R.F., Bose, S.K. (1994): J. Phys. Condens. Matter **6**, 6197; Erratum (1995) **7**, 2375
Sabiryanov, R.F., Bose, S.K., Burkov, S.E. (1995): J. Phys. Condens. Matter **7**, 5437
Sadoc. A., Belin, E., Dankhazi, Z., Flank, A.M. (1993): J. Non-Cryst. Solids **153-154**, 338
Shastri, A., Borsa, F., Torgeson, D.R., Goldman, A.I. (1994a): Phys. Rev. B **50**, 4224
Shastri, A., Borsa, F., Torgeson, D.R., Shield, J.E., Goldman, A.I. (1994b): Phys. Rev. B **50**, 16651

Shastri, A., Baker, D.B., Conradi, M.S., Borsa, F., Torgeson, D.R. (1995): Phys. Rev. B **52**, 12 681
Shechtman, D., Blech, I., Gratias, D., Cahn, J.W. (1984): Phys. Rev. Lett. **53**, 1951
Shih, C.K., Feenstra, R.M., Kirtley, J.R., Chandrashekhar, G.V. (1989): Phys. Rev. B **40**, 2682
Singh, D.J., Mazin, I.I. (1998): Phys. Rev. B **57**, 14 352
Shoenberg, D. (1984): Magnetic Oscillations in Metals. Cambridge University Press, New York
Smith, A.P., Ashcroft, N.W. (1987): Phys. Rev. Lett. **59**, 1365
Soda, K., Nozawa, K., Yanagida, Y., Morita, K., Mizutani, U., Yokoyama, Y., Note, R., Inoue, A., Ishii, H., Tezuka, Y., Shin, S., (1998a): J. Electron Spectrosc. Relat. Phenom. **88-91**, 415
Soda, K., Mizutani, U., Yokoyama, Y., Note, R., Inoue, A., Fujisawa, M., Shin, S., Suga, S., Sekiyama, A., Susaki, T., Konishi, T., Matsushita, T., Miyahara, T. (1998b): in Proceedings of the 6th International Conference on Quasicrystals, Takeuchi, S., Fujiwara, T. (eds). World Scientific, Singapore, p 676
Stadnik, Z.M. (1996): in Mössbauer Spectroscopy Applied to Magnetism and Materials Science, Vol. 1, Long, G.J., Grandjean, F. (eds). Plenum, New York, p 125
Stadnik, Z.M., Stroink, G., (1993): Phys. Rev. B **47**, 100
Stadnik, Z.M., Purdie, D. (1997): unpublished
Stadnik, Z.M., Zhang, G.W., Tsai, A.-P., Inoue, A. (1995a): Phys. Rev. B **51**, 4023
Stadnik, Z.M., Zhang, G.W., Tsai, A.-P., Inoue, A. (1995b): in Proceedings of the 5th International Conference on Quasicrystals, Janot, C., Mosseri, R. (eds). World Scientific, Singapore, p 249
Stadnik, Z.M., Zhang, G.W., Akbari-Moghanjoughi, M., Tsai, A.-P., Inoue, A. (1995c): in Proceedings of the 5th International Conference on Quasicrystals, Janot, C., Mosseri, R. (eds). World Scientific, Singapore, p 530
Stadnik, Z.M., Zhang, G.W., Tsai, A.-P., Inoue, A. (1995d): Phys. Rev. B **51**, 11 358
Stadnik, Z.M., Purdie, D., Garnier, M., Baer, Y., Tsai, A.-P., Inoue, A., Edagawa, K., Takeuchi, S. (1996): Phys. Rev. Lett. **77**, 1777
Stadnik, Z.M., Purdie, D., Garnier, M., Baer, Y., Tsai, A.-P., Inoue, A., Edagawa, K., Takeuchi, S., Buschow, K.H.J. (1997): Phys. Rev. B **55**, 10 938
Stadnik, Z.M., Purdie, D., Garnier, M., Baer, Y., Tsai, A.-P., Inoue, A., Edagawa, K., Takeuchi, S. (1998): in Proceedings of the 6th International Conference on Quasicrystals, Takeuchi, S., Fujiwara, T. (eds). World Scientific, Singapore, p 563
Stroscio, J.A., Kaiser, W.J. (eds) (1993): Scanning Tunneling Microscopy. Academic, San Diego
Takahashi (1989): in Strong Correlation and Superconductivity, Fukuyama, H., Maekawa, S., Malozemoff, A.P. (eds). Springer-Verlag, Berlin, p 311
Takeuchi, S. (1994): Mater. Sci. Forum **150-151**, 35
Takeuchi, T., Mizutani, U. (1995): Phys. Rev. B **52**, 9300
Tang, X.-P., Hill, E.A., Wonnell, S.K., Poon, S.J., Wu, Y. (1997): Phys. Rev. Lett. **79**, 1070
Terauchi, M., Tanaka, M., Tsai, A.-P., Inoue, A., Masumoto, T. (1996): Philos. Mag. B **74**, 107
Terauchi, M., Ueda, H., Tanaka, M., Tsai, A.P., Inoue, A., Masumoto, T. (1998a): in Proceedings of the 6th International Conference on Quasicrystals, Takeuchi, S., Fujiwara, T. (eds). World Scientific, Singapore, p 587
Terauchi, M., Ueda, H., Tanaka, M., Tsai, A.-P., Inoue, A., Masumoto, T. (1998b): Philos. mag. Lett. **77**, 351

Terauchi, M., Ueda, H., Tanaka, M., Tsai, A.-P., Inoue, A., Masumoto, T. (1998c): Philos. Mag. B **77**, 1625
Trambly de Laissardière, G., Fujiwara, T. (1994a): Phys. Rev. B **50**, 5999
Trambly de Laissardière, G., Fujiwara, T. (1994b): Phys. Rev. B **50**, 9843
Trambly de Laissardière, G., Fujiwara, T. (1994c): Mater. Sci. Eng. A **181-182**, 722
Trambly de Laissardière, G., Dankházi, Z., Belin, E., Sadoc, A., Nguyen Manh, D., Mayou, D., Keegan, M.A., Papaconstantopoulos, D.A. (1995a): Phys. Rev. B **51**, 12 035
Trambly de Laissardière, G., Nguyen Manh, D., Magaud, L., Julien, J.P., Cyrot-Lackmann, F., Mayou, D. (1995b): Phys. Rev. B **52**, 7920
Tsai, A.-P., Inoue, A., Masumoto, T. (1989a): Mater. Trans., Jpn. Inst. Met. **30**, 463
Tsai, A.-P., Inoue, A., Masumoto, T. (1989b): Mater. Trans., Jpn. Inst. Met. **30**, 666
Vaks, V.G., Kamyshenko, V.V., Samolyuk, G.D. (1988): Phys. Lett. A **132**, 131
Volkov, P., Poon, S.J. (1995): Phys. Rev. B **52**, 12 685
Windisch, M., Krajčí, M., Hafner, J (1994): J. Phys. Condens. Matter **6**, 6977
Wälti, Ch., Felder, E., Chernikov, M.A., Ott, H.R., de Boissieu, M., Janot, C. (1998): Phys. Rev. B **57**, 10 504
Wolf, E.L. (1985): Principles of Electron Tunneling Spectroscopy. Oxford University Press, New York
Wu, X., Homes, C.C., Burkov, S.E., Timusk, T., Pierce, F.S., Poon, S.J., Cooper, S.L., Karlow, A.M. (1993): J. Phys. Condens. Matter **5**, 5975
Wu, X., Kycia, S.W., Olson, C.G., Benning, P.J., Goldman, A.I., Lynch. D.W. (1995): Phys. Rev. Lett. **75**, 4540
Yeh, J.-J. (1993): Atomic Calculations of the Photoionization Cross-Sections and Asymmetry Parameters. Gordon and Breach, New York
Zhang, G.W., Stadnik, Z.M. (1995): in Proceedings of the 5th International Conference on Quasicrystals, Janot, C., Mosseri, R. (eds). World Scientific, Singapore, p 552
Zhang, G.W., Stadnik, Z.M., Tsai, A.-P., Inoue, A. (1994): Phys. Rev. B **50**, 6696
Zhang, G.W., Stadnik, Z.M., Tsai, A.-P., Inoue, A., Miyazaki, T. (1995): Z. Phys. B **97**, 439

9. Magnetic Properties of Quasicrystals

Kazuaki Fukamichi

9.1 Introduction

After the discovery of an icosahedral (I) Al-Mn phase (Shechtman et al. 1984), there was immediately great interest in determining whether similar phases would form in other alloy systems. A number of quasicrystals (QCs), especially Al-based QCs with a wide variety of elements, have now been discovered. In addition, Mg-Zn-Rare Earth (RE) QCs with a large localized magnetic moment have recently been found by Niikura et al. (1994).

The magnetic properties of QCs have been actively studied because they are very sensitive to the local atomic structure. The results obtained up to 1990 were reviewed by O'Handley et al. (1991). The data presented in this chapter have been limited to two categories of QCs, Al-based and Mg-RE-Zn QCs.

In Sect. 9.2, the magnetic properties of Al-based QCs are discussed. Since the magnetic properties of $3d$ transition metal (TM) alloys are affected by the local atomic environment, we first review paramagnetism, the effective magnetic moment, and the fraction of magnetic Mn atoms in the QCs. We then discuss how the small fraction of magnetic Mn atoms is associated with the spin-glass behavior and low-temperature specific heat. Furthermore, these magnetic properties of the QCs are compared with those of their amorphous (A) counterparts. The difference between the magnetic properties of the I phase and the decagonal (D) phase, as well as anisotropy in the susceptibility due to the anisotropic atomic structure in the D QCs, are demonstrated. Next, the magnetic properties of several Al-based compounds are discussed in connection with the Pauling valence (Pauling 1960). Moreover, it is pointed out that the phason plays a role in the change of the susceptibility of samples subjected to different heat treatments. The positive temperature dependence of the magnetic susceptibility at high temperatures is explained by the Pauli paramagnetism related to a pseudogap at the Fermi surface. Finally, it is suggested that peculiar ferromagnetic properties, such as a relatively high Curie temperature with a low magnetization and a large coercivity for the Al-based QCs, can be explained in terms of phasons.

Section 9.3 deals with the magnetic properties of Mg-RE-Zn QCs. First we discuss the susceptibility and spin-glass behavior in the localized $4f$ electron systems of the Mg-RE-Zn QCs. Second, we show how the low-temperature specific heat is closely correlated with the spin-glass behavior, and with the symmetry of the RE atomic site. Finally, some comments on the long-range antiferromagnetic order are given. The chapter ends with a brief summary in Sect. 9.4.

9.2 Al-Based Quasicrystals

The pair distribution functions in the *I* and *A* phases of the $Pd_{58.8}U_{20.6}Si_{20.6}$ alloy are similar up to the second-nearest neighbors at about 6 Å (Kofalt et al. 1986). Furthermore, the local *I* order in *A* alloys (Widom 1988) and the structural similarity between the $Al_{75}Cu_{15}V_{10}$ *I* QC and its *A* counterpart (Matsubara et al. 1988) have been discussed. However, detailed structural analyses of Al-Mn alloys in the quasicrystalline (*Q*) and *A* states have been made using EXAFS measurements and the difference between the pair distribution function in these two states has been demonstrated (Sadoc and Dubois 1989). Therefore, the difference between the magnetic properties of QCs and *A* alloys will occur because magnetic properties of 3*d* TMs are sensitive to the local atomic structure.

In this section, we first show the difference between various magnetic properties of the Al-based QCs and their *A* counterparts. We discuss the effective magnetic moment, the saturation magnetization, and the fraction of the magnetic Mn atoms in the Al-Mn, Al-Cu-Mn and Al-Pd-Mn systems. Next, we point out that there is a difference between the magnetic properties of the *I* and *D* phases of the Al-Pd-Mn QCs. Since the *D* QCs have an anisotropic atomic structure, the *D*-Al-Ni-Co QCs are expected to exhibit anisotropy in the magnetic susceptibility. Spin-glass behavior in the Al-Mn and Al-Pd-Mn QCs and their *A* alloys is presented, and the characteristic behavior of spin-glass and the strength of the RKKY interaction are discussed using low-temperature specific heat data. In connection with the environment effect, i.e., the atomic distance and the coordination number, the magnetism of several Al-based compounds is discussed in terms of the Pauling valence. Furthermore, the change in the susceptibility induced by heat treatments is correlated with the phason. The temperature dependence of magnetic susceptibility at high temperatures is interpreted in terms of the Pauli paramagnetism, which is caused by a pseudogap in the density of states (DOS) at the Fermi surface (E_F). Finally, the relationship between several peculiar ferromagnetic properties and phasons, are discussed in reference to magnetic clusters induced by plastic deformation in Fe-Al crystalline alloys.

9.2.1 Paramagnetism, Effective Magnetic Moment and Saturation Magnetization

Figures 9.1a and 9.1b show, respectively, the temperature dependence of the dc susceptibility of Al-Cu-TM and Al-Pd-TM QCs (Fukamichi et al. 1993).

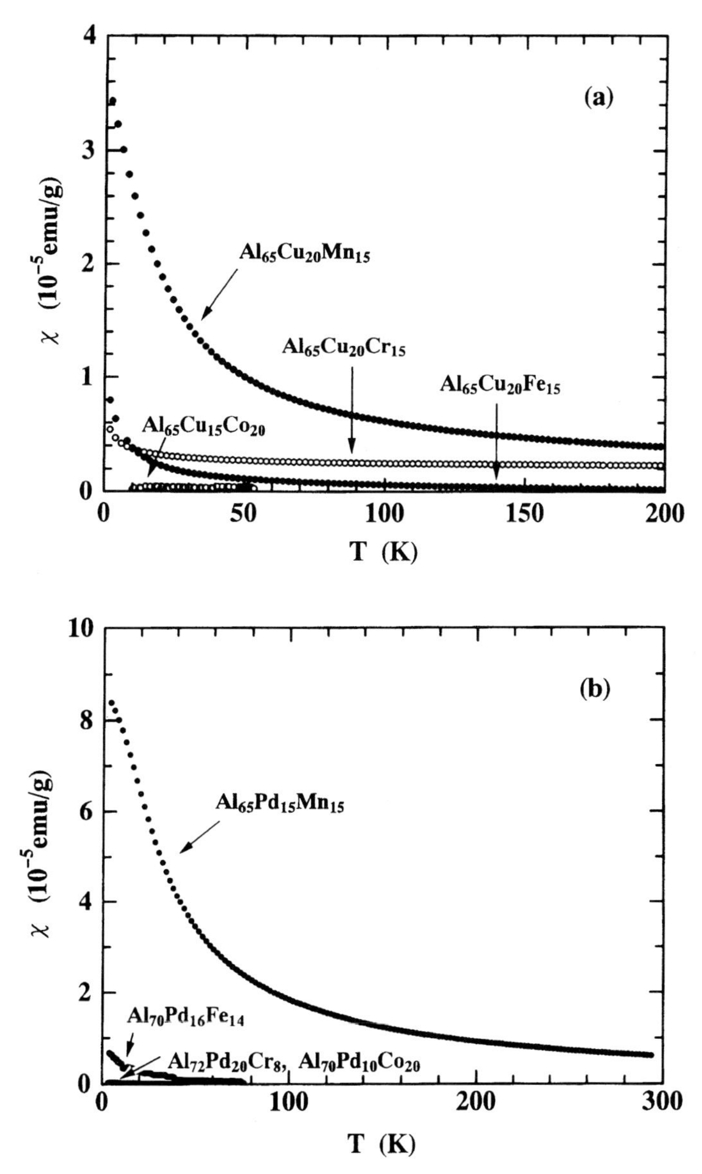

Fig. 9.1. Temperature dependence of the dc susceptibility of Al-based QCs (a) Al-Cu-TM and (b) Al-Pd-TM. After Fukamichi et al. (1993).

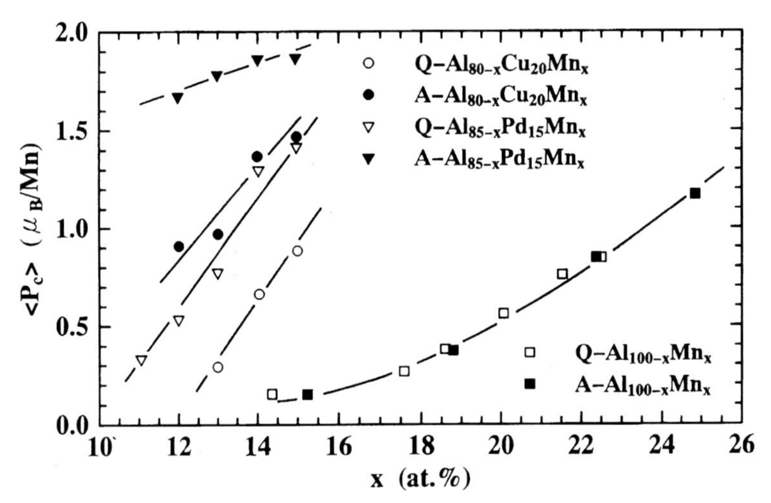

Fig. 9.2. Concentration dependence of the average local magnetic moment per Mn atom $\langle P_c \rangle$ for $Al_{100-x}Mn_x$, $Al_{80-x}Cu_{20}Mn_x$, and $Al_{85-x}Pd_{15}Mn_x$ QCs and *A* alloys. After Fukamichi et al. (1993).

The dc susceptibility χ contains a temperature independent contribution χ_0 and a Curie-Weiss contribution

$$\chi - \chi_0 = C/(T - \Theta_p), \tag{9.1}$$

where C is the Curie constant and Θ_p is the paramagnetic Curie temperature. The value of Θ_p is negative, indicating antiferromagnetic interactions. The effective magnetic moment, P_{eff}, is deduced from C. A relatively large value of χ for the $Al_{65}Cu_{20}Cr_{15}$ QC might be due to surface oxidation as CrO_2 is ferromagnetic.

Shown in Fig. 9.2 is the concentration dependence of the average local magnetic moment per Mn atom, $\langle P_c \rangle$, for $Al_{100-x}Mn_x$, $Al_{80-x}Cu_{20}Mn_x$, and $Al_{85-x}Pd_{15}Mn_x$ in the Q and A states. The value of $\langle P_c \rangle$ is deduced from the value of C by assuming $g = 2$ in the expression (Goto et al. 1988)

$$C = N\mu_B^2 \langle P_c \rangle (\langle P_c \rangle + 2)/3k_B, \tag{9.2}$$

with

$$\langle P_c \rangle = -1 + \sqrt{1 + P_{\text{eff}}^2}.$$

Here N is the number of Mn atoms, μ_B is the Bohr magneton, and k_B is the Boltzmann constant. The value of $\langle P_c \rangle$ is much smaller for the QCs than for the A alloys (Fukamichi et al. 1991, Fukamichi and Goto 1991). It is worth noting that the $Al_{80-x}Cu_{20}Mn_x$ and $Al_{85-x}Pd_{15}Mn_x$ alloys exhibit large values of $\langle P_c \rangle$ at lower Mn concentrations, as compared with the values for $Al_{100-x}Mn_x$ alloys.

The M-H curves are at first convex in shape and are difficult to saturate even at 300 kOe. Using a 1/H plot, the values of the saturation magnetization,

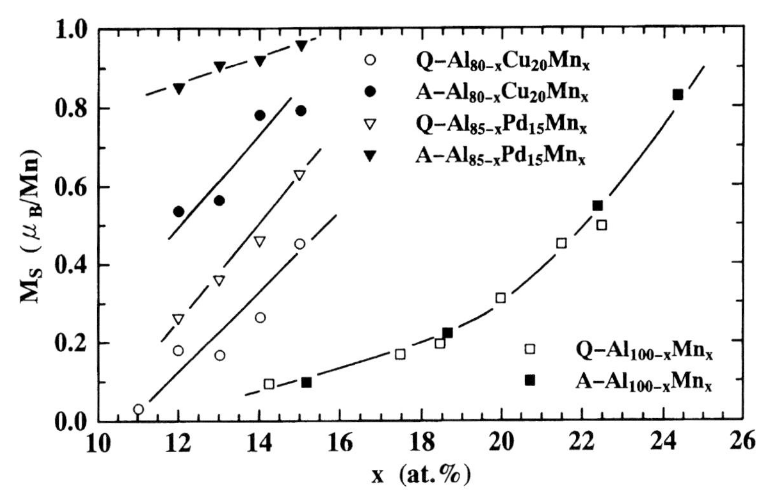

Fig. 9.3. Concentration dependence of the average saturation magnetization per Mn atom M_s for $Al_{100-x}Mn_x$, $Al_{80-x}Cu_{20}Mn_x$ and $Al_{85-x}Pd_{15}Mn_x$ QCs and A alloys. After Fukamichi et al. (1993).

M_s, of QCs are estimated by scaling with the data of Cu-Mn crystalline dilute alloys (Bellissent et al. 1987). The concentration dependence of the average M_s per Mn is presented in Fig. 9.3 (Fukamichi et al. 1993). The important point in Figs. 9.2 and 9.3 is that M_s is much smaller than $\langle P_c \rangle$ in both phases. We note that a linear extrapolation of the magnetization from low fields, which is often applied to the magnetization curves of ferromagnets, gives an incorrect result (Chernikov et al. 1993).

9.2.2 Fraction of Magnetic Mn Atoms and Giant Magnetic Moment

The data presented above imply that not all Mn sites are magnetic in both QCs and A alloys. Using the values of C and M_s, we can evaluate the fraction of magnetic Mn atoms, x_m, and the average local moment, P_c, of the magnetic Mn sites from the following relations

$$M_s(M_s x/x_m + 2\mu_B) = 3k_B C/N \tag{9.3}$$

$$P_c = (x/x_m)M_s/\mu_B. \tag{9.4}$$

The concentration dependence of x_m/x of three alloy systems is presented in Fig. 9.4 (Fukamichi et al. 1993). In the $Al_{80-x}Cu_{20}Mn_x$ and $Al_{85-x}Pd_{15}Mn_x$ systems, the value of x_m/x is much smaller for the QCs than for the A alloys. A slight difference in x_m/x is observed for the $Al_{100-x}Mn_x$ QCs and A alloys (Goto et al. 1988). The concentration dependence of P_c calculated from Eq. (9.4) is shown in Fig. 9.5 (Fukamichi et al. 1993). For the $Al_{65}Cu_{20}Mn_{15}$ QC, the value of P_c is about one half that of the A counterpart.

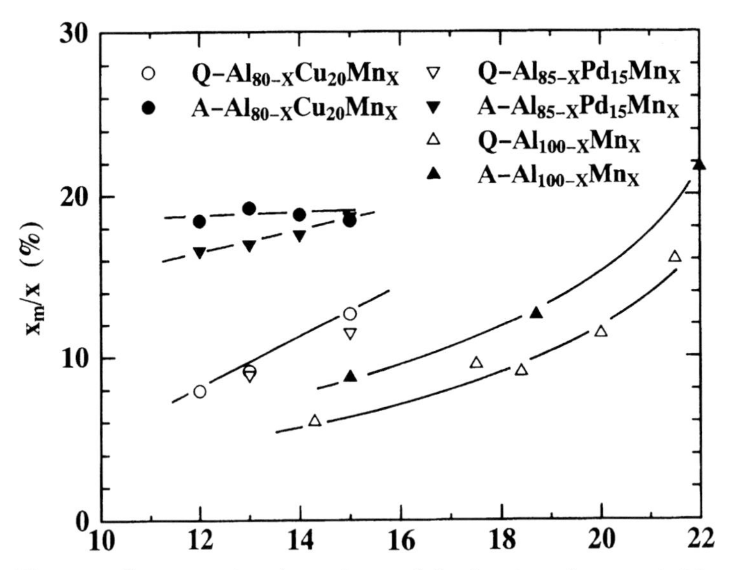

Fig. 9.4. Concentration dependence of the fraction of magnetic Mn moments x_m/x for $Al_{100-x}Mn_x$, $Al_{80-x}Cu_{20}Mn_x$, and $Al_{85-x}Pd_{15}Mn_x$ QCs and A alloys. After Goto et al. (1988) and Fukamichi et al. (1993).

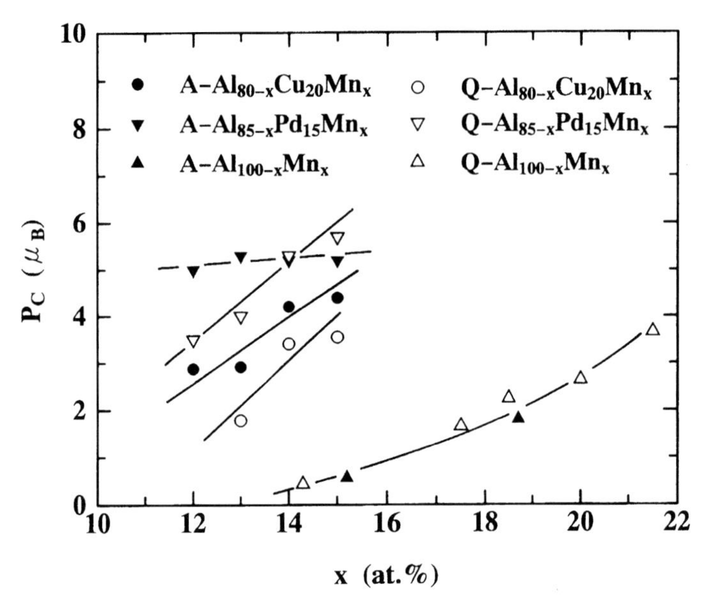

Fig. 9.5. Concentration dependence of the average local magnetic moment of magnetic Mn sites P_c for $Al_{100-x}Mn_x$, $Al_{80-x}Cu_{20}Mn_x$, and $Al_{85-x}Pd_{15}Mn_x$ QCs and A alloys. After Fukamichi et al. (1993).

In the case of the $Al_{70}Pd_{15}Mn_{15}$ QC, P_c exceeds the maximum value of $5\mu_B$ of the bare moment on the Mn atoms. Therefore, it is concluded that this alloy has a giant magnetic moment. As is well known, giant magnetic moments have been observed in crystalline dilute alloys, such as Pd-Co, Pd-Fe, and Pd-Mn. Such giant magnetic moments in Pd dilute alloys are the sum of the localized magnetic moment plus an induced moment in the surrounding host metal (Mydosh and Nieuwenhuys 1980). However, the $Al_{70}Pd_{15}Co_{15}$ and $Al_{70}Pd_{15}Fe_{15}$ QCs carry no such large magnetic moment, as seen from Fig. 9.1b. Mn atoms only in the Al-Pd-Mn alloys have a giant magnetic moment.

In the crystalline state, the Fermi energy E_F of Al is about 12 eV and the band width Δ of the virtual bound state is about 4 eV. On the other hand, E_F and Δ are respectively about 7 eV and 2 eV in Cu (Gubanov et al. 1992). The condition for the formation of localized magnetic moments of magnetic impurities is expressed as (Gubanov et al. 1992)

$$\Delta < p\Delta E, \tag{9.5}$$

where ΔE is the energy gain for a pair of electrons changing their spin orientation from antiparallel to parallel (its value is estimated to be in the range 0.6–0.8 eV) and p is the maximum number of unpaired d electrons in the outer shell of the impurity atom. Thus it is clear that the formation of localized magnetic moments in an Al host is more difficult than in a Cu host. This may explain why in the Al-Mn system the magnetic properties are not so sensitive to the different local environment in the Q and A states. On the contrary, in Cu- and Pd-based crystals, many magnetic TMs carry a localized magnetic moment. Therefore, the slight difference in the local environment in the Q and A state would be sensitively reflected in the magnetic properties, even though the studied alloys contain only 20 at.% Cu or 15 at.% Pd.

9.2.3 Difference Between Magnetic Moments in Icosahedral and Decagonal Phases

The magnetic properties of the Al-Pd-Mn system in the Q and A states differ significantly (Figs. 9.4 and 9.5). Here, we compare the magnetic properties of QCs of a different structure but of the same concentration. The $Al_{70}Pd_{30-x}Mn_x$ I QCs were prepared by melt quenching whereas the D QCs were obtained by annealing the I phase at 800°C for 15 h (Satoh et al. 1994). The magnetic susceptibility of these $Al_{85-x}Pd_{15}Mn_x$ QCs follows the Curie-Weiss law over a wide temperature range. Small negative values of θ_P, ranging from -20 to -10 K, are observed. This suggests that the Mn-Mn exchange interaction in these QCs is antiferromagnetic.

Figures 9.6a and 9.6b show the concentration dependence of M_s and $\langle P_c \rangle$ for the I and D phases (Satoh et al. 1994, Hattori et al. 1994). The value of M_s per Mn for the I phase is larger than that for the D phase, and M_s for both phases increases with increasing Mn content. The values of $\langle P_c \rangle$ for the

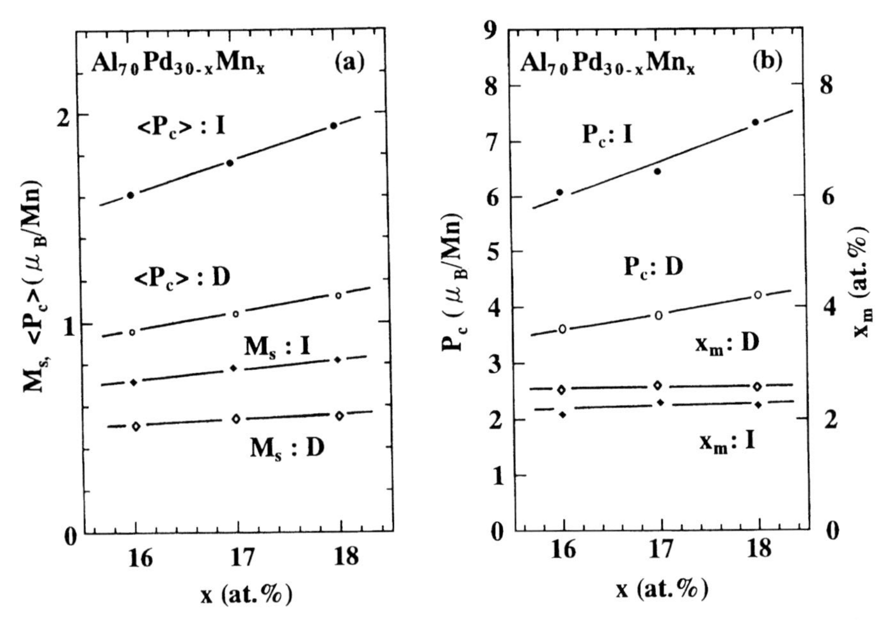

Fig. 9.6. (a) Concentration dependence of the average saturation magnetization per Mn atom M_s and the average local magnetic moment per Mn atom $\langle P_c \rangle$. (b) Concentration dependence of the average local magnetic moment of magnetic Mn sites P_c and the fraction of magnetic Mn atoms x_m, for $Al_{70}Pd_{30-x}Mn_x$ QCs in the *I* and *D* phases. After Satoh et al. (1994) and Hattori et al. (1994).

I phase are about twice as large as those for the *D* phase. The concentration dependence of P_c is proportional to the Mn content, and P_c for the *I* phase is about twice that of the *D* phase, although the fraction x_m in both phases is not so different, being about 2–2.5% of the total atomic content, as seen from Fig. 9.6b. These results imply that the magnetic properties of the Al-Pd-Mn QCs are significantly affected by the atomic distance, the coordination number (Sect. 9.2.6), and/or the kinds of atoms around the Mn atom. It should be emphasized that the concentration dependence of P_c is proportional to the Mn content. For the *I* phase, P_c is about twice that of the *D* phase (Fig. 9.6b). The values of P_c for the *I* phase are much higher than the bare Mn moment of 5 μ_B. This supports the notion of the presence of a giant magnetic moment in this alloy system (Satoh et al. 1994, Hattori et al. 1994).

Anisotropic magnetic properties are expected in the *D* QCs, because the *D* QCs have an anisotropic atomic structure with a crystalline axis perpendicular to a *Q* plane. Figure 9.7 shows the temperature dependence of the differential susceptibility χ_1, which is defined as $\Delta M/\Delta H$, for the *D*-$Al_{70}Ni_{15}Co_{15}$ single-grain QC. The data are very insensitive to temperature between about 30 and 250 K, resulting in a small anisotropy. The average values estimated for this temperature range are $\chi_{1\perp} = 8.2\times10^{-7}$ and $\chi_{1\|} = 7.1\times10^{-7}$ emu/g (Markert et al. 1994). Many ring contrasts in the c

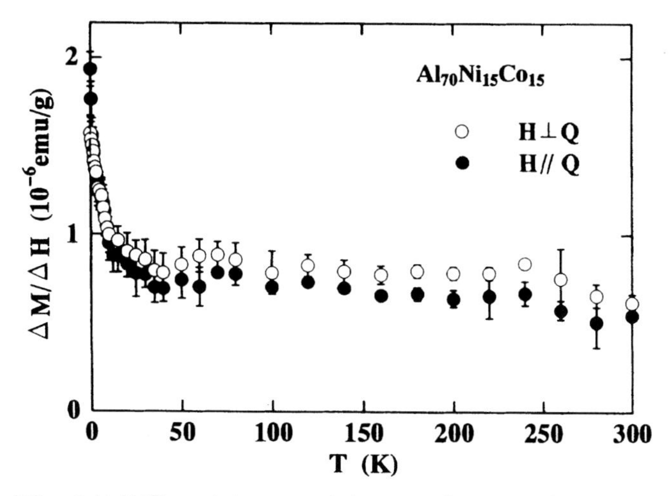

Fig. 9.7. Differential susceptibility as a function of temperature in magnetic fields perpendicular and parallel to the quasicrystallographic direction in a single-grain $Al_{70}Ni_{15}Co_{15}$ *D* phase. After Markert et al. (1994).

plane of the *D*-$Al_{72}Ni_{12}Co_{16}$ QC were observed in electron microscopy studies (Yamada et al. 1998). If the ring has one or more electrons which can freely move in it, it could be the origin of the anisotropic magnetic properties (Yamada et al. 1998), in analogy to the situation in benzene.

9.2.4 Spin-Glass Behavior

As mentioned earlier, most of the Mn sites in QCs are nonmagnetic. The model pair potential calculations indicate the existence of a characteristic preferred spacing of 4.7 Å between Mn atoms in Al-TM systems (Zou and Carlsson 1993). Therefore, magnetic Mn atoms act mainly as isolated atoms. One can construct a physical picture in a direct analogy to magnetically dilute spin-glasses. As is well known, the spin-glass behavior is created by the RKKY interaction in magnetically dilute alloys. As shown in Fig. 9.8, the temperature dependence of the ac susceptibility of Al-Mn QCs and *A* alloys exhibits a distinct cusp in low-temperature ranges. In addition, the dc magnetic susceptibility curve shows hysteresis after magnetic cooling (Fukamichi et al. 1987, Fukamichi and Goto 1989). These data indicate that the Al-Mn QCs and *A* alloys exhibit a spin-glass behavior at low temperatures.

Figure 9.9a shows the concentration dependence of the spin-freezing temperature, T_f, of $Al_{100-x}Mn_x$ (Hauser et al. 1986, Fukamichi et al. 1987, 1989, Godinho et al. 1990), $Al_{80-x}Cu_{20}Mn_x$, and $Al_{85-x}Pd_{15}Mn_x$ QCs and their *A* counterparts (Fukamichi et al. 1993). It is clear that the magnitude of T_f strongly depends on the alloy system. Note that T_f in the *A* state is higher

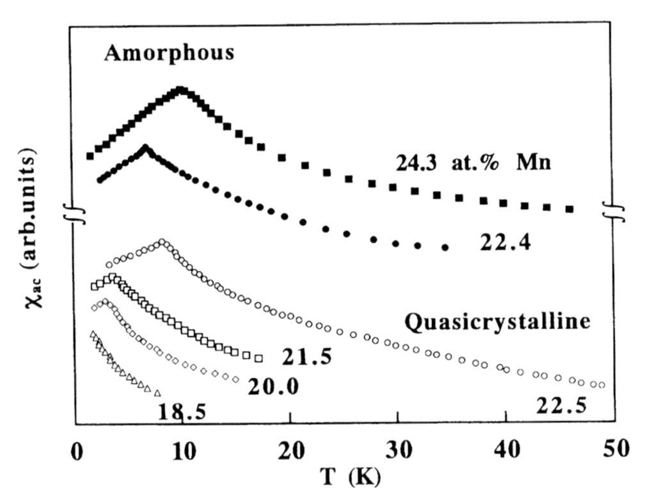

Fig. 9.8. Temperature dependence of the ac susceptibility of $Al_{100-x}Mn_x$ QCs and *A* alloys. After Fukamichi and Goto (1989).

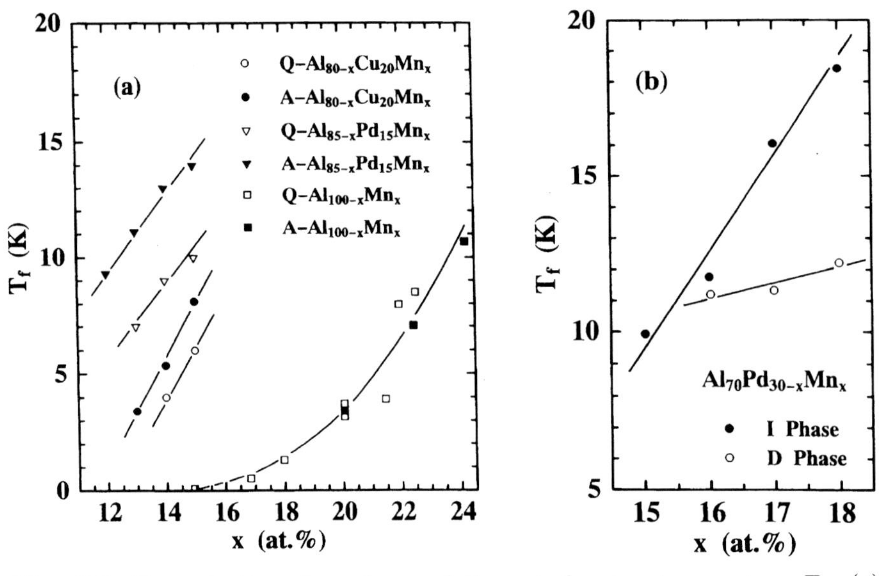

Fig. 9.9. Concentration dependence of the spin-freezing temperature T_f. (a) $Al_{100-x}Mn_x$ [after Hauser et al. (1986), Fukamichi et al. (1987), (1989), and Godinho et al. (1990)], $Al_{80-x}Cu_{20}Mn_x$, and $Al_{85-x}Pd_{15}Mn_x$ [after Fukamichi et al (1993)] QCs and *A* alloys. (b) $Al_{70}Pd_{30-x}Mn_x$ QCs in the *I* and *D* phases [after Hattori et al. (1994) and Satoh et al. (1994)].

Table 9.1. Magnetic state of 3*d* TMs in host crystalline metals and in Al-based QCs and *A* alloys. After Mydosh and Nieuwenhuys (1980), Fukamichi and Goto (1991), and Fukamichi et al. (1993).

Metal/alloy	Cr	Mn	Fe	Co	Ni
Cryst. Al	N	N	N	N	N
Q-Al-TM	N	SG	N	N	N
A-Al-TM	N	SG	–	–	–
Cryst. Cu	Y	Y	Y	Y	N
Q-Al-Cu-TM	N	Y	Y	Y	N
A-Al-Cu-TM	Δ	SG	Δ	N	–
Cryst. Pd	Y	SG+GM	GM	GM	Y
Q-Al-Pd-TM	N	SG+GM	Δ	N	–
A-Al-Pd-TM	N	SG+GM?	Δ	N	–

Y: localized magnetic moment N: no localized magnetic moment
SG: spin-glass state GM: giant magnetic moment
Δ: weak temperature dependence of susceptibility
- : no experimental data available

than in the *Q* state for the $Al_{80-x}Cu_{20}Mn_x$ and $Al_{85-x}Pd_{15}Mn_x$ alloy systems, but is essentially the same in the $Al_{100-x}Mn_x$ system. Table 9.1 summarizes the occurrence of a localized moment, a spin-glass behavior, and a giant magnetic moment in Al-Mn, Al-Cu-Mn, and Al-Pd-Mn QCs and *A* alloys (Fukamichi et al. 1991, Fukamichi and Goto 1991). For comparison, the data for magnetic impurities in Al, Cu, and Pd in the very dilute concentration range are also included (Mydosh and Nieuwenhuys 1980). One can observe (Table 9.1) that Mn atoms in the Al-Pd-Mn alloys behave just as those in the Pd-Mn crystalline dilute alloys, showing the appearance of a giant magnetic moment and a spin-glass behavior. Much remains to be investigated to explain the reason why the magnetic properties of Al-Pd-Mn alloys are different to those of other alloys.

In the same concentration range, we can see that QCs with different structures have different spin-freezing temperature T_f. Shown in Fig. 9.9b is the concentration dependence of T_f determined from the ac susceptibility for the *I* and *D* phases of $Al_{70}Pd_{30-x}Mn_x$ (Satoh et al. 1994). For the *I* phase, T_f is higher and the concentration dependence is steeper than for the *D* phase. The results in Figs. 9.6a, 9.6b, and 9.9b mean that the magnetic properties in QCs are very sensitive not only to the alloying elements, but also to the local atomic structure. The anisotropic atomic arrangement in the *D* phase would bring about such a difference.

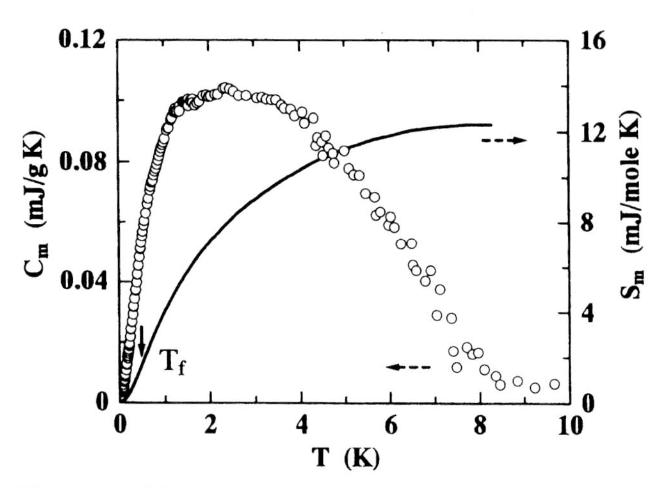

Fig. 9.10. Magnetic contribution C_m to the total specific heat and magnetic entropy S_m of $Al_{70}Pd_{21}Mn_9$ I QC as a function of temperature. After Chernikov et al. (1993).

9.2.5 Low-Temperature Specific Heat

The low-temperature specific heat of magnetic alloys may be expressed by

$$C = \gamma T + \beta T^3 + C_m(T), \tag{9.6}$$

where γ and β are, respectively, the electronic and lattice coefficients and C_m is the magnetic contribution. In the case of a long-range magnetic transition, the magnetic entropy is obtained from C_m as

$$S_m = \int (C_m/T)dT \tag{9.7}$$

$$= R\ln(2S+1), \tag{9.8}$$

where R is the gas constant. In Eq. (9.8), the spin quantum number S is used in place of the total angular quantum number J because of quenching of the orbital angular momentum in $3d$ magnetic materials.

Figure 9.10 shows the temperature dependence of C_m and S_m of the I-$Al_{70}Pd_{21}Mn_9$ QC (Chernikov et al. 1993). A broad maximum in C_m is observed near 2 K. This temperature is four times higher than T_f. The temperature dependence of magnetic entropy reveals that 85% of $R\ln(2S+1)$ is developed above T_f. This was interpreted as an indication that freezing at T_f removes only a small contribution to S_m and that a considerable short-range magnetic order exists above T_f. For the I-$Al_{80}Mn_{20}$ QC, the experimental value of S_m at the spin-freezing temperature is approximately 20% of $R\ln(2S+1)$, using the parameters of the assumed giant magnetic clusters present in this alloy (Machado et al. 1987). This was interpreted as evidence for the presence of considerable magnetic disorder even at temperatures close to 0 K.

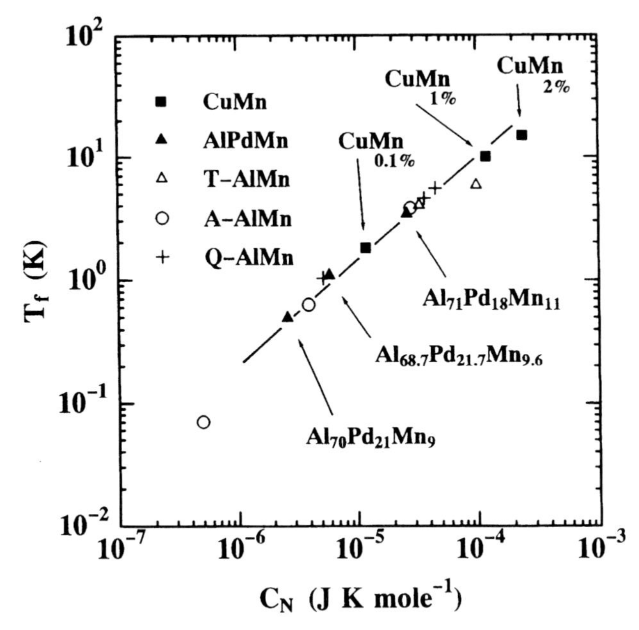

Fig. 9.11. Freezing temperature T_f vs amplitude of the nuclear hyperfine term C_N of the specific heat for Al-based QCs and A alloys. T stands for the D phase. After Lasjaunias et al. (1995).

At temperatures below 0.6 K, which is well below the T_f of all QCs in Fig. 9.11, the electronic and lattice contributions to the specific heat, C_P, become negligibly small. At such low temperatures, C_P is due to a nucler hyperfine contribution proportional to T^{-2} and to a magnetic (spin-glass) contribution proportional to T^n with n larger than 1 (Chernikov et al. 1993, Lasjaunias et al. 1995)

$$C_P = C_N T^{-2} + BT^n. \tag{9.9}$$

A plot of T_f vs the amplitude of the nuclear hyperfine term C_N for Al-based QCs and A alloys is shown in Fig. 9.11 (Lasjaunias et al. 1995). All data in Fig. 9.11 lie on a unique straight line, independent of alloy structure and composition. This implies that the RKKY interactions have the same strength in these structurally different alloys.

The nuclear hyperfine contribution to the specific heat C_N at very low temperatures is given by

$$C_N = x_m N k_B [(I+1)/3I](\mu_n H_{\text{eff}}/k_B T)^2, \tag{9.10}$$

where H_{eff} is the effective magnetic field at the nuclei of magnetic Mn atoms, N the Avogadro number, and I and μ_n are, respectively, the spin and the nuclear magnetic moments of ^{55}Mn nuclei. The value of x_m for the $Al_{75}Mn_{20}Si_5$ QC estimated from the measured C_N is about 1% (Lasjaunias et al. 1990), or x_m/x is about several %. This is in accord with the values found for the Al-Mn system (Fig. 9.4). Therefore, the measurement of the nuclear hyperfine contribution to specific heat also indicates clearly that only a small number of Mn atoms have a magnetic moment in the Al-based QCs.

9.2.6 Model for Magnetism and Pauling Valence

In theoretical calculations of the magnetism of Mn atoms in Al, the local environment (McHenry et al. 1988, Jagannathan and Schulz 1997), *sp-d* hybridization (McHenry et al. 1988), and the local symmetry (de Coulon et al. 1993, Liu et al. 1993) are taken into consideration. However, the individual atomic sites in QCs have not yet been definitely determined. Therefore, experimental results do not give definitive information on how the magnetic moment varies with the Mn site.

The magnitude of the magnetic moment of many Mn alloys and compounds is related to the atomic distance and the coordination number. The values of localized magnetic moments in a number of Mn crystalline compounds have been explained by using the concept of the Pauling valence (Môri and Mitsui 1968, Chikazumi 1997). According to Pauling's empirical expression (Pauling 1960), the bond length $D(n)_{ij}$ for i and j atoms is

$$D(n)_{ij} = D(1)_{ij} - 0.60 \log_{10} n, \tag{9.11}$$

where $D(1)_{ij}$ is the bond length for a single bond between i and j atoms and n is the bond number. The value of n is defined by the following relation

$$n = PV/CN, \tag{9.12}$$

where CN is the coordination number and PV is the Pauling valence. The Pauling valence at the site i of a Mn atom is given by the expression

$$(PV)_i = \sum_j^n (CN)_{ij} \cdot \exp\{[D(1)_{ij} - D(n)_{ij}]/0.26\}. \tag{9.13}$$

By using the single bond radii of Al (1.248 Å) and Mn (1.171 Å), the value of PV of o-Al_6Mn is calculated to be 6.1.

The values of the Mn content, CN, PV, and P_{eff} of several Al-based compounds are given in Table 9.2 (Fukamichi and Goto 1989). The magnetic moment of Mn decreases with increasing PV, and becomes almost zero at $PV = 6.5$ (Môri and Mitsui 1968). The values of the effective magnetic moment of o-Al_6Mn (Table 9.2) and the saturation magnetization of the $Al_{85.6}Mn_{14.4}$ QC (Fig. 9.3) can be understood in terms of the Pauling valence. The important point to note is that the Mn content of ϕ-$Al_{10}Mn_3$ is lower than that of δ-$Al_{11}Mn_4$, but the P_{eff} of the former is larger than that of the latter. This is in accordance with the smaller value of the PV for ϕ-$Al_{10}Mn_3$ than for δ-$Al_{11}Mn_4$. These facts imply that the Mn moments in QCs are sensitive to the structure parameters such as the atomic distance and the

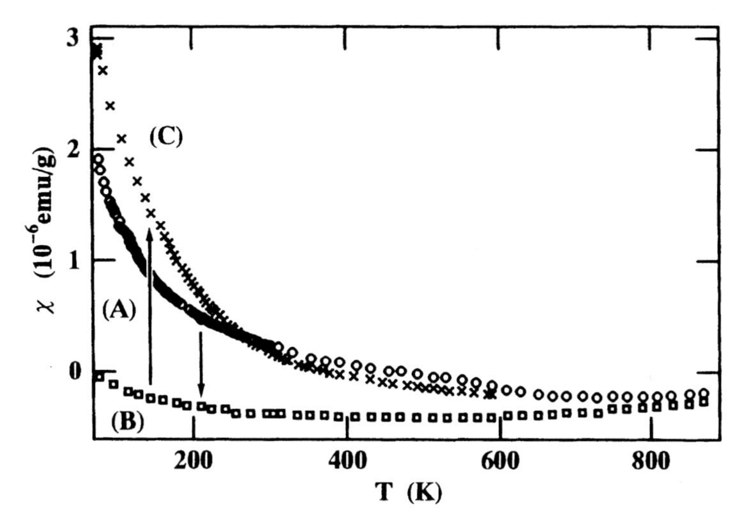

Fig. 9.12. Temperature dependence of magnetic susceptibility of the F-type $Al_{70}Pd_{21.5}Mn_{8.5}$ QC. After Matsuo et al. (1997).

coordination number. The magnetic and nonmagnetic Mn atoms in the QCs and their *A* counterparts, as well as the different magnetic properties among the *I*, *D*, and *A* phases could be understood in terms of the Pauling valence.

Table 9.2. Mn concentration, Pauling valence (*PV*), coordination number (*CN*) and effective magnetic moment P_{eff} of several Al-based compounds. After Fukamichi and Goto (1989).

	o-Al_6Mn	ϕ-$Al_{10}Mn_3$	δ-$Al_{11}Mn_4$	α-$Al_9Mn_2Si_{1.8}$	β-Al_9Mn_3Si
Mn (at.%)	14.3	23.1	26.7	17.4	23.1
PV	6.1	4.7	5.2	5.6	5.7
CN		8	Mn(1) 12	Mn(1) 10	10
			Mn(2) 12	Mn(2) 9	
$P_{\text{eff}}(\mu_B)$	0.56	1.28	0.63	0	1.09

9.2.7 Phasons, Diamagnetism, and Pauli Paramagnetism

Figure 9.12 shows the temperature dependence of magnetic susceptibility of the F-type $Al_{70}Pd_{21.5}Mn_{8.5}$ *I*QC (Matsuo et al. 1997). The specimen quenched from 1079 K exhibits a typical Curie-Weiss behavior [curve (A)], whereas the specimen annealed at 872 K for 100 h shows a significant decrease in the magnitude of susceptibility [curve (B)]. The specimen annealed again at 1080 K exhibits a recovery in susceptibility [curve (C)]. The change

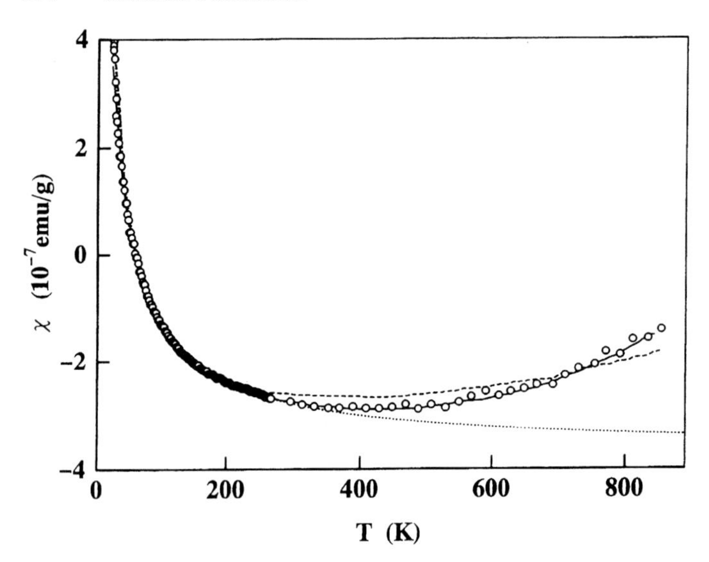

Fig. 9.13. Temperature dependence of the magnetic susceptibility for P-type $Al_{71}Pd_{20.5}Mn_{8.5}$ QC . Data: open circles; Curie-Weiss fit: dotted line; Curie-Weiss + T^2-term fit: dashed line; Curie-Weiss + T^2 term + T^4-term fit: solid line. After Kobayashi et al. (1997).

of the magnetic susceptibility upon annealing may be explained by the phason relaxation. The change in slope of the temperature dependence of the susceptibility for the D-$Al_{65}Cu_{20}Co_{15}$ and D-$Al_{70}Ni_{15}Co_{15}$ QCs was associated with a condensation of phasons in phason walls (Lück and Kek 1993). Therefore, the phasons play an important role in the change of the susceptibility. The change in the fraction of the magnetic Mn atoms determined from the magnetization measurements (Matsuo et al. 1997) is consistent with the results shown in Fig. 9.12. It is thus confirmed that the change in Curie-Weiss paramagnetism is mainly associated with the number of magnetic Mn sites.

Al-Pd-Mn QCs with low Mn concentration exhibit diamagnetic behavior. This might be an intrinsic property of high-quality QCs. These QCs have two types of structure: F-type (high-temperature phase F1) and P-type (low-temperature phase F2). Figure 9.13 shows the temperature dependence of χ together with several fits between 2 and 870 K for the $Al_{71}Pd_{20.5}Mn_{8.5}$ P-type QC (Kobayashi et al. 1997). An increase of χ with increasing temperature can be accounted for by a temperature dependence of the Pauli paramagnetism which is related to a pseudogap in the DOS at E_{F} (Wilson 1965). The temperature dependence of χ is given by

$$\chi = \chi_0 + C/(T - \theta_p) + AT^2, \tag{9.14}$$

with

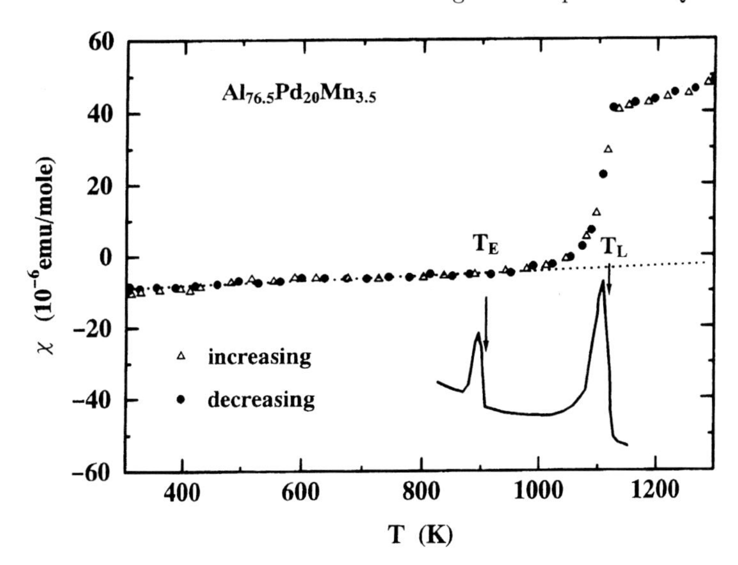

Fig. 9.14. Temperature dependence of magnetic susceptibility and a differential thermal analysis curve of $Al_{76.5}Pd_{20}Mn_{3.5}$ QC. After Hippert et al. (1996).

$$A = \frac{\mu_B^2 N(E_\mathrm{F})(\pi k)^2}{3} \left\{ \frac{1}{N(\zeta)} \left(\frac{d^2 N(\zeta)}{d\zeta^2} \right) - \left(\frac{1}{N(\zeta)} \frac{dN(\zeta)}{d\zeta} \right)^2 \right\}_{\zeta = E_\mathrm{F}} . \tag{9.15}$$

Here $N(\zeta)$ denotes the electronic DOS. By including in Eq. (9.14) a higher-order term proportional to T^4, good agreement with the experimental data was obtained (Kobayashi et al. 1997) and this was taken as evidence for the presence of the pseudogap in the DOS at E_F (Kobayashi et al. 1997).

In general, the magnetic susceptibility of approximant crystals is smaller than that of their Q counterparts. Moreover, the magnetic properties of QCs are significantly different from those of their A counterparts (Figs. 9.2–9.5). In the A state, Mn atoms are randomly distributed with a finite probability of being located as the first- and second-nearest neighbors. One can expect a similar behavior to occur in a liquid state. The magnetic susceptibility and neutron-scattering measurements for Al-Pd-Mn alloys indicate that a large localized magnetic moment appears on the Mn atoms in the liquid state and disappears in the solid state of either I or approximant phases (Hippert et al. 1996). Figure 9.14 shows the temperature dependence of χ of the $Al_{76.5}Pd_{20}Mn_{3.5}$ QC (Hippert et al. 1996). The dotted line is a fit to $\chi = \chi(T = 0) + A_2 T^2$. The temperatures T_E and T_L correspond, respectively, to the melting point of the ternary eutectic and to the liquidus temperature, as determined from a differential thermal analysis curve (Fig. 9.14). Because of the formation of a localized magnetic moment, χ increases in the liquid state. A marked increase of χ on melting has also been observed in the D-$Al_{65}Cu_{20}Co_{15}$ and D-$Al_{70}Ni_{15}Co_{15}$ QCs (Lück and Kek 1993). The increase of χ in the liquid state with increasing temperature is regarded as a strong indication of chemical short-range order in the melt, which is destroyed upon further temperature increase (Lück and Kek 1993). These data indicate again that the magnetic character of Mn atoms critically depends on their local environment.

9.2.8 Ferromagnetism

Ferromagnetic properties have been reported for several kinds of QCs (Tsai et al. 1988, Zhao et al. 1988, Dunlap et al. 1989, Yokoyama et al. 1994). The characteristic features of these magnetic QCs are a low magnetization, a relatively high Curie temperature T_c, and a large coercivity. The origin of these magnetic characteristics has been investigated in the Fe-doped QCs by Mössbauer spectroscopy (Stadnik and Stroink 1991, Nasu et al. 1992). It

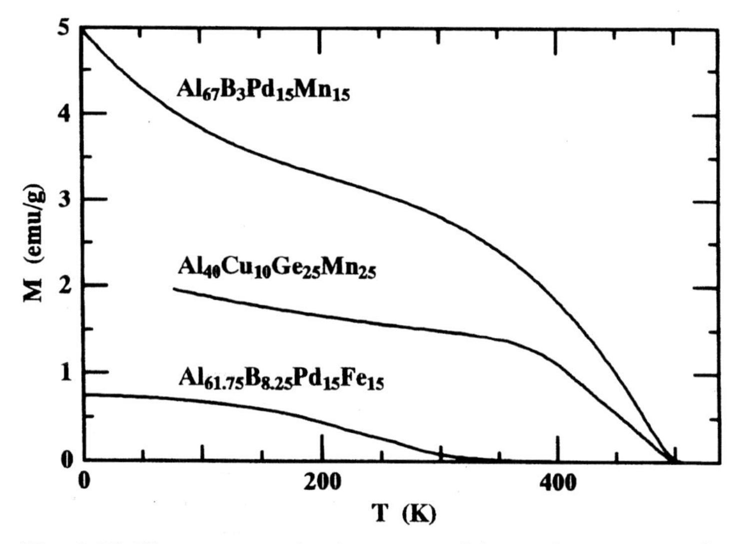

Fig. 9.15. Thermomagnetization curves of the specimens reported as ferromagnetic QCs. After Tsai et al. (1988), Yokoyama et al. (1994), and Lyubutin et al. (1997).

was shown that these magnetic properties arise from a small amount of ferromagnetic inclusions or magnetic clusters present in the QCs. However, one needs to bear in mind that there is a tendency to confuse a spin glass with a ferromagnet from the data obtained in strong magnetic fields (Fukamichi et al. 1988, Chatterjee et al. 1990).

Figure 9.15 shows the thermomagnetization curves for the specimens reported as ferromagnetic QCs (Tsai et al. 1988, Yokoyama et al. 1994,

Lyubutin et al. 1997). Assuming uniform ferromagnetism, the ratio of the number of magnetic carriers obtained from the saturation magnetization and from the Curie-Weiss law (Wohlfarth 1978) is estimated to be about 3 for the $Al_{64}Pd_{15}Mn_{15}B_6$ QC. This indicates that this QC is strongly itinerant, in contrast to the ferromagnetic materials with the same T_c. NMR results (Shinohara et al. 1993) indicate that the Al-Pd-Mn-B QCs are not ferrimagnetic (Yokoyama et al. 1994) but ferromagnetic, which is closely correlated with the heterogeneous atomic distribution, although the x-ray diffraction pattern and electron microscopy indicate a single *i* phase. This heterogeneity manifests itself via the coexistence of ferromagnetic and nonmagnetic Mn atoms in this system (Shinohara et al. 1993).

Magnetic ordering of Fe atoms was reported in the *I*-Al-B-Pd-Fe QCs (Lyubutin et al. 1997). Figure 9.16 shows the dependence of M_s on the B

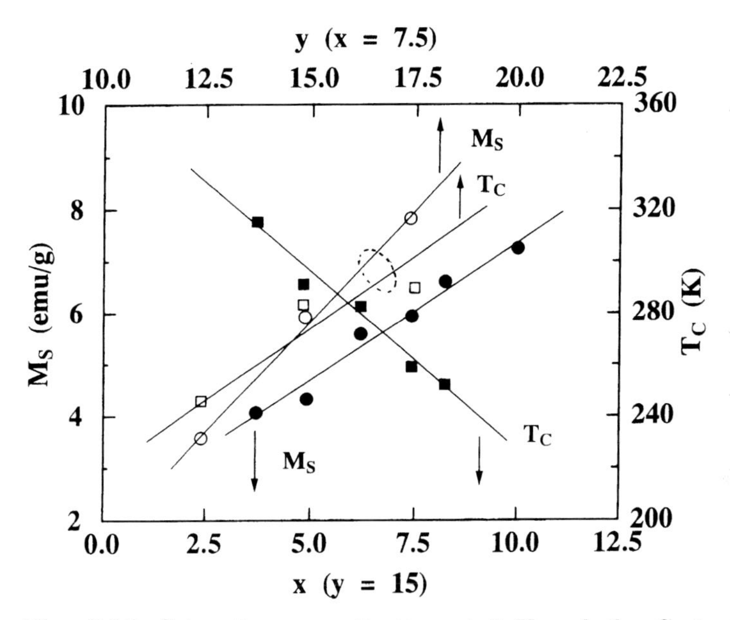

Fig. 9.16. Saturation magnetization at 5 K and the Curie temperature of $Al_{70-x}B_xPd_{30-y}Fe_y$ QCs as a function of the B and Fe concentration at the fixed concentration of $x = 7.5$ and $y = 15$. After Lyubutin et al. (1997).

and Fe concentration, at the fixed concentration of $x = 7.5$ and $y = 15$, for the *I*-$Al_{70-x}B_xPd_{30-y}Fe_y$ QCs at 5 K (Lyubutin et al. 1997). The value of M_s increases with the concentration of B and Fe, whereas the value of T_c decreases with increasing B concentration, but increases with increasing Fe concentration. It has been pointed out (Lyubutin et al. 1997) that the temperature dependence of the magnetization is very complicated and it is difficult to find a proper fitting model which is valid throughout the measured temperature range, suggesting the presence of complex magnetic phases. One may notice that an increase of the B content yields an increase of M_s but a decrease of T_c, which is analogous to what is observed in various Invar alloys (Fukamichi et al. 1983). It has been suggested (Lyubutin et al. 1997) that the ferromagnetic state is due to large magnetic clusters with a size of 20–30 nm. This was ascribed to bonding between Fe and B, similar to the situation in the *A* Fe-B alloys. In the Al-B-Pd-Mn QCs, on the other hand, it has been suggested that the role of B is to enhance the magnetic coupling between the Mn-Mn pairs, bringing about ferromagnetic properties (Shinohara et al. 1993, Yokoyama et al. 1994). It may be recalled that MnB (Lundquist et al 1962) and MnB_2 (Cadeville 1966) are ferromagnetic in the crystalline state.

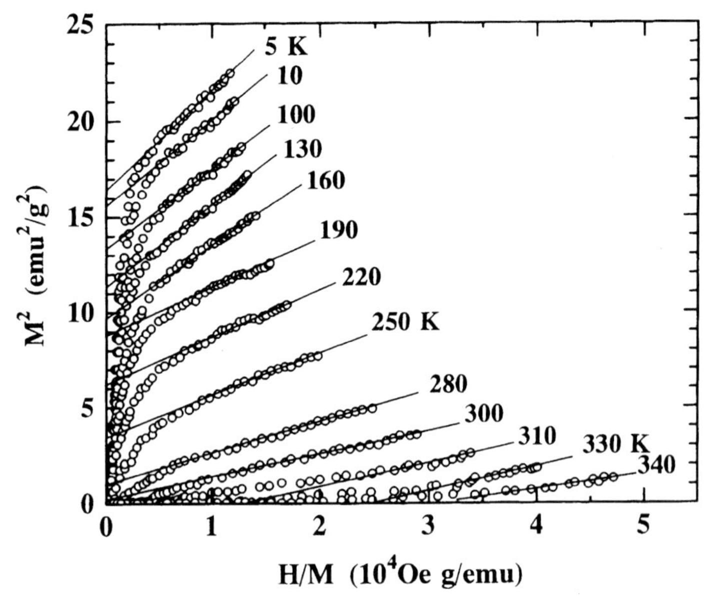

Fig. 9.17. Arrott plots of $Al_{62.5}B_{7.5}Pd_{15}Fe_{15}$ QC. After Lin et al. (1995).

The Arrott plots M^2 vs H/M for the $Al_{62.5}B_{7.5}Pd_{15}Fe_{15}$ QC are shown in Fig. 9.17 (Lin et al. 1995). The linear relationship is not strictly obeyed at low fields, and M^2 tends to be curved downward below 300 K, while it tends to be curved upward above 300 K. This has been attributed to the spatial fluctuation of the magnetization, resulting from an inhomogeneous magnetic state (Lin et al. 1995). In addition, this QC exhibits very strong temperature dependence of coercivity, which changes from 1.2 kOe at 5 K, to 40 Oe at 300 K (Lin et al. 1995).

It should be pointed out that uniform ferromagnetism of the Al-based QCs is in conflict with the formation of QCs in that high DOS at E_F is necessary for the onset of uniform ferromagnetism, as given by the Stoner criterion (Stoner 1938) or Shimizu condition (Shimizu 1965), whereas low DOS at E_F is expected for the formation of a QC via the Hume-Rothery mechanism. In this context, it is interesting to mention the following work. Fe-Al alloys containing between 35 and 50 at.% Al are paramagnetic, but ferromagnetic properties are induced by plastic deformation (Takahashi et al. 1996). The ferromagnetic clusters are created along the antiphase boundary between superpartial dislocations. Therefore, the ferromagnetic state coexists with the paramagnetic state in the crystallographically single-phase alloy. These clusters are induced by the change of the atomic arrangement in the vicinity of the antiphase boundary. The cluster size is estimated to be several tens nm (Takahashi et al. 1996), which is comparable with the size of the magnetic clusters in the $Al_{62.5}B_{7.5}Pd_{15}Fe_{15}$ QC. More noteworthy is that the clusters exhibit a strong magnetic anisotropy, which brings about a large coercivity, and the easy direction of magnetization is consistent with that of the roll-induced anisotropy in the Fe_3Al alloy. In addition, one often observes ferromagnetic properties even in antiferromagnets in the non-stoichiometric state, such as NiO (Shimomura et al. 1954), $BaVS_3$ (Massenet et al. 1979), and $TiFe_2$ (Nakamichi 1968). In these compounds, the ferromagnetic state can also be induced by a change of environment. Needless to say, just as in ferromagnetic QCs, the spontaneous magnetization per volume or gram for these compounds is small because there is only a small fraction of the ferromagnetic component in the specimens.

It thus seems that the ferromagnetic clusters in many QCs are induced by phasons, which change the local atomic arrangement. A large coercivity associated with the anisotropy would appear in ferromagnetic QCs with a low magnetization, in analogy to the Fe-Al alloys. A coexistence of ferromagnetic, diamagnetic, and superconducting states in the D-$Al_{70}Ni_{15}Co_{15}$ QC (Markert et al. 1994) could be explained in the framework of the changes of atomic arrangement induced by phasons.

9.3 Mg-RE-Zn Quasicrystals

The Frank-Kasper-type $Mg_{42}RE_8Zn_{50}$ (RE=Y,Tb,Dy,Ho,Er) QCs with a highly ordered face-centered I lattice can be prepared by conventional solidification (Niikura et al. 1994). The magnetic properties of these QCs are expected to be different from those of the Al-based QCs containing $3d$ TMs because the magnetic carriers are the $4f$ electrons located deep inside the RE atoms. Since heavy RE elements have a large localized magnetic moment, the spin-glass state develops. Because of the Schottky-type specific heat, the low-temperature specific heat of RE compounds exhibit an anomalous behavior. Therefore, the specific heat data gives useful information on the symmetry of the RE site, as well as of the spin-glass behavior.

In this section, we first show the temperature dependence of the ac magnetic susceptibility and the magnetic cooling effect, and discuss the susceptibility and spin-glass behavior. The effective magnetic moment and the saturation magnetic moment of Gd are compared with the values of the free Gd^{3+} ion. Next, the low-temperature specific heat data are given in order to elucidate the spin-glass behavior and the atomic structure in the $Mg_{42}RE_8Zn_{50}$ QCs. Moreover, the small electronic specific heat coefficient for the $Mg_{42}Y_8Zn_{50}$ QC is related to the pseudogap at E_F. Finally, we comment on the long-range antiferromagnetic ordering in the $Mg_{42}RE_8Zn_{50}$ QCs by using the low-temperature specific heat and the muon spin relaxation data.

9.3.1 Susceptibility and Spin-Glass Behavior

Figure 9.18 shows the temperature dependence of the dc magnetic susceptibility χ_{dc} for the high-quality $Mg_{42}Gd_8Zn_{50}$, $Mg_{42}Tb_8Zn_{50}$, and $Mg_{42}Dy_8Zn_{50}$ QCs. The diffraction peaks of these QCs are very sharp and the deviation of the reflections from the ideal I symmetric positions due to phason strains is very small (Hattori et al. 1995c). All of them exhibit a Curie-Weiss-type temperature dependence and the values of P_{eff} are close to those of the free RE^{3+} ions, exhibiting no concentration dependence, in contrast to the Al-Mn, Al-Cu-Mn, and Al-Pd-Mn QCs. Other magnetic properties, however, are similar to those of Al-based QCs mentioned above. That is, the $Mg_{42}RE_8Zn_{50}$ QCs

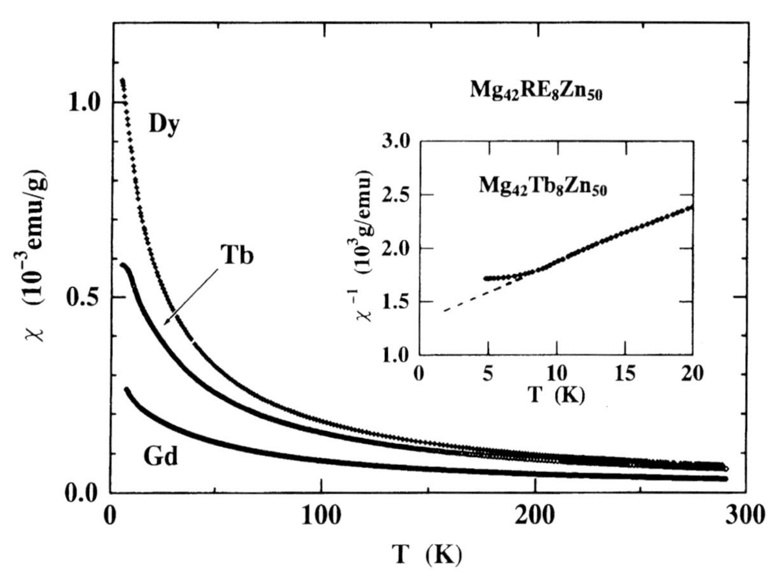

Fig. 9.18. Temperature dependence of the magnetic susceptibility for the Frank-Kasper-type (F-type) $Mg_{42}RE_8Zn_{50}$ QCs. The inset shows the inverse susceptibility of $Mg_{42}Tb_8Zn_{50}$ QC at low temperatures. After Hattori et al. (1995c).

have a negative value of θ_p, showing that the RE-RE exchange interactions are antiferromagnetic. The inverse susceptibility of the $Mg_{42}Tb_8Zn_{50}$ QC (inset in Fig. 9.18) deviates upward at low temperatures, suggesting strong magnetic interactions.

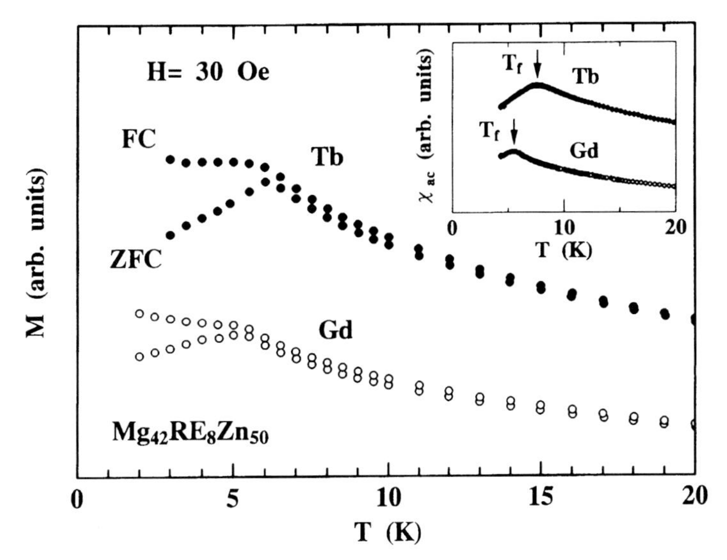

Fig. 9.19. Magnetic cooling effect in 30 Oe for the Frank-Kasper-type (F-type) $Mg_{42}Tb_8Zn_{50}$ and $Mg_{42}Gd_8Zn_{50}$ QCs. The inset shows the temperature dependence of the ac susceptibility. After Hattori et al. (1995c).

Figure 9.19 displays the magnetic cooling effect of the F-type $Mg_{42}Gd_8Zn_{50}$ and $Mg_{42}Tb_8Zn_{50}$ QCs. A marked hysteresis is observed between the curves cooled in zero-field (ZFC) and in 30 Oe (FC) at low temperatures, in analogy with the Al-based QCs discussed in Sect. 9.2.3. The inset in Fig. 9.19 shows the temperature dependence of the ac susceptibility χ_{ac} for the F-type $Mg_{42}Gd_8Zn_{50}$ and $Mg_{42}Tb_8Zn_{50}$ QCs measured at a frequency of 80 Hz in a field with an amplitude of 10 Oe. The data clearly exhibit a characteristic cusp corresponding to the spin-freezing temperature at 5.5 K and 7.6 K, respectively. Because the spin freezing is very sensitive to the strength of the applied dc magnetic field, the cusp temperature of the ZFC magnetization in the dc magnetic field is lower than that of the ac magnetic susceptibility. The de Gennes factor ξ is given by

$$\xi = c(g-1)^2 J(J+1), \tag{9.16}$$

where c is the concentration of the RE atom and g is the g factor. Various magnetic transition temperatures of heavy RE compounds may be described by the de Gennes factor. However, it is worth noting that T_f of 7.6 K for the $Mg_{42}Tb_8Zn_{50}$ is higher than that of 5.5 K for the $Mg_{42}Gd_8Zn_{50}$, which is in contrast to the magnitude of ξ found in some crystalline compounds such as RE_2Zn_{17} (Stewart and Coles 1974) and REAl (Bécle et al. 1970). This characteristic behavior may be associated with the appearance of anisotropic interactions (Bécle et al. 1970). The F-type $Mg_{42}Dy_8Zn_{50}$, $Mg_{42}Ho_8Zn_{50}$, and $Mg_{42}Er_8Zn_{50}$ QCs exhibit no spin-glass behavior above 2 K.

The effective Gd magnetic moment evaluated from the Curie constant is about 8 μ_B, which is very close to the free Gd^{3+} ion value of 7.98 μ_B. The saturation magnetic moment evaluated from the high-field magnetization curve up to 300 kOe is about 7.5 μ_B per Gd atom (Saito et al. 1997), in fair agreement with 7.55 μ_B for the saturation magnetic moment of pure Gd atom. Similar results are obtained for other F-type Mg-RE-Zn QCs (Hattori et al. 1995c).

The P-type $Mg_{30}Gd_xZn_{70-x}$ (x = 7–10) QCs are prepared by rapid quenching from the melt (Saito et al. 1997). The temperature dependence of the dc magnetic susceptibility for these QCs exhibits a negative value of θ_p, indicating that the Gd-Gd exchange interactions are antiferromagnetic. Moreover, the magnitude of θ_p increases with increasing Gd content. This indicates that the antiferromagnetic interaction becomes stronger with increasing Gd content, following the de Gennes factor given by Eq. (9.16). The effective and saturation magnetic moments of Gd for the P-type phase are the same as those for the F-type phase. In contrast to the 3d TMs such as Mn, the carriers of magnetism in the heavy RE metals are the 4f electrons located deep inside the atoms. Therefore, no change in the magnetic moment occurs regardless of the structural change in the QCs.

9.3.2 Low-Temperature Specific Heat

Figure 9.20 displays the temperature dependence of the low-temperature specific heat for the $Mg_{42}RE_8Zn_{50}$ (RE=Gd,Tb,Y) QCs (Hattori et al. 1995b). The arrows indicate the spin-freezing temperature T_f determined by the ac magnetic susceptibility measurements. Since Y atoms have no magnetic moments, the specific heat of the F-type $Mg_{42}Y_8Zn_{50}$ paramagnetic QC is fitted to the equation

$$C = \gamma T + \beta T^3 + \delta T^5, \tag{9.17}$$

where β and δ are the lattice specific heat coefficients, and the third term represents the deviation from the Debye model. The resultant values of γ, β, and δ are 0.63 mJ mole^{-1} K^{-2}, 5.59$\times$10^{-2} mJ mole^{-1} K^{-4}, and 2.16$\times$10^{-4} mJ mole^{-1} K^{-6}, respectively. The Debye temperature θ_D is obtained from β using the expression

$$\theta_D = (12\pi^4 R/5\beta)^{1/3}. \tag{9.18}$$

The magnitude of θ_D for the $Mg_{42}Y_8Zn_{50}$ QC is estimated to be 325 K. The γ value is comparable to the values found for the Mg-Al-Zn QCs

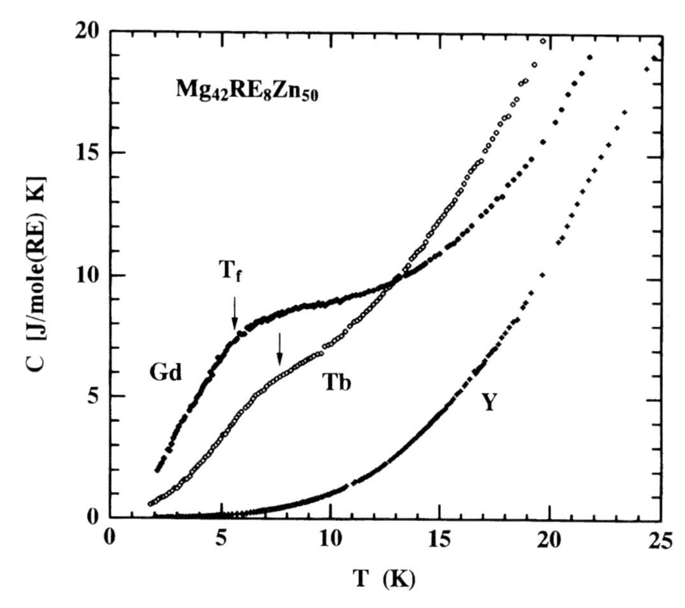

Fig. 9.20. Low-temperature specific heat as a function of temperature for the Frank-Kasper-type (F-type) $Mg_{42}RE_8Zn_{50}$ QCs. After Hattori et al. (1995b).

which have a similar Debye temperature (Matsuda et al. 1989). The experimental value of $\gamma_{\rm exp}$ is less than $\gamma_{\rm free}$ estimated from the free-electron model. Such a small value is closely correlated with the existence of a pseudogap in the DOS at $E_{\rm F}$. The relation between the small γ value and the phase stability of *I*-Al-Cu-Fe, *I*-Mg-Ga-Zn , and *I*-Al-Li-Cu QCs has been discussed in connection with the Fermi surface-Brillouin zone interaction (Wagner et al. 1989). The ratio $\gamma_{\rm exp}/\gamma_{\rm free}$ as a function of the valence concentration e/a for some Frank-Kasper-type QCs of the *sp*-electron systems becomes smaller with decreasing e/a, regardless of the alloy systems (Mizutani et al. 1990, Kimura et al. 1989).

Representative low-temperature specific heat data for the Frank-Kasper-type QCs are listed in Table 9.3 (Hattori et al. 1995b, Mizutani et al. 1990,

Table 9.3. Low-temperature specific heat data of some Frank-Kasper-type QCs. After Hattori et al. (1995b), Mizutani et al. (1990), and Kimura et al. (1989).

	$\gamma_{\rm exp}$ (mJ mole^{-1} K^{-2})	$\gamma_{\rm exp}/\gamma_{\rm free}$	e/a	θ_D (K)
$Mg_{42}Y_8Zn_{50}$	0.63	0.7	2.08	325
$Al_{55}Li_{35.8}Cu_{9.2}$	0.318	0.39	2.09	341
$Mg_{33.5}Zn_{40}Ga_{26.5}$	0.91	1.05	2.265	353

Kimura et al. 1989). The value $\gamma_{\text{exp}}/\gamma_{\text{free}}$ for the $Mg_{42}Y_8Zn_{50}$ QC is much lower than that for the $Mg_{33.5}Ga_{40.0}Zn_{26.5}$ QC with a higher e/a value.

The magnetic specific heat curve of the $Mg_{42}Gd_8Zn_{50}$ QC exhibits a broad maximum above the spin-freezing temperature T_f as shown in Fig. 9.21. The magnetic contribution does not vanish even above 20 K. Such a behavior is very similar to that observed in conventional spin-glass systems and also in the Al-Pd-Mn QC (Fig. 9.10). The inset in Fig. 9.21 shows the temperature dependence of the magnetic entropy S_m obtained from Eq. (9.7) for the $Mg_{42}Gd_8Zn_{50}$ QC. The value of S_m is about 30% of $R \ln 8$ at the freezing temperature, which is similar to the values found for other crystalline spin-glass systems (Mydosh and Nieuwenhuys 1980). The magnetic contribution to the specific heat for the $Mg_{42}Tb_8Zn_{50}$ QC seems to be smaller than that for the $Mg_{42}Gd_8Zn_{50}$ QC (Fig. 9.20), although the total angular momentum of Tb^{3+} ($J = 6$) is larger than that of Gd^{3+} ($J = 7/2$).

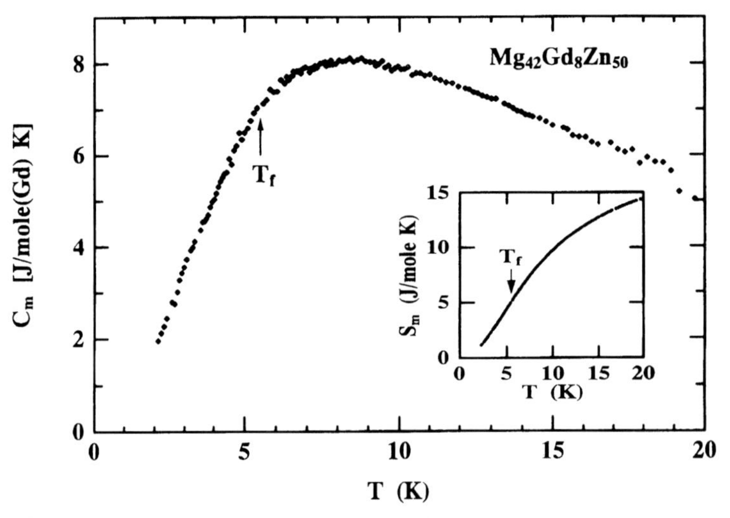

Fig. 9.21. Temperature dependence of the magnetic contribution to the specific heat for the Frank-Kasper-type (F-type) $Mg_{42}Gd_8Zn_{50}$ QC. The inset shows the temperature dependence of the magnetic entropy. After Hattori et al. (1995b).

As compared with $Mg_{42}Gd_8Zn_{50}$, the specific heat of $Mg_{42}Tb_8Zn_{50}$ gradually varies with temperature, indicating the appearance of a Schottky-type specific heat C_{Schottky} which arises from the splitting of the J multiplet energy level due to the electrostatic field. From the data in Fig. 9.20, the sum of S_m and S_{Schottky} of the $Mg_{42}Tb_8Zn_{50}$ QC is estimated to be about 3.6 $\text{J}\,(\text{mole Tb})^{-1}\,\text{K}^{-2}$. The value of S_m is also estimated to be about 30% of $R \ln(2J + 1)$ from Fig. 9.21. Consequently, $(2J + 1)$ should be 2 or 3 because 3.6 $\text{J}\,(\text{mole Tb})^{-1}\,\text{K}^{-2}$ is equal to about 30% of $R \ln 4$. In other words, if $(2J + 1)$ is 4, S_{Schottky} disappears. Therefore, the ground state of Tb should be a doublet or a triplet, implying the low symmetry of the Tb site in the $Mg_{42}Tb_8Zn_{50}$ QC (Hattori et al. 1995b). The structural analysis of the single-grain $Mg_{42}Y_8Zn_{50}$ QC indicates that the Y atom does not occupy the center of the I cluster and the electron density map shows a shift of atomic positions from the ideal positions in three-dimensional space (Yamamoto et al. 1996). In view of the similarity of the structures of the $Mg_{42}Tb_8Zn_{50}$ and $Mg_{42}Y_8Zn_{50}$ QCs, a small degeneracy of the Tb ion is consistent with the result of the structural analysis.

Typical spin-glass behavior is caused by the competition between the ferromagnetic and antiferromagetic interactions due to the fluctuating RKKY interaction (Mydosh and Nieuwenhuys 1980). Random anisotropies may destroy the long-range magnetic order, resulting in spin-glass-like behavior (Goldschmidt 1992). In the latter, one may observe a large coercive field at low temperatures (Hattori et al. 1995a). This behavior is caused by a single-ion anisotropy. In contrast to the anisotropy of Tb^{3+}, Gd^{3+} is an S-state ion and consequently single-ion anisotropy is negligible or very small. However, the spin-glass behavior is observed not only in the $Mg_{42}Tb_8Zn_{50}$ QC but also in the $Mg_{42}Gd_8Zn_{50}$ QC (Fig. 9.19). In the long-range magnetic order transition in RE compounds, the magnetic entropy S_m is given by $R\ln(2J+1)$. The value of S_m for magnetically dilute crystalline alloys, such as AuFe and CuMn, is about 20–30% of $R\ln(2S+1)$ at the spin-freezing temperature T_f (Wenger et al. 1976). The value S_m at T_f of the $Mg_{42}Gd_8Zn_{50}$ QC becomes about 30% of $R\ln(2J+1)$, which is comparable with the result for the $Al_{70}Pd_{21}Mn_9$ QC (Fig. 9.10). On the other hand, the value S_m at T_f for an A alloy system with a large random magnetic anisotropy, such as Er-Ni, is about 50% of $R\ln(2J+1)$ (Hattori et al. 1995a). These results imply that the fluctuating RKKY interaction is responsible for the spin-glass behavior in the Mg-RE-Zn QCs.

9.3.3 Antiferromagnetism

In $Mg_{42}Tb_8Zn_{50}$ a long-range antiferromagnetic order at 20 K is expected because the neutron diffraction peaks due to the magnetic order vanish above 20 K and the inverse susceptibility deviates downward at this temperature from the straight line extrapolated from high temperatures. This magnetic ordering is characterized by the propagation vector $\boldsymbol{Q} = (1/4, 0, 0, 0, 0, 0)$ in the six-dimensional notation (Charrier and Schmitt 1997, Charrier et al. 1997). In contrast to these reports, another experiment shows no such a deviation of the inverse susceptibility (inset in Fig. 9.18). The deviation is upward, in analogy with that of Al-based QCs in the spin-glass state. Furthermore, no specific heat anomaly is observed in the vicinity of 20 K (Fig. 9.20).

In zero applied field, the fast-relaxing signal could be fitted in the paramagnetic regime above the spin-freezing temperature T_f by the following power-exponential relaxation

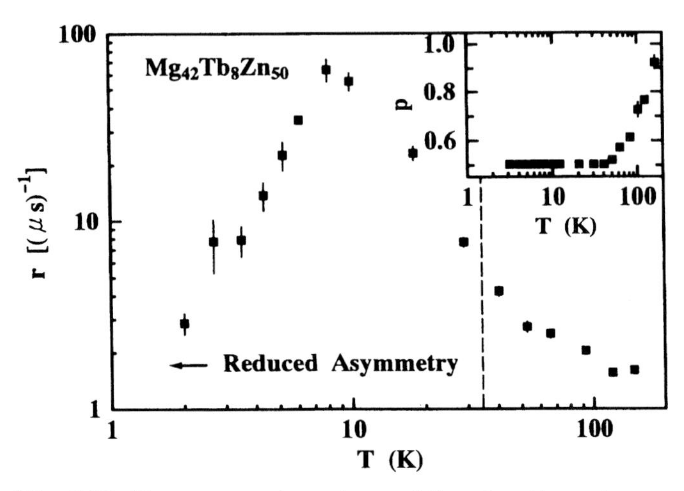

Fig. 9.22. Temperature dependence of the relaxation rate r on a double logarithmic scale for $Mg_{42}Tb_8Zn_{50}$ QC. The inset is the temperature dependence of the power p on a semi-logarithmic scale. After Noakes et al. (1998).

$$A(t) = A_0 \exp[-(rt)^p], \tag{9.19}$$

where r is the relaxation rate and p is the power. Figure 9.22 shows the temperature dependence of the relaxation rate of the $Mg_{42}Tb_8Zn_{50}$ QC on a double logarithmic scale. The inset shows the temperature dependence of the power p on a semi-logarithmic scale. The power p decreases from a simple exponential ($p \approx 1$) at 180 K to a root-exponential ($p = 1/2$) in the vicinity of 40 K. The vertical dashed line indicates the onset of loss of asymmetry at low temperatures. The relaxation rate begins to decrease as the temperature decreases below about 10 K, which is associated with spin freezing. In contrast to the neutron diffraction data (Charrier et al. 1997), there is no indication of the magnetic phase transition at 20 K. Slower spin dynamics and more prompt spin freezing in the $Mg_{42}Tb_8Zn_{50}$ QC, as compared with the the $Mg_{42}Gd_8Zn_{50}$ QC, is due to splitting of the crystalline electric field in the low symmetry of the Tb site (Noakes et al. 1998). These results are consistent with the low-temperature specific data (Hattori et al. 1995b) and the structural analysis (Yamamoto et al. 1996) mentioned in Sect. 9.3.2. Recently, a $Mg_{16}Ho_{16}Zn_{68}$ crystalline phase was found in the $Mg_{42}Ho_8Zn_{50}$ QC by annealing under the same conditions as those used by Charrier et al. (1997). The neutron diffraction peak positions and intensities of this compound (Sato et al. 1998) coincide with the results obtained by Charrier et al. (1997). Therefore, the establishment of long-range antiferromagnetic order in the $Mg_{42}RE_8Zn_{50}$ QCs remains an open question.

9.4 Summary

The magnetization and low-temperature specific heat measurements for Al-based QCs reveal that only a small number of Mn atoms carry the magnetic moment and that these magnetically dilute QCs exhibit spin-glass behavior. In the Al-Mn, Al-Cu-Mn and Al-Pd-Mn alloy systems, the fraction of magnetic Mn atoms and the effective magnetic moment of QCs are smaller than those of their *A* counterparts. In addition, the spin-freezing temperature of these QCs is lower. A localized magnetic moment is created on melting in the diamagnetic Al-based QCs. These facts imply that the structural disorder in *A* and liquid states gives rise to a change in the distribution of the Mn atoms, changing the magnetic and nonmagnetic states.

The magnitude of the effective magnetic moment of several kinds of Al-based compounds can be explained in terms of the Pauling valence, indicating that the magnetic properties are mainly governed by the local atomic environment, i.e., the atomic distance and the coordination number. Consequently, for the same compositions of $Al_{70}Pd_{30-x}Mn_x$ QCs, clear differences are observed between the magnetic properties of the *I* and *D* phases. Moreover, an anisotropic magnetic susceptibility is found in the *D*-Al-Ni-Co QCs due to the anisotropic atomic structure.

The positive temperature dependence of the magnetic susceptibility for the Al-Pd-Mn QC is attributed to the Pauli paramagnetism associated with the pseudogap at the Fermi surface. The peculiar ferromagnetic properties of the Al-based QCs, such as a relatively high Curie temperature with a low magnetization, a large coercivity and magnetic heterogeneity, could be correlated with the magnetic clusters created by phasons.

The magnetic properties of the Mg-RE-Zn QCs differ significantly from those of the Al-Mn, Al-Cu-Mn, and Al-Pd-Mn QCs and their *A* counterparts. There is no concentration dependence for the effective magnetic moment or the saturation magnetic moment, which reflect the localized magnetic moment of the $4f$ electrons.

The small γ value of nonmagnetic QC is attributed to the formation of a pseudogap at the Fermi level. The pseudogap explains the temperature dependence of the magnetic susceptibility of the $Al_{71}Pd_{20.5}Mn_{8.5}$ QC. The magnetic entropy at the spin-freezing temperature for the QCs is much smaller than the theoretical value for the long-range magnetic transition. Furthermore, the specific heat associated with the Schottky anomaly for the $Mg_{42}Tb_8Zn_{50}$ QC indicates that the Tb atom does not occupy the center of the *I* cluster, which is consistent with the structural analysis and the muon spin relaxation measurements.

Magnetic properties are very sensitive to the quality of QCs. In order to discuss the long-range antiferromagnetic order in the $Mg_{42}RE_8Zn_{50}$ QCs, various additional magnetic data from high-quality specimens are needed.

Acknowledgments

The author would like to acknowledge fruitful collaboration with Professor T. Goto of the Institute for Solid State Physics, the University of Tokyo. Grateful acknowledgements are made to Professor T. Fujiwara of the University of Tokyo and Dr. A.-P. Tsai of the National Research Institute of Metals for stimulating discussions and helpful information.

References

Bécle, C., Lemaire, R., Paccard, D. (1970): J. Appl. Phys. **41**, 855.
Bellissent, R., Hopper, F., Mond, P., Vigneron, F. (1987): Phys. Rev. B **36**, 5540
Cadeville, M.C. (1966): J. Phys. Chem. Solids **27**, 667
Charrier, B., Schmitt, D. (1997): J. Magn. Magn. Mater. **171**, 106
Charrier, B., Ouladdiaf, B., Schmitt, D. (1997): Phys. Rev. Lett. **78**, 4637
Chatterjee, R., Dunlap, R.A., Srinivas, V., O'Handley, R.C. (1990): Phys. Rev. **42**, 2337
Chernikov, M.A., Bernasconi, A., Beeli, C., Schilling, A., Ott, H.R. (1993): Phys. Rev. B **48**, 3058
Chikazumi, S. (1997): Physics of Ferromagnetism. Oxford University Press, Oxford, p 223
de Coulon, V., Reuse, F.A., Khanna, S.N. (1993): Phys. Rev. B **48**, 814
Dunlap, R.A., Lawther, D.W., Lloyd, D.J. (1988): Phys. Rev. B **38**, 3649
Dunlap, R.A., McHenry, M.E., Srinivas, V., Bahadur, D., O'Handley, R.C. (1989): Phys. Rev. B **39** 4808
Fujiwara, T. (1993): J. Non-Cryst. Solids **156-158**, 865
Fujiwara, T., Yokoyama, T. (1991): Phys. Rev. Lett. **66**, 333
Fukamichi, K. (1983): in Amorphous Metallic Alloys, Luborsky, F.E. (ed). Butterworths, London, p 317
Fukamichi, K., Goto, T. (1989): Sci. Rep. Res. Inst. Tohoku Univ. A **34**, 267
Fukamichi, K., Goto, T. (1991): Sci. Rep. Res. Inst. Tohoku Univ. A **36**, 143
Fukamichi, K., Goto, T., Masumoto, T., Sakakibara, T., Oguchi, M., Todo, S. (1987): J. Phys. F **17**, 743
Fukamichi, K., Goto, T., Wakabayashi, H., Bizen, Y., Inoue, A., Masumoto, T. (1988): Sci. Rep. Res. Inst. Tohoku Univ. A **34**, 93
Fukamichi, K., Kikuchi, T., Hattori, Y., Tsai, A.-P., Inoue, A., Masumoto, T., Goto, T. (1991): in Quasicrystals, Kuo, K.H., Ninomiya, T. (eds). World Scientific, Singapore, p 256
Fukamichi, K., Hattori, Y., Nakane, N., Goto, T. (1993): Mater. Trans., Jpn. Inst. Met. **34**, 122
Godinho, M., Berger, C., Lasjaunias, J.C., Hasselbach, K., Bethoux, O. (1990): J. Non-Cryst. Solids **117-118**, 808
Goldschmidt, Y.Y. (1992): in Recent Progress in Random Magnets, Ryan, D.H. (ed). World Scientific, Singapore, p 151
Goto, T., Sakakibara, T., Fukamichi, K. (1988): J. Phys. Soc. Jpn. **57**, 1751
Gubanov, V.A., Liechtenstein, A.I., Postnikov, A.V. (1992): in Magnetism and the Electronic Structure of Crystals, Springer-Verlarg, Berlin, p 24
Hattori, Y., Fukamichi, K., Chikama, H., Aruga-Katori, H., Goto, T. (1994): J. Phys. Condens. Matter **6**, 10 129

Hattori, Y., Fukamichi, K., Suzuki, K., Aruga-Katori, H., Goto, T. (1995a): J. Phys. Condens. Matter **7**, 4193
Hattori, Y., Fukamichi, K., Suzuki, K., Niikura, A., Tsai, A.-P., Inoue, A., Masumoto, T. (1995b): J. Phys. Condens. Matter **7**, 4183
Hattori, Y., Niikura, A., Tsai, A.-P., Inoue, A., Masumoto, T., Fukamichi, K., Aruga-Katori, H., Goto, T. (1995c): J. Phys. Condens. Matter **7**, 2313
Hauser, J.J., Chen, H.S., Waszczak, J.V. (1986): Phys. Rev. B **33**, 3577
Hippert, F., Audier, M., Klein, H., Bellissent, R., Boursier, D. (1996): Phys. Rev. Lett. **76**, 54
Jagannathan, A., Schulz, H.J. (1997): Phys. Rev. B **55**, 8045
Kimura, K., Iwahashi, H., Hashimoto, T., Takeuchi, S., Mizutani, U., Ohashi, S., Itoh, G. (1989): J. Phys. Soc. Jpn. **58**, 2472
Kobayashi, A., Matsuo, S., Ishimasa, T., Nakano, H. (1997): J. Phys. Condens. Matter **9**, 3205
Kofalt, D.D., Nanao, S., Egami, T., Wong, K.M., Poon, S.J. (1986): Phys. Rev. Lett. **57**, 114
Lasjaunias, J.C., Godinho, M., Berger, C. (1990): Physica B **165-166**, 187
Lasjaunias, J.C., Sulpice, A., Keller, N., Préjean, J.J., de Boissieu, M. (1995): Phys. Rev. B **52**, 886
Lin, C.R., Lin, C.M., Lin, S.T., Lyubutin, I.S. (1995): Phys. Lett. A **196**, 365
Liu, F., Khanna, S.N., Magaud, L., Jena, P., de Coulon V., Reuse, F., Jaswal, S.S., He, X.-G., Cyrot-Lackmann, F. (1993): Phys. Rev. B **48**, 1295
Lundquist, N., Myers, H.P., Westin, R. (1962): Philos. Mag. **7**, 1187
Lück, R., Kek, S. (1993): J. Non-Cryst. Solids **153-154**, 329
Lyubutin, I.S., Lin, Ch.R., Lin, S.T. (1997): Sov. Phys. JETP **84**, 800
Machado, F.L.A., Clark, W.G., Azevedo, L.J., Yang, D.P., Hines, W.A., Budnick, J.I., Quan, M.X. (1987): Solid State Commun. **61**, 145
Matsubara, E., Waseda, Y., Tsai, A.-P., Inoue, A., Masumoto, T. (1988): Z. Naturforsch. a **43**, 505
Markert, J.T., Cobb, J.L., Bruton, W.D., Bhatnagar, A.K., Naugle, D.G., Kortan, A.R. (1994): J. Appl. Phys. **76**, 6110
Massenet, O., Since, J.J., Mercier, J., Avignon, M., Buder, R., Nguyen, V.D., Kelber, J. (1979): J. Phys. Chem. Solids **40**, 573.
Matsuda, T., Ohara, I., Sato, H., Ohashi, S., Mizutani, U. (1989): J. Phys. Condens. Matter **1**, 4087
Matsuo, S., Ishimasa, T., Nakano, H. (1997): Solid State Commun. **102**, 575
McHenry, M.E., Vvedensky, D.D., Eberhart, M.E., O'Handley, R.C. (1988): Phys. Rev. B **37**, 10887
Mizutani, U., Sakabe, Y., Matsuda, T. (1990): J. Phys. Condens. Matter **2**, 6153
Môri, N., Mitsui, T. (1968): J. Phys. Soc. Jpn. **25**, 82
Mydosh, J.A., Nieuwenhuys, G.J. (1980): in Ferromagnetic Materials, Vol. 1, Wohlfarth, E.P. (ed). North-Holland, Amsterdam, p 71
Nakamichi, T. (1968): J. Phys. Soc. Jpn. **25**, 1189
Nasu, S., Miglieriani, M., Kuwano, T. (1992): Phys. Rev. B **45**, 12 778
Niikura, A., Tsai, A.-P., Inoue, A., Masumoto, T. (1994): Philos. Mag. Lett. **69**, 351
Noakes, D.R., Kalvius, G.M., Wäppling, R., Stronach, C.E., White Jr., M.F., Saito, H., Fukamichi, K. (1998): Phys. Lett. A **238**, 197
O'Handley, R.C., Dunlap, R.A., McHenry, M.E. (1991): in Handbook of Magnetic Materials, Vol. 6, Buschow, K.H.J. (ed). North-Holland, Amsterdam, p 453
Pauling, L. (1960): The Nature of the Chemical Bond. Cornell University Press, Ithaca, p 221
Sadoc, A., Dubois, J.M. (1989): J. Phys. Condens. Matter **1**, 4283

Saito, H., Fukamichi, K., Goto, T., Tsai, A.-P., Inoue, A., Masumoto, T. (1997): J. Alloys Comp. **252**, 6
Sato, T.J., Takakura, H., Tsai, A.-P., Shibata, K. (1998): Phys. Rev. Lett. **81**, 2364
Satoh, M., Hattori, Y., Kataoka, N., Fukamichi, K., Goto, T. (1994): Mater. Sci. Eng. A **181-182**, 801
Shechtman, D., Blech, I., Gratias D., Cahn J.W. (1984): Phys. Rev. Lett. **53**, 1951
Shimizu, M. (1965): Proc. Phys. Soc. **86**, 147
Shimomura, Y., Tsubokawa, I., Kojima, M. (1954): J. Phys. Soc. Jpn. **9**, 521
Shinohara, T., Yokoyama, Y., Sato, M., Inoue, A., Masumoto, T. (1993): J. Phys. Condens. Matter **5**, 3673
Stadnik, Z.M., Stroink, G. (1991): Phys. Rev. B **43**, 894
Stewart, A.M., Coles, B.R. (1974): J. Phys. F **4**, 458
Stoner, E.C. (1938): Proc. Roy. Soc. A **165**, 372
Takahashi, S., Li, X.G., Chiba, A. (1996): J. Phys. Condens. Matter **8**, 11 243
Tsai, A.-P., Inoue, A., Masumoto, T., Kataoka, N. (1988): Jpn. J. Appl. Phys. **27**, L2252
Wagner, J.L., Wong, K.M., Poon, S.J. (1989): Phys. Rev. B **39**, 8091
Wenger, L.E., Keesom, P.H. (1976): Phys. Rev. B **13**, 4053
Widom, M. (1988): in Introduction to Quasicrystals, Jarić, M.V. (ed). Academic Press, Boston, p 59
Wilson, A.H. (1965): The Theory of Metals. Cambridge University Press, London, p 151
Wohlfarth, E.P. (1978): J. Magn. Magn. Mater. **7**, 113
Yamada, Y., Yokoyama, Y., Matomo, K., Fukaura, K., Sunada, H. (1998): Jpn. J. Appl. Phys. in press
Yamamoto, A., Weber, S., Sato, A., Kato, K., Ohashima, K., Tsai, A.-P., Niikura, A., Hiraga, K., Inoue, A., Masumoto, T. (1996): Philos. Mag. Lett. **73**, 247
Yokoyama, Y., Inoue, A., Masumoto, T. (1994): Mater. Sci. Eng. A **181-182**, 734
Zhao, J.G., Yang, L.Y., Fu, Q., Guo, H.Q. (1988): Mater. Trans., Jpn. Inst. Met. Suppl. **29**, 497
Zou, J., Carlsson, A.E. (1993): Phys. Rev. Lett. **70**, 3748

10. Surface Science of Quasicrystals

Patricia A. Thiel, Alan I. Goldman, and Cynthia J. Jenks

10.1 Introduction

10.1.1 Background

Some of the interesting properties of quasicrystals (QCs), such as low friction and "non-stick" character, involve surface phenomena. Furthermore, in most proposed applications, QCs would be in the form of low-dimensional solids – coatings, thin films, precipitates, or composites – where the ratio of surface area to volume is high, and where surface or interface properties could be crucial to the success of the application (Dubois et al. 1993, Dubois et al. 1994a, Dubois 1997). These facts motivate fundamental studies of structure, composition, and chemical reactivity of QC surfaces. A main goal is to determine whether and how the interesting properties of the surface may relate to the unusual atomic structure of the bulk.

In order to pursue the goal mentioned above, one must make comparisons. For instance, one must compare the surface properties of a bulk QC with the surface properties of a crystalline material having similar chemical composition. This will reveal whether the interesting surface properties result fortuitously from chemical composition. One must also compare the properties of oxidized surfaces with those of clean surfaces, in order to determine which properties are due to the chemistry of the ubiquitous oxide, and which are intrinsic to the metallic substrate. The latter comparison obviously requires control over the reaction of the surface with its environment.

The use of ultrahigh vacuum (UHV) provides the ability to prepare a surface of desired composition, such as an oxidized surface or a clean surface. UHV also enables the use of surface-sensitive spectroscopies, such as x-ray photoelectron spectroscopy (XPS), Auger electron spectroscopy (AES), and low-energy electron diffraction (LEED). These techniques involve electrons, which have much shorter mean free paths in air than in vacuum.

The use of UHV, and associated spectroscopies, distinguishes recent and ongoing investigations of surface properties of QCs from earlier work. This current research is providing a wealth of new experimental information about surface structure and properties. Previous work was largely theoretical, and focussed on the surface phenomena responsible for the beautiful growth facets exhibited so often by QCs (Ho 1991).

In the early 1990's, a group of experimentalists at AT&T Bell Laboratories ushered in the current generation of surface work, with the first LEED and scanning tunneling microscopy (STM) experiments on clean, single-grain surfaces of decagonal (d) Al-Co-Cu (Kortan et al. 1990, Becker and Kortan 1991, Kortan et al. 1992). Subsequent work focussed mainly upon icosahedral (i) Al-Pd-Mn, primarily because of availability of high quality, large samples (typically, a sample must have minimum surface dimensions of 2 mm$\times$2 mm for such studies). At the present time, surface work is branching out to the d phase of Al-Ni-Co and i-Al-Cu-Fe, because of recent advances in the growth of these materials. Thus, all work to date concerns Al-rich alloys. Consequently, the present article is restricted to these materials.

10.1.2 Outline

This chapter deals first with oxidized surfaces, and our current understanding of them. This information is relevant to any application of QCs that might take place in an oxidizing environment. A discussion of surface energy, which is typically measured only for oxidized surfaces and which may depend critically upon the nature of the oxide, is presented in Sect. 10.3. Clean (non-oxidized) surfaces of QCs and related alloys are discussed next. Clean surfaces are relevant to growth, and also to fracture, friction, and wear processes in which clean surfaces may be created instantaneously even if they cannot survive. Finally, friction is discussed in Sect. 10.5. Friction is an area where the information gathered for both oxidized and clean surfaces is necessary for the types of comparisons described as a major goal in Sect. 10.1.1, although much work remains to be done.

10.2 Oxidized Surfaces

10.2.1 Overview

Abundant evidence now shows that a thin, protective layer of Al oxide tends to form on the Al-rich QCs (Chang et al. 1995a, 1995b, Kang and Dubois 1995, Chang et al. 1996, Suzuki et al. 1996, Chevrier et al. 1997, Jenks et al. 1997, Pinhero et al. 1997, Rouxel et al. 1997, Wehner and Köster 1997, Gavatz et al. 1998, Jenks et al. 1998c, Pinhero et al. 1998a, 1998b, Wehner and Köster 1998). This is based upon data both for i-Al-Pd-Mn and i-Al-Cu-Fe. At least for i-Al-Pd-Mn, the oxide forms independently of whether the sample is prepared by cleaving, polishing, or sputter-annealing (Jenks et al. 1998c). A comparison of single-grain versus multi-grain i-Al-Cu-Fe samples shows that it also does not depend upon sample type (Pinhero et al. 1998a). Passivation by an Al oxide layer is common among alloys of Al (Hoffman et al. 1988, Mesarwi and Ignatiev 1992, Libuda et al. 1994), although not universal (Ahmed and Smeltzer 1986, Splinter and McIntyre 1994). The phenomenon is driven by the relatively large and negative heat of formation of the oxide of this element, -1676 to -1657 kJ/mol of Al_2O_3, or -1117 to -1105 kJ/mol of O_2 (Wagman et al. 1982). Hence, with respect to the basic formation of a passivating Al oxide layer, QCs do not appear to be distinctive.

10.2.2 Oxide Composition

Representative data supporting this conclusion are shown in Fig. 10.1. Here, several Al-Cu-Fe alloys were prepared by gas atomization of powders, followed by hot-isostatic-pressing. This yields monoliths, consisting of many individual grains, and without detectable porosity. Surfaces of these materials were cleaned by sputter-annealing in UHV (see Sect. 10.3), then exposed to air at room temperature (Figs. 10.1a–c). Comparisons were also made to pure Al, Cu and Fe. Figure 10.1d shows that, in air, all three of the pure elements oxidize, based upon established signatures of oxidation (Moulder et al. 1992). However, these signatures fail to appear for any element except Al, when the alloys are oxidized in air. Furthermore, measurements of the surface composition of the alloys indicate enrichment of Al under these conditions. The enrichment in Al is apparent even from visual inspection of Figs. 10.1a–c, which reveals the attenuation of the Cu and Fe peaks relative to Al after oxidation. Together, both observations support the formation of a pure, or nearly-pure, oxide of Al at the surface. However, this oxide is thin enough that it does not obscure the underlying Fe and Cu, still present in the metallic substrate.

10.2.3 Oxide Depth

The depth of the oxide, formed by air at room temperature, has been measured in several ways. Analysis of XPS data relies upon the relative attenuation of the metallic Al signal, relative to the oxide signal, and upon estimates of the inelastic mean free paths of the photoelectrons both in the oxide and the metal. Such evaluations consistently yield a depth of 19–26 Å for *i*-Al-Cu-Fe (Pinhero et al. 1998a, 1998b), and 22–29 Å for *i*-Al-Pd-Mn (Chang et al. 1995b, 1996, Pinhero et al. 1997, Jenks et al. 1998c). These are to be compared with XPS measurements of approximately 30 Å for pure Al (Chang et al. 1996, Pinhero et al. 1998b). A different technique, x-ray reflectivity and diffuse scattering, yields 23 Å for *i*-Al-Pd-Mn (Gu et al. 1997), in good agreement with the XPS values. These depths are consistent with the fact that Fe and Cu are still visible after oxidation as shown in Fig. 10.1, since XPS typically probes about the top 100 Å of surface material (Strohmeier 1990).

However, there are at least three variables that can affect the oxide layer's characteristics. The first, which probably contributes some variability to the measurements described above, is humidity. Water affects strongly the oxide thickness and composition, with more water presenting a more aggressive environment (Jenks et al. 1997, Pinhero et al. 1997, 1998b). For instance,

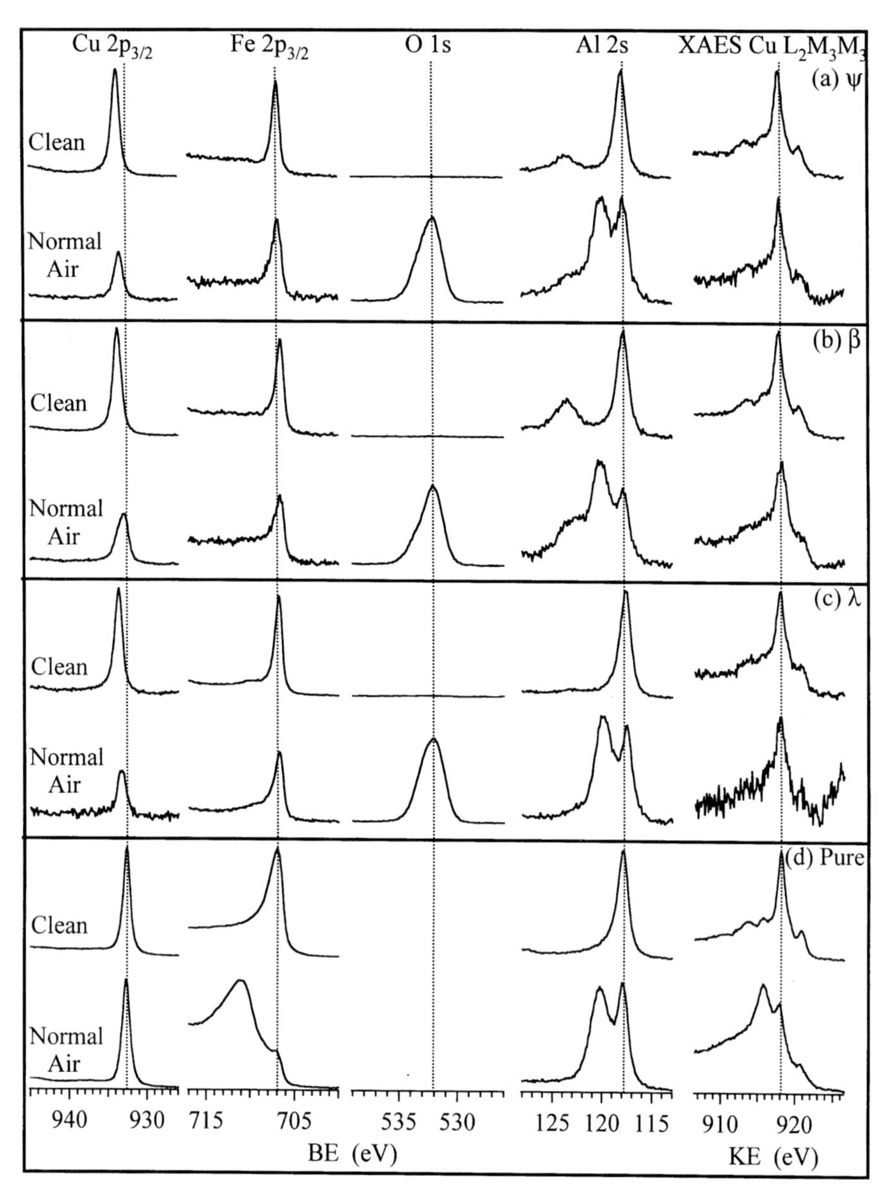

Fig. 10.1. XPS and x-ray induced Auger electron spectra (XAES) of pressed powders of Al-Cu-Fe alloys after various treatments (Pinhero et al. 1998b). Vertical dashed lines show the average peak positions for the clean surfaces of the pure metal constituents, measured in our laboratory. In air, all three pure elements oxidize, based on the appearance of new peaks which are several eV lower in kinetic energy (higher binding energy), compared to the clean surface. These new peaks are evident in the Fe $2p_{3/2}$, Al $2s$, and Cu $L_2M_3M_3$ lines in (d). In the alloys, only Al oxidizes, as shown in (a), (b), and (c). Small shifts in the Cu $2p_{3/2}$ line between the pure element and the alloys do not signal oxidation, but rather intermetallic bonding in the alloy. Reprinted with permission of Taylor and Francis.

experiments such as those represented in Fig. 10.1 show that in vacuum, using very pure oxygen at room temperature, the oxide layer saturates at 4–8 Å thickness on *i*-Al-Cu-Fe (Jenks et al. 1997, Pinhero et al. 1997, Jenks et al. 1998c, Pinhero et al. 1998b). The analogous value for *i*-Al-Pd-Mn is 4–5 Å (Chang et al. 1996, Jenks et al. 1997, 1998c). For both alloys, this is essentially the same as the depth on pure Al under the same conditions. This is much thinner than the oxide formed in air, presumably due to the presence of some water in the air. At the other extreme, immersion in water leads to a much deeper oxide than in air, and causes oxidation of other elements in the alloy (Jenks et al. 1997, Pinhero et al. 1997, Jenks et al. 1998c, Pinhero et al. 1998b).

A second variable is temperature: higher temperatures produce thicker oxides, all other conditions being equal, at least up to very high temperatures where oxide evaporation and oxygen dissolution may occur (Chang et al. 1996, Rouxel et al. 1997, Gavatz et al. 1998). This has been investigated thoroughly by Rouxel and coworkers for *i*-Al-Cu-Fe (Rouxel et al. 1997, Gavatz et al. 1998). Some of their data are shown in Fig. 10.2a, which illustrates the surface Al content, derived from AES, after saturation with oxygen in vacuum (using very pure oxygen at 4×10^{-7} Torr). The increase in Al content represents enhanced oxidation. It can be seen that 670 K is a critical point; only above 670 K does oxidation depend strongly upon temperature. Presumably, diffusion of the Al (and/or O) becomes significantly faster above this temperature. Similarly, an AES-based study of *i*-Al-Pd-Mn showed that oxidation was not temperature-dependent in the range 100–500 K, but was enhanced at 870 K (Chang et al. 1995a). XPS also shows that the limiting depth on *i*-Al-Pd-Mn increases from approximately 5 to 10 Å, upon going from 300 to 870 K, after oxidation in vacuum. Pure Al also shows a temperature-dependence (Lauderback and Larson 1990, Chang et al. 1996).

A third variable is pressure. Comparing Figs. 10.2a and 10.2b shows the effect of pressure. There is an abrupt increase in Al content at 670 K only if the pressure is "high", as in Fig. 10.2a; there is no change up to 1070 K when the pressure is an order of magnitude lower, as in Fig. 10.2b. This pressure dependence is attributed to a competition between surface oxidation and diffusion of oxygen into the bulk. High pressure favors oxidation over bulk diffusion (Gavatz et al. 1998).

10.2.4 Comparison to Crystalline Materials

Is there anything "special" about oxidation of the QCs? There are two sets of contradictory evidence on this point. XPS measurements of oxide thickness and oxide composition, in general, show no difference between the Al-rich QCs and pure Al, nor between the QCs and their crystalline counterparts. For instance, Fig. 10.1 compares data for the *i* (Ψ) phase of Al-Pd-Mn, with data for two crystalline alloys (the λ and β phases) which are very close in

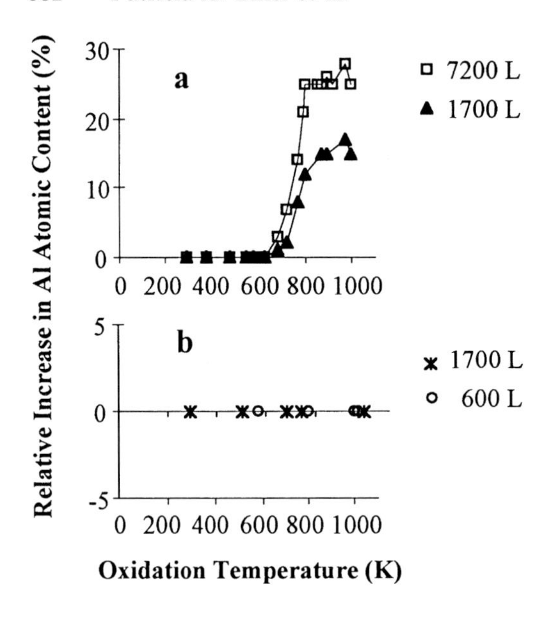

Fig. 10.2. Change in surface and near-surface Al concentration, measured with AES after high exposures of O_2 in UHV, as a function of oxidation temperature (Gavatz et al. 1998). (a) Pressure during exposure at 3.8×10^{-7} Torr; (b) Pressure during exposure at 1.5×10^{-8} Torr.

the phase diagram. There are no significant differences in the spectra, except for those expected from the differences in bulk alloy composition. This is for oxidation at room temperature. On the other hand, Fig. 10.3 shows weight gain during oxidation at elevated temperature (Sordelet et al. 1997). The alloys being compared are the same as those in Fig. 10.1, although the physical form of the alloys is different: In Fig. 10.3 they are gas-atomized powders, the precursors to the monoliths used in Fig. 10.1. The data of Fig. 10.3 show parabolic behavior, indicative of passivation, but the Ψ phase takes up less oxygen than the crystalline phases. This would indicate a thinner oxide layer, and/or less oxygen dissolution, for the Ψ phase at these elevated temperatures. For i-Al-Pd-Mn, measurements of oxide depth via sputtering (Chevrier et al. 1997) have been interpreted to mean that the depth on the QC is about half that on elemental Al, after air oxidation at room temperature (Dubois 1998). Finally, a limited comparison between a quasicrystalline and crystalline phase in the Al-Pd-Mn system suggested that the quasicrystalline phase is more effective at protecting Mn from oxidation, than is the crystalline CsCl analog (Jenks et al. 1997).

At present, we believe that a critical examination of all available data favors the conclusion that the oxidation chemistry of these alloys is determined by their Al-rich composition, and is not related to the aperiodic structure, at least for oxidation at room temperature. The data presently available, however, mainly address the average oxide thickness and average composition of the saturated layer at room temperature. Systematic comparisons between QCs and related alloys have not been made as a function of temperature, nor have the kinetics, morphology, or heterogeneity of oxidation been addressed

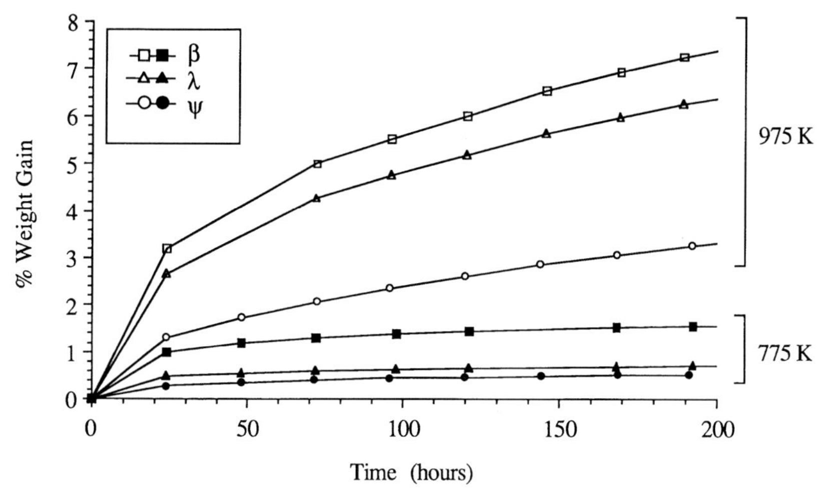

Fig. 10.3. Weight gain vs time for initially single-phase crystalline and quasicrystalline powders oxidized at 775 and 975 K. After Sordelet et al. (1997).

via such comparisons. The QCs may still exhibit unusual behavior when these other topics are studied, as suggested by Fig. 10.3.

10.2.5 Oxide Structure

Preliminary data on the structure of the oxide layer suggest that this is indeed an intriguing area for exploration. Figure 10.4 shows an STM micrograph of a grain of *i*-Al-Cu-Fe prepared by polishing and oxidizing in air (Machizaud et al. 1997). The dark areas, which may represent oxide islands, are on the order of 10 Å wide and 5 Å high. Although measurements of height from STM are, in general, somewhat questionable, this estimate of roughness (10 Å) is in good agreement with a previous measurement of 15 Å for air-oxidized Al-Pd-Mn, using x-ray scattering techniques (Gu et al. 1997). Such structural features were not observed in STM studies of air-oxidized Al (Szklarczyk et al. 1990). Whether and how the quasicrystalline substrate dictates the structure in this oxide should be investigated. Such investigation could elucidate the relationship, if any, between the roughness of this oxide and the low surface energy of QCs (see Sect. 10.3) (Rivier 1997, Dubois et al. 1998).

10.2.6 Oxidation-Induced Phase Transformations

The importance of studying oxidation is underscored by the fact that oxidation can move the surface, and even bulk, composition out of the range of quasicrystalline stability. If the temperature is sufficiently high, this can cause transformations to crystalline phases (Kang and Dubois 1995, Sordelet et al. 1997, Wehner and Köster 1997, 1998). This was suggested first by a

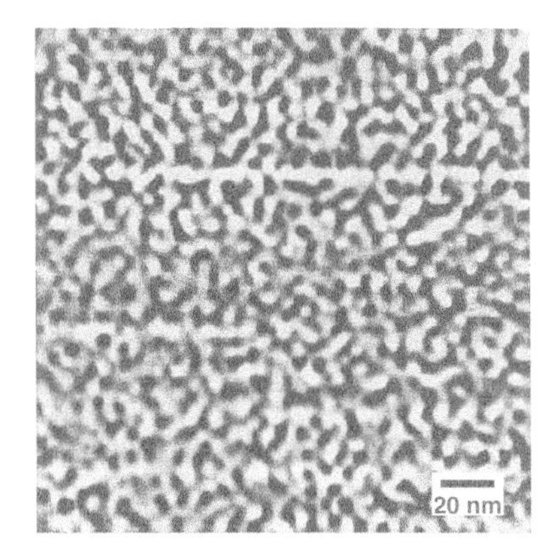

Fig. 10.4. Scanning tunneling micrograph of a grain of *i*-Al-Cu-Fe in air. The grain prepared by polishing and oxidizing is part of a larger monolith prepared by crushing and sintering. From data of F. Machizaud, B. Vigneron, J. M. Dubois, and J. P. Dufour (Jenks and Thiel 1997).

study of the influence of annealing atmosphere on stability of Al-Cu-Fe samples (Kang and Dubois 1995). Phase transformation occurs because diffusion of Al to the surface depletes Al in the underlying material (Sordelet et al. 1997), and also because oxygen diffuses into the bulk (Kang and Dubois 1995, Wehner and Köster 1997, 1998). All observations to date show that, if oxidation triggers transformation, the β phase (CsCl structure) is the end result in the Al-Cu-Fe system (Kang and Dubois 1995, Sordelet et al. 1997, Wehner and Köster 1997, 1998).

Such phase transformations are likely to be particularly important in situations where the ratio of surface area to bulk volume is high. For instance, Wehner and Köster (1997) have shown that oxidation of very thin films can transform the substrate to a cubic phase, whereas oxidation of thicker samples under similar conditions does not (Wehner and Köster 1998). Similarly, in the experiments represented by Fig. 10.3, x-ray diffraction (XRD) shows that at 950 K the quasicrystalline bulk begins to transform into a crystalline phase which is deficient in Al, relative to the i phase (Sordelet et al. 1997). The kinetics depend upon particle size, faster kinetics being observed for smaller particles. Hence, the surface chemistry dictates the phase stability, particularly of low-dimensional materials, and will be important for applications in which the relevant properties of quasicrystalline and crystalline phases differ significantly.

10.3 Surface Energies

The surface energy is an example of an important, fundamental physical property that can only be measured directly for oxidized QC surfaces. Surface energy can be derived from contact angle measurements, in which the so-called sessile drop method is used often. This involves placing a drop of liquid on top of a substrate and measuring the angle the drop makes at the point of contact with the surface. The angle measured in these experiments is shown in Fig. 10.5. By definition, a liquid is said to not wet a surface if the contact angle is greater than 90°. If the contact angle is zero, then the liquid is said to wet the surface. The angle formed at the point of contact depends upon a complex set of intermolecular interactions such as dipole interactions between the solid and the liquid and is very sensitive to small changes at the liquid/solid interface (Bose 1993). For example, changes in the roughness of a surface or the presence of chemical inhomogeneities will alter the resulting contact angle. Therefore, surfaces must be carefully prepared because of the sensitivity of the results; the surface preparation needs to be identical for each sample being examined, and the appearance of gross chemical inhomogeneities must be ruled out.

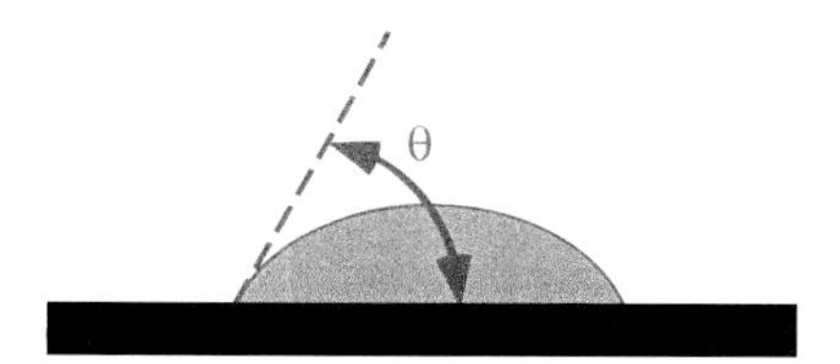

Fig. 10.5. Schematic of a drop of liquid in contact with a surface. The angle Θ represents the angle measured in the contact angle experiments discussed in the text.

Surface energy is related to the contact angle via the well-known Young equation:

$$\gamma_{LV}\cos(\Theta) = \gamma_{SV} - \gamma_{SL}, \tag{10.1}$$

where L denotes liquid, S - solid, V - vapor, Θ is the contact angle, and γ is the interfacial free energy. The resulting surface energy can be broken down into two components, a polar component and a dispersive component. The polar component takes into account induced polarization and permanent dipoles, while the dispersion component results from van der Waals forces (Adamson 1976). By measuring the contact angle using liquids with varying degrees of polarizability, these two components can be separated (Dubois et al. 1998).

Recent surface energy measurements which compare QCs to other materials show very interesting results (Dubois et al. 1998). Such measurements were made by analyzing, *under atmospheric conditions*, the contact angle of various liquids with different surfaces including several Al-based QCs, oxidized Al, Al oxide, transition metals, and related crystalline phases. Despite the presence of an overlayer composed predominately of Al oxide (Chang et al. 1995a, 1995b, Kang and Dubois 1995, Chang et al. 1996, Suzuki et al. 1996, Chevrier et al. 1997, Jenks et al. 1997, Pinhero et al. 1997, Rouxel et al. 1997, Wehner and Köster 1997, Gavatz et al. 1998, Jenks et al. 1998c, Pinhero et al. 1998a, 1998b, Wehner and Köster 1998), the surfaces energies of *i*-Al-Pd-Mn and *i*-Al-Cu-Fe do not equal that of Al oxide. For *i*-Al-Pd-Mn under ambient conditions, the total surface energy is 28 mJ/m^2, which is only slightly higher than that of Teflon, 18 mJ/m^2. This is in contrast to Al_2O_3, which has a surface energy of 47 mJ/m^2. Freshly polished Al (which forms an oxide with an average thickness virtually identical to that of the QC) has a total surface energy of 33 mJ/m^2 (Dubois et al. 1998). These values can be compared to surface energies of clean metals, which are typically in the range of a few thousand mJ/m^2 (Somorjai 1994).

When the polar and dispersive components are examined, it is found that quasicrystalline substrates tend to behave more like Teflon (which has a small polar component to the surface energy) than like typical metals (Dubois 1998, Dubois et al. 1998). In fact, the more perfect the quasicrystalline sample, the lower the polar component. Polished Al exposed to ambient air has a polar component larger than that of the QCs but smaller than that of bulk Al_2O_3. Al_2O_3 has an extremely high polar component compared to the other materials studied. Based on these contact angle measurements, the quasicrystalline surfaces behave much more like covalently bound materials than like metals.

The polar contribution to the surface energy correlates with the density of states at the Fermi level; materials with lower polar components have lower electron densities at the Fermi level, as shown in Fig. 10.6. The suppression in the density of states bridging the Fermi level is called the pseudogap (Belin and Mayou 1993, Belin et al. 1994, Belin-Ferré et al. 1997, Belin-Ferré 1998). The pseudogap is a bulk property of all QCs studied thus far, and it is also seen in related crystalline materials. However, the gap is larger for QCs than for related crystalline materials (Belin-Ferré 1998), which is consistent with the wetting results. The pseudogap at the Fermi level has been related to the poor electrical and thermal conductivity of QCs and their approximants relative to their metallic constituents, as well as the small electronic contribution to the specific heat. In addition, a pseudogap near the Fermi level may stabilize the structure by lowering the energy according to Hume-Rothery arguments.

If the pseudogap persists in the surface and near-surface region, then the low surface polarizability correlating to the reduced density of states at the Fermi level may be understandable. In the language of surface science, one expects a surface pseudogap to reduce the ability of the metal to screen a charged or highly polar adsorbate by forming an image dipole. In the contact angle measurements, the drop (as shown in Fig. 10.5) always rests atop an oxidized surface, yet is apparently sensitive to the electronic properties of the metal beneath the oxide layer. Hence, it is important to consider whether and how the pseudogap of the bulk is maintained at the oxide interface, an issue which (to our knowledge) has not been addressed. More consideration of the surface pseudogap, as it relates to the chemistry of the *non-oxidized* surface, is given in Sect. 10.4.4.

The low surface energy may also be related to a larger surface roughness on the microscopic or mesoscopic length scale for the quasicrystalline material (Rivier 1997), such as the roughness illustrated in Fig. 10.4.

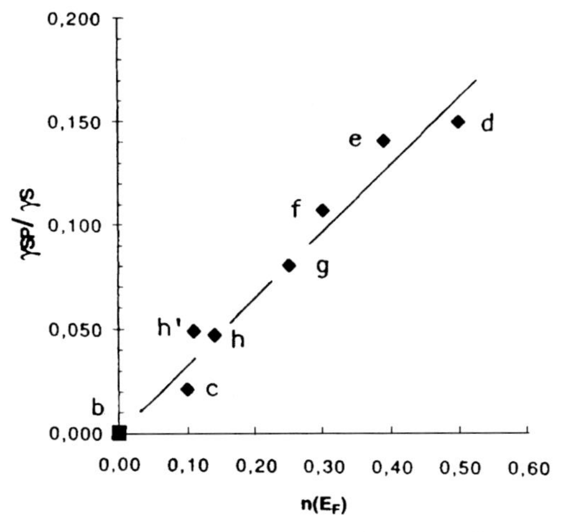

Fig. 10.6. Linear correlation between the ratio of the polar component of the solid surface energy to total solid surface energy and the density of states at the Fermi level (Dubois et al. 1998). The letters correspond to the following materials: b is Teflon, c is *i*-Al-Pd-Mn, d is a typical polycrystalline fcc metal, e is cubic Al-Cu-Fe, f is a cubic approximant to *i*-Al-Cu-Fe, namely $Al_{55}Si_7Cu_{25.5}Fe_{12.5}$, g is amorphous $Al_{62}Cu_{25.5}Fe_{12.5}$, h is a *i*-Al-Cu-Fe sample with many defects, and h′ is a highly perfect *i*-Al-Cu-Fe sample.

10.4 Clean Surfaces

The ability to produce clean surfaces of QCs provides an opportunity for unprecedented study of the fundamental properties of these surfaces. Questions abound regarding to the nature of the clean surface. For example, is it crystalline or quasicrystalline? Does the pseudogap which exists in the bulk extend to the surface? Do these complex alloys exhibit surface segregation? In this section, we review methods for preparing clean surfaces. In addition, we show results of studies of the surface structure of clean surfaces and their reactivity measured within UHV.

10.4.1 Methods of Clean Surface Preparation

Growth and preparation of clean QCs pose special challenges, compared with elemental metals (Goldman and Thiel 1998). Our approach to preparing these samples for surface experiments is described in detail elsewhere (Jenks et al. 1996a, 1998d, Jenks and Thiel 1998b). The possible existence of secondary phases, and of porosity, deserves special attention since these features may be critical to the interpretation of surface measurements. We recommend screening quasicrystalline surfaces routinely after growth and polishing, using a combination of scanning electron and scanning Auger microscopies, and energy-dispersive spectroscopy. These techniques are nondestructive, are able to scan large sample areas, and have an areal detection limit better than 1% for secondary phases. These present advantages over Laue XRD, powder XRD, and transmission electron microscopy, which are used more traditionally to screen for secondary phases. Variability in phase purity and porosity can occur within a given growth, which provides the motivation to screen every sample used.

In order to obtain a clean surface in UHV, one must usually deal with the residue from air exposure. We find that air always leaves oxygen (Jenks et al. 1998c), as well as carbon- and sulfur-containing compounds, just as would be the case for an elemental metal. It is sometimes thought that preparation of a surface in an "inert" atmosphere, such as a glove box filled with rare gas, might circumvent these problems. However, this is not true. Contaminant levels in a glove box are typically on the order of 1 ppm, which corresponds to a partial pressure of about 10^{-3} Torr and an incident flux of about 10^3 monolayers per second. Hence, if the glove box contains even 1 ppm of water, and if the surface is reasonably reactive, the surface will oxidize very quickly in the "inert" environment.

Commonly, surface impurities are removed in UHV by sputtering with noble gas ions, typically Ar^+. Because this damages the surface, the sample is usually heated to restore structural and chemical integrity. Another source of contamination is trace impurities in the bulk, some of which may have been incorporated from the original ores. Impurities such as P, S, Si, and/or B are often observed in crystalline samples, where they usually segregate to the surface of a fresh sample after heating (Musket et al. 1982). We have detected such impurities at the surface even when they were not detected by highly sensitive bulk techniques, including inductively-coupled plasma atomic emission spectroscopy. Usually, bulk contaminants can be depleted from the near-surface region to an acceptable level by repeated cycles of sputtering and annealing.

The sputter-annealing sequence is traditional for metal surfaces. Its advantage is that it can regenerate a clean surface repeatedly in UHV, from a single sample. Its disadvantage is that it is chemically and structurally disruptive. Chemical changes at the surface can occur via two routes: (1) preferential sputtering, and (2) preferential evaporation at high temperature.

Simple momentum-transfer arguments lead to the expectation that the lightest element in an alloy will be sputtered preferentially. Schaub et al. were the first to report that Ar^+ sputtering of an Al-rich QC, *i*-Al-Pd-Mn, leads to preferential loss of Al, the lightest element (Schaub et al. 1995). This observation has since been confirmed in other laboratories (Chang et al. 1995a, Chang et al. 1996, Suzuki et al. 1996, Naumović et al. 1998). Similar observations of Al depletion upon Ar^+ sputtering have been reported also on two other Al-rich alloys, *i*-Al-Cu-Fe (Rouxel et al. 1997, Shen et al. 1997b) and *d*-Al-Ni-Co (Zurkirch et al. 1998). However, a composition close to that of the bulk can be restored by annealing to temperatures around 700–900 K, for *i*-Al-Pd-Mn and *i*-Al-Cu-Fe.

The other problem, preferential evaporation, can be observed at temperatures in excess of about 900 K, at least for *i*-Al-Pd-Mn. Preferential evaporation in this alloy usually leads to Pd enrichment at the expense of the other two metals (Jenks et al. 1996a, Suzuki et al. 1996, Jenks et al. 1998d). This can be reversed by annealing again at 800 K (Jenks et al. 1996a). The change in composition is illustrated in Fig. 10.7, which shows that compositional changes due to preferential evaporation start at 870–920 K (600–650 °C). Preferential evaporation may begin even earlier, but if it is sufficiently slow that the losses can be replaced by diffusion from the bulk, it would not necessarily cause compositional changes at the surface. Chevrier et al. have demonstrated that Mn is the predominant metal lost from *i*-Al-Pd-Mn at 1000 K (Chevrier et al. 1998), consistent with the depletion of Mn evident in Fig. 10.7.

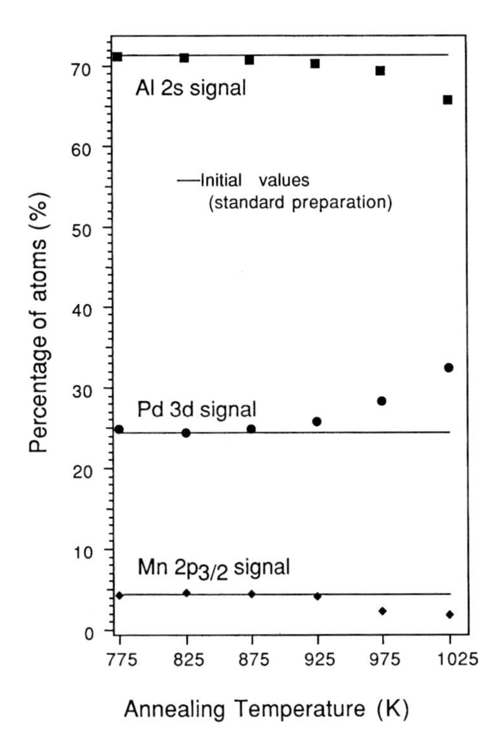

Fig. 10.7. Surface compositions of an $Al_{70.3}Pd_{21.4}Mn_{8.3}$ sample provided by Y. Calvayrac (CECM-CNRS, Vitry-sur Seine, France) in UHV. Before each annealing the crystal was sputtered for 1 h at decreasing values of discharge potential (between 1.0 and 0.5 kV). Then the crystal was annealed for 30 min at the given temperatures between 775 and 1025 K (every 50 K). After each preparation surface characteristics were measured with LEED, XPS, and x-ray photoelectron diffraction (XPD). Changes were observed in LEED and XPD at 1025 K due to preferential evaporation. Courtesy of Dusanka Naumović.

Additionally, we have observed that heating some samples of *i*-Al-Pd-Mn to elevated temperatures can cause precipitation of secondary phases, leading to irreversible sample degradation. For this reason, we suggest that heating above 900 K, and perhaps even 800 K, be avoided (Jenks et al. 1998a). This is a noteworthy correction to our previous recommendation (Jenks et al. 1996a).

Because of the chemical and structural damage which accompanies sputter-annealing, alternative approaches have been investigated. In order to avoid sputtering, Chevrier et al. report that one can transfer a sample from air to UHV and heat it to high temperature, 1000 K for *i*-Al-Pd-Mn (Chevrier et al. 1997, 1998). The oxide layer simply evaporates, although one is potentially left with contaminant segregation from the bulk, which may necessitate sputtering after all. Furthermore, preferential evaporation can occur, although this might be reversed by annealing at lower temperature (Jenks et al. 1996a). Little work has been done to characterize clean surfaces produced by this intriguing method, except that STM images reveal facetted voids that may be remnants of metal evaporation (Chevrier et al. 1997).

Another approach is to fracture the sample. This approach clearly causes the least chemical damage. However, if fracture is carried out at relatively low temperature, there is no diffusion, and hence no possibility for equilibration. Annealing in UHV may allow the surface to equilibrate, but may also induce segregation of contaminants from the bulk. An even larger problem is that fracture consumes samples irreversibly.

In summary, a major challenge to the field at present is determining which preparation route is most appropriate for understanding global surface properties. Part of the solution must be provided by comparisons between properties, in UHV, of surfaces prepared via different routes. Thus far, no such comparisons have been made except for comparisons of surface structure (see Sect. 10.4.3).

10.4.2 Surface Composition

A basic question is whether the surface of a QC exhibits thermodynamically-driven segregation of individual alloy components. Many studies of surface composition have been reported, primarily using XPS and AES, which provide a depth-weighted average over approximately the top 100 Å of material. All such studies have been on sputter-annealed surfaces, and most have been done on *i*-Al-Pd-Mn and *i*-Al-Cu-Fe samples. With one exception, these XPS and AES studies are probably only accurate to within ±3–5 at.%, a level of uncertainty which exceeds the range of compositional stability of the typical bulk QC.

The reason why accuracy is not better lies in the need for sensitivity factors: all studies to date (with one exception) have utilized values for the pure elements. However, sensitivity factors determined from chemically-similar alloys should provide much better accuracy, probably in the range of ±1 at.%.

The best way, and perhaps the only way, to prepare an alloy surface with known composition (and without composition gradients over the depth probed by XPS or AES) is via fracture at relatively low temperature (≤ 300 K). This is an experiment for which many UHV instruments are not equipped. An example of the importance of such calibration, however, was first provided by Rouxel et al. (1997), who fractured a poly-grain sample of *i*-Al-Cu-Fe in

UHV and used scanning electron microscopy to identify a flat, smooth area in the fracture front for analysis. They found that the sensitivity factor for Fe in the alloy differed significantly from that for elemental Fe. More recent work has shown that there can even be differences between AES sensitivity factors of *i* samples and related crystalline alloys (Bloomer et al. 1998). This could arise because of differences in the structural and chemical environment of each element.

For *i*-Al-Pd-Mn, XPS and AES studies based on elemental sensitivity factors show that the fivefold surface is deficient in Mn after heating to 800 K, relative to bulk compositions. One could argue that this arises because the findings are not properly calibrated. However, low energy ion scattering (LEIS) supports these results. LEIS probes the topmost surface layers exposed to vacuum as opposed to the several atomic layers probed by XPS and AES. LEIS itself is not free from problems associated with calibration (Gierer et al. 1998). However, as discussed in the next section, the maximum Mn concentration possible is still below the minimum value expected based on LEED structural analysis. Additionally, Mn depletion at the surface is also supported by a recent XRD study of the fivefold surface plane which showed that the surface plane contains no detectable Mn (Capitan et al. 1998). Thus, several techniques suggest in concert that the sputter/annealed fivefold surface is depleted in Mn. The surface free energies of the various elements point to Al as the likely element to replace Mn at the surface.

10.4.3 Surface Structure and Topography

Interest abounds in the surface structure and topography of these materials. From the studies that have been reported it is quite clear that the surface structure depends strongly on how the surface is prepared. In this section, we discuss surfaces prepared by sputter-annealing and fracturing. The sputter-annealed results are divided into two categories: (1) results for which the surface is annealed below 700 K, and (2) results for which the surface is annealed between 700 and 900 K.

As discussed previously, sputtering results in a depletion of Al at the surface relative to the bulk composition. Generally speaking, the Al-based phase diagrams containing quasicrystalline structures indicate the presence of a bcc cubic crystalline phase at Al contents below that of the quasicrystalline structure. This depletion thus shifts the surface composition away from the quasicrystalline region of the phase diagram and towards that of cubic structure (Rouxel et al. 1997). This structure is typically B2 (CsCl). This structure has been observed after sputtering, with and without low temperature anneals, for several Al-based QCs using a number of surface sensitive techniques. These systems include *d*-Al-Ni-Co, studied by secondary-electron imaging (SEI) (Zurkirch et al. 1998), *i*-Al-Cu-Fe, studied by LEED (Shen et al. 1997b, 1998a, Shi et al. 1998), and *i*-Al-Pd-Mn, studied by SEI (Bolliger et al. 1998), LEED (Shen et al. 1997a, 1998a), and by XPD (Naumović et al. 1998). These techniques probe approximately the first 10–100 Å and all indicate that the crystalline phase at the surface, under these preparation conditions, is at least this deep. Furthermore, results for fivefold (tenfold) surfaces of *i* (*d*) samples, all agree that the cubic alloy adopts an orientation that is structurally related to that of the bulk; in each case the [110] crystalline axis is parallel to a bulk fivefold (tenfold) axis. The CsCl structure also forms when *i*-Al-Pd-Mn is heated to very high temperatures, such that Pd enrichment again results, albeit from preferential evaporation (Naumović 1996).

For the fivefold surface of Al-Cu-Fe, the cubic overlayer has also been examined by LEED-IV analysis (Shi et al. 1998). These studies also support the conclusion of a B2 overlayer, with Al occupying the corners of the cubic unit cell, and Cu and Fe randomly occupying the center (the electron scattering properties of Cu and Fe are similar, so distinguishing between the two is difficult with LEED). These studies also show that within the outermost surface layer the Al atoms buckle out 0.128 Å and Cu buckles out 0.020 Å relative to the bulk positions, as shown in Fig. 10.8. For bulk B2 crystals, such as NiAl (Yalisove and Graham 1987, Davis and Noonan 1988b, Davis and Noonan 1988a, Mullins and Overbury 1988) and CoAl (Blum et al. 1996), outward buckling of Al has also been observed but to an even larger degree.

In LEED, a number of rotational domains are observed within these crystalline overlayers (Shen et al. 1998a). On a fivefold surface of *i*-Al-Pd-Mn and *i*-Al-Cu-Fe, at first glance, the surface appears to have 10-fold symmetry at certain energies, but truly consists of five (110) domains oriented 72° apart. On a twofold surface of *i*-Al-Pd-Mn, two (110) domains are observed (Shen et al. 1997a). On the threefold surface of *i*-Al-Pd-Mn, one (111) domain is observed. The domain sizes are estimated to be 35–60 Å based on LEED results. This is in contrast to a surface which has been annealed higher; the domain size observed in this case is greater than 850 Å (see paragraphs that follow) (Shen et al. 1998c). We have developed a model to explain the crystalline orientation with respect to the bulk (Shen et al. 1998a). This model explains not only the surface orientation with respect to the bulk orientation, but also the domains present in the CsCl overlayer. In this model, the structural relationship between cubic closed packing (ccp) and *i* packing (ip) plays a dominant role. The relationship can be visualized best by comparing clusters of spheres packed with appropriate symmetry, as shown in Fig. 10.9. Both packings involve a sphere surrounded by 12 nearest neighbors. The ccp cluster can be transformed to the ip cluster, essentially, by altering the middle layer. To achieve ip, the middle six spheres in ccp are rotated by 30°, with three rotated up by roughly 20% and the others down by 20%. Based on this, the relationship between the various symmetry axes become apparent. The threefold axes of the *i* and the cubic closed packing are identical, thus explaining the observation of the (111) oriented crystalline overlayer on the threefold *i*-Al-Pd-Mn sample. [110] axes are parallel to the twofold axes of the

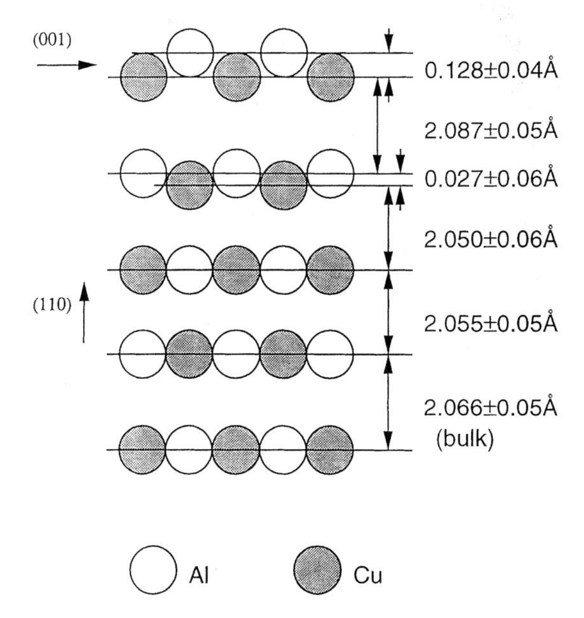

Fig. 10.8. Schematic top view of the β-Al(Cu)$_{1-x}$Fe$_x$(110) surface, showing the top two layers. For clarity, the atom sizes are reduced from touching-sphere radii. Reprinted from Surf. Sci., Shi et al. (1998), The Surface Structure of a β-Al(Cu)$_{1-x}$Fe$_x$-(110) Film Formed on an AlCuFe Quasicrystal Substrate, Analyzed by Dynamical LEED, in press. © 1998, with permission from Elsevier Science.

i packing. Other [110] axes are approximately parallel to the fivefold axes. In summary, maximum alignment between the high symmetry axes of the substrate and the crystalline overlayers is sought in these systems. An alternative transformation is possible which also preserves some high-symmetry axes, but this transformation is not consistent with the data for surfaces of *i*-Al-Pd-Mn and *i*-Al-Cu-Fe (Shen et al. 1998a).

The observation of such a transition has also been observed in a number of electron-microscopy and crystal-growth studies. Ion bombardment of *i*-Al-Cu-Fe samples and subsequent examination with electron microscopy reveals CsCl-type structures (Wang et al. 1993, Zhang et al. 1993, Wang et al. 1995, Yang et al. 1996). During crystal growth, CsCl structures have been observed together with *i* structures (Dong et al. 1987a, 1987b, 1992). Thus, their formation under the conditions presented above are not entirely surprising.

Upon heating to higher temperatures, a different structure forms. Upon cooling, the crystalline structure mentioned above does not reappear. Thus, this new structure is thermodynamically stable relative to the crystalline surface layer which forms upon annealing below about 700 K. Within the ex-

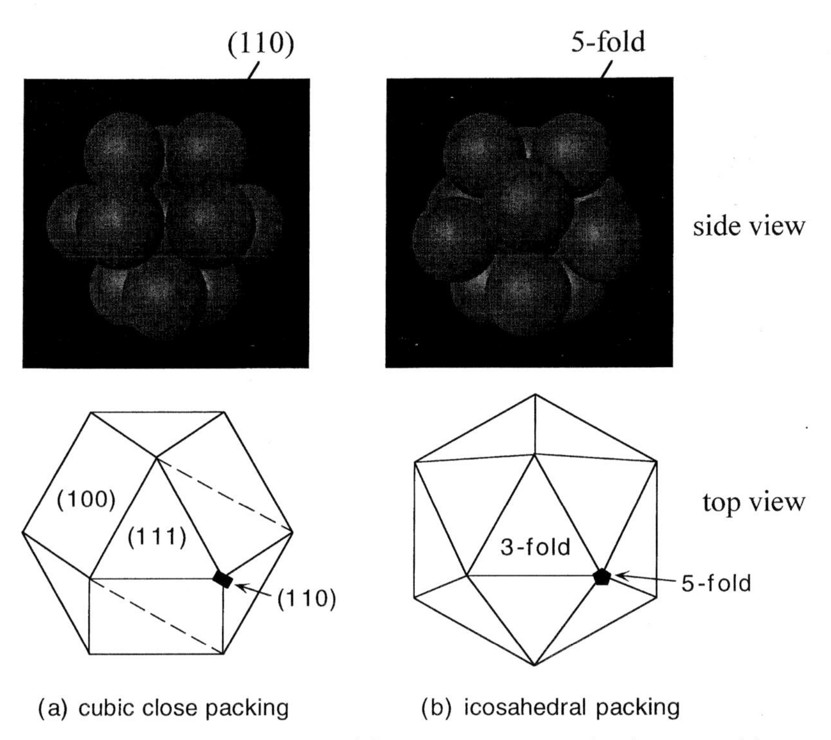

Fig. 10.9. Structure models of (a) cubic close packed (ccp) cluster; (b) icosahedral packed (ip) cluster. Top row is a side view, and bottom row is a top view. After Shen et al. (1998a). © 1998 the American Physical Society.

perimental limitations of all the techniques used to study this new structure to date, it appears that the surface formed at higher annealing temperatures (greater than about 800 K) has a bulk-like quasicrystalline structure. Furthermore, it is distinguished from the lower-temperature cubic phase by a particular distinctive lineshape in XPS, which may be attributable to the electronic structure (pseudogap) of the QC (Shen et al. 1997a, Jenks et al. 1996b). However, the exact structure of this surface is the subject of ongoing debates, as will be discussed. Furthermore, the structure differs greatly from the fracture surface, leading to discussions of the true nature of the quasicrystalline surface.

All quasicrystalline materials prepared by sputter-annealing exhibit the terrace-step structure which is familiar for crystalline metals. This is based upon data from techniques capable of nm-scale resolution, namely STM or atomic force microscopy. Thus far, these techniques have been used to study several alloys, namely, Al-Co-Cu (Kortan et al. 1990, Becker and Kortan 1991, Becker et al. 1991, Kortan et al. 1992), Al-Cu-Fe (Becker et al. 1991), and Al-Pd-Mn (Schaub et al. 1994a, 1994b, 1995, 1996, Chevrier et al. 1997, Raberg 1998, Shen et al. 1998c). The terraces on all these surfaces tend to be large (greater than 75 Å and commonly several hundred Å). This result is corroborated by LEED data. All three high-symmetry surfaces of sputter-annealed *i*-Al-Pd-Mn, for instance, yield narrow diffraction spots, consistently showing an average terrace size greater than 150 Å (the apparent instrumental limit of the low-resolution optics) (Shen et al. 1997a, Jenks et al. 1998d, Shen et al. 1998a, 1998b). A single high-resolution LEED measurement on fivefold Al-Pd-Mn revealed an average terrace width of about 900 Å (Shen et al. 1998b) (note that the LEED experiments average over an area of about 1–2 mm^2). An XRD measurement on this same type of surface yielded an average terrace dimension of 380 Å (Capitan et al. 1998). One would expect the terrace width to depend upon the sample history, but in most cases the widths are as large or larger than those expected for a typical crystalline metal with similar sample history.

The features on the terraces revealed by STM are quite intriguing. For the case of *d*-Al-Co-Cu, the surface structure does not undergo reconstruction to a crystalline or twinned termination (Kortan et al. 1990, Becker and Kortan 1991, Becker et al. 1991, Kortan et al. 1992). In fact it appears to have a bulk-like termination. The corrugation on this surface is 0.1 Å, also suggesting that the surface is not strongly reconstructed. Structural arrangements consistent with tenfold symmetry are evident in the normalized power spectrum of the surface, as shown in Fig. 10.10. Surface features form lines which are 72° apart. The lines are continuous across terraces, suggesting a strong structural relationship between terraces. The single-valued step height of 1.94 Å is also noteworthy. Given that the bulk *d* structure is periodic in the direction perpendicular to the surface, regularity in the step height is expected. However, high resolution XRD of the same alloy reveals a repeat unit of 8.26 Å. The approximate step height is about $\frac{1}{4}$ of the periodicity of the bulk axis, and is explained as representing a cut along one of the stacking planes that form the periodic axis.

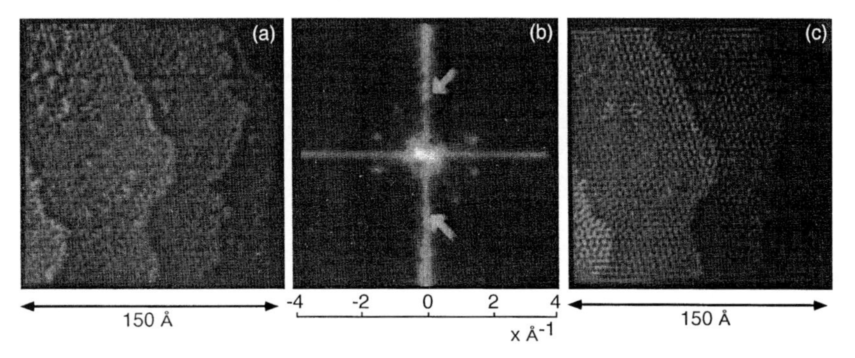

Fig. 10.10. A scanning tunneling micrograph of the tenfold surface of *d*-Al-Co-Cu (Kortan et al. 1990). (a) The lateral scale is indicated and the gray scale is derived from the local height. (b) Normalized power spectrum of the image. (c) Enhanced image of surface. © 1990 the American Physical Society.

The structural features of the terraces of two i materials have also been examined by STM. For Al-Cu-Fe, limited results show 12 Å wide tenfold features on the surface which are distinct from those on d-Al-Co-Cu (Becker et al. 1991). Also, the corrugation is only 0.25 Å across a 125 Å terrace. Unfortunately, step heights are not reported. A more extensive database exists for fivefold surfaces of i-Al-Pd-Mn annealed at high temperatures. Schaub et al. demonstrate that the step heights and sequence of step heights are consistent with known aspects of the bulk structure (Schaub et al. 1994a, 1994b, 1995, 1996). Two types of steps are observed, denoted high (H) and low (L), with heights of 6.78 Å and 4.22 Å, respectively. The ratio of these two heights is 1.61, equal within experimental error to the golden mean $\tau = (1 + \sqrt{5})/2$. Micrographs from these studies are shown in Fig. 10.11. The step heights are consistent with the distances between dense atomic planes in the bulk structure model of Boudard et al. (1992). Furthermore, the sequence of these heights in a given image forms a portion of a Fibonacci chain, namely HHLHHLHLHH. The terrace corrugation is 0.5 Å, and the terraces contain fivefold features, as shown in Fig. 10.11b. Autocorrelation functions of STM images show long-range spatial correlation. The implication of all these results is that the quasicrystallinity of the bulk extends up to the surface.

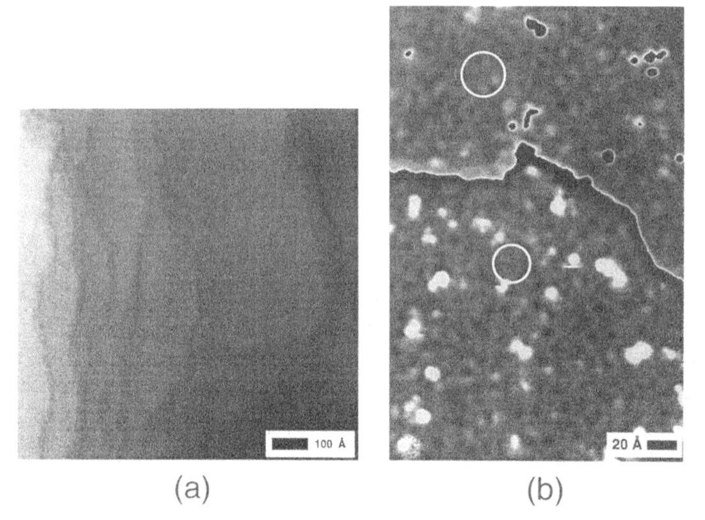

Fig. 10.11. Scanning tunneling micrographs at two different magnifications of a fivefold surface of i-Al-Pd-Mn (Schaub et al. 1995). (a) The terrace-step-kink morphology. (b) The fine structure present on the terraces, with fivefold features highlighted by circles. A 4 Å step cuts across the upper middle of (b). © 1995 Springer-Verlag.

Our own recent results are largely consistent with those of Schaub et al., even though they were obtained under very different tunneling conditions (nA vs pA). Interestingly, we demonstrated that the structural features observed at high resolutions on the terraces are not atomic in origin (Shen et al. 1998c). Rather, the scale of the features is comparable to the separation between intact pseudo-Mackay icosahedra. It is well known that metallic clusters are prevalent in the *i* alloys. There is good evidence this particular type, the pseudo-Mackay icosahedron (PMI), exists in *i*-Al-Pd-Mn, although this conclusion is not uncontested (Elser 1998). Photoelectron diffraction also gives evidence of PMI's in the surface and subsurface region of this alloy (Naumović et al. 1997, 1998). The PMI contains 51 atoms and is approximately 10 Å in diameter.

Janot and de Boissieu (1994) have postulated that these PMI's are, in a sense, the "atoms" of the three-dimensional structure. Electrons are confined in the deep potential well of these clusters, but are capable of tunneling through the potential barrier. Furthermore, these clusters consist of "magic numbers" of electrons which give rise to successive clustering and overlapping of the PMI's to form a self-similar hierarchy of aggregates which maintains the overall density and keeps the average number of electrons equal to this "magic number". As the number of inflations increases, the ability of the electrons to tunnel across the barrier increases. The end result is a structure which is quite robust and a model that can explain certain peculiar aspects of the thermal and electrical conductivity of QCs (Janot and de Boissieu 1994, Janot 1996, Janot and Dubois 1998). According to the bulk structure model of Boudard et al. (1992), any planar termination of the bulk must necessarily cut through some of these clusters (Dubois 1998). This contradicts the notion that the clusters are inviolate, which has prompted a suggestion that the surface planar terminations produced by sputter-annealing cannot be bulk like in their structure (Dubois 1998).

These ideas are illustrated in Fig. 10.12, which shows a twofold projection of the bulk structure model with the arrow indicating a fivefold zone axis and the solid lines indicating possible planar terminations of the surface. Intact PMI's are indicated by circles. The solid lines are, in fact, the terminations suggested by Gierer et al. (1997, 1998) from the LEED surface structural analysis discussed below. It can be seen that these favored planes are tangent to many intact PMI's, while other possible planes (dashed lines, for example) would not preserve any PMI's in the cut. This may be a factor which stabilizes the terminations symbolized by the solid lines, relative to others which nature could select. The tops of the clusters touch the top of the plane in the terminations indicated by solid lines. Dubois (1998) proposes that the fivefold surface is actually a pentagonal phase, related closely in structure to the QC but periodic along a twofold axis (it would be aperiodic in the other two dimensions). It should be noted that no such phase is known in the bulk phase diagram of *i*-Al-Pd-Mn, but Dubois (1998) argues that since two

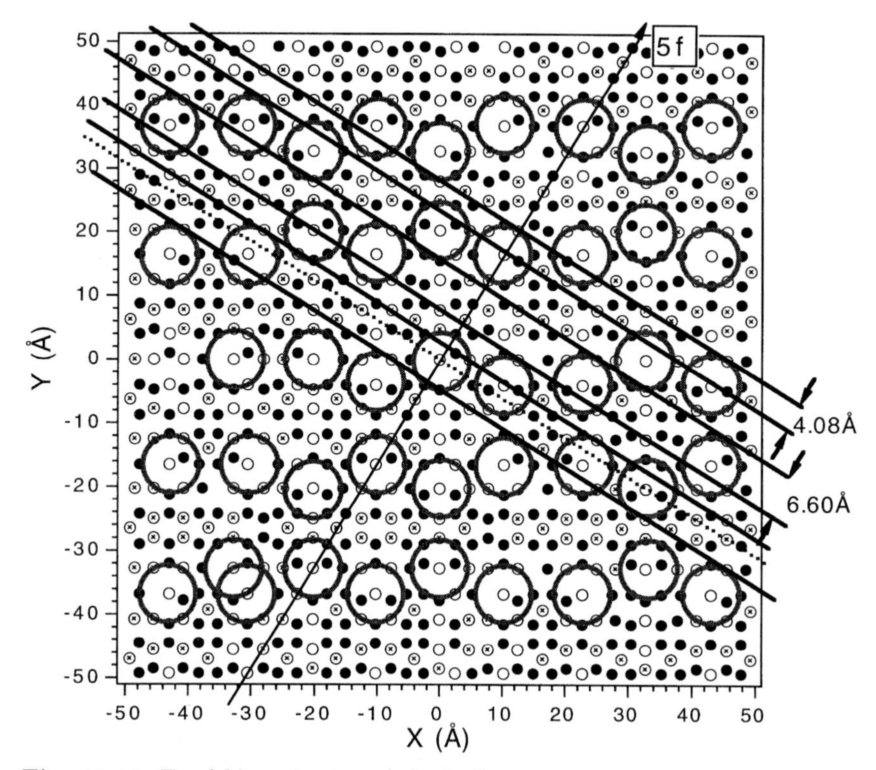

Fig. 10.12. Twofold projection of the bulk structure model of *i*-Al-Pd-Mn. The arrow shows a fivefold zone axis and the solid and dashed lines show possible planar terminations of the fivefold surface (Goldman and Thiel 1998). Circles are the cross-sections of complete PMI's cut by this particular projection. Solid lines preserve some PMI's, dashed lines preserve none.

have been discovered in *i*-Al-Cu-Fe (Menguy et al. 1993, Quiquandon et al. 1996), they most likely exist in *i*-Al-Pd-Mn as well. In any case, it is possible that surface phases can differ from those which exist in the bulk; surface alloys of crystalline metals are certainly known which have no bulk counterpart (Stevens and Hwang 1995). In the language of surface science, such a phase could be regarded as a subtle surface relaxation (defined as perpendicular deviations of atomic planes from bulk positions), although also possibly involving some reconstruction (defined as lateral, or in-plane, deviations of atoms from bulk-like positions) (Thiel and Estrup 1995). The proposal of the pentagonal phase is also based partly on the surface deviation from the bulk stoichiometry (Mn depletion, Sect. 10.4.2). This interpretation highlights the need to compare QCs with closely-related phases. It is unfortunate that, like the quasicrystalline materials, crystals of related compounds can be quite difficult to grow; extensive efforts are underway by a number of groups to alleviate this problem.

LEED studies have also been done on quasicrystalline surfaces, including *i*-Al-Cu-Fe and *i*-Al-Pd-Mn. The results on both surfaces are consistent with what one would expect for an unreconstructed surface, based on bulk x-ray scattering data and given the resolution of the technique. It should be noted though that the results would also be consistent with an approximant with a large unit cell. LEED results for *i*-Al-Pd-Mn are shown in Fig. 10.13. SEI data also provide evidence for *i* symmetry at the surface and within the near-surface region (Erbudak et al. 1994, Zurkirch et al. 1997). A full dynamical structure analysis of LEED intensity-voltage (IV) data for fivefold *i*-Al-Pd-Mn (Gierer et al. 1997, 1998) has also been performed. In order to carry out such an analysis, a new type of approximation had to be developed because of a lack of periodicity. This approximation is the so-called "average neighborhood approximation," in which the scattering properties of all atoms in a plane are assumed to be equal, and the average neighborhood of an average atom in a given plane is described by a kind of radial distribution function. Using the bulk structure model of Boudard et al.(1992) as a starting point, this study showed that a mix of several closely similar, relaxed, bulk-like lattice terminations is favored, all of which have a dense Al-rich layer on top followed by a layer with a composition of about 50% Al and 50% Pd. The interlayer spacing between these two topmost planes is contracted from the bulk value by 0.1 Å, to a final value of 0.38 Å, and the lateral density of the two topmost layers taken together is similar to that of an Al(111) surface. The vertical separation between similar terminations must correspond physically to the step height (separation between terraces), and indeed this was found true upon comparison with the STM data of Schaub et al. (1994a, 1994b, 1995). LEIS of the surface was consistent with the composition of the outermost layer predicted by LEED (except for Mn depletion) (Gierer et al. 1998). Some of the LEED results were also corroborated by a subsequent XRD study (Capitan et al. 1998). The LEED and XRD results thus indicated that, of the many possible planes which could terminate the bulk structure, nature selects those which are dense and Al-rich. These factors, density and high Al content, should both maximize stability, based upon principles known to govern stabilities of surfaces of crystalline materials.

Turning now to fracture surfaces, Ebert et al. (1996, 1998) cleaved single grains along either the fivefold or twofold axes within UHV. The resulting cleaved surfaces are within $\pm 5°$ of the expected plane. The results are fundamentally different from those for the sputter-annealed surfaces. The cleaved surfaces are comparatively rough with an average corrugation of about 10–15 Å (versus < 1 Å for the sputter-annealed surfaces). Furthermore, the surface is made up of clusters with the smallest cluster 8–10 Å in diameter. This cluster size is consistent with the diameter of a PMI (9.6 Å) and the roughness is consistent with the expected roughness if the PMI's remain intact (Janot 1997, Janot and Dubois 1998). This roughness is supported qualitatively by molecular dynamics simulations of crack propagation (Mikulla et

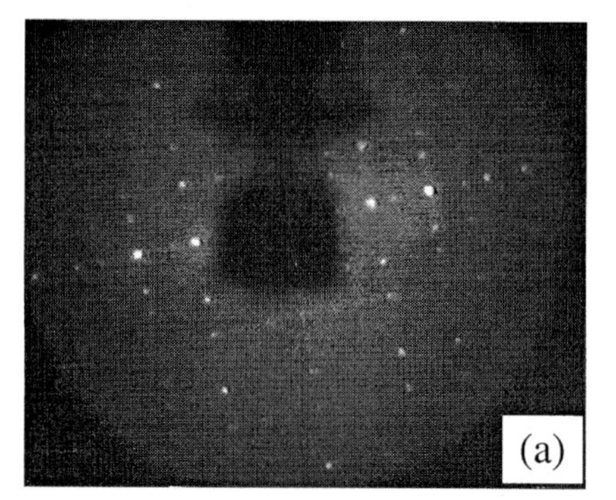

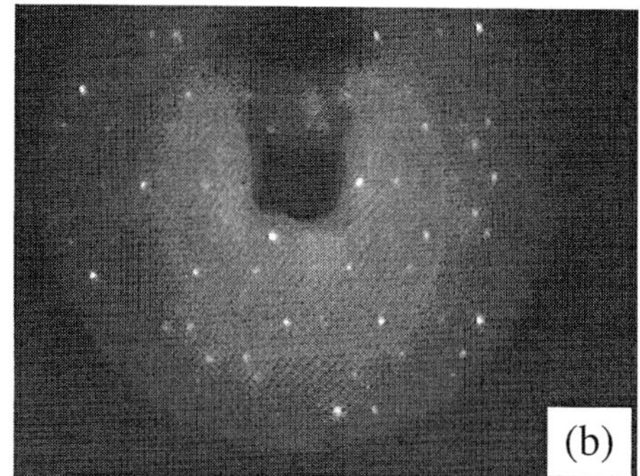

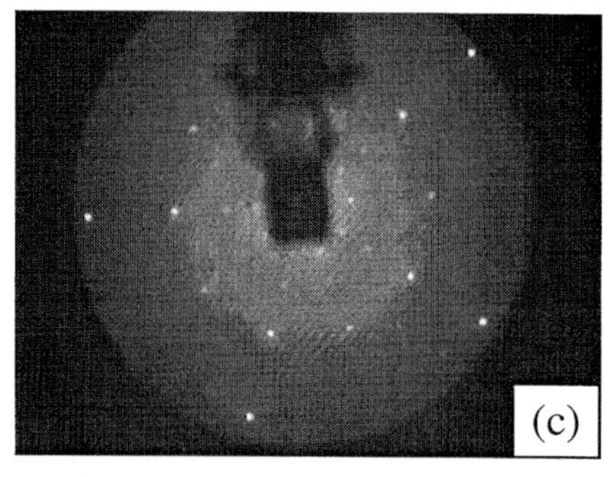

Fig. 10.13. LEED patterns of three high-symmetry surfaces of *i*-Al-Pd-Mn (Goldman and Thiel 1998). (a) Twofold surface at 50 eV, after annealing at 900 K. (b) Threefold surface at 50 eV, after annealing at 700 K. (c) Fivefold surface at 50 eV, after annealing at 750 K.

al. 1997). It is notable, however, that cleaved crystalline metals are rough on the length scale probed by optical and electron microscopies (Marder 1997). To our knowledge, no higher-resolution studies have been conducted on cleaved metal crystals. STM has revealed ball-like structures for metallic glasses (Schaub et al. 1996).

Interestingly, the clusters on the fracture surfaces of *i*-Al-Pd-Mn seem to form the self-similar hierarchical structure first described by Janot and de Boissieu (1994). If one examines relatively large scale micrographs the surface appears to be made up of clusters of various sizes and when the surface is examined with higher resolution these clusters appear to be made up of smaller clusters, the smallest being the 8–10 Å clusters mentioned above. This is illustrated in Fig. 10.14.

These results generate several questions. What is the true nature of the quasicrystalline surface? Does one method of surface preparation lead to a thermodynamically stable structure? Is the sputter-annealed surface aperiodic in three or in two dimensions? Which surface is relevant to applications

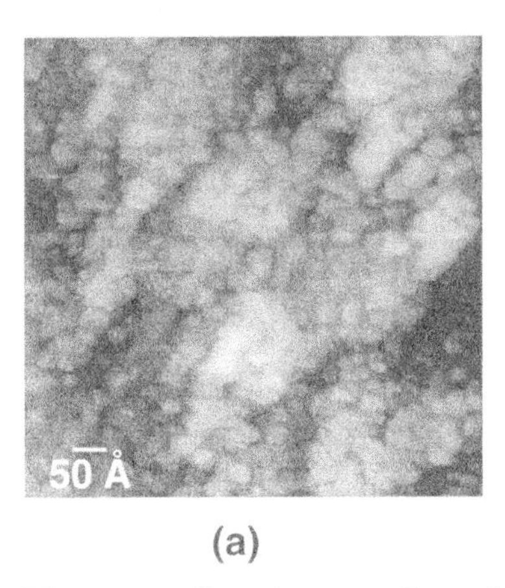

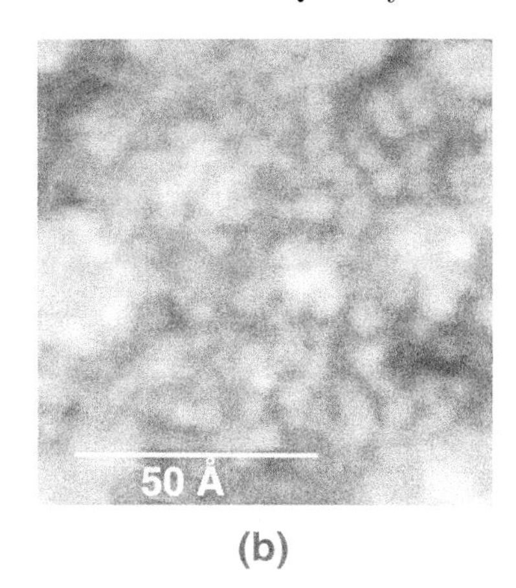

Fig. 10.14. Scanning tunneling micrographs showing an *i*-Al-Pd-Mn surface at two different magnifications. The surface was prepared by cleaving along a fivefold plane at room temperature (Ebert et al. 1996). © 1996 the American Physical Society.

of these materials? All these questions are the subject of current investigation within the scientific community.

One approach to help address the issue of thermodynamic stability is to test the effect of heating on fracture surfaces. Ebert et al. (1998) have heated cleaved surfaces to 940 K and find variable results. In particular, portions of the samples become matté whereas other portions maintain high optical reflectivity. The matté portions are found to consist of faceted holes. The faceted holes appear to arise from evaporation. Interestingly, terraces up to 700 Å large are in this area, with less than 1 Å of corrugation. The other portions of the surface still show a cluster-subcluster structure, as discussed previously, with the smallest clusters being 3–6 Å in diameter rather than the 8–10 Å without annealing. The corrugation of the surface is still $>$ 10 Å. It is not clear what causes the differences; non-detectable compositional variations are proposed by the authors.

While many studies of clean surfaces have been conducted in the last several years, open questions remain and new ones have been asked. There is much yet to be learned.

10.4.4 Surface Chemistry

Another fundamental surface property is chemical reactivity. Studies under various environmental conditions suggest that QCs are relatively unreactive and/or corrosion-resistant (Massiani et al. 1993, Dubois et al. 1994b, Massiani et al. 1995). As mentioned in Sect. 10.3, it has been proposed (Rivier 1993, 1997) that the surface chemistry of QCs should be dominated by the suppression of the density of states at the Fermi level, the "pseudogap", which is characteristic of bulk QCs.

The key question here is whether the pseudogap persists at the clean surface. There is, for example, evidence from high resolution photoemission measurements (Stadnik et al. 1996, Neuhold et al. 1998) that quasicrystalline surfaces are more metallic than the bulk. That is, the electronic density of states at Fermi level in the surface region is enhanced relative to that of the bulk. This may result, in part, from the reduced symmetry introduced by the discontinuity of the surface itself, as is true for periodic crystals with dangling bonds at the surface. Some theoretical results, however, suggest that the pseudogap can extend to the surface as a result of localization effects (Janssen and Fasolino 1998). Further, Rivier (1997) has argued that the pseudogap in QCs arises from the inflation symmetry of these structures, an idea that is also implicit to the hierarchical cluster model of Janot and de Boissieu (1994). Cut or cleaved surfaces that preserve these clusters should then preserve the pseudogap at the surface. A metallic-like surface could result from disruption of these clusters at the surface. In this context, the nature of the pseudogap at the QC surface may depend upon surface structure and preparation history. In any case, this is an issue which is far from resolved.

We recently studied the reactivity of fivefold *i*-Al-Pd-Mn, prepared by sputter-annealing, towards a number of simple molecules, namely H_2, CO, CH_3OH, and two iodoalkanes (Jenks et al. 1998b, Jenks and Thiel 1998a). Temperature programmed desorption (TPD) was employed. This technique probes molecule-surface bond energies, desorption product distributions, and surface kinetics. We find that the reactivity is dominated by the presence of Al at the surface. That is, the reactivity is comparable to that of pure Al. For example, H_2 is found to not dissociate on *i*-Al-Pd-Mn. The same is true for pure Al. As discussed previously, LEED-IV and LEIS studies suggest that the topmost surface layers are about 85 at.% Al so such a result might be expected. Presumably, any Pd or Mn atoms do not form large enough ensembles to significantly perturb the chemistry of the Al. Pd in particular is known to be fairly aggressive towards bond cleavage. Our result is perhaps also not too surprising given that the chosen molecules form surface bonds which are primarily covalent in nature. As discussed in Sect. 10.3, the effect of a surface pseudogap is expected to be less pronounced for covalently-bonding adsorbates than for highly polar or ionic adsorbates. From these studies we conclude that the clean surface is not particularly “inert”. Of course, different types of adsorbates or other quasicrystalline materials may behave differently.

10.5 Friction

Hard, wear resistant materials, in particular metal-based materials, with low coefficients of friction, have potential for a variety of applications. It is important to note that friction is not strictly a materials property, but rather a number which characterizes an entire set of measurement parameters (Rigney and Hammerberg 1998). Studies have shown that indeed quasicrystalline materials exhibit low coefficients of friction, are wear resistant, and are hard. In this section, we discuss friction studies of various Al-based quasicrystalline materials.

Friction is defined as the resistance of two bodies in contact to tangential motion (Quinn 1991). The coefficient of friction is the ratio of the tangential (shear) force to the normal force. Two methodologies for determining friction coefficients of QCs have been used. One involves rubbing a pin or ball slider made of various materials against a quasicrystalline surface under atmospheric conditions, and the other involves examining the tribology of QCs under UHV conditions where truly clean surfaces of single grains can be prepared, as discussed earlier. These clean surfaces can subsequently be exposed to common atmospheric contaminants, such as oxygen and water.

In slider experiments under atmospheric conditions, coatings and bulk materials comprised of Al-Cu-Fe, Al-Cu-Fe-Cr, and Al-Cu-Fe-Cr-Si QCs show an average friction coefficient of 0.15 with a steel indentor versus 0.11–0.18 for diamond indentor, depending on its size (Dubois et al. 1993, Kang et al. 1993, Dubois et al. 1994a). In these studies, a constant normal load of 20 N is used. In comparison, diamond on diamond is known to have a friction coefficient of 0.1 (Weast 1986-87). More typical for metals, such as low-carbon steel, is about 0.3–0.4 (Kang et al. 1993). Both single pass and multipass studies have been performed. The effect of multiple passes is to increase the friction coefficient. AES studies of Al-Cu-Fe-B QC samples, after sliding friction studies, show that Al and iron oxides transfer to the WC slider used (Singer et al. 1998), suggesting that the low friction of these materials relative to most metals is a result of the low energy necessary to remove the oxide film.

However, recent work by Ko et al. (1998) suggests that the low friction coefficients may be, at least in large part, intrinsic to the QC structure or its mechanical properties. Here, a UHV tribometer (Gellman 1992) was used to measure the friction coefficient of two single grain Al-Pd-Mn QCs sliding against one another. A load of 10–50 mN was employed. One grain was polished flat while the other was rounded to avoid edge effects. Results show an average static friction coefficient of 0.6 for clean surfaces undergoing a single pass slide across one another where static friction is defined as the force necessary to initiate motion of two objects in contact. This contrasts clean metal surface, such as Cu, Fe and Ni, which exhibit stick-slip behavior and friction coefficients > 2. These data are illustrated in Fig. 10.15. Since the sputtered-annealed surface of an Al-Pd-Mn QC as prepared in these studies is predominately Al, as shown by low energy electron diffraction intensity versus voltage (LEED-IV) calculations and ion surface scattering (Gierer et al. 1997, 1998), a comparison was also made to clean Al. In comparison to clean Al sliding against clean Al, the friction coefficient is much lower for the

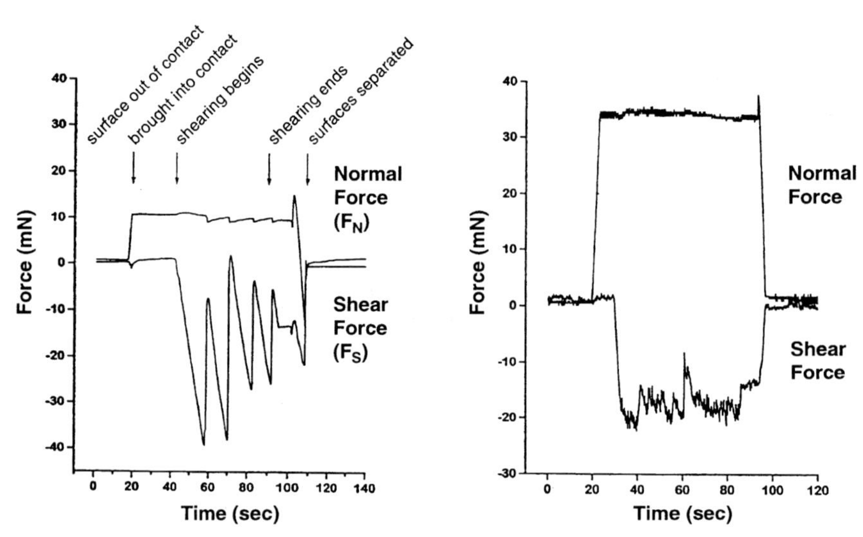

Fig. 10.15. Measurements of static friction coefficients between clean, self-similar surfaces in UHV. Left panel: Clean Cu(111) on Cu(111). The static friction coefficient, defined as the ratio of the shear force to the normal force (F_S/F_N), is about 4. Stick-slip behavior is evident in the large variation of F_S between 40 and 100 s. Reprinted with permission from McFadden and Gellman (1995). © 1995 the American Chemical Society. Right panel: Clean fivefold *i*-Al-Pd-Mn sliding on same. The static friction coefficient is 0.60. The shear speed is 20 μm/s and T = 300 K (Ko et al. 1998).

QCs in contact (1.85 vs 0.6). Stick-slip behavior did not occur in either set of studies.

Upon oxidation (using water or O_2) in vacuum, the static friction coefficient for Al-Pd-Mn decreases slightly, to 0.35. As discussed in Sect. 10.2, oxidation of *i*-Al-Pd-Mn yields a surface comprised almost entirely of oxidized Al. Similarly, oxidation of pure Al in vacuum does not greatly alter its friction coefficient. Furthermore, sulfidation or surface crystallization do not affect the friction coefficient of the QCs. These results suggest that surface modifications to the surface chemistry are not important, at least under the conditions of these measurements.

Another possible contribution to friction is the mechanical response of the bulk material. Interestingly, the value for the oxidized QCs in contact is actually close to that found in pin-on-disk friction measurements of Al_2O_3 versus Al_2O_3 (Blau 1991). This may be due to the relative hardness of the Al-Pd-Mn samples and Al_2O_3 versus Al. Because Al is relatively soft, the bulk may deform plastically under conditions where the other two materials do not. Further work is need to distinguish the contributions of mechanical, chemical, and structural properties in the interesting friction results.

10.6 Concluding Remarks

The surface science of QCs is a fascinating and fertile area. The intense work of the past several years has settled some questions but also generated many new ones. Future progress will rest upon careful, systematic comparisons, in addition to the few which have been made already. For instance, there is a need to establish or broaden our understanding of the relationships between properties of oxidized and clean surfaces; sputter-annealed and fractured surfaces; crystalline and quasicrystalline surfaces; quasicrystalline surfaces of different symmetry; and QCs of different chemical compositions. In this light, the expansion of surface research into quasicrystalline alloys which are *not* Al-rich is a particularly interesting frontier, and one that is currently untouched. We hope that the present article is useful to those who are working in, or curious about, this rich field.

Acknowledgments

The authors wish to acknowledge useful discussions with, and ssistance from, T. Lograsso, M. Kramer, D. Sordelet, I. Fisher, P. Canfield, P. Ebert, K. Urban, D. Rouxel, D. Naumović, T. Schaub, J. Chevrier, and J. M. Dubois. This work was supported by the Ames Laboratory, which is operated for the U.S. Department of Energy by Iowa State University under Contract No. W-7405-Eng-82.

References

Adamson, A.W. (1976): Physical Chemistry of Surfaces. John Wiley and Sons, New York

Ahmed, H.A., Smeltzer, W.W. (1986): J. Electrochem. Soc.: Solid-State Sci. Technol. **133**, 212

Becker, R.S., Kortan, A.R. (1991): in Quasicrystals, The State of the Art, DiVincenzo, D.P., Steinhardt, P.J. (eds). World Scientific, Singapore, p 111

Becker, R.S., Kortan, A.R., Thiel, F.A., Chen, H.S. (1991): J. Vac. Sci. Technol. B **9**, 867

Belin-Ferré, E. (1998): in An Introduction to Structure, Physical Properties and Application of Quasicrystalline Alloys, Suck, J.-B., Schreiber, M., Häussler, P. (eds). Springer-Verlag, Berlin, in press

Belin, E., Mayou, D. (1993): Phys. Scr. T **49**, 356

Belin, E., Miyoshi, Y., Yamada, Y., Ishikawa, T., Matsuda, T., Mizutani, U. (1994): Mater. Sci. Eng. A **181-182**, 730

Belin-Ferré, E., Fournée, V., Dubois, J.-M. (1997): in New Horizons in Quasicrystals, Goldman, A.I., Sordelet, D.J., Thiel, P.A., Dubois, J.M. (eds). World Scientific, Singapore, p 9

Blau, P.J. (1991): Wear **151**, 193

Bloomer, T.E., Jenks, C.J., Kramer, M.J., Lograsso, T.A., Delaney, D.W., Sordelet, D.J., Besser, M.F., Thiel, P.A. (1998): J. Am. Chem. Soc., in press
Blum, V., Rath, C., Castro, G.R., Kottcke, M., Hammer, L., Heinz, K. (1996): Surf. Rev. Lett. **3**, 1409
Bolliger, B., Erbudak, M., Vvedensky, D.D., Zurkirch, M., Kortan, A.R. (1998): Phys. Rev. Lett. **80**, 5369
Bose, A. (1993): in Wettability, Berg, J.C. (ed). M. Dekker, New York, p 149
Boudard, M., de Boissieu, M., Janot, C., Heger, G., Beeli, C., Nissen, H.-U., Vincent, H., Ibberson, R., Audier, M., Dubois, J.M. (1992): J. Phys. Condens. Matter **4**, 10 149
Capitan, M.J., Alvarez, J., Joulaud, J.L., Calvayrac, Y. (1998): unpublished
Chang, S.-L., Chin, W.B., Zhang, C.-M., Jenks, C.J., Thiel, P.A. (1995a): Surf. Sci. **337**, 135
Chang, S.-L., Zhang, C.-M., Jenks, C.J., Anderegg, J.W., Thiel, P.A. (1995b): in Proceedings of the 5th International Conference on Quasicrystals, Janot, C., Mosseri, R. (eds). World Scientific, Singapore, p 786
Chang, S.-L., Anderegg, J.W., Thiel, P.A. (1996): J. Non-Cryst. Solids **195**, 95
Chevrier, J., Cappello, G., Comin, F., Palmari, J.P. (1997): in New Horizons in Quasicrystals, Goldman, A.I., Sordelet, D.J., Thiel, P.A., Dubois, J.M. (eds). World Scientific, Singapore, p 144
Chevrier, J., Cappello, G., Schmithüsen, F., Déchelette, A., Comin, F., Stierle, A. (1998): Abstracts of the GDR-SPQK Joint Colloquium at Strasbourg
Davis, H.L., Noonan, J.R. (1988a): Phys. Rev. Lett. **54**, 566
Davis, H.L., Noonan, J.R. (1988b): in The Structure of Surfaces II, van der Veen, J.F., Van Hove, M.A. (eds). Springer-Verlag, Berlin, p 152
Dong, C., Chattopadhyay, K., Kuo, K.H. (1987a): Scr. Metall. **21**, 1307
Dong, C., Kuo, K.H., Chattopadhyay, K. (1987b): Mater. Sci. Forum **22-24**, 555
Dong, C., Dubois, J.M., Kang, S.S., Audier, M. (1992): Philos. Mag. B **65**, 107
Dubois, J.M. (1997): in New Horizons in Quasicrystals, Goldman, A.I., Sordelet, D.J., Thiel, P.A., Dubois, J.M. (eds). World Scientific, Singapore, p 208
Dubois, J.M. (1998): in An Introduction to Structure, Physical Properties and Application of Quasicrystalline Alloys, Suck, J.-B., Schreiber, M., Häussler, P. (eds). Springer-Verlag, Berlin, in press
Dubois, J.-M., Kang, S.S., Massiani, Y. (1993): J. Non-Cryst. Solids **153-154**, 443
Dubois, J.-M., Kang, S.S., Perrot, A. (1994a): Mater. Sci. Eng. A **179-180**, 122
Dubois, J.M., Proner, A., Bucaille, B., Cathonnet, P., Dong, C., Richard, V., Pianelli, A., Massiani, Y., Ait-Yaazza, S., Belin-Ferre, E. (1994b): Ann. Chim. Fr. **19**, 3
Dubois, J.M., Plaindoux, P., Belin-Ferre, E., Tamura, N., Sordelet, D.J. (1998): in Proceedings of the 6th International Conference on Quasicrystals, Takeuchi, S., Fujiwara, T. (eds). World Scientific, Singapore, p 733
Ebert, P., Feuerbacher, M., Tamura, N., Wollgarten, M., Urban, K. (1996): Phys. Rev. Lett. **77**, 3827
Ebert, P., Yue, F., Urban, K. (1998): Phys. Rev. B **57**, 2821
Elser, V. (1998): in Proceediings of the 6th International Conference on Quasicrystals, Takeuchi, S., Fujiwara, T. (eds). World Scientific, Singapore, p 19
Erbudak, M., Nissen, H.-U., Wetli, E., Hochstrasser, M., Ritsch, S. (1994): Phys. Rev. Lett. **72**, 3037
Gavatz, M., Rouxel, D., Claudel, D., Pigeat, P., Weber, B., Dubois, J.M. (1998): in Proceedings of the 6th International Conference on Quasicrystals, Takeuchi, S., Fujiwara, T. (eds). World Scientific, Singapore, p 765
Gellman, A.J. (1992): J. Vac. Sci. Technol. A **10**, 180

Gierer, M., Van Hove, M.A., Goldman, A.I., Shen, Z., Chang, S.-L., Jenks, C.J., Zhang, C.-M., Thiel, P.A. (1997): Phys. Rev. Lett. **78**, 467
Gierer, M., Van Hove, M.A., Goldman, A.I., Shen, Z., Chang, S.-L., Pinhero, P.J., Jenks, C.J., Anderegg, J.W., Zhang, C.-M., Thiel, P.A. (1998): Phys. Rev. B **57**, 7628
Goldman, A.I., Thiel, P.A. (1998): in Quasicrystals, The State of the Art, DiVincenzo, D.P., Steinhardt, P.J. (eds). World Scientific, Singapore, in press
Gu, T., Goldman, A.I., Pinhero, P., Delaney, D. (1997): in New Horizons in Quasicrystals, Goldman, A.I., Sordelet, D.J., Thiel, P.A., Dubois, J.M. (eds). World Scientific, Singapore, p 165
Ho, T.-L. (1991): in Quasicrystals, The State of the Art, DiVincenzo, D.P., Steinhardt, P.J. (eds). World Scientific, Singapore, p 403
Hoffman, A., Maniv, T., Folman, M. (1988): Surf. Sci. **193**, 57
Janot, C. (1996): Phys. Rev. B **53**, 181
Janot, C. (1997): J. Phys. Condens. Matter **9**, 1493
Janot, C., de Boissieu, M. (1994): Phys. Rev. Lett. **72**, 1674
Janot, C., Dubois, J.-M. (1998): in An Introduction to Structure, Physical Properties and Application, Suck, J.-B., Schreiber, M., Häussler, P. (eds). Springer-Verlag, Berlin, in press
Janssen, T., Fasolino, A. (1998): in Proceedings of the 6th International Conference on Quasicrystals, Takeuchi, S., Fujiwara, T. (eds). World Scientific, Singapore, p 757
Jenks, C.J., Thiel, P.A. (1997): MRS Bull. **22**, 55
Jenks, C.J., Thiel, P.A. (1998a): J. Molec. Catal. A **131**, 301
Jenks, C.J., Thiel, P.A. (1998b): Langmuir **14**, 1392
Jenks, C.J., Delaney, D.W., Bloomer, T.E., Chang, S.-L., Lograsso, T.A., Shen, Z., Zhang, C.-M., Thiel, P.A. (1996a): Appl. Surf. Sci. **103**, 485
Jenks, C.J., Chang, S.-L., Anderegg, J.W., Thiel, P.A., Lynch, D.W. (1996b): Phys. Rev. B **54**, 6301
Jenks, C.J., Pinhero, P.J., Chang, S.-L., Anderegg, J.W., Besser, M.F., Sordelet, D.J., Thiel, P.A. (1997): in New Horizons in Quasicrystals, Goldman, A.I., Sordelet, D.J., Thiel, P.A., Dubois, J.M. (eds). World Scientific, Singapore, p 157
Jenks, C.J., Lograsso, T.A., Delaney, D.W., Pinhero, P.J., Anderegg, J.W., Goldman, A.I., Islam, A.H.M.Z., Thiel, P.A. (1998a): unpublished
Jenks, C.J., Lograsso, T.A., Thiel, P.A. (1998b): J. Am. Chem. Soc., in press
Jenks, C.J., Pinhero, P.J., Bloomer, T.E., Chang, S.-L., Anderegg, J.W., Thiel, P.A. (1998c): in Proceedings of the 6th International Conference on Quasicrystals, Takeuchi, S., Fujiwara, T. (eds). World Scientific, Singapore, p 761
Jenks, C.J., Pinhero, P.J., Shen, Z., Lograsso, T.A., Delaney, D.W., Bloomer, T.E., Chang, S.-L., Zhang, C.-M., Anderegg, J.W., Islam, A.H.M.Z., Goldman, A.I., Thiel, P.A. (1998d): in Proceedings of the 6th International Conference on Quasicrystals, Takeuchi, S., Fujiwara, T. (eds). World Scientific, Singapore, p 741
Kang, S.-S., Dubois, J.-M. (1995): J. Mater. Res. **10**, 1071
Kang, S.S., Dubois, J.M., von Stebut, J. (1993): J. Mater. Res. **8**, 2471
Ko, J.S., Gellman, A.J., Lograsso, T.A., Jenks, C.J., Thiel, P.A. (1998): unpublished
Kortan, A.R., Becker, R.S., Thiel, F.A., Chen, H.S. (1990): Phys. Rev. Lett. **64**, 200
Kortan, A.R., Becker, R.S., Thiel, F.A., Chen, H.S. (1992): in Physics and Chemistry of Finite Systems: From Clusters to Crystals, Vol. I, Jena, P., Khanna, S.N., Rao, K. (eds). Kluwer Academic, Dordrecht, p 29
Lauderback, L.L., Larson, S.A. (1990): Surf. Sci. **234**, 135
Libuda, J., Winkelmann, F., Bäumer, M., Freund, H.-J., Bertrams, T., Neddermeyer, H., Müller, K. (1994): Surf. Sci. **318**, 61
Machizaud, F., Vigneron, B., Dubois, J.M., Dufour, J.P. (1997): unpublished

Marder, M. (1997): Science **277**, 647
Massiani, Y., Yaazza, S.A., Crousier, J.P. (1993): J. Non-Cryst. Solids **159**, 92
Massiani, Y., Ait Yaazza, S., Dubois, J.-M. (1995): in Proceedings of the 5th International Conference on Quasicrystals, Janot, C., Mosseri, R. (eds). World Scientific, Singapore, p 790
McFadden, C.F., Gellman, A.J. (1995): Langmuir **11**, 273
Menguy, N., Audier, M., Guyot, P., Vacher, M. (1993): Philos. Mag. B **68**, 595
Mesarwi, A., Ignatiev, A. (1992): J. Appl. Phys. **71**, 1943
Mikulla, R., Krul, F., Gumbsch, P., Trebin H.-R. (1997): in New Horizons in Quasicrystals, Goldman, A.I., Sordelet, D.J., Thiel, P.A., Dubois, J.M. (eds). World Scientific, Singapore, p 200
Moulder, J.F., Stickle, W.F., Sobol, P.E., Bomben, K.D. (1992): Handbook of X-ray Photoelectron Spectroscopy. Perkin-Elmer Corporation, Eden Prairie, Minnesota
Mullins, D.R., Overbury, S.H. (1988): Surf. Sci. **199**, 141
Musket, R.G., McLean, W., Colmenares, C.A., Makowiecki, D.M., Siekhaus, W.J. (1982): Appl. Surf. Sci. **10**, 143
Naumović, D. (1996): unpublished
Naumović, D., Aebi, P., Schlapbach, L., Beeli, C. (1997): in New Horizons in Quasicrystals, Goldman, A.I., Sordelet, D.J., Thiel, P.A., Dubois, J.M. (eds). World Scientific, Singapore, p 86
Naumović, D., Aebi, P., Schlapbach, L., Beeli, C., Lograsso, T.A., Delaney, D.W. (1998): in Proceedings of the 6th International Conference on Quasicrystals, Takeuchi, S., Fujiwara, T. (eds). World Scientific, Singapore, p 749
Neuhold, G., Barman, S.R., Horn, K., Theis, W., Ebert, P., Urban, K. (1998): Phys. Rev. B **58**, 734
Pinhero, P.J., Chang, S.-L., Anderegg, J.W., Thiel, P.A. (1997): Philos. Mag. B **75**, 271
Pinhero, P.J., Anderegg, J.W., Sordelet, D.J., Besser, M.F., Thiel, P.A. (1998a): unpublished
Pinhero, P.J., Anderegg, J.W., Sordelet, D.J., Besser, M.F., Thiel, P.A. (1998b): Philos. Mag. B, in press
Quinn, T.F.J. (1991): Physical Analysis for Tribology. Cambridge University Press, New York
Quiquandon, M., Quivy, A., Devaud, J., Faudot, F., Lefèbvre, S., Bessière, M., Calvayrac, Y. (1996): J. Phys. Condens. Matter **8**, 2487
Raberg, W. (1998): Ph.D. Thesis, Universität Bonn
Rigney, D.A., Hammerberg, J.E. (1998): MRS Bulletin **23**, 32
Rivier, N. (1993): J. Non-Cryst. Solids **153-154**, 458
Rivier, N. (1997): in New Horizons in Quasicrystals, Goldman, A.I., Sordelet, D.J., Thiel, P.A., Dubois, J.M. (eds). World Scientific, Singapore, p 188
Rouxel, D., Gavatz, M., Pigeat, P., Weber, B., Plaindoux, P. (1997): in New Horizons in Quasicrystals, Goldman, A.I., Sordelet, D.J., Thiel, P.A., Dubois, J.M. (eds). World Scientific, Singapore, p 173
Schaub, T.M., Bürgler, D.E., Güntherodt, H.-J., Suck, J.-B. (1994a): Z. Phys. B **96**, 93
Schaub, T.M., Bürgler, D.E., Güntherodt, H.-J., Suck, J.B. (1994b): Phys. Rev. Lett. **73**, 1255
Schaub, T.M., Bürgler, D.E., Güntherodt, H.-J., Suck, J.B., Audier, M. (1995): Appl. Phys. A **61**, 491
Schaub, T.M., Bürgler, D.E., Schmidt, C.M., Güntherodt, H.-J. (1996): J. Non-Cryst. Solids **205-207**, 748

Shen, Z., Jenks, C.J., Anderegg, J., Delaney, D.W., Lograsso, T.A., Thiel, P.A., Goldman, A.I. (1997a): Phys. Rev. Lett. **78**, 1050

Shen, Z., Pinhero, P.J., Lograsso, T.A., Delaney, D.W., Jenks, C.J., Thiel, P.A. (1997b): Surf. Sci. Lett. **385**, L923

Shen, Z., Kramer, M.J., Jenks, C.J., Goldman, A.I., Lograsso, T., Delaney, D., Heinzig, M., Raberg, W., Thiel, P.A. (1998a): Phys. Rev. B, in press

Shen, Z., Raberg, W., Heinzig, M., Jenks, C.J., Gierer, M., Van Hove, M.A., Lograsso, T., Cai, T., Thiel, P.A. (1998b): unpublished

Shen, Z., Stoldt, C., Jenks, C., Lograsso, T., Thiel, P.A. (1998c): unpublished

Shi, F., Shen, Z., Delaney, D.W., Goldman, A.I., Jenks, C.J., Kramer, M.J., Lograsso, T., Thiel, P.A., Van Hove, M.A. (1998): Surf. Sci. **411**, 86

Singer, I.L., Dubois, J.M., Soro, J.M., Rouxel, D., von Stebut, J. (1998): in Proceedings of the 6th International Conference on Quasicrystals, Takeuchi, S., Fujiwara, T. (eds). World Scientific, Singapore, p 769

Somorjai, G.A. (1994): Introduction to Surface Chemistry and Catalysis. John Wiley and Sons, New York

Sordelet, D.J., Gunderman, L.A., Besser, M.F., Akinc, A.B. (1997): in New Horizons in Quasicrystals, Goldman, A.I., Sordelet, D.J., Thiel, P.A., Dubois, J.M. (eds). World Scientific, Singapore, p 296

Splinter, S.J., McIntyre, N.S. (1994): Surf. Sci. **314**, 157

Stadnik, Z.M., Purdie, D., Garnier, M., Baer, Y., Tsai, A.-P., Inoue, A., Edagawa, K., Takeuchi, S. (1996): Phys. Rev. Lett. **77**, 1777

Stevens, J.L., Hwang, R.Q. (1995): Phys. Rev. Lett. **74**, 2078

Strohmeier, B.R. (1990): Surf. Interface Anal. **15**, 51

Suzuki, S., Waseda, Y., Tamura, N., Urban, K. (1996): Scr. Mater. **35**, 891

Szklarczyk, M., Minevski, L., Bockris, J.O.M. (1990): J. Electroanal. Chem. **289**, 279

Thiel, P.A., Estrup, P.J. (1995): in The Handbook of Surface Imaging and Visualization, Hubbard, A.T. (ed). CRC Press, Boca Raton, Florida, p 407

Wagman, D.D., Evans, W.H., Parker, V.B., Schumm, R.H., Halow, I., Bailey, S.M., Churney, K.L., Nuttall, R.L. (1982): J. Phys. Chem. Ref. Data **11**, Suppl. 2

Wang, R., Yang, X., Takahashi, H., Ohnuki, S. (1995): J. Phys. Condens. Matter **7**, 2105

Wang, Z., Yang, X., Wang, R. (1993): J. Phys. Condens. Matter **5**, 7569

Weast, R.C. (ed). (1986-87): CRC Handbook of Chemistry and Physics. Chemical Rubber Company, Boca Raton, Florida

Wehner, B.I., Köster, U. (1997): in New Horizons in Quasicrystals, Goldman, A.I., Sordelet, D.J., Thiel, P.A., Dubois, J.M. (eds). World Scientific, Singapore, p 152

Wehner, B.I., Köster, U. (1998): in Proceedings of the 6th International Conference on Quasicrystals, Takeuchi, S., Fujiwara, T. (eds). World Scientific, Singapore, p 773

Yalisove, S.M., Graham, W.R. (1987): Surf. Sci. **183**, 556

Yang, X., Wang, R., Fan, X. (1996): Philos. Mag. Lett. **73**, 121

Zhang, Z., Feng, Y.C., Williams, D.B., Kuo, K.H. (1993): Philos. Mag. B **67**, 237

Zurkirch, M., Erbudak, M., Hochstrasser, M., Kortan, A.R. (1997): Surf. Rev. Lett. **4**, 1143

Zurkirch, M., Erbudak, M., Kortan, A.R. (1998): in Proceedings of the 6th International Conference on Quasicrystals, Takeuchi, S., Fujiwara, T. (eds). World Scientific, Singapore, p 67

11. Mechanical Properties of Quasicrystals

Knut Urban, Michael Feuerbacher, Markus Wollgarten, Martin Bartsch, and Ulrich Messerschmidt

11.1 Introduction

The mechanical behavior, i.e., the response of a material to an externally applied mechanical load, is of basic importance for any kind of structural application. It can be described largely in terms of the material properties that govern plastic deformation and fracture. Macroscopically, these properties can be expressed in terms of materials parameters, which can usually be measured without detailed knowledge of the microscopic origin of these properties. Microscopically, these properties are related to processes on a scale ranging from atomic dimensions to the typical dimensions of the morphology of the material.

In classical crystalline materials, since the first attempts towards a microscopic understanding of mechanical behavior in the 1920s and 1930s, great progress has been made in establishing experimentally and theoretically the microscopic mechanisms and relating them to the properties that can be observed macroscopically. The key concepts (see, e.g., Seeger 1958, Hirth and Lothe 1982, Mughrabi 1993) are based on the observation that plastic deformation occurs by shear on densely packed crystallographic slip planes. Along a slip plane, shear occurs not uniformly but in a very localized manner due to the movement of dislocations, i.e., one-dimensional (1D) lattice defects. It is the mobility and the multiplication of these dislocations which controls the yield stress and plastic flow. The interaction of dislocations with other lattice defects or second-phase particles reduces their mobility, and the mutual interaction of dislocations plays an important part in work hardening, i.e., the observation that the stress required to continue plastic flow, the flow stress, increases with increasing strain.

In the classical microscopic treatment of plasticity lattice periodicity is implicitly assumed in one way or another. Therefore, the mechanical behavior of quasicrystalline materials, whose structure combines long-range order with the absence of translational symmetry, poses an intriguing problem. In fact, the fundamental questions are: Which of the basic concepts developed for crystalline materials are universal enough to be also applicable to quasicrystal (QC) deformation? Provided a defect-related mechanism brings about and controls plastic behavior, how can these defects be described in a quasiperiodic lattice? Are there specific defects or deformation modes which are a direct consequence of the quasiperiodic structure?

The mechanical properties of quasicrystalline materials were studied some time ago (Chen et al. 1985, Bhaduri and Sekhar 1987). However, systematic investigations were hampered by difficulties in obtaining material of reasonable quality. The progress in materials preparation in recent years has triggered new activity in this field. As in conventional alloys, experiments under well-defined conditions are required which can serve as a basis for understanding the intrinsic mechanical properties of QCs. Such studies are now increasingly possible after techniques have been developed to grow large single QCs up to a few cm in size directly from the melt (Yokoyama et al. 1992, Boudard et al. 1995).

The data on the mechanical and plastic properties of polyquasicrystalline and, in a few cases, also on singlequasicrystalline (sq) materials indicate a high hardness at room temperature which, unfortunately, is combined with a low fracture toughness or, in other words, a very high brittleness. This rules out the direct use of these materials as structural materials. However, it has been demonstrated that coating of a ductile technical material with a thin quasicrystalline surface layer produces very interesting mechanical properties of the composite, in particular, a substantial reduction of friction and wear (Dubois et al. 1991, Kang et al. 1993). On the other hand, a number of very promising QC-hardened alloys have been developed in which the particular ageing properties of quasicrystalline precipitates in an Al-rich matrix (Inoue et al. 1995, Inoue et al. 1997, Kita et al. 1997, Büchler et al. 1997) or in steels (Liu et al. 1995, Liu and Nilsson 1997) have turned out to offer technically interesting advantages.

Most of our detailed and quantitative, microscopic knowledge about QC plasticity comes from experiments on single QCs of icosahedral (i) Al-Pd-Mn. Therefore, a major part of this chapter deals with the data obtained from experiments on Al-Pd-Mn. Since the mechanical behavior of quasicrystalline alloys is to a great extent determined by a brittle-to-ductile transition at about 70% of the absolute melting temperature, it is useful to discuss mechanical properties with reference to appropriately defined low-temperature and high-temperature ranges.

11.2 Low-Temperature Mechanical Properties

11.2.1 Mechanical Property Data

At low or ambient temperatures quasicrystalline materials are relatively hard. For example, the Vickers hardness of decagonal (d) $Al_{62}Co_{15}Cu_{20}Si_3$ measured at room temperature is 9.5 GPa (HV 950) (Wittmann et al. 1991) and that of i-$Al_{72}Pd_{20}Mn_8$ is 7.8 GPa (Tsai et al. 1992). For comparison, hardness values for steels range from 1.74 for 0.36C1.2Mn (BS120M36) to 9.12 GPa for 9-3-3-9 WMoVCo (Smithells 1992). The QC values compare well to the hardness of singlecrystalline silicon (10 GPa). The values obtained for

Young's modulus at room temperature range from 87 GPa for $Al_{5.1}Li_3Cu$ (Reynolds et al. 1990) to 168 GPa in $Al_{65}Cu_{20}Fe_{15}$ (Tanaka et al. 1996) and to 194 GPa in $Al_{70.5}Pd_{21}Mn_{8.5}$ (Feuerbacher et al. 1996).

Typical values for the room-temperature fracture toughness K_{Ic} are 1.0 MPa m$^{1/2}$ in *d*-$Al_{62}Co_{15}Cu_{20}Si_3$ (Wittmann et al. 1991) and 1.6 MPa m$^{1/2}$ in *i*-$Al_{64}Cu_{22}Fe_{14}$ (Köster et al. 1993). These are quite low compared to the values of 22 to 37 MPa m$^{1/2}$ typical of 2000 series Al alloys or 40 to 83 MPa m$^{1/2}$ for maraging steels (Bardes 1978). Again, with respect to fracture toughness, QCs behave quite similar to singlecrystalline Si (0.7 MPa m$^{1/2}$). In fact, this combination of high hardness with high brittleness at room temperature is a property shared also with many intermetallic phases. For instance, the bcc λ phase $Al_{13}Fe_4$ combines a hardness value of 10.7 GPa with a K_{Ic} value of 1.03 MPa m$^{1/2}$ and the respective values for the θ phase Al_2Cu are 6 GPa and 1.1 MPa m$^{1/2}$ (Köster et al. 1993). These values are characteristic of alloys with directed bonds of highly covalent nature.

Detailed inspection of Vickers pyramid indentations and also scratching experiments point to some very limited room-temperature ductility of QCs. However, cross-sectional electron microscopy of the area under Vickers indentations did not reveal any evidence for dislocations but demonstrated the presence of a high density of shear cracks (Urban et al. 1993), Wollgarten and Urban 1994, Wollgarten and Saka 1997). In cases where extended shear-crack areas could be imaged along the crack plane it could be observed that the cracks frequently change direction and thus avoid, at least for small strains, combination with others to form supercritical cracks which would induce immediate catastrophic failure.

11.2.2 Fracture

Experiments have shown that *sq* *i*-$Al_{70.5}Pd_{21}Mn_{8.5}$ can be cleaved along the most densely packed quasicrystallographic planes, i.e., the fivefold, threefold, and twofold planes. Scanning tunneling microscopy (STM) of the cleavage planes thus obtained shows extended flat areas and a corresponding low density of cleavage steps. However, the condition of "flatness" only holds on a scale of the order of a few ten nm. On a nm scale, the surfaces are rough. In fact, the finer-scale details seen in STM supply information not only on the micromechanics of fracture but also on a locally varying elastic strength of QCs which is closely related to their particular structure (Ebert et al. 1996).

In the theory of Mode I brittle fracture it is assumed that fracture is caused by the initial presence and subsequent widening of cracks on the surface or within the interior of the material. The crack, under the applied load, produces a stress concentration at its tip. Fracture occurs when the maximum stress at the tip equals the theoretical strength of the material, and the crack will propagate if the increase in surface energy is overcompensated by a reduction of the elastic strain energy of the volume around the crack

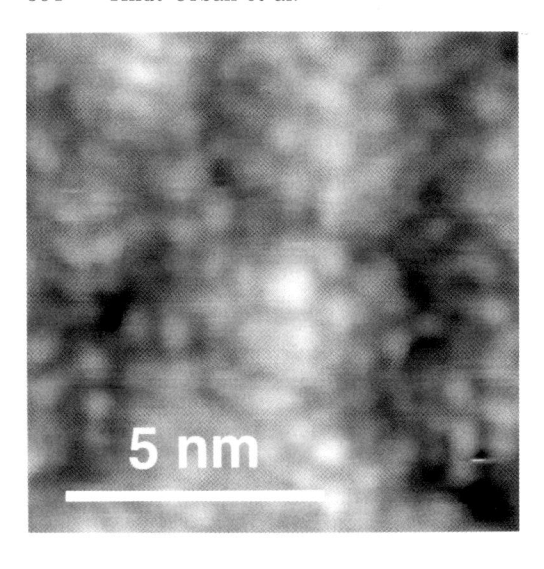

Fig. 11.1. STM image of a fracture surface obtained by cleavage along a twofold plane of an $Al_{70.5}Pd_{21}Mn_{8.5}$ single QC. The circular contrast features originate from calotte-shaped sections of Mackay-type clusters sticking out from the background. After Ebert et al. (1996).

tip. This condition is expressed by the Griffith condition for the stress σ_F required for fracture

$$\sigma_F = \left(\frac{2\gamma E}{\pi c}\right)^{1/2}, \tag{11.1}$$

where γ denotes the surface energy, E is Young's modulus, and c is the initial crack length (see, e.g., Courtney 1990).

Let us now imagine that a crack advances towards an obstacle provided by a volume area with a locally higher bond strength. This can be formally characterized by a higher spring constant between atoms and gives rise to an increased value for E. This, in turn, leads to a locally increased σ_F. As a result, the crack advances in the "softer" areas around the obstacle while it may be hindered to do so in the "harder" area represented by the obstacle. Consequently the local stress level in the obstacle area increases. At a certain level, this will allow the more rigid bonds to break. In general, before this happens, the crack will temporarily change direction thus overcoming the hard obstacle. The resulting overall path of the crack will create a cleavage surface which, within certain limits, can be considered a "map" of the local elastic strength. Figure 11.1 shows an STM image of a fracture surface obtained by cleavage along a twofold plane of an i-$Al_{70.5}Pd_{21}Mn_{8.5}$ single QC. It is obvious that the structure is rough and consists of a size hierarchy of clusters. The circular contrast features in Fig. 11.1 originate from calotte-shaped sections of an elementary cluster sticking out from the background. The diameter of this cluster of the order of 1 nm fits well to the diameter of the Mackay-type cluster (Mackay 1962, Boudard et al. 1992). This cluster contains 52 atoms in three concentric shells and occurs as the basic element in current models of the Al-Pd-Mn i QC lattice. Based on the above discussion of crack propagation one can draw conclusions from such pictures not only about the existence of these clusters but also about their particular mechanical strength. This observation is of fundamental importance for much of our understanding of the microplasticity of *i* QCs, an account of which is given below.

11.3 Dislocations in Quasicrystals

11.3.1 Background

The concept of dislocation-mediated plastic flow of periodic crystals, as introduced in 1934 by Orowan, Polanyi, and Taylor (see, e.g., Mughrabi 1993), contains as an essential feature the fact that at any time during slip the corresponding atom rearrangements are concentrated in a very narrow region. Let us consider the simple case in which an edge dislocation of Burgers vector **b** is created in a primitive cubic lattice by insertion of an extra lattice half-plane of atoms. To accommodate the dislocation, there is a disregistry of atomic coordination across the slip plane. The width of the zone within which this disregistry is larger than a certain minimum value defines the core of the dislocation. Outside the core, the atomic lattice is, apart from the presence of an elastic strain-field, unaffected by the presence of the dislocation. In more complicated lattices where the dislocations can split into partials, these are separated by a narrow ribbon of stacking fault. In ordered alloys the superdislocations of the superstructure can split into partials which are the elementary dislocations of the underlying matrix. In all these cases, however, due to translational symmetry, the lattice far away from the dislocations is undisturbed and, in principle, the movement of a dislocation through the crystal can take place without producting any permanent damage.

In a QC, which combines long-range order with the absence of translational symmetry, the situation is much more complicated. It is not possible to create a dislocation by simply inserting an additional quasilattice half-plane without creating a fault all along the inserted plane, and it cannot be expected that a dislocation can simply move in a plane of the quasiperiodic lattice introducing slip in the conventional way without leaving behind some sort of planar fault. Nevertheless, dislocations do occur in QCs. The first evidence was obtained by Hiraga and Hirabayashi (1987) in electron microscopic lattice-fringe pictures of *i*-Al-Mn-Si. Later, dislocations exhibiting conventional diffraction contrast behavior could be identified and analyzed by Bragg-diffraction contrast electron microscopy in *d*-$Al_{65}Cu_{20}Co_{15}$ (Zhang and Urban 1989) and *i*-$Al_{65}Cu_{20}Fe_{15}$ (Ebalard and Spaepen 1989, Zhang et al. 1990).

11.3.2 Dislocations in a Quasilattice

The first theoretical treatments of dislocations were carried out for 2D quasilattices within the density-wave approach by Levine et al. (1985), Lubensky et al. (1985), and by Socolar et al. (1986). In the following, a brief account will be given of the treatment of strain in QC lattices and the introduction and quantitative characterisation of dislocations. The quasilattice model (Levine and Steinhardt 1984) is adopted in this discussion although it has been shown that dislocations can also be defined without any principal difficulty in the random tiling model of QCs (Henley 1987).

An adequate characterization of lattice strain and of extended defects in *i* QCs can only be achieved on the basis of the one-to-one correspondence of the lattice geometry in 3D space and that in 6D reference space $\mathbf{R}^6$ where full periodicity is recovered.

The lattice nodes of the primitive hypercubic lattice $\mathbf{Z}^6$ in $\mathbf{R}^6$ are given by position vectors $\mathbf{X}$ with components $\mathbf{x}_\mathrm{i} = An_\mathrm{i}\mathbf{e}_\mathrm{i}$ ($i = 1...6$, n_i are integers, $\mathbf{e}_\mathrm{i}$ are orthonormal basis vectors, A is the hyperlattice parameter) along the six coordinate axes of the 6D Cartesian coordinate system $\mathbf{C}_1$. For a construction of the QC lattice in 3D physical space a second 6D coordinate system $\mathbf{C}_2$ is introduced which is rotated with respect to the original one. With respect to $\mathbf{C}_2$ two 3D subspaces of $\mathbf{R}^6$ can be defined by grouping its orthonormal basis vectors $\mathbf{e}_\mathrm{j}$ into two sets. The first subspace $E_{\|}$, for $j = 1, 2, 3$, represents parallel or physical space. Here the position vectors are denoted $\mathbf{x}_{\|}$ with components x_1, x_2, x_3. The orientation of $E_{\|}$ with respect to $\mathbf{C}_1$ is chosen such that the projections of the six axes of $\mathbf{C}_1$ lie along the six fivefold axes of an icosahedron. This has the consequence that the rotation matrix relating $\mathbf{C}_2$ to $\mathbf{C}_1$ has irrational entries (see, e.g., Cahn et al. 1986). The second 3D subspace $E_\perp$, for $j = 4, 5, 6$, is perpendicular to $E_{\|}$. The respective position vectors are $\mathbf{x}_\perp$ with components x_4, x_5, x_6. The particular orientation of $\mathbf{C}_2$ with respect to $\mathbf{C}_1$ ensures that the *i* point-group symmetry of the QC lattice is constructed. The lattice nodes of the original hypercubic lattice are decorated by 3D atomic hypersurfaces having only extension parallel to $E_\perp$. The lattice nodes of the QC lattice $\mathbf{Z}^3$ in $E_{\|}$ are then constructed as the cut points between these hypersurfaces and $E_{\|}$. This determines not only the position of the individual lattice points of the QC structure but also the occupation of these by the different atomic species (see, e.g., Steurer 1990, Janot 1992).

Strain can be introduced into the 3D QC lattice by applying a corresponding strain field on the hyperlattice in $\mathbf{R}^6$. This is done by displacing hyperlattice points from positions $\mathbf{X}_0$ to positions $\mathbf{X}_1 = \mathbf{X}_0 + \mathbf{U}(\mathbf{X})$, where $\mathbf{U}(\mathbf{X})$ is a 6D displacement vector. It can be decomposed into two components $\mathbf{u}_{\|}$ and $\mathbf{u}_\perp$ in physical and perpendicular space, respectively. The following special cases illustrate the effect of these two displacement components in $\mathbf{R}^6$ on the resulting atom arrangement in $E_{\|}$. Performing the cut in the same way as before on a lattice of hyperlattice points shifted by $\mathbf{u}_{\|} = \mathbf{u}_{\|}^0 = const$ and

$\mathbf{u}_\perp = 0$, yields an identical quasilattice but with all lattice points shifted by $\mathbf{u}_\parallel^0$. Locally varying $\mathbf{u}_\parallel$ leads to classical or *phonon strain* in the QC lattice. On the other hand, for $\mathbf{u}_\parallel = 0$ and $\mathbf{u}_\perp = \mathbf{u}_\perp^0 = const$, some of the hyperlattice points of $\mathbf{Z}^6$, whose positions in the perfect lattice were such that the appertaining atomic hypersurfaces did intersect the 3D cut surface, may now lie outside the range of the cut. In consequence, the corresponding QC lattice points disappear. On the other hand, hyperlattice points originally situated such that the appertaining atomic hypersurfaces did not intersect the cut surface may now be shifted inside the range of the cut. As a result, in the QC lattice in physical space lattice points appear which are absent in the original lattice. For locally varying $\mathbf{u}_\perp$ we obtain *phason strain*. This means that QC-lattice point arrangements occur which, in the tiling picture, violate the matching rules and are therefore not found in ideal QCs.

Dislocations can be introduced into the QC lattice by introducing a corresponding hyperdislocation into the hypercubic lattice in $\mathbf{R}^6$. As discussed by Kléman (1988), Lubensky (1988), Bohsung and Trebin (1989), Kléman and Sommers (1990), and Wollgarten et al. (1991), this can be done by a generalisation of the Volterra process (Cottrell 1953). This procedure, which in reality is quite involved, has been realized on the basis of the hyperspace model of Boudard et al. (1992) for the construction of dislocations in *i*-Al-Pd-Mn by Yang et al. (1998b).

A hyperlattice edge dislocation is created by insertion or removal of a 5D hyperlattice half-plane. The 6D Burgers vector $\mathbf{B}$ is defined as the resulting net displacement of the hyperlattice points for any closed circuit C around the dislocation line in $\mathbf{R}^6$, i.e.,

$$\mathbf{B} = \oint_C \mathrm{d}\mathbf{U}, \tag{11.2}$$

or within the subspaces

$$\mathbf{b}_\parallel = \oint_C \mathrm{d}\mathbf{u}_\parallel \quad \text{and} \quad \mathbf{b}_\perp = \oint_C \mathrm{d}\mathbf{u}_\perp, \tag{11.3}$$

where $\mathbf{b}_\parallel$ and $\mathbf{b}_\perp$ denote the Burgers vector components in physical and perpendicular space, respectively.

When a defect-free QC lattice is constructed by the described cut method, shifts of the cut surface in a direction perpendicular to it, i.e., by any vector in $E_\perp$, produce equivalent lattices belonging to the same local isomorphism class (Levine and Steinhardt 1984, Socolar and Steinhardt 1986, Steinhardt and Ostlund 1987). By analogy, it is necessary to postulate, in order that a shift of the cut surface parallel to a vector in $E_\perp$ produces structurally equivalent dislocations in the QC lattice, that the displacement function $\mathbf{U}$ of the hyperdislocation in $\mathbf{R}^6$ must only depend on the coordinates of physical space $\mathbf{x}_\parallel$. As a consequence, the result of the Burgers circuit is independent of the location in perpendicular space and it suffices to perform this circuit in $\mathbf{R}^6$ with respect to the physical-space coordinates only. In order to clarify the term "line" for a hyperdislocation in $\mathbf{R}^6$ we recall that for dislocations in ordinary crystals the line is characterized by the condition that the displacement field does not change along any path parallel to the line direction. Applying this condition in an equivalent form to $\mathbf{R}^6$ we find that there the dislocation line is a 4D object spanned by the 3D perpendicular space together with the 1D line in physical space. The reason for this is that, as just discussed, the displacement field is independent not only of the line coordinate in $E_{\|}$ but also of the coordinates in $E_{\perp}$.

Finally, we note that since $E_{\|}$ and $E_{\perp}$ have an irrational orientation with respect to the basic coordinate system in $\mathbf{R}^6$, the Burgers vector of a QC dislocation always has finite $\mathbf{b}_{\|}$ and $\mathbf{b}_{\perp}$. This means that a QC dislocation is always characterized by the presence of both a phonon and a phason strain field. The phonon part is characterized by $\mathbf{b}_{\|}$. It is equivalent to the strain and Burgers vector of dislocations in ordinary crystals and characterizes the elementary slip step in the QC lattice. The phason part of the strain field characterized by $\mathbf{b}_{\perp}$ is a measure of the density of phason defects, i.e., matching-rule violations associated with the dislocation. For a characterisation of how the strain is distributed into phonon and phason parts it is useful to define a *strain accommodation parameter* ζ (Feuerbacher et al. 1997) according to

$$\zeta = \frac{|\mathbf{b}_{\perp}|}{|\mathbf{b}_{\|}|} . \tag{11.4}$$

Provided that the same metrics apply to the Burgers vectors and the axes of C_2 in hyperspace, the ζ values are defined unambiguously allowing a comparison of values appertaining to different dislocations and Burgers vectors.

A dislocation of mixed character produced in i-$Al_{70.5}Pd_{21}Mn_{8.5}$ by performing the Volterra construction in 6D hyperspace is shown in Fig. 11.2. The figure demonstrates the presence of both phonon and phason strain (matching-rule violations). It also shows that dislocations produced in the way described in this section exhibit the important property that both types of strain decrease with distance from the dislocation core. This means that far away from the dislocation the QC lattice is undisturbed.

11.3.3 Dislocation Analysis

Since a QC dislocation is always characterized by the simultaneous presence of a phonon and a phason strain field a quantitative analysis requires the measurement of both. This is equivalent to the determination of the 6D Burgers vector of the reference dislocation in 6D hyperspace. Two techniques for the quantitative analysis of QC dislocations in transmission electron microscopy have been developed. These are essentially extensions of the corresponding classical techniques employed for ordinary crystals.

Wollgarten et al. (1991) treated the Bragg-diffraction contrast of QC dislocations within the framework of the kinematical theory of electron diffrac-

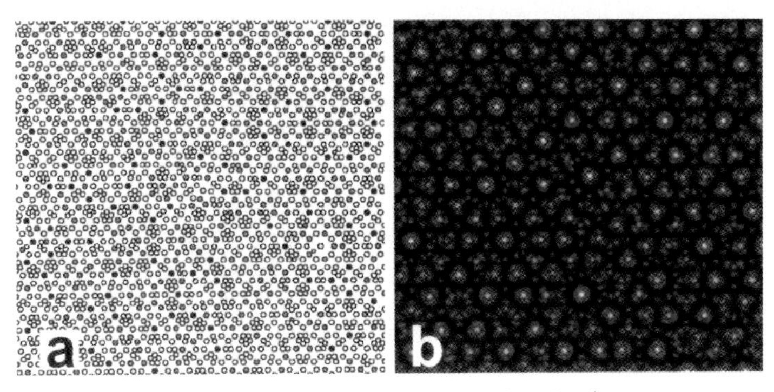

Fig. 11.2. A dislocation of a mixed edge/screw type produced in an $Al_{70.5}Pd_{21}Mn_{8.5}$ *i* QC by performing the Volterra construction in 6D hyperspace on the basis of the lattice model of Boudard et al. (1992). (a) Lattice-point presentation (Al, Pd, and Mn are represented respectively by white, grey, and black circles). (b) Simulated high-resolution transmission electron micrograph employing the multi-slice technique for the solution of the Schrödinger equation. Regarding the figure at a glancing angle, the presence of both phonon and phason strain can be recognized. The latter expresses itself in the gradual disappearance of atomic point contrast along certain lattice lines while new but displaced parallel lines of atomic points emerge. After Yang et al. (1998b).

tion and the column approximation. This allows conditions to be derived under which, in two-beam Bragg-diffraction contrast imaging, the dislocation contrast becomes extinct. Systematic experimental investigation of these contrast extinctions allows the *direction* of the 6D Burgers vector to be determined. The theory yields the result that the classical criterion $\mathbf{g}\,\mathbf{b} = 0$ for contrast extinction of ordinary crystal dislocations (see, e.g., Hirsch et al. 1960, Reimer 1989) has to be replaced by

$$\mathbf{G}\,\mathbf{B} = \mathbf{g}_{\parallel}\,\mathbf{b}_{\parallel} + \mathbf{g}_{\perp}\,\mathbf{b}_{\perp} = 0. \tag{11.5}$$

Here $\mathbf{g}$ is the imaging reciprocal lattice vector and $\mathbf{G}$ is a reciprocal lattice vector in $\mathbf{R}^6$ whose components in $E_{\parallel}$ and $E_{\perp}$ are $\mathbf{g}_{\parallel}$ and $\mathbf{g}_{\perp}$, respectively. This condition can be divided into two cases,

$$\text{(a) } \mathbf{g}_{\parallel}\,\mathbf{b}_{\parallel} = \mathbf{g}_{\perp}\,\mathbf{b}_{\perp} = 0 \quad \text{and (b) } \mathbf{g}_{\parallel}\,\mathbf{b}_{\parallel} = -\mathbf{g}_{\perp}\,\mathbf{b}_{\perp} \neq 0. \tag{11.6}$$

Condition (a) is called *strong extinction condition* (SEC), while condition (b) is termed *weak extinction condition* (WEC). Experimentally, for the SEC no dislocation contrast is observed for any reflection with a reciprocal lattice vector parallel to $\mathbf{b}_{\parallel}$. This means that if a single $\mathbf{g}_{\parallel}$ is found which leads to contrast extinction, this also holds for all reflections of the appertaining systematic row of reflections through the origin of the diffraction pattern. This condition is therefore equivalent to the classical extinction condition for periodic crystals. The WEC is an additional condition only occurring in quasiperiodic lattices. It only holds for a single reflection of a systematic row for which the contrast becomes extinct, while the dislocation remains in contrast for the other reflections. If two non-equivalent imaging conditions can be found for which the SEC holds, the direction of the Burgers vector in the subspaces $E_{\|}$ and $E_{\perp}$ can be determined. If, additionally, the WEC can be observed, the orientation of the dislocation Burgers vector can be fixed within the 6D hyperlattice. Applications of this technique of dislocation-contrast analysis can be found in Wollgarten et al. (1992), Wollgarten et al. (1993a), Wang et al. (1994), Baluc et al. (1995), and Gastaldi et al. (1995).

Bragg-diffraction contrast analysis allows a large number of dislocations to be investigated in a given area at the same time. On the other hand, as in conventional crystals, it is not possible to determine the magnitude of the strain fields characterized by the lengths of the respective Burgers vectors. However, for individual dislocations, the modulus and the direction of the 6D Burgers vector can be determined by means of the defocus convergent beam electron diffraction (d-CBED) technique which was developed for QCs by Wang and Dai (1993). Here, instead of the parallel illumination employed in ordinary diffraction experiments, a convergent electron beam is used which is focused just above or below the specimen plane and a small, circular region is illuminated. As a consequence, the reflections of the diffraction pattern are expanded to discs. If the sample is properly oriented, there is a set of beams within the illumination cone which fulfil a specific Bragg condition for diffraction. All these beams point along the outside of a cone which is defined by the wave vectors $\mathbf{k}_0$ and $\mathbf{k}$ fulfilling the Bragg condition $\mathbf{k} - \mathbf{k}_0 = \mathbf{g}$, where $\mathbf{g}$ defines the cone's symmetry axis. In the diffraction pattern, the set of diffracted beams is seen as a bright line. In the bright zero-disc, a dark deficiency line is observed. Since the crossover of the illumination is placed outside the specimen plane, each beam direction belongs to a different position in the sample. Therefore, the diffraction discs provide a mapping of Bragg plane orientations. A defect can distort the lattice plane orientations, thereby locally shifting the areas where the Bragg condition is fulfilled. In turn, the deficiency line and the corresponding excess line are distorted and split. The number of splittings, n, can be calculated according to the generalized Cherns-Preston rule (Cherns and Preston 1986, Wang and Dai 1993):

$$\mathbf{GB} = \mathbf{g}_{\|}\,\mathbf{b}_{\|} + \mathbf{g}_{\perp}\,\mathbf{b}_{\perp} = n. \tag{11.7}$$

To determine all components of the Burgers vector one has to observe six lines for which the corresponding reciprocal lattice vectors have to be linearly independent. Experimental results employing this technique for *sq* *i*-$Al_{70.5}Pd_{21}Mn_{8.5}$ are given below. We note that throughout this paper the indexing scheme of Cahn et al. (1986) is employed.

11.4 High-Temperature Plastic Deformation

11.4.1 Background

The mechanical properties of quasicrystalline materials change drastically with increasing temperature. In $Al_{72}Pd_{20}Mn_8$, for example, the Vickers hardness decreases continuously from 7.8 GPa at room temperature to about 0.2 GPa at 700 °C (Tsai et al. 1992), a temperature at which this material can be easily deformed plastically. A material's plastic response to mechanical loading can be investigated by means of a tension or compression test. In such a test the material is stretched or compressed at a given strain rate $\dot{\varepsilon}$, i.e., the relative length change per unit time. The force to be exerted by the testing machine to maintain this strain rate is measured. The force divided by the specimen cross-section, the stress σ, is then plotted as a function of the strain ε.

The first compression tests on *i* polyquasicrystalline Al-Cu-Ru were performed by Shibuya et al. (1990), on Al-Cu-Co-Si, Al-Cu-Fe-Cr-Si-(B), and Al-Cu-Fe-(B) alloys by Kang and Dubois (1992), on Al-Cu-Fe by Bresson and Gratias (1993), and on Al-Pd-Mn by Takeuchi and Hashimoto (1993). These experiments demonstrated substantial ductility in the temperature range $\sim$ 0.7–0.9 T_{m}, where T_{m} denotes the absolute melting temperature. They also provided first evidence for the phenomenon of *deformation softening* at higher strain values, which, independent of the alloy system and on whether the material is single- or polyquasicrystalline, is a characteristic property of QC plasticity. We note that this is in pronounced contrast to the behavior observed in conventional crystals where, due to work hardening, the flow stress increases with strain.

The first high-temperature experiments on Czochralski-grown single-phase single QCs were carried out by Wollgarten et al. (1993b) on *i*-Al-Pd-Mn. These experiments yielded direct evidence of a dislocation-based mechanism for plastic deformation. It was found that upon deformation, the dislocation density increased by orders of magnitude. Furthermore, in high-temperature *in-situ* straining experiments in a high-voltage electron microscope, the motion of dislocations could be directly observed during deformation (Wollgarten et al. 1995).

In the following, we will concentrate on single-QC experiments carried out in *i*-$Al_{70.5}Pd_{21}Mn_{8.5}$. The aim is to determine the intrinsic deformation parameters and to develop an understanding of the microscopic processes governing QC plasticity.

11.4.2 Theoretical

The following brief treatment of the theory of plasticity serves a more quantitative description of QC plastic behavior. In particular, it provides the basis for the design and interpretation of specific deformation experiments which allow us to determine the thermodynamic activation parameters. This treatment follows the guidelines of the theory for periodic crystals (for a review, see Evans and Rawlings 1969). The dislocation Burgers vector $\mathbf{b}$ occurring in the relations derived in this section is taken to be identical with $\mathbf{b}_{\|}$, i.e., the component of the 6D Burgers vector in physical space.

The contribution of a dislocation to slip depends on the slip plane and the slip direction, which together characterize the *slip system.* In an uniaxial test, the resolved shear stress τ is related to the axial stress σ by

$$\tau = m_{\mathrm{S}}\sigma, \tag{11.8}$$

where the *Schmid factor* is given by $m_{\mathrm{S}} = \cos\phi\cos\lambda$. Here ϕ is the angle between the specimen axis and the normal to the slip plane, λ is the angle between the load axis and the slip direction, and $0 \leq m_{\mathrm{S}} \leq 0.5$. For a given $\dot{\varepsilon}$, there is a corresponding *slip velocity* $\dot{a}$ in the slip plane. If slip is induced by the movement of straight dislocations occurring at a density ρ and moving at a velocity v, the slip velocity is given by the *Orowan relation*

$$\dot{a} = \rho b v, \tag{11.9}$$

where b is the modulus of the Burgers vector.

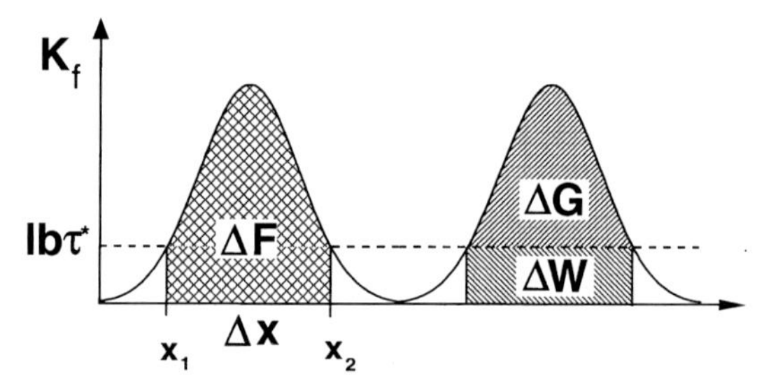

Fig. 11.3. Schematic force-distance diagram at an obstacle to be overcome by a moving dislocation. K_{f} is the force acting on the obstacle, x is the distance, ΔG is the Gibbs free energy of thermal activation, ΔF is the Helmholtz free energy, and ΔW is the work term. For details see text.

The effective shear stress τ^* acting on the dislocation (Seeger 1958) is given by

$$\tau^* = \tau + \tau_{\mathrm{i}}, \tag{11.10}$$

where τ_{i} originates from the *internal stress* resulting from the superposition of the stress fields of the dislocations present in the material. The effective shear stress induces a perpendicular force K acting on a line element l of the dislocation of magnitude $K = \tau^* lb$. This is counteracted by a friction force $K_{\mathrm{f}} = \tau_{\mathrm{f}} lb$, where τ_{f}, the line glide resistance, is the shear stress equivalent to friction. The dislocation can move through the lattice if $\tau^* > \tau_{\mathrm{f}}$. If the dislocation meets an obstacle (Fig. 11.3) for which locally $\tau^* < \tau_f$, this obstacle cannot be overcome. Instead, the dislocation takes a stable position at position x_1 in front of the obstacle. At finite temperature the obstacle can be overcome by means of thermal fluctuations provided that a Gibbs free energy of activation

$$\Delta G = \int_{x_1}^{x_2} (\tau_f - \tau^*) lb dx \tag{11.11}$$

is supplied. The total work done in overcoming the obstacle is given by the Helmholtz free energy (Gibbs 1967)

$$\Delta F = \int_{x_1}^{x_2} \tau_f lb dx. \tag{11.12}$$

According to Fig. 11.3,

$$\Delta G = \Delta F - \Delta W, \tag{11.13}$$

where ΔW, the *work term*, is given by the work done by the effective shear stress in overcoming the obstacle

$$\Delta W = \tau^* lb \Delta x. \tag{11.14}$$

Here $\Delta x = x_2 - x_1$ is the distance covered during the activation step.

For plastic deformation controlled by thermally activated processes, the temperature dependence of the plastic strain rate is given by the Arrhenius expression

$$\dot{\varepsilon} = \dot{\varepsilon}_0 \exp\left(-\frac{\Delta G(\tau)}{k_B T}\right), \tag{11.15}$$

where Boltzmann's constant is denoted k_B and T is the absolute temperature. The pre-exponential factor $\dot{\varepsilon}_0$ is proportional to ρ. Since ΔG depends on temperature and the effective shear stress, we can write the Gibbs equation as $d(\Delta G) = -\Delta S dT - V^* d\tau^*$ with

$$\Delta S \equiv -\left.\frac{\partial(\Delta G)}{\partial T}\right|_{\varepsilon^*} \tag{11.16}$$

and

$$V^* \equiv -\left.\frac{\partial(\Delta G)}{\partial \tau^*}\right|_{T}, \tag{11.17}$$

where ΔS denotes the activation entropy related to ΔG by $\Delta G = \Delta H - T\Delta S$. Here ΔH is the activation enthalpy and V^* is the activation volume. The activation volume can be written as (Kocks et al. 1975)

$$V^* = lb\Delta x = b\Delta A, \tag{11.18}$$

where ΔA is the activation area which, for the case of the interaction of a dislocation with a localized obstacle, is illustrated by means of Fig. 11.4. The work term is related to the activation volume by

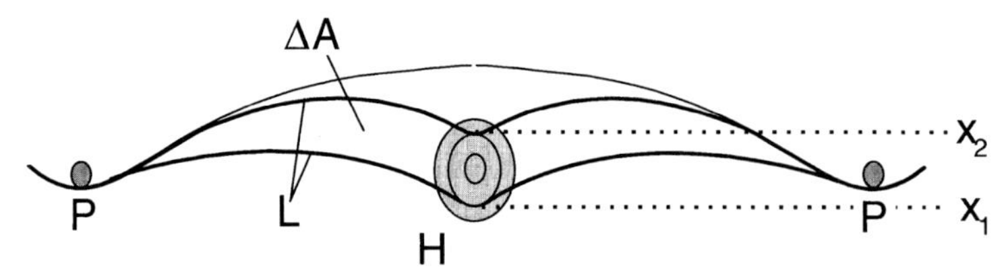

Fig. 11.4. Schematic representation of a dislocation overcoming an obstacle H. The activation area is denoted by ΔA. The dislocation line is shown for two cases: for the beginning of the thermal activation step, where the line is positioned at x_1 at H, and at the end of activation, where the line has moved to x_2.

$$\Delta W = \tau^* V^* . \tag{11.19}$$

The values of the thermodynamic parameters defined above are characteristic of the active microscopic deformation mechanisms. Unfortunately, they cannot be measured directly but have to be determined from the measurable experimental parameters. These are strain, strain rate, applied stress, and temperature.

The activation volume can be determined by means of *stress-relaxation* experiments. In these the sample is at first deformed at a constant rate up to a total strain ε_t. The testing machine is then stopped and, while the sample is held at constant ε_t, the stress is measured as a function of time. The total strain consists of an elastic part ε_e and the plastic part ε (the latter is subject to the definitions in the previous paragraphs), i.e., $\varepsilon_\mathrm{t} = \varepsilon_\mathrm{e} + \varepsilon$. During stress relaxation the elastic part is transformed into additional plastic strain. Since $\varepsilon_\mathrm{t} = const$,

$$\dot{\varepsilon_\mathrm{t}} = \dot{\varepsilon} + \dot{\varepsilon_\mathrm{e}} = 0. \tag{11.20}$$

According to Hooke's law, $\dot{\varepsilon_\mathrm{e}} = C\dot{\sigma}$, where C is the total elastic compliance of the specimen and the testing machine. With Eq. (11.20),

$$\dot{\varepsilon} = -C\dot{\sigma}. \tag{11.21}$$

Taking the logarithm of Eq. (11.15), differentiating with respect to τ, and taking into account Eqs. (11.8) and (11.21), we obtain

$$V = \frac{k_\mathrm{B}T}{m_\mathrm{S}} \frac{\partial \ln(-\dot{\sigma})}{\partial \sigma}\bigg|_T , \tag{11.22}$$

which allows the *experimental* activation volume V to be derived from plots of the negative time derivative of the observed stress as a function of stress. It is important to differentiate between the experimental activation volume V and the theoretical activation volume V^* (Evans and Rawlings 1969, Gibbs 1967). An important prerequisite for $V \approx V^*$ is that, during the experiment, the microstructure remains unchanged.

Phenomenologically the plastic strain rate can be related to the stress by the simple power law

$$\dot{\varepsilon} \propto \sigma^m. \tag{11.23}$$

The stress-relaxation experiments allow us to determine the *stress exponent* m using the relation

$$m = \left. \frac{\partial \ln(-\dot{\sigma})}{\partial ln\sigma} \right|_T. \tag{11.24}$$

We note that the Gibbs free energy of thermal activation cannot be measured directly. Instead, the activation enthalpy can be obtained using the following relations. Taking the logarithm and differentiating Eq. (11.15) with respect to T yields

$$\begin{aligned} \left. \frac{\partial \ln \dot{\varepsilon}}{\partial T} \right|_T &= -\left. \frac{\partial \ln \dot{\varepsilon}}{\partial \sigma} \right|_T \left. \frac{\partial \sigma}{\partial T} \right|_{\dot{\varepsilon}} = -\frac{1}{k_B T} \left. \frac{\partial(\Delta G)}{\partial T} \right|_\tau + \frac{\Delta G}{k_B T^2} \\ &= \frac{T\Delta S + \Delta G}{k_B T^2} = \frac{\Delta H}{k_B T^2}. \end{aligned} \tag{11.25}$$

With Eq. (11.22),

$$\Delta H = -m_S T V \left. \frac{\partial \sigma}{\partial T} \right|_{\dot{\varepsilon}} \tag{11.26}$$

is obtained. Thus, the activation *enthalpy* can be determined by the combination of stress relaxations and temperature changes. The determination of the Gibbs free energy requires additional information on the nature of the entropy terms which is also not directly available from experiments (see below).

11.4.3 Results of Mechanical Testing

Figure 11.5 shows the stress/strain curve for *sq* i-$Al_{70.5}Pd_{21}Mn_{8.5}$ samples deformed in compression along a twofold direction at 760 and 800 °C at a strain rate of 10^{-5} s^{-1}. At 769 °C, after a region of Hooke-type linear elastic behavior, the sample yields plastically at about 180 MPa. There is an *upper yield point* at 270 MPa and 0.6% strain and a *lower yield point* at 220 MPa and a strain of 1%. From there on, the stress decreases continuously with increasing strain. This behavior was observed up to 20% plastic strain without any evidence for saturation. This means that the more the material has already been deformed, the easier it will be to compress it further. This demonstrates the most prominent special feature of QC plasticity, the phenomenon of *deformation softening*.

Figure 11.6 shows the temperature dependence of the maximum flow stress σ_{max}. It decreases from about 750 MPa at 680 °C to 120 MPa at 800 °C. At about 600 °C the samples break due to brittle fracture.

Figure 11.7 shows an example of the evaluation of data obtained in stress-relaxation experiments carried out at 760 °C in terms of the activation volume and the stress exponent by means of Eqs. (11.22) and (11.24). A value of

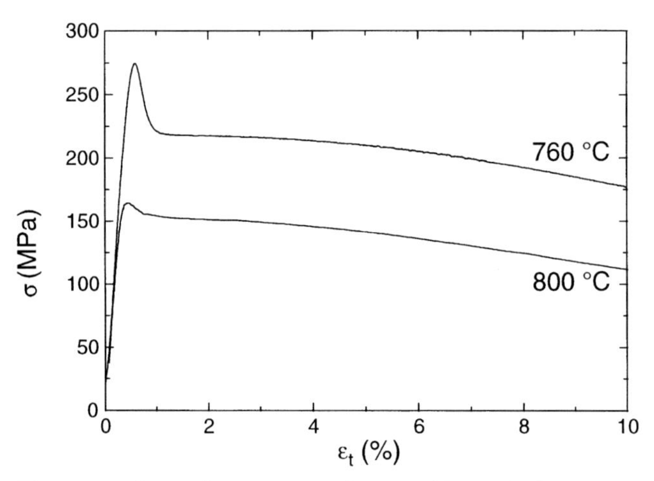

Fig. 11.5. Stress/strain curve for *sq* i-$Al_{70.5}Pd_{21}Mn_{8.5}$ samples deformed in compression along the twofold [0/0, 0/0, 0/2] direction at 760 and 800 °C at a strain rate $\dot{\varepsilon} = 10^{-5}\ s^{-1}$. The cuboid-shaped samples about 7 mm in length and $2{\times}2\ mm^2$ in cross-section were oriented with the long axis (compression axis) parallel to the twofold direction. After Feuerbacher et al. (1997).

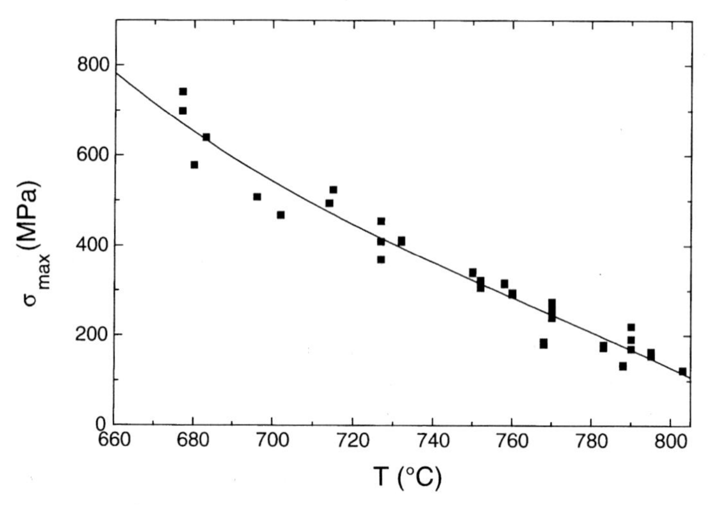

Fig. 11.6. Temperature dependence of the maximum flow stress σ_{max} for deformation along a twofold direction. After Feuerbacher (1996).

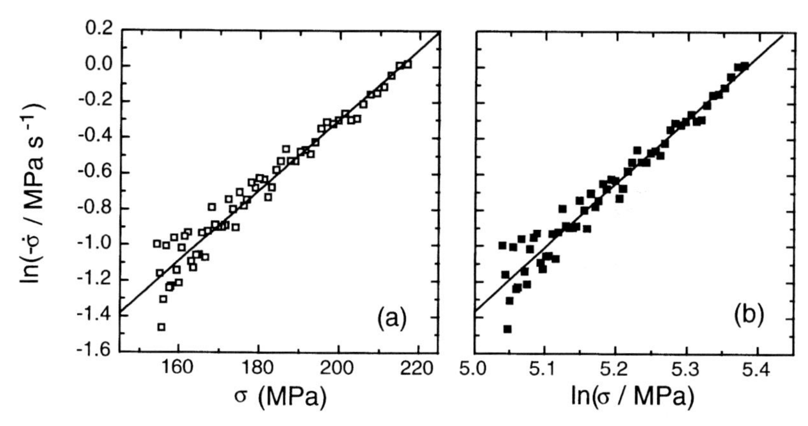

Fig. 11.7. Plot of $\ln(-\dot{\sigma})$ vs σ for the determination of the experimental activation volume (a) and of $\ln(-\dot{\sigma})$ vs $\ln \sigma$ for the determination of the stress exponent (b) on the basis of the data obtained in a stress-relaxation experiment at 760 °C. The stress-relaxation experiment was carried out after the sample was deformed at $\dot{\varepsilon} = 10^{-5}$ to $\varepsilon \approx 0.6\%$ total strain corresponding to the stress maximum in Fig. 11.5. After Feuerbacher (1996).

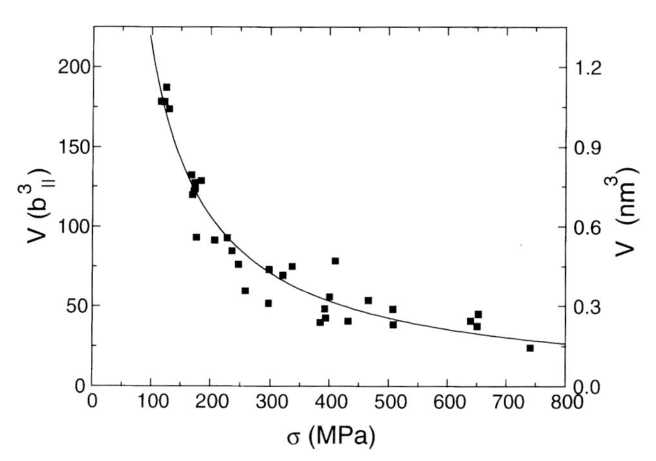

Fig. 11.8. The activation volume as a function of stress determined from stress-relaxation experiments. After Feuerbacher et al. (1997).

$m_S = 0.5$ is employed, an assumption justified by the high isotropy of the i lattice. The results are $V = 0.56\ \mathrm{nm}^3$ and $m = 3.6$. In order to express the activation volume in terms of the modulus of the Burgers vector we take the value of $b_\| = 0.183$ nm, a typical value most frequently found in plastically deformed $Al_{70.5}Pd_{21}Mn_{8.5}$ (see below). We find $V = 92b_\|^3$.

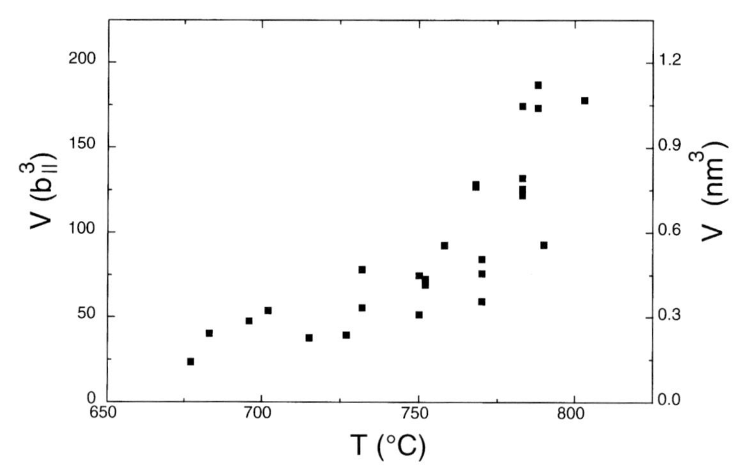

Fig. 11.9. The temperature dependence of the activation volume. After Feuerbacher (1996).

The stress and the temperature dependence of the activation volume is depicted in Figs 11.8 and 11.9, respectively. The stress dependence follows a hyperbolic function. This is to be expected within the framework of classical theory. Inserting Eqs. (11.13) and (11.14) into Eq. (11.15) yields

$$\dot{\varepsilon} \propto \exp\left(\frac{\tau^* V^*}{k_B T}\right). \tag{11.27}$$

With Eq. (11.8)

$$V^* = \frac{k_B T}{m_S} \frac{\partial \ln \dot{\varepsilon}}{\partial \sigma}, \tag{11.28}$$

and with Eq. (11.23) we find $V^* = const(T)/\sigma$.

The values of the activation volume range from 0.2 to 1.2 nm^3. This means that it is about two orders of magnitude larger than $b_\|^3$. The stress exponent decreases linearly with temperature from 5 at 680 °C to 3 at 800 °C.

The stress/strain curve obtained in a typical temperature-change and stress-relaxation experiment is shown in Fig. 11.10. The data were evaluated employing Eq. (11.26). For V, the mean of the value obtained before and after a temperature change, was used. The results are depicted in Fig. 11.11. The activation enthalpy is about 7 eV, which is rather high. The work term also depicted in Fig. 11.11 was calculated according to Eqs. (11.8), (11.19), and (11.22).

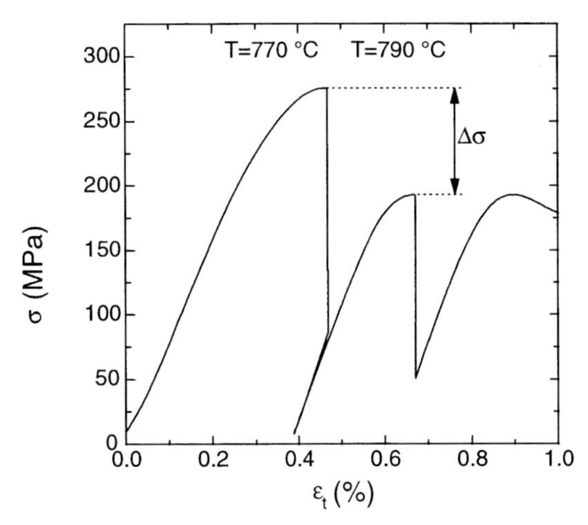

Fig. 11.10. Stress/strain curve recorded during a temperature-change and stress-relaxation experiment. The sample was deformed at a strain rate of 10^{-5} s^{-1} up to the stress maximum at a total strain of 0.47%, after which the stress relaxation was carried out which was followed by deloading and a subsequent temperature increase. At the new stress maximum a second stress relaxation was carried out. After Feuerbacher (1996).

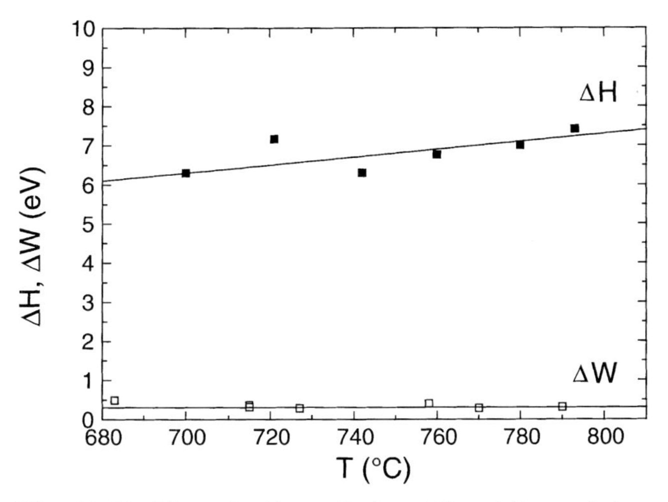

Fig. 11.11. The activation enthalpy ΔH and the work term ΔW as a function of temperature. After Feuerbacher et al. (1997).

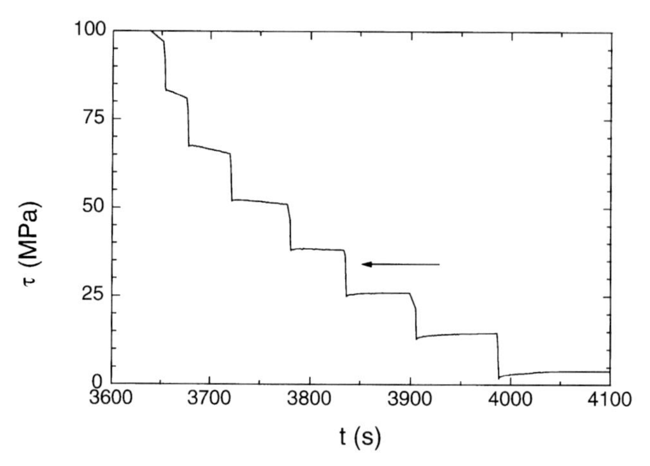

Fig. 11.12. The shear stress as function of time recorded during a stress transient dip test. The stress decreases with time for $\tau > 33$ MPa but increases for smaller values, indicating that $\tau_\mathrm{i} \approx 33$ MPa. After Feuerbacher (1996).

In order to obtain, according to Eq. (11.10), the effective shear stress τ^*, the internal stress level τ_i has to be determined. This can be done by the technique of *stress-transient dip tests* (MacEwen et al. 1969). The shear stress (calculated from the applied stress and employing $m_\mathrm{S} = 0.5$) as a function of time measured in an experiment carried out at 760 °C is shown in Fig. 11.12. For the dip test, in a stress-relaxation experiment the relaxation is stopped at short intervals, the stress is reduced by a certain amount, and the relaxation is continued. As long as $\tau_\mathrm{i} < \tau$, the stress decreases during relaxation. For $\tau_\mathrm{i} > \tau$, it increases. From the change in slope we can derive $\tau_\mathrm{i} \approx 33$ MPa. This means that the internal stress level amounts to about 25% of the maximum shear stress at 760 °C. We note that the measured τ_i in Eq. (11.10) enters at negative sign. This means that, apart from special cases, the long-range stress field resulting from the entire dislocation structure of the sample reduces the effective shear stress (Seeger 1958, Haasen 1986).

11.4.4 Microscopic Observations

At the present state-of-the-art of $Al_{70.5}Pd_{21}Mn_{8.5}$ single-QC growth by the Czochralski technique, the density of grown-in dislocations is of the order of a few 10^7 cm^{-2}. During plastic deformation in the high-temperature region the density increases typically by about two orders of magnitude. Figure 11.13 shows the dislocation arrangement in a sample after deformation at 760 °C to a total strain of 6%. The long dislocation segments lie in the (0/0,0/1,1/0) slip plane. The fact that the dislocations are indeed mobile during deformation is demonstrated by Fig. 11.14 which shows successive video frames recorded during straining *in situ* in a high-voltage electron microscope at 750 °C under

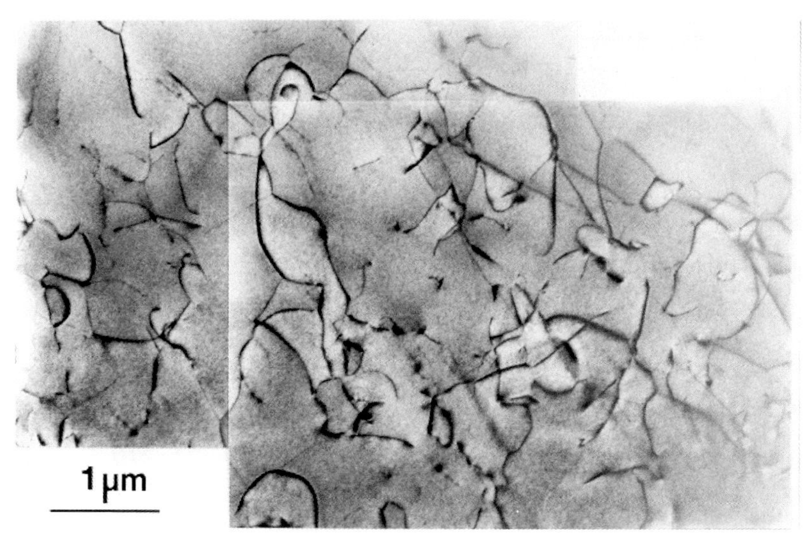

Fig. 11.13. Dislocation arrangement in a sample after deformation at 760 °C to a total strain of 6%. The electron-microscopic sample was prepared in such a way that the (0/0,0/1,1/0) slip plane is parallel to the sample plane. After Feuerbacher (1996).

direct observation. These experiments indicate a viscous, i.e., a smooth and spacially uniform motion of the dislocation lines.

If during the *in-situ* experiments the specimen is oriented for suitable Bragg-diffraction contrast conditions, a bright or dark fault contrast can be recognized which is drawn out by the dislocations during their motion. This is shown in Fig. 11.15. This contrast fades with time and disappears after a time of the order of 10 min. Figure 11.16 shows dislocations in a specimen which was produced from a bulk sample and which was deformed at about 700 °C and subsequently rapidly quenched to room temperature. The fault contrast visible in Fig. 11.16a disappears during a heat treatment of about 10 min at 750 °C in the hotstage of a high-voltage electron microscope (Fig. 11.16b). This observation provides an explanation for the fact that in samples taken from material deformed in the bulk the fault contrast is only seen occasionally. It is only expected for dislocations which have moved shortly before the deformation was stopped and the sample was cooled to room temperature.

Measurements of the dislocation density in *sq* $Al_{70.5}Pd_{21}Mn_{8.5}$ as a function of temperature and strain were carried out by Schall (1998). Figure 11.17a shows stress/strain curves recorded at temperatures between 695 °C and 820 °C. The dislocation density in the undeformed state was about 8×10^7 cm^{-2}. The density at the lower yield point as a function of temperature is depicted in Fig. 11.17b. It drops from about 9×10^9 cm^{-2} at 695 °C to 5×10^8 cm^{-2} at 820 °C. Detailed measurements of the development of the disloca-

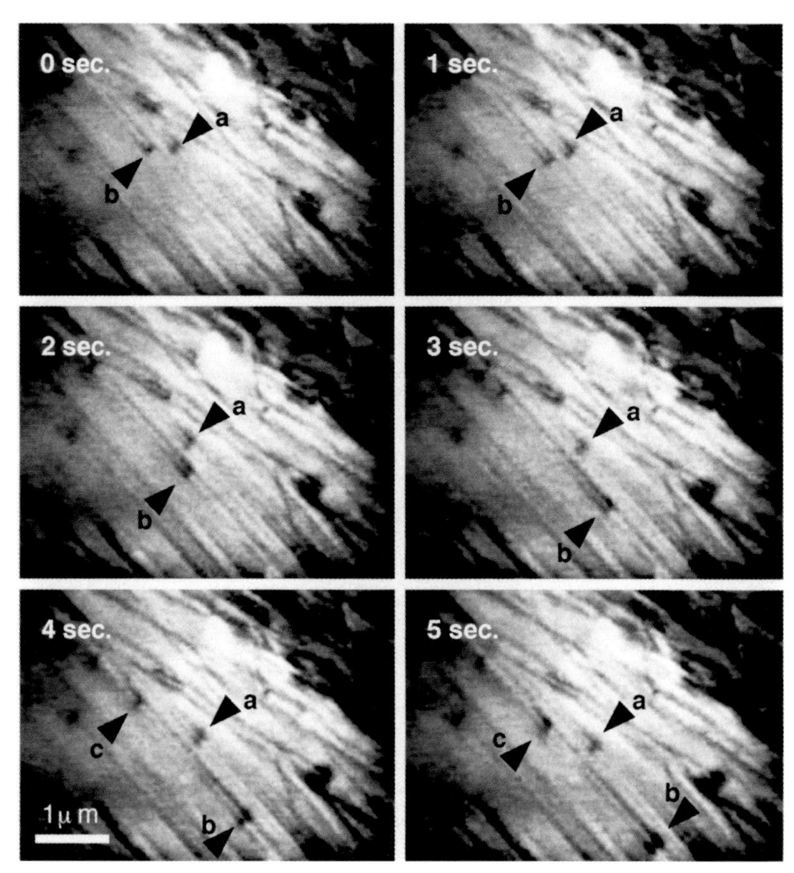

Fig. 11.14. Successive video frames recorded during straining of *sq* $Al_{70.5}Pd_{21}Mn_{8.5}$ *in situ* in a 1 MV high-voltage electron microscope at 750 °C. After Wollgarten et al. (1995).

tion density as a function of strain were carried out at 730 °C (Fig. 11.18). The density rises to a maximum value of 1.2×10^9 cm^{-2} at a plastic strain of 6%, after which it drops to about 6×10^8 cm^{-2} at 12% plastic strain. This indicates that the increase of the dislocation density continues to relatively high strain values compared to the position of the stress maximum (400 MPa) which, at 730 °C, occurs at about 0.8% plastic strain.

It was already noted by Wollgarten et al. (1993b) that dislocations anneal out at temperatures in the range used in the plastic deformation experiments. This effect was investigated in detail by Schall (1998). Figure 11.19 depicts the results of annealing experiments carried out at 730 °C in samples which were predeformed at the same temperature and then quenched to room temperature. Within 30 min the dislocation density falls by about 60% of the original value.

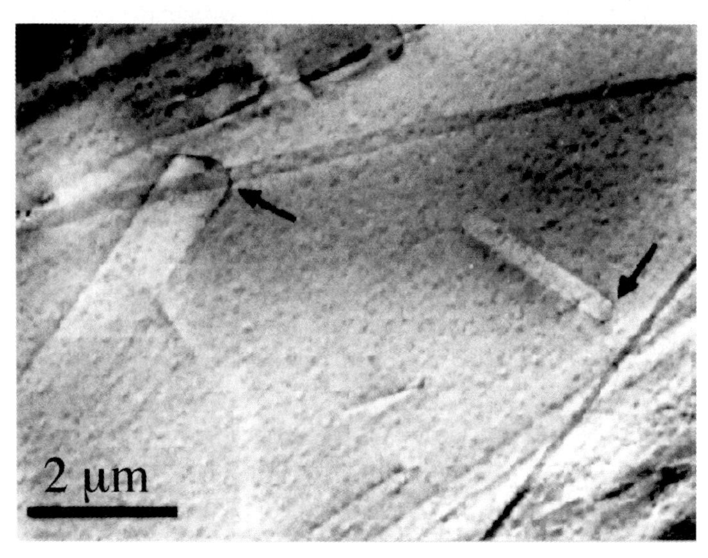

Fig. 11.15. Dislocations during plastic deformation of *sq* $Al_{70.5}Pd_{21}Mn_{8.5}$ in the high-voltage electron microscope at 750 °C. The dark and narrow inclined bands indicate fault contrast drawn out by dislocations which have crossed the field of view. The dislocations marked by arrows exhibit a "tail" of bright contrast which fades with increasing distance from the dislocation line. After Feuerbacher et al. (1997).

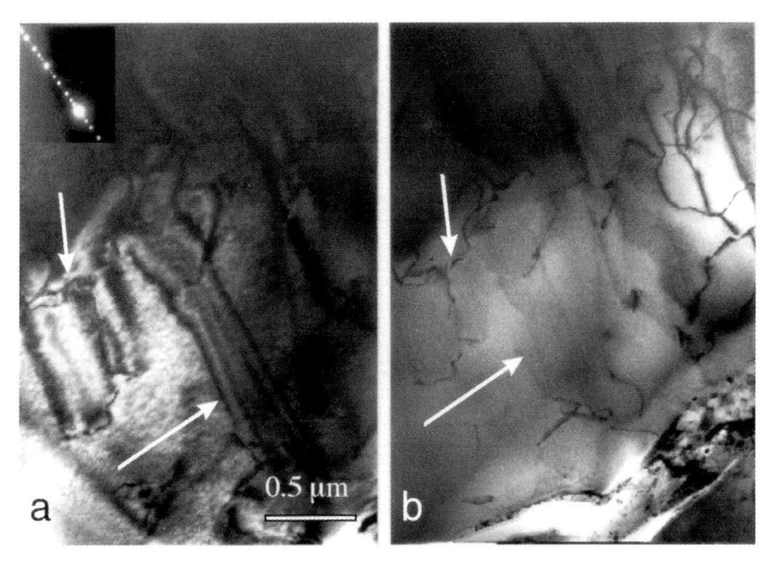

Fig. 11.16. Dislocations in a specimen produced from a bulk sample deformed at about 700 °C and quenched to room temperature. The fault contrast visible in (a) disappears during heat treatment for about 10 min at 750 °C *in situ* in a high-voltage electron microscope (b). After Yang et al. (1998a).

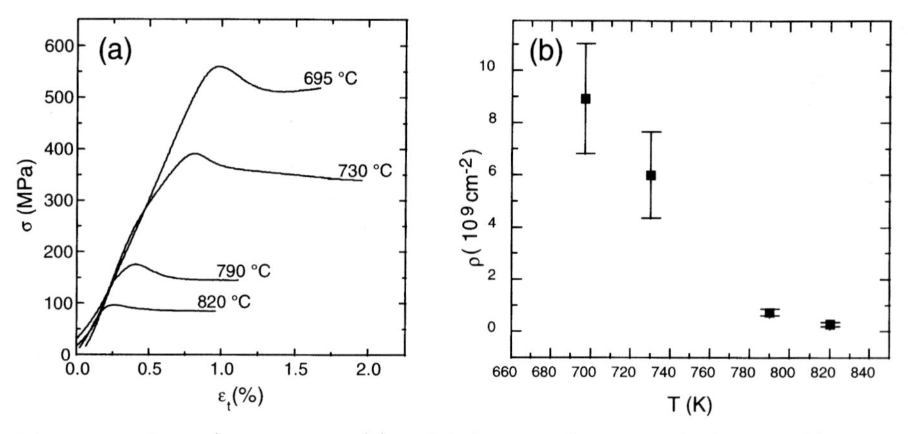

Fig. 11.17. Stress/strain curves (a) and dislocation density at the lower yield point (b) in *sq* $Al_{70.5}Pd_{21}Mn_{8.5}$ as a function of temperature. The samples were deformed at $\dot{\varepsilon} = 10^{-5}$ s^{-1}. After Schall (1998).

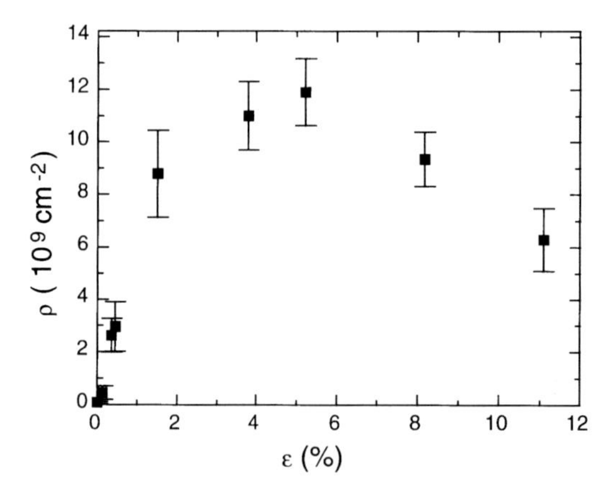

Fig. 11.18. Dislocation density as a function of strain for deformation at $\dot{\varepsilon} = 10^{-5}$ s^{-1} at 730 °C. After Schall (1998).

A detailed analysis of dislocations with respect to Burgers vector and slip system was carried out by Rosenfeld (1994), Metzmacher (1995), and Rosenfeld et al. (1995) in *sq* $Al_{70.5}Pd_{21}Mn_{8.5}$ deformed at 760 and 800 °C. The authors used the d-CBED technique for Burgers vector measurement and stereo analysis for determination of the habit plane. The results obtained from material in the as-grown state and after plastic deformation are compiled in Table 11.1. For a given symmetry direction several types of Burgers vectors exist. These are characterized by different values of the strain accommodation parameter ζ. For the tabulated dislocations this parameter can be expressed in powers of τ, i.e.,

$$\zeta = \tau^n . \tag{11.29}$$

For dislocations with Burgers vectors parallel to twofold QC lattice directions (referred to as *twofold* Burgers vectors), the exponent n is an odd integer

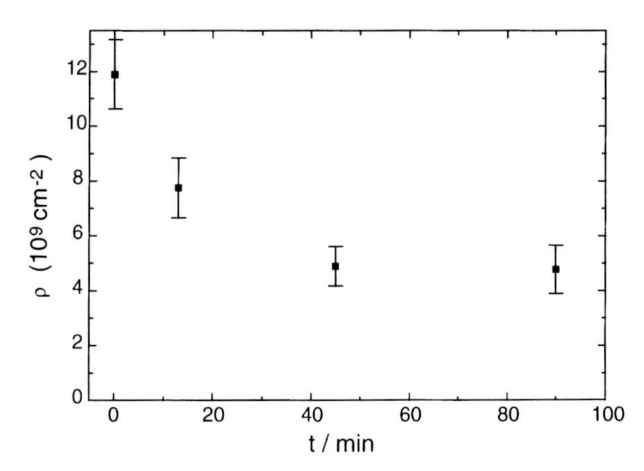

Fig. 11.19. Reduction of the dislocation density during annealing at 730 °C of an $Al_{70.5}Pd_{21}Mn_{8.5}$ sample which was predeformed at the same temperature and quenched to room temperature. After Schall (1998).

Table 11.1. Burgers vectors of dislocations observed in i-$Al_{70.5}Pd_{21}Mn_{8.5}$. Ψ denotes the direction in physical space to which the Burgers vector is parallel. $\mathbf{B}$ is the Burgers vector of the 6D reference dislocation and $\mathbf{b}_{\parallel}$ its Burgers vector component in physical space. $A = 0.645$ nm denotes the lattice parameter of the 6D hyperlattice, $a = A/\sqrt{4+2\tau} = 0.24$ nm, ζ denotes the strain accommodation parameter, and f gives the frequency of occurrence in deformed and as-grown material (in brackets). For a small number of analyzed additional dislocations (not tabulated), the slip systems could not be attributed to simple low-index quasicrystallographic planes and directions. After Metzmacher (1995), Rosenfeld et al. (1995), and Feuerbacher (1996).

Ψ	$\mathbf{B}/A$	$\mathbf{b}_{\parallel}/a$	$\lvert\mathbf{b}_{\parallel}\rvert$/nm	ζ	$f(\%)$
2-fold	$<\bar{1}\,1\,1\,0\,0\,\bar{1}>$	$<\bar{1}/1, 2/\bar{1}, 1/0>$	0.296	$\tau^3 \approx 4.2$	10(22)
2-fold	$<\bar{2}\,1\,1\,0\,0\,\bar{2}>$	$<\bar{2}/1, 3/\bar{2}, 1/\bar{1}>$	0.183	$\tau^5 \approx 11.1$	61(55)
2-fold	$<\bar{3}\,2\,2\,0\,0\,\bar{3}>$	$<\bar{3}/2, 5/\bar{3}, 2/\bar{1}>$	0.113	$\tau^7 \approx 29.0$	18(10)
2-fold	$<\bar{5}\,3\,3\,0\,0\,5>$	$<\bar{5}/3, 8/\bar{5}, 3/\bar{2}>$	0.070	$\tau^9 \approx 76.0$	0(5)
3-fold	$\frac{1}{2}<\bar{3}\,3\,\bar{1}\,3\,1\,\bar{1}>$	$<\bar{3}/2, 0/0, 1/\bar{1}>$	0.159	$\tau^5 \approx 11.1$	2.5(0)
3-fold	$\frac{1}{2}<\bar{5}\,5\bar{1}\,5\,1\,\bar{1}>$	$<\bar{5}/3, 0/0, 2/\bar{1}>$	0.098	$\tau^7 \approx 29.0$	0.6(2.5)
5-fold	$\frac{1}{2}<3\,\bar{1}\,\bar{1}\,\bar{1}\,\bar{1}\,\bar{1}>$	$<2/\bar{1}, \bar{1}/1, 0/0>$	0.174	$\tau^4 \approx 6.9$	2.5(0)
5-fold	$\frac{1}{2}<\bar{4}\,2\,2\,2\,2\,\bar{2}>$	$<\bar{3}/2, 2/\bar{1}, 0/0>$	0.108	$\tau^6 \approx 17.9$	1.9(0)
5-fold	$\frac{1}{2}<7\,\bar{3}\,\bar{3}\,\bar{3}\,\bar{3}\,\bar{3}>$	$<5/\bar{3}, \bar{3}/2, 0/0>$	0.067	$\tau^8 \approx 47.0$	0.6(0)

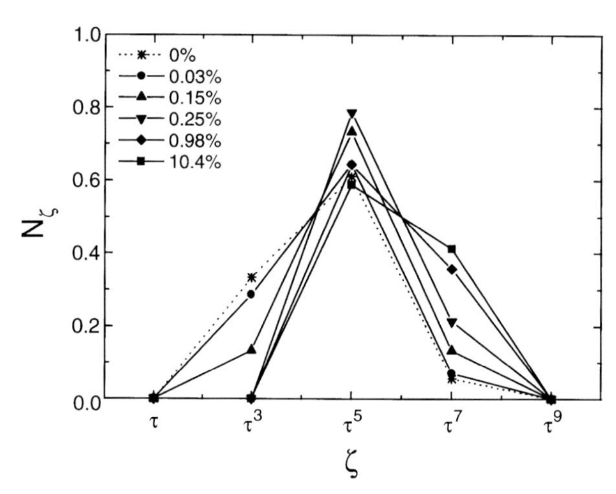

Fig. 11.20. Fraction of dislocations exhibiting a certain value of ζ for different total strain for $Al_{70.5}Pd_{21}Mn_{8.5}$ samples deformed at 760 °C. After Rosenfeld et al. (1995).

whereas in the case of fivefold and twofold Burgers vectors n is even. Both perfect and partial dislocations are observed. The former are characterized by Burgers vectors that are vectors of the F-type hyperlattice (Ebalard and Spaepen 1989, Tsai et al. 1990, Boudard et al. 1992). The latter can be recognized in Table 11.1 by the prefactor 1/2.

Regarding the frequency of occurrence of the individual types of dislocations, it is obvious that the majority has Burgers vector components in physical space which are parallel to twofold QC lattice directions. Among these the frequency of the type of dislocation characterized by $\mathbf{B} = \mathrm{A} < \bar{2}\,1\,1\,0\,0\,\bar{2} >$, $\mathbf{b}_{\|} = \mathrm{a} < \bar{2}/1, 3/\bar{2}, 1/\bar{1} >$, $|\mathbf{b}_{\|}| = 0.183$ nm, and $\zeta = \tau^5$ is by far the highest, i.e., 61% (a is the component of the hyperlattice constant A parallel to physical space and is given by $a = A/\sqrt{4+2\tau}$). Correlating the measured direction of the Burgers vector with the line direction $\mathbf{l}$ obtained by stereo analysis yielded angles between $\mathbf{b}_{\|}$ and $\mathbf{l}$ between close to 0 and 90 °. The dislocations predominantly exhibit a mixed character between screw and edge type (Rosenfeld et al. 1995).

Figure 11.20 shows the fraction of dislocations exhibiting a certain value of ζ for different total strain for samples deformed at 760 °C. The weighted mean $\bar{\zeta}$ vs ε is depicted in Fig. 11.21. It is obvious that the frequency at which dislocations exhibiting high ζ values are observed increases with plastic strain. This means that the character of the dislocations changes with strain. This effect has been investigated in detail both theoretically and experimentally by Wollgarten et al. 1997 and Wang et al. 1998. It was shown that dislocations with elevated ζ values can arise from dislocation reactions and from a novel type of dissociation of perfect dislocations into two other also perfect dislocations. These processes, since they lead to an increase of

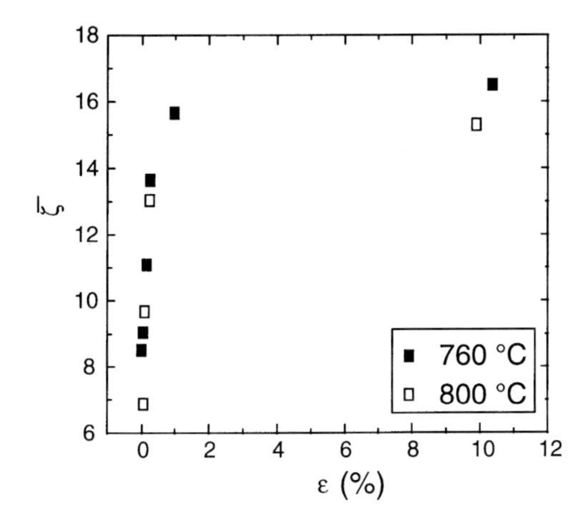

Fig. 11.21. Weighted mean $\bar{\zeta}$ as a function of total strain. After Feuerbacher et al. (1997).

the phason-strain and a corresponding decrease of the phonon-strain component, are energetically favorable if the total strain energy of a QC dislocation increases less strongly with $|\mathbf{b}_\perp|$ than with $|\mathbf{b}_\parallel|$.

The dislocation slip-plane normals $\mathbf{n}$ can be calculated according to $\mathbf{n} = \mathbf{l} \times \mathbf{b}_\parallel$. It is found that slip occurs on low-index planes. Data of an analysis of the slip systems are compiled in Table 11.2. At a twofold $[0/0, 0/0, 0/2]$ sample compression direction, the slip systems most frequently observed consist of a fivefold {0/0,1/0,0/1} slip plane with a twofold $< 1/0, \bar{1}/\bar{1}, 0/1 >$ slip direction and a fivefold {0/1,0/0,1/0} slip plane with a twofold $< 0/\bar{1}, 1/0, 1/1 >$ slip direction, both exhibiting a Schmid factor of 0.43. Other slip systems with twofold and threefold slip planes, having smaller Schmid factors, are found less frequently (Rosenfeld 1994, Metzmacher 1995, Feuerbacher 1996).

Table 11.2. Slip systems observed in *sq* $Al_{70.5}Pd_{21}Mn_{8.5}$ deformed along the twofold $[0/0, 0/0, 0/2]$ direction at 760 °C. Ψ denotes the direction in physical space to which the slip-plane normal is parallel, m_S is the Schmid factor, and f gives the frequency of occurrence. After Rosenfeld et al. (1995) and Feuerbacher (1996).

Ψ	slip plane	slip direction	m_S	$f(\%)$
5-fold	$\{0/0, 1/0, 0/1\}$	$< 1/0, \bar{1}/\bar{1}, 0/1 >$	0.43	11.7
5-fold	$\{0/1, 0/0, 1/0\}$	$< 0/\bar{1}, 1/0, 1/1 >$	0.43	11.7
5-fold	$\{0/0, 1/0, 0/1\}$	$< 1/1, 0/\bar{1}, 1/0 >$	0.26	13.3
5-fold	$\{0/1, 0/0, 1/0\}$	$< \bar{1}/0, 1/1, 0/1 >$	0.26	1.7
2-fold	$\{0/1, 1/0, 1/1\}$	$< \bar{1}/0, \bar{1}/\bar{1}, 0, 1 >$	0.41	16.7
2-fold	$\{0/1, 1/0, 1/1\}$	$< \bar{1}/\bar{1}, 0/1, 1/0 >$	0.25	8.3
3-fold	$\{0/1, 0/0, 1/2\}$	$< \bar{1}/\bar{1}, 0/1, 1/0 >$	0.30	13.3
3-fold	$\{0/0, 1/2, 0/1\}$	$< 0/1, \bar{1}/0, 1/1 >$	0.30	6.7
3-fold	$\{1/1, 1/1, 1/1\}$	$< 1/0, \bar{1}/\bar{1}, 0/1 >$	0.29	3.3

The presence of phason strain in plastically deformed material can be studied by means of quantitative electron diffraction experiments in the transmission electron microscope. In the diffraction pattern, the presence of phason defects leads to deviations of the diffraction peaks from their ideal positions for a perfect lattice (Horn et al. 1986, Bancel and Heiney 1986, Zhang and Kuo 1990). Franz (1997) and Franz et al. (1998) measured for deformed *sq* $Al_{70.5}Pd_{21}Mn_{8.5}$ the shifts of reflections with high values of the perpendicular-space component. This allowed a derivation of the so-called *phason matrix* which relates the observable shifts $\Delta\mathbf{g}_{\parallel}$ to the perpendicular-space component $\mathbf{g}_{\perp}$ of a reflection. We note that, due to the irrational orientation of $\mathbf{E}_{\parallel}$ and $\mathbf{E}_{\perp}$ with respect to $\mathbf{Z}^6$ in $\mathbf{R}^6$, there is a definite relation between $\mathbf{g}_{\parallel}$ and $\mathbf{g}_{\perp}$. This means that after the reflections of an observed diffraction pattern have been indexed in terms of $\mathbf{g}_{\parallel}$, the appertaining $\mathbf{g}_{\perp}$ can be calculated (Cahn et al. 1986). By appropriately placing the selected-area diffraction aperture, the authors used areas of the deformed specimen which were free of dislocations for their measurements (nevertheless, it can be anticipated that these areas were crossed by dislocations during deformation). Figure 11.22 shows the shift of the reflections from the ideal positions in the straight systematic row passing through the incident-beam position. Quantitative results are depicted in Fig. 11.23. While, as expected, the plot of the reflection shift $|\Delta\mathbf{g}_{\parallel}|$ vs $|\mathbf{g}_{\parallel}|$ does not indicate any systematic behavior (left-hand side of Fig. 11.23), the plot of the shift vs $|\mathbf{g}_{\perp}|$ (right-hand side) shows the linear increase with $|\mathbf{g}_{\perp}|$ predicted by phason-strain theory. In addition, we find that the shift increases with strain. This provides direct evidence for the accumulation of phason defects in the QC lattice during deformation.

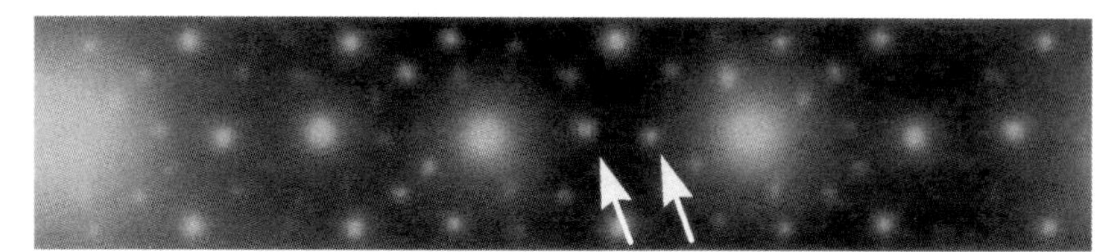

Fig. 11.22. Shift of reflections from the ideal positions in the straight systematic row passing through the incident-beam position (left) in a diffraction pattern recorded from a dislocation-free area of *sq* $Al_{70.5}Pd_{21}Mn_{8.5}$ plastically deformed at 760 °C. After Franz (1997) and Feuerbacher (1996).

11.5 Discussion

From the observations on *sq* *i*-$Al_{70.5}Pd_{21}Mn_{8.5}$ it can be directly inferred that plastic deformation in the ductile temperature regime is based on thermally

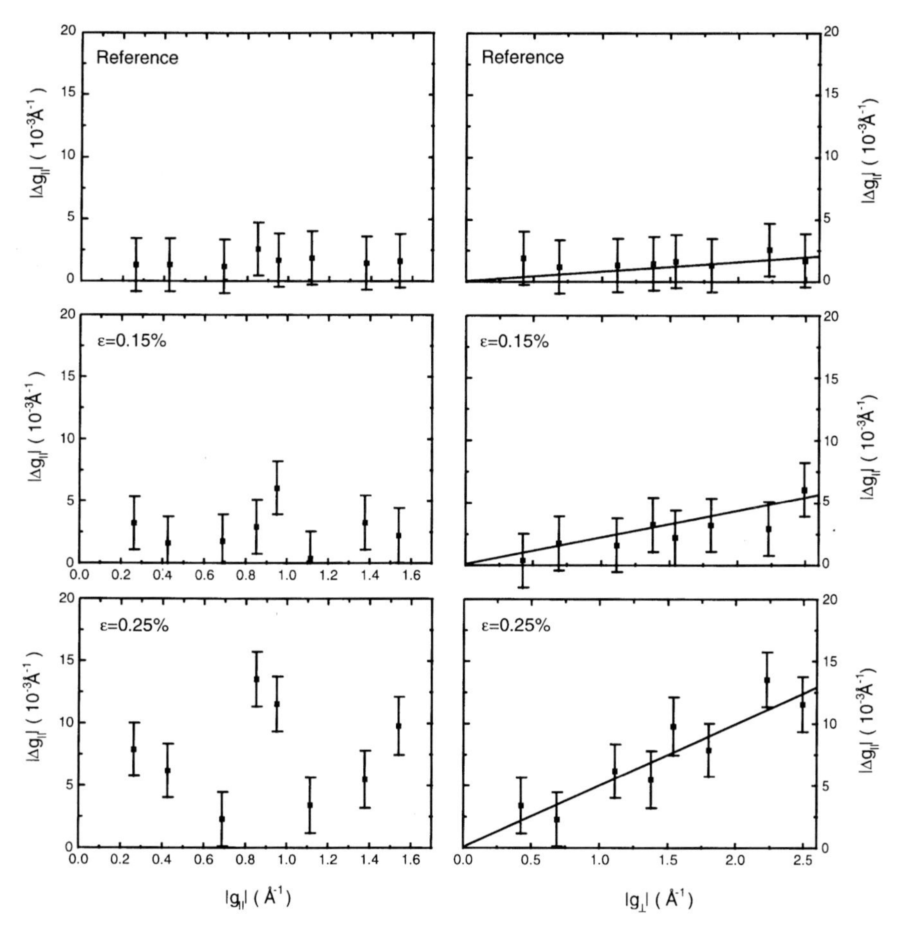

Fig. 11.23. Plot of the measured reflection shift $|\Delta \mathbf{g}_{\|}|$ vs $|\mathbf{g}_{\|}|$ (left) and vs $|\mathbf{g}_{\perp}|$ (right-hand side) for 0.15 and 0.25% plastic strain. After Franz (1997) and Feuerbacher (1996).

activated dislocation motion. As in conventional crystals, dislocations move on low-index planes of the QC lattice (see below). These are fivefold, twofold, and threefold planes which, in this order, also correspond to the densest planes in the Al-Pd-Mn structure.

The pronounced yield phenomenon (Fig. 11.5), i.e., the drop in stress immediately after yielding, is a phenomenon also observed in other materials [see, e.g., Hull and Bacon (1984)]. This can be understood qualitatively using the Orowan relation [Eq. (11.9)]. If the initial dislocation density is low, a high dislocation velocity is required to bring about the strain rate enforced by the testing machine. The velocity depends on stress. As a consequence, the stress in the specimen rises and, as it does so, both the velocity and the density of dislocations increase. This is because above a certain stress level, dislocation sources and multiplication processes become activated. Since multiplication continues even after the dislocations can eventually accommodate the externally applied strain rate, the velocity and therefore the stress required to induce further strain decreases. Introducing Eqs. (11.13), (11.19) and (11.8) into Eq. (11.15), we can derive the following relation for the flow stresses and dislocation densities at the upper (index u) and lower (index l) yield point:

$$\ln \frac{\rho_{\mathrm{l}}}{\rho_{\mathrm{u}}} = \frac{V^*}{k_{\mathrm{B}}T} \Delta\sigma_{\mathrm{y}}, \tag{11.30}$$

where $\Delta\sigma_{\mathrm{y}} = \sigma_{\mathrm{u}} - \sigma_{\mathrm{l}}$. At 760 °C the experimental values are $\Delta\sigma = 57\,\mathrm{MPa}$ and $V = 0.5\ \mathrm{nm}^3$, yielding a dislocation-density ratio of 3.8 which is in excellent agreement with the observed value of 3.6 (Feuerbacher 1996).

The experimental activation volume is a very important parameter with respect to an interpretation of the experimental data in terms of a microscopic mechanism controlling dislocation motion. For a simple Peierls-type lattice friction mechanism, an activation volume comprising a few atomic volumes, i.e., a few $|\mathbf{b}_{\|}|^3$, is expected (e.g., Krausz and Eyring 1975). In contrast, the experimental results (Figs. 11.8 and 11.9) range from 1.2 nm^3 to 0.15 nm^3, i.e., 200 to 25 $|\mathbf{b}_{\|}|^3$ (using the modulus 0.183 nm of the most frequently occurring dislocations with Burgers vector component $\mathrm{a} < \bar{2}/1, 3/\bar{2}, 1/\bar{1} >$ in Table 11.1), depending on stress. This indicates that dislocation motion is not controlled by a Peierls mechanism but by the interaction with obstacles.

The activation enthalpy of 7 eV is a rather high value. Since it is not ΔH but ΔG which determines the temperature dependence of the deformation rate, the contribution of the deformation entropy has to be estimated. For this an approach suggested for crystals by Schöck (1965) can be employed. Making the assumption that the entropy term arises solely from the temperature dependence of the elastic shear modulus, Schöck (1965) derived the following expression

$$\Delta G = \frac{\Delta H + C\tau V^*}{1 - C}, \tag{11.31}$$

where (μ - shear modulus)

$$C = -\frac{T}{\mu} \frac{\partial \mu}{\partial T}. \tag{11.32}$$

With the data for $\mu(T)$ of Tanaka et al. (1996), we arrive for, e.g., 730 °C at $\Delta G \approx 4\mathrm{eV}$, which is still a relatively high value. It therefore appears necessary to discuss the conditions under which the combined stress-relaxation and temperature-change experiments are carried out.

The homologous temperatures, $0.7T_m$ to $0.8T_m$, at which plastic behavior of QCs is observed, are very high. In fact, in Al alloys, creep conditions apply at homologous temperatures higher than about $0.54\,T_{\mathrm{m}}$ (Nabarro and de Villiers 1995). In addition, with self-diffusion activation energies of

1.99 eV (Mn) (Zumkley et al. 1996) and 1.2 eV and 2.3 eV (Pd) (Blüher et al. 1997), substantial long-range atomic motion is expected at $T > 600$ °C in $Al_{70.5}Pd_{21}Mn_{8.5}$. Indeed, the heat-treatment experiments carried out by Schall (1998) in the very temperature range in which the deformation experiments are carried out indicate a drop (Fig. 11.19) in the dislocation density by about 60% within 30 min of annealing at 730 °C. This means that during plastic deformation both dislocation production and thermally induced annihilation processes take place simultaneously. At best, this results in a dynamic equilibrium, and transients are expected during changes of the experimental conditions as employed in the measurements of the thermodynamic parameters. On the other hand, in deriving the theoretical relations in Sect. 11.4.2 and applying them for the evaluation of the experimental data, the assumption is made that the microstructure remains constant. This means that to ensure an at least approximate applicability of these relations experimental conditions have to be chosen for which annealing effects are limited.

Based on the observation by Schall (1998) that during annealing the dislocation density adopts a quasi-stationary value after about 30 min (Fig. 11.19), Geyer et al. (1998) investigated the effect of repeating a stress-relaxation experiment. For this purpose, after the first relaxation (R_1), the specimen was reloaded to stress values lower than the stress at which the R_1 experiment was begun. As expected, the two relaxation curves did not coincide. The curves of the second relaxation (R_2) exhibited smaller strain rates (proportional to the negative stress rates) for the same stresses and therefore yielded smaller values for the activation volume. This can be explained by the reduction of the dislocation density due to annealing during R_1. Reloading to the starting stress of R_2 and performing a third relaxation yielded the same result as R_2, indicating an essentially stable dislocation structure. Geyer et al. (1998) therefore only employed the R_2 data for the determination of ΔH. In addition, they performed their measurements just behind the lower yield point rather than at the upper yield point (Feuerbacher et al. 1995, 1997) in order to reduce the effect of the temperature-dependent yield phenomenon ($\Delta\sigma_y$ decreases with temperature, see Fig. 11.17) on the results. In this way, lower activation enthalpy values were obtained, i.e., about 4.5 eV for temperatures between 680 and 800 °C. Using Eq. (11.31) yields $\Delta G = 2.3$ eV at 735 °C.

With about 33 MPa the internal stress value obtained by the dip tests carried out at 760 °C (Feuerbacher 1996) amounts to about 12% of the total flow stress after the lower yield point (Fig. 11.5). Calculating the long-range athermal component of the flow stress according to the Taylor formula (Taylor 1934)

$$\tau_i = \alpha\mu b\sqrt{\rho}, \tag{11.33}$$

with $b = b_{\|} = 0.183\,\mathrm{nm}$, $\mu = 50\,\mathrm{GPa}$ (Tanaka et al. 1996), $\rho = 6\times10^9$ cm^{-2} (Schall 1998), and $\alpha \approx 0.3$ yields $\tau_i \approx 36\,\mathrm{GPa}$. This indicates that the observed dislocation density can account for the observed internal-stress value.

A particularly important result of the *in-situ* deformation experiments (Figs. 11.15 and 11.16) and of the diffraction analysis of deformed samples (Fig. 11.23) is the production of a fault contrast behind the moving dislocations and an accumulation of phason strain in the samples. On the other hand, the fault contrast is much less frequently observed in the *ex-situ* contrast analysis of dislocations in samples prepared from material deformed in the bulk condition. From this it can be concluded that QC dislocations in general do not move in the ideal configuration derived above in Sect. 11.3.2. The applied shear stress enforces slip by rapid motion of the dislocations under conditions where they cannot maintain their "equilibrium" configuration with respect to their phason strain field. As a consequence, the dislocations leave behind a *phason layer* in their plane of motion. At the high temperature of deformation which, as already discussed, lies well above the self-diffusion temperature, the related phason defects can diffuse apart and eventually, after a very long time, anneal out.

There is a number of studies in which the motion of dislocations in 2D and 3D quasiperiodic model lattices is investigated by computer simulation. The results are in excellent agreement with the experimental observations. Molecular-statics calculations indicate that the dislocations glide on well-defined planes, producing a layer of a high concentration of phason defects along their glide plane (Dilger et al. 1997, Takeuchi et al. 1997). In a 3D simulation, the energy of the phason layers was estimated to be about 58% of the surface energy of a QC (Dilger et al. 1997). In 2D molecular-dynamics simulations the dislocations were observed to stop in front of local atomic configurations of high coordination (2D atom clusters). At low temperature, the phason layers remain flat, while at elevated temperature the defects are observed to diffuse away from the glide plane (Mikulla et al. 1998a).

11.5.1 Model of Dislocation Friction in Quasicrystals

According to the structure model of Boudard et al. (1992) for *i*-Al-Pd-Mn, the structure can be described by an arrangement of two types of Mackay-type clusters. These clusters of 52 atoms exhibit *i* symmetry and consist of a central Mn atom, a core consisting of Al atoms surrounded by an inner icosahedron of either Mn and Al atoms or Mn and Pd, and an outer icosidodecahedron of either Pd and Al atoms or Al atoms only. Their diameter is about 0.9 nm. Perfect Mackay-type clusters comprise about 60% of the atoms while the rest is arranged in patterns which can also be described on the basis of the Mackay-type clusters but taking into account overlapping. The physical properties of this cluster-based structure have been discussed by Janot and de Boissieu (1994) and Janot (1995) who arrived at the conclusion that, due to their particular electronic properties, the Mackay-type clusters have to be considered as energetically very stable entities. This is corroborated by the results of the investigation of cleavage surfaces of *sq* $Al_{70.5}Pd_{21}Mn_{8.5}$ discussed in Sect. 11.2.2. Complete clusters are preserved at the surface (Fig.

11.1). This indicates that cracks circumvent the Mackay-type clusters, which demonstrates that they are mechanically strong entities capable of deflecting the propagating cracks.

We note that Mikulla et al. (1998a, 1998b) studied crack propagation as a function of strain in a 2D binary model QC by molecular-dynamics simulations. At low load and low strain values, the crack tip proceeds with constant velocity along an "easy" line until it hits an obstacle. In the model, this is a centre of high coordination and strength corresponding to the 3D clusters just mentioned. The crack tip then stops and emits a dislocation along another easy lattice line along which the shear stress is highest. The dislocation core moves until it is stopped itself by another obstacle. This dislocation has left a track of high phason-defect concentration in its wake. There the cohesive strength of the material is reduced and, after a period of further straining, the crack propagates along this line. The result of crack propagation is a rough cleavage line whose profile shows the 2D cluster obstacles. These observations are in very good agreement with the STM results reported in Sect. 11.2.2.

Figure 11.24 shows the arrangement of complete Mackay-type clusters calculated for sections of the model of Boudard et al. (1992) projected along twofold direction. That this arrangement indeed corresponds to the STM images can be realized by rotating the image of Fig. 11.1 counterclockwise by about 30 °, which aligns it parallel to the cluster arrangement shown in Fig. 11.24 for the twofold plane. Regarding, e.g., the arrangement calculated for the view along the twofold direction we find in Fig. 11.24 that there are planes where the clusters above and below the plane exhibit, along the viewing direction, very little overlap. These planes are the twofold (inclined in the figure), threefold, and fivefold planes (the latter arranged horizontally in the figure). As will be discussed below, these are good candidates for dislocation glide.

It was suggested by Feuerbacher et al. (1997) that the Mackay-type clusters act as obstacles to the dislocation motion, and that the resulting friction controls the plastic deformation. The main argument in favour of this *cluster-friction model* is based on the value of the activation volume which is larger by one to two orders of magnitude than the few atomic volumes expected for a simple Peierls model in which dislocation motion would be controlled by double-kink formation [e.g., Hirth and Lothe (1982)]. Indeed, the mutual distance as well as the volume of the Mackay-type clusters fit this length scale.

We note that in the model of Boudard et al. (1992) the atoms are aligned in sets of quasiperiodically spaced planes. Certainly, due to its large diameter, a given Mackay-type cluster comprises atoms localized on many different sets of parallel atomic planes. As a consequence, when a dislocation moves on these planes it can, in principle, cut through the Mackay-type clusters at many different azimuthal angles. However, on account of the high mechanical strength of the clusters demonstrated by the STM experiments and by the

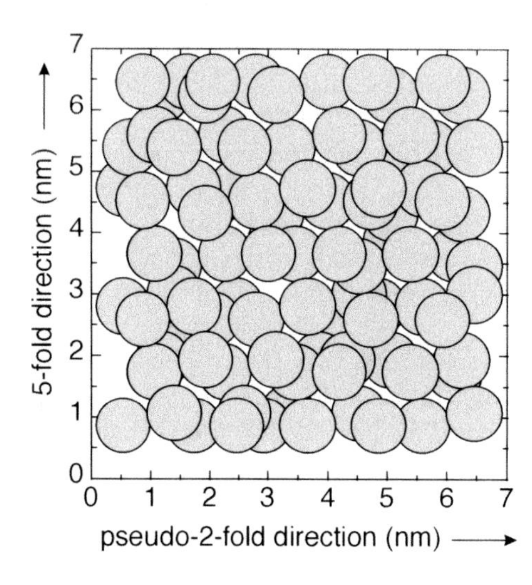

Fig. 11.24. Arrangement of complete Mackay-type clusters (diameter 0.9 nm) calculated for a section of the model of Boudard et al. (1992) projected along a twofold direction. Note that the complete clusters comprise only 60% of the atoms. After Yang et al. (1998c).

corresponding results of the computer simulations (Mikulla et al. 1995, 1998a) it appears obvious to assume that the dislocations avoid as far as possible the central parts of the clusters. This defines certain planes out of a full set of parallel atomic planes which are especially favorable for glide. In Fig. 11.24 these are the planes of small overlap of the clusters along the viewing direction which we have already discussed. QCs therefore differ from ordinary simple crystals for which any plane of a set of parallel atomic planes is suitable as a glide plane.

The cluster-friction model has been worked out in detail by Messerschmidt et al. (1998). Assuming that the rate-controlling event is, as sketched in Fig. 11.4, the overcoming of individual obstacles, a maximum value of the activation volume can be estimated. For this we assume that the three obstacles indicated in the figure are neighbouring Mackay-type clusters. Inserting in Eq. (11.18) both for l and Δx the cluster diameter, and for $b = b_{\|} = 0.183$ nm, yields $V^* = 0.15\,\mathrm{nm}^3$. This value has to be compared with the maximum measured value of about $1.1\,\mathrm{nm}^3$. As can be understood on the basis of. Eq. (11.18) and the definitions in Fig. 11.3, the maximum of the activation volume is expected for the lowest stress values. It is found that the measured value is about an order of magnitude larger than the activation volume estimated for a process where Mackay-type clusters are overcome individually. Indeed, due to line-energy arguments, it appears unlikely that a dislocation bows out between two neighbouring individual clusters in the way anticipated for this estimate.

In order to find an explanation for this discrepancy Messerschmidt et al. (1998) investigated the conditions for the applicability of the theory developed by Labusch (1988) and Labusch and Schwarz (1992) for so-called *extended* obstacles. This theory treats the situation where a relatively rigid dislocation moves in the field of relatively weak extended obstacles. For certain conditions this theory yields the simultaneous activation of several obstacles acting as one *effective* obstacle with considerably enlarged activation volume and activation energy. This is the case if the characteristic dimension of the obstacle interaction profile is not small compared to the dimension of the region of bending of the dislocation between obstacles. The extended-obstacle condition holds if the normalized obstacle width η_0 is not small compared to 1, where

$$\eta_0 = d_0 \sqrt{\frac{2c\Gamma}{F_0}}. \tag{11.34}$$

Here d_0 denotes the effective Δx value at zero stress, c is the effective concentration of obstacles in the slip plane, F_0 is the maximum force the obstacles can sustain, and Γ is the dislocation line tension. According to Labusch and Schwarz (1992), F_0 can be calculated if the athermal flow stress τ_0 related to the obstacle arrangement is known. From the computer calculations of Mikulla et al. (1998b), $\tau_0 \approx 0.03\,\mu$ can be derived. The calculations of Messerschmidt et al. (1998) for a wide range of parameter values yield the result $0.3 \leq \eta_0 \leq 0.9$. This means that η_0 is not small compared to 1 and therefore the extended-obstacle situation holds.

Applying the numerical data of Labusch and Schwarz (1992), it is found that the *experimental* value for the activation volume is about a factor of 10 larger than that appertaining to a single-obstacle process. Furthermore, the Gibbs free energy of activation is about 2.5 times larger than the value appertaining to the overcoming of an individual cluster. Using the aforementioned value for ΔG of 2.3 eV determined at 735 °C by Geyer et al. (1998), a value for the activation energy appertaining to the overcoming of an individual obstacle of about 0.9 eV is obtained. This is of a reasonable order of magnitude.

In conclusion, the cluster-friction model allows us to qualitatively understand the rate-controlling process of QC plasticity. However, our discussion has shown that a decisive element of this model is that dislocation interacts simultaneously with groups of Mackay-type clusters. With respect to the thermodynamic activation parameters, these groups are acting as one effective obstacle.

According to the model of Boudard et al. (1992), the QC is an ordered alloy which owes its particular structure, lattice symmetry, and physical properties, including its mechanical strength, to the particular arrangement of the Mackay-type clusters (Janot 1996). The production of phason defects, i.e., matching-rule violations in the structure of the QC lattice, by the motion of dislocations has already been discussed. These defects are equivalent to chemical and structural disorder. Taking into account the diameter of the Mackay-type clusters of about 0.9 nm, it is obvious that any macroscopic shear induced by the moving dislocations will cut these clusters apart all along the glide plane if a given plane, as observed in the *in-situ* experiments

(Wollgarten et al. 1995), is used by a higher number of dislocations following each other. This means that the original QC structure is successively replaced by a substantially disordered structure. Even if, as described above, annealing effects are taken into account at the high temperatures at which the experiments are carried out, the structure cannot recover completely. This is demonstrated by the diffraction experiments of Franz (1997) and Franz et al. (1998) in which an increase in the density of phason defects with strain was found.

The modification and eventual destruction of the cluster-based framework of the QC lattice can be considered responsible for the *deformation softening* observed as a general feature of QC high-temperature deformation. If the material is weakened in the course of plastic deformation a reduced stress is needed to maintain a given dislocation velocity. As a consequence, in experiments at constant strain rate, this leads, as observed, to a decrease of the flow stress with increasing plastic strain. The density of dislocations contributing to shear is the result of a dynamic equilibrium between dislocation production in sources and simultaneous annealing, as observed experimentally by Schall (1998). According to the Orowan relation [Eq. (11.9)] an increasing dislocation velocity requires a lower steady-state value of the dislocation density to maintain the constant strain rate imposed by the testing machine.

11.6 Concluding Remarks

Plastic deformation of QCs is only possible in a range of very high homologous temperatures which in other materials have mainly been studied under the aspects of creep. For this reason the applicability of the theoretical relations of Sect. 11.4.2 has to be carefully investigated. The experimental results available today allow us to draw a qualitative picture of QC plasticity. However, much more work will be required if the picture is to become more quantitative. In addition, many aspects of QC plasticity have not been studied yet. Examples are the detailed understanding of the complete absence of work hardening and the technically relevant problem of the plastic behavior of well-defined polyquasicrystalline materials. With respect to the field of work to be covered, this review is brief, and omissions are unavoidable. We have essentially concentrated on the plastic behavior of *sq* *i*-$Al_{70.5}Pd_{21}Mn_{8.5}$ since this material can be considered a model system due to the unique quality of the single QCs obtained by the Czochralski technique. Only recently have equivalent studies become possible in Al Ni Co *d* single QCs (Feuerbacher et al. 1997). In these studies the experimental results suggested the existence of an equivalent friction process based on the interaction of dislocations with columnar clusters as the rate-controlling mechanism for dislocation motion. A number of plastic deformation experiments have been carried out on polyquasicrystalline *i*-Al-Cu-Fe (Bresson and Gratias 1993, Shield et al. 1993, Bresson 1994, Shield 1997, Köster et al. 1998). However, no evidence has been found for a dislocation mechanism in this material. We should note the very recent work on the measurement of the thermodynamic activation parameters of singlequasicrystalline Al-Li-Cu (Semadeni et al. 1998) and of polyquasicrystalline i-$Mg_{36}Zn_{56}Y_8$ and i-$Mg_{36}Zn_{56}Gd_8$ (Takeuchi et al. 1998). It is obvious that materials preparation is the key issue for any type of physical-property investigation. Progress in our understanding of the mechanical properties of QCs will therefore largely depend on the success of the efforts under way to produce better defined materials both with respect to structure and chemical composition.

Acknowledgments

A major part of this chapter was written while one of the authors (K.U.) was a guest professor at the Institute of Advanced Materials Processing at Tohoku University, Sendai, Japan. Thanks are due to Prof. Y. Waseda for his hospitality and the Japanese Society for the Promotion of Science for a Japanese-German Research Award. Thanks are also due to Prof. S. Suzuki for his invaluable support during this stay. The authors are grateful to V. Franz, D. Geyer, C. Metzmacher, R. Rosenfeld, and Dr. N. Tamura for stimulating discussions.

References

Baluc, N., Yu, D.P., Kléman, M. (1995): Philos. Mag. Lett. **72**, 1

Bancel, P.A., Heiney, P.A. (1986): J. Phys. (Paris) **C3**, 341

Bardes, B.P. (ed) (1978): Metals Handbook, Vol. 1, 9th ed. American Society of Metals, Metals Park

Bhaduri, S.B., Sekhar, J.A. (1987): Nature **327**, 609

Blüher, R., Scharwaechter, P., Frank, W., Kronmüller, H. (1998): Phys. Rev. Lett. **80**, 1014

Bohsung, J., Trebin, H.-R. (1989): in Introduction to the Mathematics of Quasicrystals, Jarić, M.V. (ed). Academic, Boston, p 183

Boudard, M., de Boissieu, M., Janot, C., Heger, G., Beeli, C., Nissen, H.-U., Vincent, H., Ibberson, R., Audier, M., Dubois, J.M. (1992): J. Phys. Condens. Matter **4**, 10 149

Boudard, M., Bourgeat-Lami, E., de Boissieu, M., Janot, C., Durand-Charre, M., Klein, H., Audier, M., Hennion, B. (1995): Philos. Mag. Lett. **71**, 11

Bresson, L., Gratias, D. (1993): J. Non-Cryst. Solids **153-154**, 468

Bresson, L. (1994): in Lectures on Quasicrystals, Hippert, F., Gratias, D. (eds). Les Editions de Physique, Les Ulis, p 549

Büchler, E.H., Watanabe, E., Kazama, N.S. (1997): Int. J. Non-Equil. Process. **10**, 35

Cahn, J.W., Shechtman, D., Gratias, D. (1986): J. Mater. Res. **1**, 13

Chen, H.S., Chen, C.H., Inoue, A., Krause, J.T. (1985): Phys. Rev. B **32**, 1940

Cherns, D., Preston, A.R. (1986): J. Electron Microsc., Suppl. **35**, 721

Cotrell, A.H. (1953): Dislocations and Plastic Flow in Crystals. Oxford University Press, London
Courtney, T.H. (1990): Mechanical Behavior of Materials. McGraw-Hill, New York
Dilger, C., Mikulla, R., Roth, J., Trebin, H.R. (1997): Philos. Mag. A **75**, 425
Dubois, J.M., Kang, S.S., von Stebut, J. (1991): J. Mater. Sci. Lett. **10**, 537
Ebalard, S., Spaepen, F. (1989): J. Mater. Res. **4**, 39
Ebert, Ph., Feuerbacher, M., Tamura, N., Wollgarten, M., Urban, K. (1996): Phys. Rev. Lett. **77**, 3827
Evans, A.G., Rawlings, R.D. (1969): Phys. Status Solidi **34**, 9
Feuerbacher, M. (1996): Doctor Thesis, Rheinisch Westfälische Technische Hochschule Aachen
Feuerbacher, M., Baufeld, B., Rosenfeld, R., Bartsch, M., Hanke, G., Beyss, M., Wollgarten, M., Messerschmidt, U., Urban, K. (1995): Philos. Mag. Lett. **71**, 91
Feuerbacher, M., Weller, M., Diehl, J., Urban, K. (1996): Philos. Mag. Lett. **74**, 81
Feuerbacher, M., Metzmacher, C., Wollgarten, M., Urban, K., Baufeld, B., Bartsch, M., Messerschmidt, U. (1997): Mater. Sci. Eng. A **233**, 103
Franz, V. (1997): Diploma Thesis, Rheinisch Westfälische Technische Hochschule Aachen
Franz, V., Feuerbacher, M., Wollgarten, M., Urban, K. (1998): Philos. Mag. Lett., in press
Gastaldi, J., Reinier, E., Grange, G., Jourdan, C., Smolsky, I. (1995): in Proceedings of the 5th International Conference on Quasicrystals, Janot, C., Mosseri, R. (eds). World Scientific, Singapore, p 287
Geyer, B., Bartsch, M., Feuerbacher, M., Urban, K., Messerschmidt, U. (1998): unpublished
Gibbs, G.B. (1967): Philos. Mag. **16**, 97
Haasen, P. (1986): Physical Metallurgy. Cambridge University Press, Cambridge
Henley, C.L. (1987): Comments Condens. Matter Phys. **13**, 59
Hiraga, K., Hirabayashi, M. (1987): Jpn. J. Appl. Phys. **26**, L155
Hirsch, P.B., Howie, A., Whelan, M.J. (1960): Philos. Trans. R. Soc. London, Ser. A **252**, 499
Hirth, J.P., Lothe, J. (1982): Theory of Dislocations. John Wiley and Sons, New York
Horn, P.M., Malzfeldt, W., DiVincenzo, D.P., Toner, J., Gambino, R. (1986): Phys. Rev. Lett. **57**, 1444
Hull, D., Bacon, D.J. (1984): Introduction to Dislocations, 3rd ed. Pergamon, Oxford
Inoue, A., Kimura, H.M., Sasamori, K., Masumoto, T. (1995): Mater. Trans., Jpn. Inst. Met. **36**, 6
Inoue, A., Kimura, H., Kita, K. (1997): in New Horizons in Quasicrystals, Goldman, A.I., Sordelet, D.J., Thiel, P.A., Dubois, J.M. (eds). World Scientific, Singapore, p 256
Janot, C. (1992): Quasicrystals, A Primer, 2nd ed. Oxford University Press, New York
Janot, C. (1996): Phys. Rev. **53**, 181
Janot, C., de Boissieu, M. (1994): Phys. Rev. Lett. **72**, 1674
Kang, S.S., Dubois, J.M. (1992): Philos. Mag. A **66**, 151
Kang, S.S., Dubois, J.M., von Stebut, J. (1993): J. Mater. Res. **8**, 2471
Kita, K., Saitoh, K., Inoue, A., Masumoto, T. (1997): Mater. Sci. Eng. A **226-228**, 1004
Kléman, M. (1988): in Quasicrystalline Materials, Janot, C., Dubois, J.M. (eds). World Scientific, Singapore, p 318

Kléman, M., Sommers, C. (1988): Acta Metall. Mater. **39**, 287
Kocks, U.F., Argon, A.S., Ashby, M.F. (1975): Prog. Mater. Sci. **19**, 1
Köster, U., Liu, W., Liebertz, H., Michel, M. (1993): J. Non-Cryst. Solids **153-154**, 446
Köster, U., Ma, X.L., Greiser, J., Liebertz, H.(1998): in Proceedings of the 6th International Conference on Quasicrystals, Takeuchi, S., Fujiwara, T. (eds). World Scientific, Singapore, p 505
Krausz, A.S., Eyring, H. (1975): Deformation Kinetics. John Wiley and Sons, New York
Labusch, R. (1988): Czech. J. Phys. B **38**, 474
Labusch, R., Schwarz, R.B. (1992): in Proceedings of the 9th International Conference on the Strength of Metals and Alloys, Brandon, D.G., Chaim, R., Rosen, A. (eds). Freund, London, p 47
Levine, D., Steinhardt, P.J. (1984): Phys. Rev. Lett. **53**, 2477
Levine, D., Lubensky, T.C., Ostlund, S., Ramaswamy, S., Steinhardt, P.J., Toner, J. (1985): Phys. Rev. Lett. **54**, 1520
Liu, P., Hultin Stigenberg, A., Nilsson, J.O. (1995): Acta Metall. Mater. **43**, 2881
Liu, P., Nilsson, J.-O. (1997): in New Horizons in Quasicrystals, Goldman, A.I., Sordelet, D.J., Thiel, P.A., Dubois, J.M. (eds). World Scientific, Singapore, p 264
Lubensky, T.C., Ramaswamy, S., Toner, J. (1985): Phys. Rev. B **32**, 7444
Lubensky, T.C. (1988): in Introduction to Quasicrystals, Jarić, M.V. (ed). Academic, Boston, p 199
MacEwen, S.R., Kupcis, O.A., Ramaswami, B. (1969): Scr. Metall. **3**, 441
Mackay, A. (1962): Acta Crystallogr. **15**, 916
Messerschmidt, U., Bartsch, M., Feuerbacher, M., Geyer, B., Urban, K. (1998): unpublished
Metzmacher, C. (1995): Diploma Thesis, Rheinisch Westfälische Technische Hochschule Aachen
Mikulla, R., Roth, J., Trebin, H.-R. (1995): Philos. Mag. B **71**, 981
Mikulla, R., Krul, F., Gumbsch, P., Trebin, H.-R. (1997): in New Horizons in Quasicrystals, Goldman, A.I., Sordelet, D.J., Thiel, P.A., Dubois, J.M. (eds). World Scientific, Singapore, p 200
Mikulla, R., Stadler, J., Gumbsch, P., Trebin, H.-R. (1998a): in Proceedings of the 6th International Conference on Quasicrystals, Takeuchi, S., Fujiwara, T. (eds). World Scientific, Singapore, p 485
Mikulla, R., Stadler, J., Krul, F., Trebin, H.R., Gumbsch, P. (1998b): preprint
Mughrabi, H. (1993): in Materials Science and Technology, Vol. 6, Mughrabi, H. (ed). VCH, Weinheim, p 1
Nabarro, F.R.N., de Villiers, H.L. (1995): The Physics of Creep. Taylor and Francis, London
Reimer, L. (1989): Transmission Electron Microscopy, 2nd ed. Springer-Verlag, Berlin
Reynolds, G.A., Golding, B., Kortan, A.R., Parsey, J.M. (1990): Phys. Rev. B **41**, 1194
Rosenfeld, R. (1994): Diploma Thesis, Rheinisch Westfälische Technische Hochschule Aachen
Rosenfeld, R., Feuerbacher, M., Baufeld, B., Bartsch, M., Wollgarten, M., Hanke, G., Beyss, M., Messerschmidt, U., Urban, K. (1995): Philos. Mag. Lett. **72**, 375
Schöck, G. (1965): Phys. Status Solidi **8**, 499
Schall, P. (1998): Diploma Thesis, Rheinisch Westfälische Technische Hochschule Aachen

Seeger, A. (1958): in Handbuch der Physik, Vol. VII/2, Flügge, S. (ed). Springer-Verlag, Berlin, p 1
Semadeni, F., Baluc, N., Bonneville, J. (1998): in Proceedings of the 6th International Conference on Quasicrystals, Takeuchi, S., Fujiwara, T. (eds). World Scientific, Singapore, p 513
Shibuya, T., Hashimoto, T., Takeuchi, S. (1990): Jpn. J. Appl. Phys. **29**, L349
Shield, J.E. (1997): in New Horizons in Quasicrystals, Goldman, A.I., Sordelet, D.J., Thiel, P.A., Dubois, J.M. (eds). World Scientific, Singapore, p 312
Shield, J.E., Kramer, M.J., McCallum, R.W. (1993): J. Mater. Res. **9**, 343
Smithells, C.J. (1992): Smithells Reference Book, 7th ed. Butterworth-Heinemann, Oxford
Socolar, J.E.S., Steinhardt, P.J. (1986): Phys. Rev. B **43**, 617
Socolar, J.E.S., Lubensky, T.C., Steinhardt, P.J. (1986): Phys. Rev. B **34**, 3345
Steinhardt, P.J., Ostlund, S. (1987): The Physics of Quasicrystals. World Scientific, Singapore
Steurer, W. (1990): Z. Kristallogr. **190**, 179
Takeuchi, S., Hashimoto, T. (1993): Jpn. J. Appl. Phys. **32**, 2063
Takeuchi, S., Shinoda, K., Fujiwara, H., Edagawa, K. (1997): preprint
Takeuchi, S., Shinoda, K., Yoshida, T., Kakegawa, T. (1998): in Proceedings of the 6th International Conference on Quasicrystals, Takeuchi, S., Fujiwara, T. (eds). World Scientific, Singapore, p 541
Tanaka, K., Mitari, Y., Koiwa, M. (1996): Philos. Mag. A **73**, 1715
Taylor, G.I. (1934): Proc. R. Soc. London, Ser. A **145**, 362
Tsai, A.-P., Inoue, A., Masumoto, T (1990): Philos. Mag. Lett. **62**, 95
Tsai, A.P., Suenaga, H., Ohmori, M., Yokoyama, Y., Inoue, A., Masumoto, T. (1992): Jpn. J. Appl. Phys. **31**, 2530
Urban, K., Wollgarten, M., Wittmann, R. (1993): Phys. Scr. T **49**, 360
Wang, R., Dai, M.X. (1993): Phys. Rev. B **47**, 15 326
Wang, R., Feuerbacher, M., Wollgarten, M., Urban, K. (1998): Philos. Mag. A **77**, 523
Wang, Z.G., Wang, R., Feng, J.L. (1994): Philos. Mag. A **70**, 577
Wittmann, R., Urban, K., Schandl, M., Hornbogen, E. (1991): J. Mater. Res. **6**, 1165
Wollgarten, M., Urban, K. (1994): in Lectures on Quasicrystals, Hippert, F., Gratias, D. (eds). Les Editions de Physique, Les Ulis, p 535
Wollgarten, M., Saka, H. (1997): in New Horizons in Quasicrystals, Goldman, A.I., Sordelet, D.J., Thiel, P.A., Dubois, J.M. (eds). World Scientific, Singapore, p 320
Wollgarten, M., Gratias, D., Zhang, Z., Urban, K. (1991): Philos. Mag. A **64**, 819
Wollgarten, M., Zhang, Z., Urban, K. (1992): Philos. Mag. Lett. **65**, 1
Wollgarten, M., Lakner, H., Urban, K. (1993a): Philos. Mag. Lett. **67**, 9
Wollgarten, M., Beyss, M., Urban, K., Liebertz, H., Köster, U. (1993b): Phys. Rev. Lett. **71**, 549
Wollgarten, M., Bartsch, M., Messerschmidt, U., Feuerbacher, M., Rosenfeld, R., Beyss, M., Urban, K. (1995): Philos. Mag. Lett. **71**, 99
Wollgarten, M., Metzmacher, C., Rosenfeld, R., Feuerbacher, M. (1997): Philos. Mag. A **76**, 455
Yang, W., Feuerbacher, M., Bartsch, M., Messerschmidt, U., Urban, K. (1998a): unpublished
Yang, W., Feuerbacher, M., Tamura, N., Ding, D, Wang, R., Urban, K. (1998b): Philos. Mag. A **77**, 1481
Yang, W., Feuerbacher, M., Urban, K. (1998c): unpublished

Yokoyama, Y., Miura, T., Tsai, A.P., Inoue, A., Masumoto, T. (1992): Mater. Trans., Jpn. Inst. Met. **33**, 97
Zhang, H., Kuo, K.H. (1990): Phys. Rev. B **41**, 3482
Zhang, Z., Urban, K. (1989): Philos. Mag. Lett. **60**, 97
Zhang, Z., Wollgarten, M., Urban, K. (1990): Philos. Mag. Lett. **61**, 125
Zumkley, T., Mehrer, H., Freitag, K., Wollgarten, M., Tamura, N., Urban, K. (1996): Phys. Rev. B **54**, R6815

12. Toward Industrial Applications

Patrick C. Gibbons and Kenneth F. Kelton

12.1 Introduction

Almost all alloys that form quasicrystals (QCs) contain as major elements Al, Ti, or (Ti_xZr_{1-x}). These are shiny metal alloys with thin surface-oxide layers that protect the bulk from further oxidation. All contain transition metals and are hard and brittle, like many intermetallic crystals. The most structurally perfect of them have surprisingly low electrical and thermal conductivities. They seem a poor set of materials for industrial applications, but some show promise.

The applications for which QCs may prove suitable are determined by their physical and chemical properties, some of which are unique. To prepare for a discussion of applications we first review these properties, many of which are presented in detail in the preceding chapters. This review is not exhaustive, but rather focuses only on those properties that are important in existing and proposed applications

12.2 The Relevant Properties of Quasicrystals

Unless otherwise indicated, alloy compositions are given in atomic percent (at%.).

12.2.1 Electronic Structure and Transport

The magnitude and temperature dependence of the electrical resistivity in QCs seemed unremarkable initially, but were found to be quite unusual as stable QCs with highly perfect structures were synthesized. Perhaps surprisingly, the greater the structural order in an icosahedral (i) QC, the greater the resistivity (Fig. 12.1). Conductivities as much as three orders of magnitude lower than Mott's nominal value (Mott and Davis 1979) near which electron localization should occur have been observed (Pierce et al. 1993, Pierce et al. 1994, Poon et al. 1995). The temperature dependence of the resistivity is metallic, with resistivity increasing with increasing temperature, in less-resistive, less well-ordered samples. It changes to semiconductor-like (Takeuchi et al. 1995, Tamura et al. 1995) or semimetallic (Haberkern and

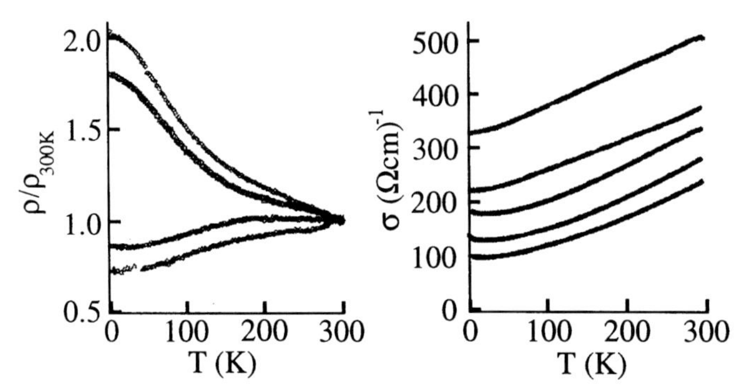

Fig. 12.1. Left: The low-temperature resistivity of 1/1 Al-Mn-Si increases with increasing structural order produced by different annealing conditions. After Poon (1992). Right: The conductivity of Al-Cu-Fe decreases with increasing order and also depends on composition. From top to bottom: $Al_{63}Cu_{24.5}Fe_{12.5}$ annealed 2 h at 600 °C, 3 h at 800 °C; $Al_{62}Cu_{25.5}Fe_{12.5}$ annealed 2 h at 600 °C, 3 h at 800 °C; $Al_{62.5}Cu_{25}Fe_{12.5}$ annealed 3 h at 800 °C. After Dubois (1993).

Fritsch 1995) behavior, with resistivity increasing with decreasing temperature, in the highly resistive, highly ordered samples. The discoverer of these phenomena has claimed an insulating QC, even though non-zero conductivity was measured at every temperature studied. The data suggest that, at non-zero temperatures lower than could be reached in the experiment, the conductivity would have become zero (Poon et al. 1995).

Spectroscopic studies have been made to estimate the energy density of states (DOS) in some of the stable, *i* QCs. X-ray absorption and photoelectron spectroscopies can probe the empty and filled electron states, respectively, near the Fermi energy E_F and ten or more eV away from it. Even the most intense and highly collimated synchrotron x-ray sources, however, have limited energy resolution, no better than about 0.3 eV, which obviously is insufficient to study thoroughly the kind of narrow, local minimum in the DOS that would be expected if a Hume-Rothery mechanism were stabilizing a highly-ordered QC, as has been proposed. The intrinsic widths of the spectral lines due to core-hole lifetimes are greater than 0.3 eV for core states bound by more than 1 keV (Belin et al. 1995, Belin-Ferré et al. 1997). Ultraviolet photoelectron spectroscopy using resonance radiation from a helium discharge can probe the filled states within a few eV of E_F with an energy resolution of 30 meV (Stadnik 1997) and more recently, 6 meV (Stadnik et al. 1996, 1998). The spectroscopic studies have confirmed that there appear to be minima in the DOS at the Fermi energies that are narrow (0.2 to 0.4 eV Lorentzian FWHM) on the filled-state side with values 0.1 to 0.4 of the values of the DOS 1 eV below the Fermi energies. The Fermi edges from the DOS minima down to zero have the energy widths expected at the measurement temperatures of 12 K to 45 K (Chap. 8). The dips in the DOS may be as broad as many eV on the empty-state side of the E_F, as shown in Fig. 12.2

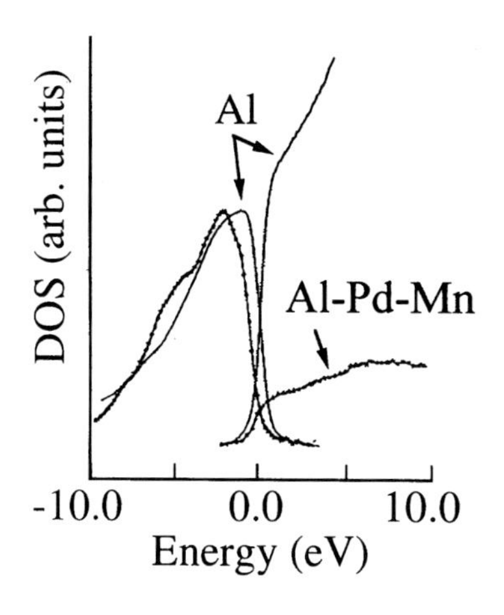

Fig. 12.2. Empty and filled Al $3p$ states in Al cross at E_F (0.0 eV) at about one-half the DOS at energies near E_F; those from *i*-$Al_{71}Pd_{19}Mn_{10}$ cross much lower and its DOS remains low above E_F. After Belin et al. (1994).

(Belin-Ferré et al. 1997, 1998), suggesting that the DOS maxima are probably produced by *d*-electron peaks only a few eV wide (Hennig and Teichler 1997).

The available evidence supports the notion of a quasiperiodic-order-induced energy gap in the DOS. The electronic transport properties and the small electronic specific heat (see Sect. 12.2.4) are then consistent with the properties of a semi-metal or narrow-gap semiconductor.

12.2.2 Visible and Infrared Optical Properties

Many QCs have reflectivities that are nearly constant at about 60% for wavelengths from 300 nm to 15 μm. In contrast, simple metals reflect nearly 100% of incident light in this range, and semiconductors have high transmission in the infrared (IR) (Eisenhammer et al. 1995, Eisenhammer 1997). Experiments with different instruments, on different samples of a number of different metastable and stable QCs have produced consistent results. After reasonable extrapolations to longer and shorter wavelengths than were included in the measurements, the phase shift upon reflection could be computed using the Kramers-Krönig relations, and from the complex reflectivity thus obtained other optical parameters could be estimated. The real part of the optical conductivity $Re\,\sigma(\omega)$ is one of the most useful of these properties to examine. In simple metals it is approximately a Lorentzian function centered at zero frequency, as the Drude theory for nearly-free electrons predicts. Most QCs studied have $Re\,\sigma(\omega)$ increasing approximately linearly from wave numbers as low as 14 cm^{-1}, about 1.7 meV, up to 4000 cm^{-1}, about 0.5 eV. Measurements to higher energies reveal that the linear rise is followed by a peak around 12 000 cm^{-1} or about 1.5 eV, after which the conductivity again decreases. The peak conductivity is comparable to the conductivities exhibited by simple metals in the visible part of the spectrum. At the low-energy end, extrapolation to zero frequency produces a conductivity that is typically in good agreement with the measured DC conductivity. This is the case for metastable Al-Mn-Si and for stable Al-Cu-Fe and Al-Pd-Mn, materials with resistivities lower than the critical value of about 100 Ω cm, above which electron localization is expected (Mott and Davis 1979). The Al-Mn-Si 1/1 approximant has optical properties very similar to those of the Al-Mn-Si i phase, so it is not quasicrystallinity *per se* that produces the unusual optical properties. Al-Pd-Re, with a resistivity that is at the critical value and that increases as temperature decreases (Basov et al. 1995, Poon et al. 1995), has an optical conductivity that is very low for a metal up to 2000 cm^{-1}, and then increases linearly up to 4000 cm^{-1}, as shown in Fig. 12.3. This looks

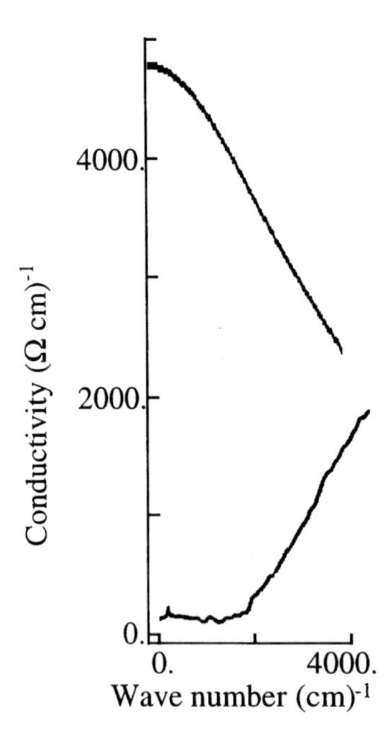

Fig. 12.3. The room-temperature optical conductivity of i-$Al_{70}Pd_{20}Re_{10}$ remains low for photon energies below 0.25 eV and then increases linearly with increasing energy, strongly suggesting a gap in the DOS at E_F; the conductivity of pure Al is higher and decreases with increasing energy (upper line, a nearly free electron calculation). After Basov et al. (1995).

like the conductivity of a material with a 0.25 eV wide pseudogap at E_F, in which the DOS is very low. The non-approximant crystal Al_2Ru, with six atoms per primitive cell, has optical reflectivity and conductivity similar to those of the Al-Pd-Re QC. Models based on weak localization and others based on Hume-Rothery stabilization mechanism predict pseudogaps and are generally consistent with the data.

12.2.3 Thermopower

For simple sample geometries, the thermopower Q is the ratio of the voltage across the sample to the temperature difference across it. In simple metals Q is expected (Ashcroft and Mermin 1976) and observed (Lin et al. 1996) to be proportional to absolute temperature T. Deviations from a linear temperature dependence can be caused by phonon drag or an electron-phonon mass enhancement (Lin et al. 1996). In careful measurements of Q as a function of T in single grains of decagonal (d) $Al_{73}Ni_{17}Co_{10}$, the temperature dependence was linear in the periodic direction but deviated from linearity in directions lying in the quasiperiodic plane. In the interpretation of these results, phonon drag was eliminated as a possible cause. A strongly anisotropic electron-phonon mass enhancement, weak in the periodic direction but strong in a quasiperiodic direction, may be the cause (Lin et al. 1996).

12.2.4 Lattice Dynamics

The quasilattice contributions to heat capacities and thermal conductivities have been measured and appear similar in magnitude to those of ordinary insulating crystals at all temperatures (Dubois et al. 1993b). Consideration of the details of their dependences on temperature and quasilattice quality suggests scattering mechanisms that are different from those in perfect crystals. In insulating crystals, in the vicinity of 20 K there is a peak in the thermal conductivity. Its low-temperature side is produced by the T^3 increase of the lattice heat capacity with increasing temperature and a temperature-independent phonon mean-free path due to surface scattering in high-quality samples. At temperatures greater than 20 K, the high-temperature side of the peak is produced by the decrease in the umklapp mean-free path of phonons as more phonon modes are occupied with increasing temperature (Kittel 1986).

In single-grain QCs with sufficiently low electrical conductivity that the electronic contributions to their thermal properties are very small below 80 K, there is a peak or plateau in the thermal conductivity at about 20 K, similar in appearance to those found in insulating glasses, as Fig. 12.4 demonstrates. The low-temperature side of the peak exhibits a T^2 temperature dependence, however. Scattering of phonons by tunneling states like those in amorphous materials has been proposed to explain this temperature dependence (Chernikov et al. 1995a, 1995b, 1997, 1998, Legault et al. 1995) although others have noted that the scattering of phonons by electrons at low temperatures also leads to a T^2 temperature dependence of thermal conductivity (Perrot and Dubois 1993, Perrot et al. 1995). That QCs with very different electrical conductivities exhibit similar thermal conductivities argues against this electronic explanation (Legault et al. 1997).

From 20 to 80 K in single-grain QCs with low electrical conductivity, the thermal conductivity decreases slightly with increasing temperature. Polygrained QC samples, still with low electrical conductivity, exhibit a constant or slowly-increasing thermal conductivity in this same temperature range, indicating that there is additional scattering of phonons by micro-voids, impurities, phasons, or other disorder (Chernikov et al. 1995a, 1995b, Legault et al. 1995, 1997). The conductivity decrease in single-grain samples may

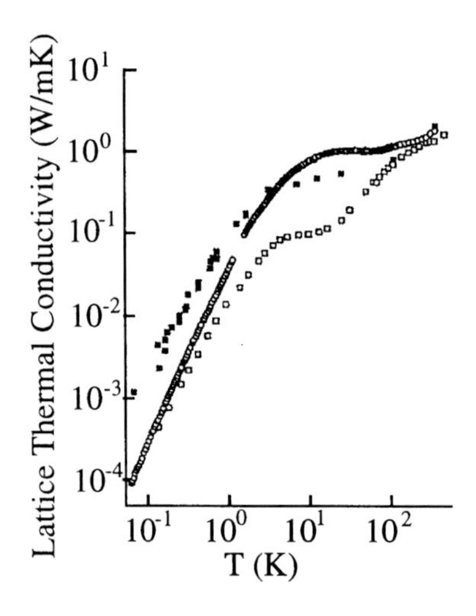

Fig. 12.4. The temperature dependence of the phonon thermal conductivity of *i*-$Al_{70}Pd_{21}Mn_9$ (closely-spaced open circles) is similar to that of amorphous Pd-Si (full squares) or SiO_2 (open squares). After Chernikov et al. (1995a).

be due in part to the decrease in the umklapp mean-free path of phonons as more phonon modes are occupied with increasing temperature, the same mechanism found in insulating crystals (Legault et al. 1997). But other scattering mechanisms are required to explain the temperature dependences of the phonon mean-free paths extracted from conductivity data from single-grain and poly-grained samples. The scatterers are rarer in single-grained samples than in poly-grained ones, but the short mean-free paths computed suggest that they cannot be the grain boundaries in the poly-grained samples. The same phasons, impurities, disorder, or two-level systems that scatter at lower temperatures may also dominate the scattering in the 20–80 K range (Legault et al. 1997).

Above about 80 K in QCs the thermal conductivity increases with increasing temperature and is proportional to T, in single- and poly-grained samples (Legault et al. 1995, Perrot et al. 1995, Archambault 1997). In this temperature range, and in particular above room temperature, the increasing thermal conductivity is attributed to electronic conduction (Perrot et al. 1995). An additional, temperature-independent term observed in poly-grained Al-Cu-Fe QCs suggests significant phonon conductivity in that material in the 80–200 K temperature range (Perrot et al. 1995). At room temperature and above the electronic contribution to the thermal conductivity is dominant, but as one would expect from the normal ratios of thermal to electronic conductivities (Perrot et al. 1995), it is very small compared to the thermal conductivity of ordinary metals. At room temperature the thermal conductivity of QCs, single-or poly-grained, is similar in magnitude to that of zirconia (Archambault 1977). At temperatures up to 600–800 °C, the thermal conductivity, with its relatively slow, linear temperature dependence, remains much closer to those of ceramic insulators than to those of pure metals (Archambault 1997).

12.2.5 Ductility

All known QCs are hard and brittle at temperatures near room temperature, quite similar to silicon (Feuerbacher et al. 1997). This could have been expected in materials with d-electron bonds that resist changes in angles between bonds as well as stretching; crystalline transition-metal intermetallic alloys are also often hard and brittle. Another reason for the hardness and lack of ductility arises from a consideration of the source of ductility in simple metals and alloys. Plastic flow occurs when dislocations move freely through the structure. The multiple, incommensurate length scales in QC structures and the lack of any small, periodically repeated units make the motion of dislocations quite difficult at room temperature.

At room temperature the phason degrees of freedom involving short-distance atomic hops are frozen; the energy barriers to atomic motions typically require temperatures of about 600 °C to activate those motions. Quenched phason strains at room temperature produce measurable effects in diffraction (Lubensky et al. 1986). The activation of local atomic hopping at high temperatures has been detected by an investigation of the quasielastic peak in inelastic neutron scattering, as a function of temperature (Coddens and Bellissent 1993). The dynamic phason modes they studied may be the mechanism by which the phase transformation from quasicrystalline to micro-crystalline structures in Al-Cu-Fe occurs. Neutron scattering did not distinguish between local hops and long-range diffusion.

The activation of atomic hopping suggests a possible mechanism for the observed plastic and super-plastic behavior at high temperature (Shibuya et al. 1990). An example is shown in Fig. 12.5. In fact, more than one mechanism has been observed. In *d*-Al-Cu-Co creep occurred in association with

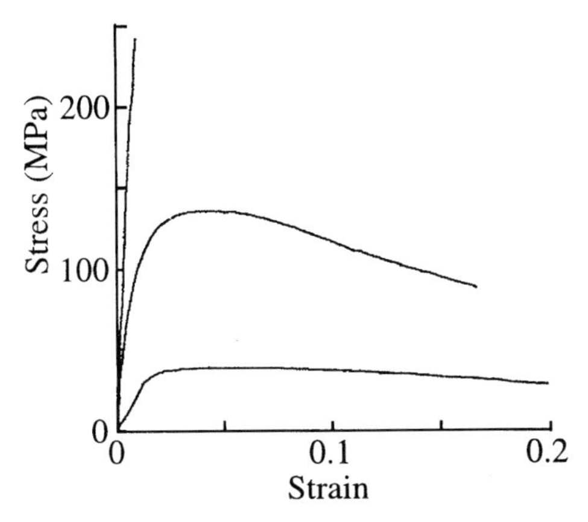

Fig. 12.5. Decagonal $Al_{63}Cu_{17.5}Co_{17.5}Si_2$ exhibits plastic flow in compression tests as temperature increases from 600 °C (top curve) to 700 °C (middle curve) and 800 °C (bottom curve). After Kang and Dubois (1992).

the formation of voids (Kang and Dubois 1992). Super-plastic deformation in compression occurs at temperatures from 600 to 750 °C in Al-Cu-Fe *i* QCs (Bresson and Gratias 1993, Dubois 1993). In this work no nucleation nor migration of linear defects, no shear bands, and no slip bands were detected. The stress-strain curves showed no evidence of hardening. Even though dislocations had been found in QCs before these measurements, they did not seem to be the mechanism of the observed plastic behavior. Rather, relying on evidence for a thermally activated process from stress relaxation measurements (Bresson and Gratias 1993), the authors concluded that long-range diffusion through phason hops was the most likely mechanism. More recent work has shown that, in *i*-Al-Cu-Fe at temperatures of 500 and 750 °C, plastic deformation under stress of 300 MPa occurs by a shear transformation similar to a martensitic transformation (Shield and Kramer 1995, Shield 1997). No evidence for a dislocation mechanism was uncovered during the electron microscopy of the deformed samples. In another study of deformation of *i* and 1/1 approximant Al-Cu-Fe, dislocation glide was the deformation mechanism in the approximant. In the QC the data suggest a localized-obstacle, strain-rate controlling mechanism, with an obstacle size of about 36 atoms (Bresson 1997). The results of Bresson and of Shield and Kramer may be consistent aspects of a single deformation mechanism in *i*-Al-Cu-Fe.

In *i*-Al-Pd-Mn dislocation motion is the mechanism of high-temperature plasticity in the 680–800 °C temperature range. Compression testing followed by transmission electron microscopy (TEM), and TEM *in situ* investigation of a sample during high-temperature straining established this. The motion of dislocations was observed (Feuerbacher et al. 1995, 1997). Consistent with the quasicrystallography, the Burgers vectors of the dislocations have non-zero components in the perpendicular or non-physical space of the *i* structure, as well as in physical space. The most prominent dislocations moved on fivefold planes with twofold Burgers vectors. Their motion was thermally activated, controlled by localized obstacles. The smooth motion, without any evidence of pinning, demonstrated that the obstacles were dense in the material. The activation volume, the size of an obstacle, was about 100 atomic volumes, which argues against diffusion control of the dislocation motion. The authors suggested that tightly bonded pseudo-Mackay clusters containing of order 50 atoms, with weak bonding between clusters, could be the obstacles. Examination of fractured surfaces showed that cleavage cracks avoid these clusters (Feuerbacher et al. 1997). Simulations of crack formation in two dimensions exhibit similar behavior (Mikulla et al. 1997). Gibbs free energies consistent with dislocation motion in the temperature range studied require that a large activation enthalpy be balanced by a substantial deformation entropy.

The room temperature properties of QCs have been improved by annealing treatments and by the addition of selected impurities. They are hard and brittle, fracturing before any plastic deformation occurs. The fracture strength in compressive testing of an $Al_{63.5}Cu_{24.5}Fe_{12}$ poly-grained quasicrystalline alloy increased from 250 MPa in the as-cast state to 690 MPa after annealing at 650 °C for 3 h (Kang and Dubois 1992). The Young's modulus and Vickers hardness remained the same, with high hardness of 715. Addition of boron produced an $(Al_{63.5}Cu_{24.5}Fe_{12})_{96.5}B_{3.5}$ poly-grained quasicrystalline alloy that, without annealing, exhibited a fracture strength of 1010 MPa and Vickers hardness of 850. Conventional, dispersion-hardened Al alloys have tensile strengths that approach 500 MPa. Electron diffraction patterns from the B-doped sample showed more diffuse scattering and greater phason strain than those from the undoped sample (Kang and Dubois 1992).

12.2.6 Surface Properties

Their hardness and strong bonding make QC surfaces resistant to wear when rubbed or scratched by other materials (von Stebut et al. 1997). A surprising result from hardness studies, however, is that under high stress ductility increases, even near room temperature (Dubois et al. 1994b). Even after the stress is removed, a stressed region of a surface is more ductile than nearby, unstressed areas. Vickers indentations were made on a single grain of *i*-Al-Pd-Mn at room temperature, after which the regions just adjacent to the indentations were examined by TEM. Below the indents were regions in which the material remained icosahedral, but had cracked and contained dislocations. The grain sizes in this poly-grained material were 10 to 500 nm, much smaller than the 2 to 7 μm sides of the indents. An observed increase in ductility with increasing load was ascribed to deformation along the grain boundaries. Nevertheless, even in regions thinned for microscopy, the fragmentation into small grains did not make the sample fall apart; the grains remained bonded together. The dislocations observed did not appear to be the primary deformation mechanism. Plastic flow under high stress at room temperature involves cracking and cold welding (Wollgarten and Saka 1997).

QC surfaces, either deposited films or boundaries of bulk material, exhibit low friction (Kang et al. 1993, Dubois et al. 1994b, Sordelet et al. 1995, von Stebut et al. 1997). This is due in part to their hardness. Further reasons offered include the atomic-scale mismatches between the structure of a QC and that of any crystalline or polycrystalline material (Rivier and Boose 1995) and the observed low interfacial energies at various liquid-QC boundaries (Dubois 1997, Rivier 1997, Dubois et al. 1998). Friction and surface structures of oxide-free QCs have been studied under ultra-high vacua (UHV) (Gelmann 1996), complementing the more application-relevant measurements made in air. The UHV work has shown that some QCs have a low coefficient of static friction (< 0.2), when clean, and highly variable friction when oxidized by exposure to O_2 or air, with the coefficient in the range 0.05–1.5.

The best-known property of QCs of possible technological importance is their reported non-stick behavior. N. Rivier has considered the observed non-wetting property of QCs from three points of view: (1) thermodynamic, examining possible surface-energy effects; (2) electronic, exploring the relation between the density of states at the Fermi energy and adhesion, and using experimental and theoretical results that suggest that the pseudogap in well-ordered QCs appears on the surface as well as in the bulk; and (3) surface roughness, the pinning of a liquid, and the hysteresis produced by the pinning (Rivier 1993, 1997, Rivier and Boose 1995). The theoretical arguments that a pseudogap exists in the bulk and at the surface of any well-ordered QC are based on inflation symmetry, and on the assertion that if a one-dimensional QC has a pseudogap because of inflation-symmetry, two- and three-dimensional QCs must also (Rivier and Boose 1995, Rivier 1997).

Experimental estimates of surface energies from measurements of contact angles between liquids and solid surfaces have shown that large contact angles and therefore non-wetting are typical for QCs and their related complex crystals (Dubois et al. 1998). These experiments were made using nine different solids and five different liquids. The solids were all ground and polished in the same manner to a finish with less than 0.1 μm roughness.

The reversible work of adhesion of a liquid L onto a solid surface S is

$$W_{SL} = \gamma_{LV}(1 - \cos(\theta)) + \gamma_{So} - \gamma_{SV},$$

where θ is the observed contact angle, γ_{LV} is the liquid-vapor interfacial energy, γ_{SV} is that for the solid in the presence of the vapor from the liquid, and γ_{So} is the interfacial free energy of the solid with no vapor present, i.e., with no adsorbed layer of the molecules from the liquid (Adamson 1990, Dubois et al. 1998). To use instead an energy difference between a solid-liquid interface and a solid-vapor interface, the last two terms are ignored: this is justified experimentally (Dubois et al. 1998). The use of the geometric-mean rule to approximate the interaction potentials between two different molecules, given interaction potentials between molecules of one type and between molecules of the second type (Adamson 1990), leads to

$$W_{SL} = 2\sqrt{\gamma_{SD}\gamma_{LD}} + 2\sqrt{\gamma_{SP}\gamma_{LD}},$$

where D and P identify dispersive and polar parts of the surface energies. Dispersive parts arise from interactions between dipolar fluctuations, and polar parts describe interactions between permanent dipoles and dipole moments induced by them (Dubois et al. 1998). The polar parts of the liquid surface energies are not included here because they describe, in fact, chemical bonding energies not expected to contribute to the interaction between different molecules (Adamson 1990). Combining these two equations leads to a graphical method for estimating γ_{SD} and γ_{SP} from contact angle data, using the known decomposition of the surface energies of the five liquids used into dispersive and polar parts (Dubois et al. 1998).

The results are that *i*-Al-Pd-Mn and PTFE (polytetrafluoroethylene or Teflon, a registered trademark of the DuPont Corporation), the widely used consumer non-stick coating for cookware, have in common very small values of γ_{SP}. *i*-Al-Cu-Fe has a larger γ_{SP} that is, nevertheless, less than one-third of the value found for fcc metals; less than one-tenth the value for alumina. The QCs have a γ_{SD} comparable to those of fcc metals and alumina, 55% higher than that of PTFE. The reversible work of adhesion for water on the various surfaces in mJ/m^2 is 111.4 for alumina, 38.9 for PTFE, 59.7 for *i*-Al-Pd-Mn, 67.5 for *i*-Al-Cu-Fe, and 81.1 for fcc metals. It is interesting to note that the thin oxide layers on the QC surfaces did not make them behave like bulk alumina in this experiment. The polar parts of the surface energies of the conducting solids were correlated with estimates of the partial density of states at the Fermi energy due to Al 3*p* electrons (Dubois et al. 1998). QCs are non-stick and non-wetting. It appears, then, that this is due to the covalent nature of their bonding, as opposed to simple-metal bonding.

12.2.7 Corrosion Resistance

Tests of the corrosion of QCs have been made in UHV exposing oxide-free surfaces to dry O_2 (Chang et al. 1995, Gavatz et al. 1998, Rouxel et al. 1997, Sordelet et al. 1997), dry air (Gu et al. 1997b), dry air after UHV cleaning (Chevrier et al. 1997, Wehner and Köster 1997), and humid air and water as well as dry air (Gu et al. 1997a, Jenks et al. 1997). These are reviewed here.

A simple, practical test of coatings for use on industrial cookware was made by boiling inside coated utensils 500 mL of a solution of 3% acetic acid in water for 1h. A subsequent chemical analysis of the solution revealed the species that were leached from the coating. The most effective variable for reducing corrosion was the composition of the coating. *i*-Al-Cu-Fe, applied by thermal-spray techniques (Sect. 12.3.1), was most resistant when Cr was added, giving $Al_{70}Cu_9Fe_{10.5}Cr_{10.5}$. Impurities were important. Zn and Mg decrease corrosion resistance in many Al alloys, and Ni and Co promote the formation of the competing cubic β Cs-Cl phase that, with a different composition from the *i* phase, generates galvanic effects between grains of the different phases (Dubois et al. 1993a, 1994a).

Polarization measurements and low-frequency impedance spectroscopy provide more information (Castle 1993, Massiani et al. 1993, 1995, Dubois et al. 1994a). Coatings of composition $Al_{70}Cu_9Fe_{10.5}Cr_{10.5}$, mostly *i* phase as deposited on an aluminum alloy and on low carbon steel substrates, were tested. The master ingot of the alloy and uncoated substrates were also measured. The master ingot was the most corrosion resistant. The coatings greatly improved the corrosion resistance of the steel, but had a lesser effect on the resistance of the aluminum alloy.

In impedance spectroscopy (McDonald 1987) an uncoated aluminum alloy and a coated sample with a small hole in the coating behaved similarly, whereas a completely coated sample behaved differently, demonstrating that the coatings are waterproof when free from cracks (Dubois et al. 1994a).

More extensive tests have been made on bulk materials by using solutions with a wide range of pH, to obtain data from surfaces without a passivating aluminum oxide that forms, on pure Al, in the range $pH = 4$ to $pH = 8$.

Solutions with $pH = 2$, 6.8, and 13 were prepared and used at room temperature. Bulk materials, having compositions near that of the coatings and i or approximant structures, were tested, as well as Al metal and single-phase crystalline $Al_{70}Cu_{20}Fe_{10}$. The differences in corrosion resistance were greatest at $pH = 13$. Impedance spectroscopy in that solution showed that all other specimens differed from pure Al in their behavior. Interpretation of the spectra produced the conclusion that, in all the materials but pure Al, an initial stage of dissolution was followed by formation of a corrosion layer that blocked further dissolution in the alkaline solution. Composition was more important than structure in determining corrosion resistance, and addition of Cr was the most effective way to reduce corrosion (Massiani et al. 1993).

Al-Pd-Mn and Al-Cu-Fe QCs and crystals of similar compositions to the QCs have been oxidized, after removal in UHV of all pre-existing oxides, and then examined by x-ray photoemission spectroscopy. Peaks produced by metal atom core levels, Al $2p$ for example, occur at resolvably different energies for oxidized and non-oxidized atoms. Thus this spectroscopy can be used to determine which elements have oxidized and which have not. The oxidizing agents used include dry O_2, dry air, humid air, and water, all at room temperature (Jenks et al. 1997). Dry O_2 oxidizes only the Al in the QCs; air exposure produces greater Al oxidation than dry O_2. In Al-Pd-Mn air exposure also produces some oxidation of the Mn, more if the air is humid than if it is dry, but only water exposure oxidizes the Pd. In Al-Cu-Fe oxidation of Cu begins with exposure to humid air. Oxidation increases with humidity, and in both QCs Al enrichment of the surfaces grows as the thickness of the oxide layer increases. This was not the case for cubic $Al_{56}Pd_{33}Mn_{11}$, in which all metals oxidized upon air exposure and, after water immersion, were almost completely oxidized at the surface (Jenks et al. 1997).

12.2.8 Hydrogen Storage

Ti-Zr-Ni QCs, stable for at least one composition (Kelton et al. 1997a, 1997b, Kim, W.J. et al. 1998a), absorb and release large amounts of hydrogen. The amount absorbed in a metal alloy can be described by the weight percent (wt%) or the atomic percent (at.%). A measure derived from the atomic percent of hydrogen present, the hydrogen-to-metal ratio H/M, is also used. H/M is the number of hydrogen atoms present per host metal atom; it exceeds 1.0 in all of the alloys considered for possible applications. The atomic percent of hydrogen in a metal hydride is $[(H/M)/(1+H/M)]100$. The weight percent is $[(H/M)/(A_M/A_H + H/M)]100$, where A_M/A_H is the ratio of the average atomic weight of the metals in the alloy to the atomic weight of hydrogen. Quasicrystalline Ti-Zr-Ni can absorb almost two hydrogen atoms per host metal atom or of order 2 wt% (Viano et al. 1995a, 1995b, 1998, Kelton et al. 1997a). Absorption has been observed by using the QC as a cathode in an electrolytic cell, and absorption and desorption have been observed through specially prepared surfaces (Kim, J.Y. et al. 1998) exposed to modest pressures of hydrogen gas at temperatures well below those that promote irreversible formation of stable crystalline hydride phases.

Hydrogen in a metal is in an ionized state, a proton surrounded by an excess of conduction electrons that screen the positive charge in a distance on the order of 1 nm. There is no one electron bound to the proton. The protons occupy interstitial sites, the holes in the lattice between metal atoms. The most favorable sites, those occupied by the first hydrogen atoms absorbed in most cases, are tetrahedral interstitials like the position $(1/4, 1/4, 1/4)$ in the conventional cubic cell of a primitive fcc lattice. Some materials continue to absorb hydrogen after all the tetrahedral interstitials are filled. In these, the hydrogens occupy octahedral interstitials, like the position $(1/2, 0, 0)$ in the conventional cubic cell of a primitive fcc lattice. In some materials, fcc Pd for example, the octahedral interstitials fill before the tetrahedral ones (Switendick 1978b). Usually not all interstitials can be occupied; generally no two hydrogen atoms can be closer than 0.21 nm (Switendick 1978a). Westlake (1980) developed a hard-sphere model for host-metal and hydrogen atoms that allows hydrogen atoms to occupy off-center interstitials sites, thus possibly increasing the separation of first-neighbor hydrogens, but forbidding overlap of hydrogen and metal atoms.

12.2.8.1 Icosahedral Quasicrystals. *i* QCs contain many tetrahedral interstitials. The Mackay two-shell atomic cluster for example, believed to be the building block of QCs and related crystal phases in Al-Mn, Ti-Cr-Si-O, and Ti-Zr-Fe alloys, contains 20 tetrahedral interstitials inside its inner shell and 60 tetrahedral interstitials and 20 octahedral interstitials between its inner and outer shells. The Bergman two-shell atomic cluster, believed to be the building block of QCs and of related crystal phases in Al-Mg-Zn, Al-Li-Cu, and Ti-Zr-Ni alloys, contains 20 tetrahedral interstitials within its inner shell, 120 between its inner and outer shells, and no octahedral interstitials. These sites are shown in Fig. 12.6. Using the Switendick criterion, only four of the sites inside the inner shells can be occupied, or eight if the hydrogen atoms are moved radially outward from the centers of the tetrahedra (Westlake 1980, Viano 1996, Hennig 1997). In the Mackay cluster only 24 of the 60 sites between the shells can be occupied, for totals of 28 or 32 hydrogen atoms within the cluster. In the Bergman cluster 40 of the sites between the shells can be occupied (Viano 1996, Hennig 1997, Shastri et al. 1998) for totals of 44 or 48 hydrogen atoms within the cluster, assuming that the tetrahedral sites fill before the octahedral sites. Recent work has suggested that this may not be the case, however. The structure of the 1/1 approximant to the Ti-Cr-Si-O *i* QC has been determined by Rietveld refinement of x-ray (XRD) and neutron powder diffraction data (Libbert et al. 1994). Neutron powder diffraction data from a crystal loaded with deuterium to $D/M = 0.3$ suggest that the tetrahedral sites inside the first shell of the Mackay clus-

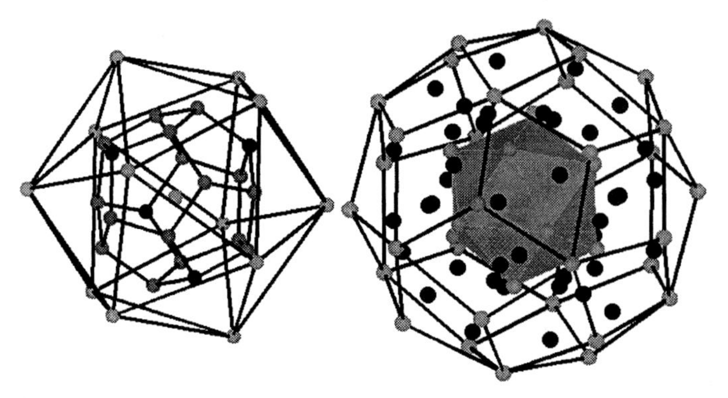

Fig. 12.6. Tetrahedral sites in the Bergman cluster within the inner icosahedron (left, inner dodecahedron) and between the two *i* shells (right, dark circles). After Viano (1996).

ter are mostly empty; that between-shells octahedra in <111> directions are 50% occupied by deuterium; that octahedra between the shells and midway between clusters in <111> directions are mostly occupied by oxygen; and that the rest of the deuterium is in tetrahedral sites between first and second shells and in the regions between the clusters (Kim, J.Y. 1997, Kim, W.J. 1997).

12.2.8.2 Icosahedral Ti-Zr-Ni. The Ti-Zr-Ni *i* phase was first formed by rapid quenching from the melt (Sibirtsev et al. 1988, Molokanov and Chebotnikov 1990, Zhang et al. 1994). It is now known to be a stable QC that can be formed by annealing as cast crystalline material, which contains mostly the C14 Laves phase (Kelton et al. 1997b). It is critical for *i* phase formation that oxygen content be minimized, both in the elemental materials and in all processing steps (Kelton et al. 1997b). Annealing in an atmosphere containing some oxygen produces the Ti_2Ni phase (Stroud et al. 1996).

It is interesting that in $Ti_{41.5}Zr_{41.5}Ni_{17}$, the QC is the low-temperature stable phase, which is not the case in Al-Cu-Fe and Al-Pd-Mn alloys. This argues against entropic stabilization of the stable Ti-Zr-Ni QC (Kelton et al. 1997a, 1998). The structure of the Ti-Zr-Ni 1/1 approximant has been determined by refinement of x-ray and neutron powder diffraction data (Kim, W.J. et al. 1997, 1998b) The approximant is isostructural with the Al-Mg-Zn 1/1 approximant, containing two-shell Bergman clusters of atoms on a bcc lattice. This indicates that the Ti-Zr-Ni QC is also likely based on the Bergman cluster, not the pseudo-Mackay cluster found in *i*-Al-Pd-Mn (Boudard et al 1992). Measured densities and atomic sizes also classify this QC in the Bergman class with Al-Li-Cu and Al-Mg-Zn (Kim, W.J. et al. 1998b, 1998c).

In addition to the tetrahedral interstitial structure of the Bergman type, this *i* phase also has favorable chemistry for hydrogen storage. Attracted to Ti and to Zr, hydrogen can be loaded into the elemental metals up to H/M of 2.0 with structural transformations to hydride phases (Beck and Mueller

1968, Mueller 1968). Binary TiZr alloys can absorb hydrogen to $H/M > 1.0$, 1.9 in one instance (Mueller 1968). The success of the $LaNi_5$ alloy as a metal hydride negative battery electrode, and other Ni-bearing alloys that store hydrogen well (Willems 1986), show that the considerable Ni present in this i phase does not necessarily spoil its hydrogen-storage potential.

12.2.8.3 Loading Hydrogen into Icosahedral Ti-Zr-Ni. Loading hydrogen into the Ti-Zr-Ni QC was first accomplished by exposing as-quenched ribbons to high pressure hydrogen gas ($\sim$ 30 atm) at temperatures near 230 °C. The amount of hydrogen was measured both directly from weight changes using an electrobalance (± 5 μg) and by the pressure change in the loading cell. The QC peaks in powder XRD shifted to lower angles upon hydrogen absorption, proving that the hydrogen went into the QC and not into some secondary phase or the grain boundaries. The calibration of XRD peak shifts with mass gains and pressure changes was obtained for each composition of QC studied. Once calibrated, XRD is the most reliable method for determining H/M.

The Surface Barrier to Hydrogenation. In all instances of hydrogen loading from the gas phase, an induction time was observed, followed by rapid loading, as shown in Fig. 12.7. The typical maximum H/M for this method ($\sim$ 1.6) is independent of the length of the induction time, which tends to increase with increasing storage time of the ribbons in air (Viano et al. 1995a, 1998, Kim, J.Y. et al. 1998), suggesting that a surface oxide layer either inhibits dissociation of molecular hydrogen or acts as a barrier to rapid diffusion of atomic hydrogen into the bulk metal. This was confirmed by depth profiling studies using Auger spectroscopy (Viano 1996, Kelton et al. 1997a, Viano et

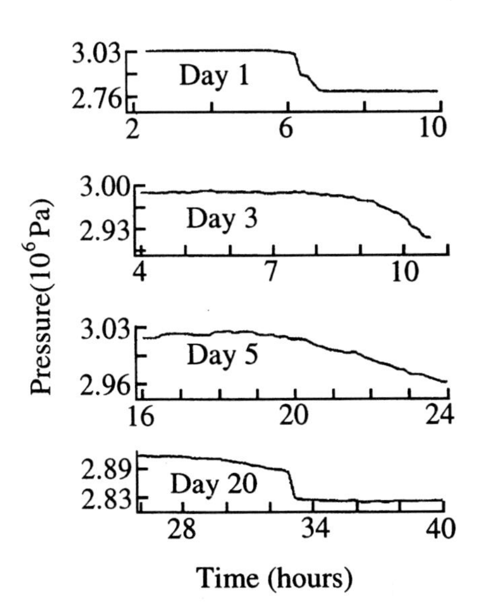

Fig. 12.7. The induction time for gas-phase loading of hydrogen at 230 °C into i-Ti-Zr-Ni previously stored in air increases from 6 h to more than 30 h as the time of air exposure increases. After Viano et al. (1998).

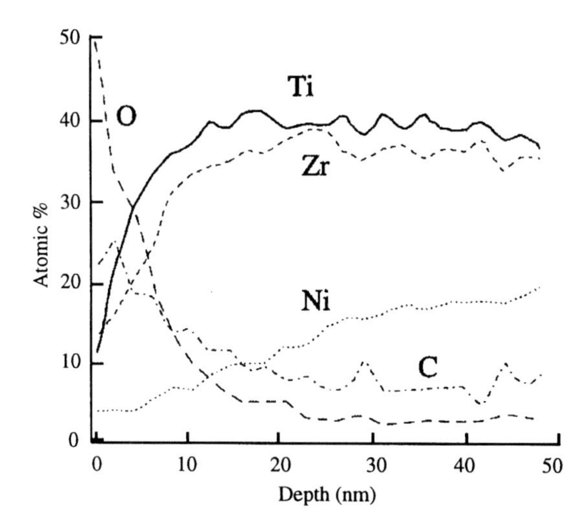

Fig. 12.8. An Auger depth profile reveals the surface oxide. After Viano et al. (1998).

al. 1998). The profiles in Fig. 12.8 show that the oxygen is confined to the top 10 nm of the sample studied, with Ni depleted and Ti and Zr enriched in that region (Viano et al. 1998).

Overcoming the Surface Barrier. The surface barrier to loading can be overcome in a number of ways. One is continuously exposing fresh surface in a hydrogen atmosphere. QC ribbons that were gently ball-milled in only a few atm of hydrogen could be fully loaded at room temperature (Viano et al. 1998). One problem is that the temperature of a freshly exposed, absorbing surface is not known, because the impacts of the ball bearings can heat the powder locally at the fresh surface, and the exothermic absorption reaction can heat a grain as it takes up hydrogen. That no crystalline hydride phases are formed during loading, however, suggests that any temperature increase is small.

The surface barrier to loading from the gas phase could also be breached by Ar plasma etching followed by the deposition of 10 to 30 nm of Pd. Pd does not oxidize and hydrogen passes easily through the air-Pd surface and through bulk Pd. Ribbons that had been stored in air for one year, after cleaning and coating, loaded to $H/M = 1.6$ with induction times of 4 h or less at *room temperature* in 17 atm of hydrogen. Without the surface treatment these ribbons had not absorbed hydrogen after 200 h at 230 °C in 22 atm of hydrogen. Further, the ribbons with the surface treatment, when loaded at 230 °C in 32 atm of hydrogen, reached a higher H/M value of 1.9. These high H/M values were obtained with no detectable second phase formation as a result of the hydrogenation (Kim, J.Y. et al. 1998). Like the surface-treated ribbons, an etched and Pd-coated ingot of the stable QC loaded after an induction time of less than 4 h in 17 atm of hydrogen at room temperature, reaching $H/M = 1.7$, but containing a small amount of the C14 hexagonal Laves phase after loading (Kelton et al. 1997b). That the i phase induction time did not go to zero suggests that a thin oxide may still be present at the Pd–Ti-Zr-Ni interface, requiring some time in the presence of hydrogen to develop paths for hydrogen to pass through. This may be due to incomplete etching or to oxygen contamination from the background gas in the diffusion-pumped chamber between etching and deposition.

The surface can also be activated and the hydrogen loaded into the QC by electrolytical means. A ribbon of QC was cathodically biased to 3.5 V in an electrolytic cell with a Pt anode and an electrolyte of 5 M KOH in deionized water. Hydrogen was produced at the cathode from OH^- ions in the electrolyte. Occasional operation for short times with reversed bias tended to remove the oxide from the QC. Ribbons were loaded repeatedly up to $H/M \sim 1.9$. Partial loading was easily achieved by controlling the total charge that flowed through the cell (Majzoub 1996).

12.2.8.4 Desorbing and Cycling Hydrogen. For any hydrogen-storage application, the ability to recover stored hydrogen and to make many store-and-recover cycles is crucial. Some tests have been made of the recovery and cycling ability of the Ti-Zr-Ni QC; systematic measurements of pressure-composition isotherms are planned. *In situ* powder XRD measurements made on loaded ribbons of QC in a temperature regulated heating stage with an oxygen-gettered atmosphere of inert gas, vacuum, or hydrogen have demonstrated reversible loading and desorption. The samples were heated in a clean vacuum until, at high temperature, the XRD peak positions indicated that all the hydrogen had left the sample. At intermediate temperatures the samples appeared to contain uniform concentrations of hydrogen, less than the starting concentration. H/M decreased from 1.54 to 0.025 as the temperature increased to 620 °C (Stroud et al. 1996). In a subsequent experiment a sample with initial H/M of 1.55, held in a gettered He atmosphere at 400 °C, released hydrogen and reached H/M of 0.71. Cooling in 1 atm hydrogen caused the sample to reabsorb hydrogen to a final H/M of 0.92 (Viano et al. 1998); a higher hydrogen pressure would likely have produced more reabsorption of hydrogen. The reabsorption occurred without any time delay in the non-oxidizing atmosphere, supporting the hypothesis that the surface oxide forms the barrier to hydrogenation.

Reversibility was demonstrated more directly by placing a sample of hydrogenated ribbon ($H/M = 1.6$) in an evacuated chamber that was heated from ambient temperature to 350 °C. The pressure in the chamber increased with increasing temperature and, upon cooling, decreased as shown in Fig. 12.9. That the heating and cooling curves nearly coincide demonstrates equilibrium between the hydrogen in the gas and that in solution in the QC. The hydrogen pressure changed much more with temperature than can be accounted for by ideal-gas expansion. The sample released and reabsorbed 25% of the stored hydrogen, with H/M varying between 1.6 and 1.2. This can be repeated for a few cycles, but growth of a crystalline metal hydride, fcc Ti-Zr, which binds hydrogen more tightly than the QC, eventually reduces the

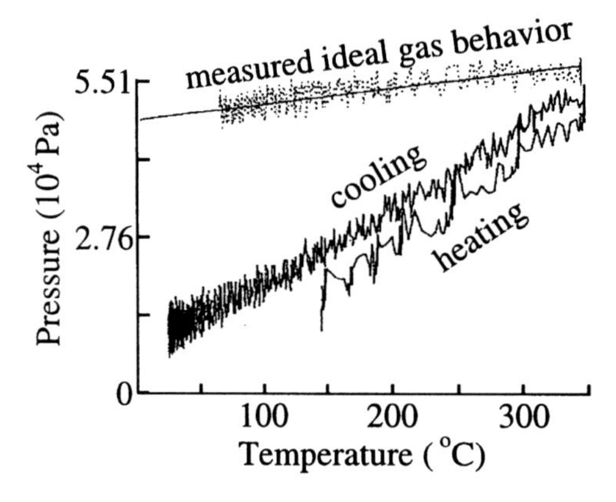

Fig. 12.9. Desorption and reabsorption of hydrogen from loaded ribbons of *i*-Ti-Zr-Ni on heating and cooling in a closed chamber. Measured ideal gas (Ar) pressure changes are also shown. After Kim, J.Y. et al. (1998).

amount of hydrogen released. After the first heating and cooling, the ribbon had broken into a fine-grained powder (Kim, J.Y. et al. 1998), likely allowing some oxidation by residual gasses and so degrading the cycling ability.

From these data the binding energy of the least tightly bound hydrogen in the QC can be estimated using a simple, one-binding-energy model (Speiser 1968) with saturation:

$$\frac{r}{s-r} = \sqrt{\frac{P}{P_0}} \exp(\frac{\Delta S}{k} - \frac{\Delta H}{kT}).$$

Here r is H/M and s is its maximum or saturation value, P is pressure, P_0 is a reference pressure, ΔS and ΔH are entropy and enthalpy differences between H_2 gas and H in solution in the QC, and k is the Boltzmann constant. The energy is (-0.10 ± 0.03) eV per hydrogen atom, an encouragingly low value. Some of the hydrogen must be more tightly bound, however. If a large mass ($\sim$ 1 g) of sample is hydrogenated at 230 °C in 22 atm of hydrogen, for example, the exothermic hydrogen absorption reaction generates enough heat to melt some of the ribbon (Kim, J.Y. 1997). Clearly the binding energy of the first hydrogen absorbed is much larger than that for the least tightly bound hydrogen that exhibits cycling.

12.2.8.5 Hydrogen as a New Characterization Tool for Quasicrystals. Probes of quasilattice structure other than diffraction, and probes of electronic structure have produced useful information about the local structures around hydrogen sites, the motions of the hydrogen in the alloy, both local and long-range, and the effect of the hydrogen on the valence electronic states of host metal atoms. Not directly relevant to hydrogen storage applications, this information is nevertheless important in selecting candidate materials for further study and development. Nuclear magnetic resonance of hydrogen and deuterium nuclei in powdered samples of the QC provide information about the hopping rate of hydrogen diffusing through the lattice, the distribution of near-neighbor H-H separations, and the symmetries of the interstitial sites occupied by D atoms. The time scale for long-range diffusion of hydrogen was found to be 17 ps (Viano et al. 1995b, Shastri et al. 1998). Inelastic neutron scattering from samples loaded with hydrogen, measured with high resolution as a function of temperature and scattering wave vector, revealed the presence of hydrogen temporarily bound in a potential with more than one minimum and hopping between minima only when assisted by the excitation of a local vibrational mode (130 meV) of a nearby hydrogen (Coddens et al. 1997).

12.3 Possible Applications

Based on the properties discussed above, several applications have been proposed for QCs. The most promising ones are discussed here.

12.3.1 Coatings

Coatings are the original and one of the most promising applications. Based on their hardness, non-stick quality, and corrosion resistance, QCs may be used in commercial cookware as PTFE is used in consumer cookware, and they may improve wear resistance and durabilty of systems through which reactive or abrasive materials are conveyed.

Thermal projection techniques, sub-sonic and super-sonic plasma spraying for example, can be used to deposit finely powdered metal alloys onto the surface of a bulk metal. The properties of the coatings depend on intrinsic properties of the coating material and on many parameters in the thermal-spray process. These include: the gases used; the speed and temperature of the projected particles; their size distribution; and the cooling rate, which can be controlled to some extent by choosing the temperature of the target metal (Dubois et al. 1994b, Besser and Sordelet 1997). The cooling rate depends on the thermal conductivity of the target material as well (Dubois et al. 1994b). Coating parameters that are important include: the thickness of the coating, typically 50 to a few hundred μm; porosity and surface roughness; and the phases present in the coating (Dubois et al. 1994b). In thermal projection the powder grains become molten, either throughout the particles or in a surface layer, and then impact the target with high kinetic energy. In plasma spraying an electric arc ionizes a flowing inert gas, usually argon, which carries powder particles along and transfers heat to them. On the target the molten particles spread out maintaining intimate contact with the target; they splat. They cool rapidly as a consequence, interlock, and also become very tightly bonded to the target (Besser and Sordelet 1997). Good thermal-spray coatings do not peel off.

Coatings formed using Al-based QCs, Al-Cu-Fe in particular, may contain a mixture of phases. In one experiment both the *i* phase and the high-temperature cubic β phase were produced. Al depletion due to selective evaporation can move the coating composition away from the range in which the i phase forms, especially if smaller powder grains, less than 50 μm in diameter, are used. With larger grains there is less depletion and the i phase may be the major phase. Annealing (700 °C in vacuum for 2 h) coatings that have enough Al to remain in the single i phase field can produce fully quasicrystalline coatings (Besser and Sordelet 1997).

The coatings produced are porous and rough. Roughness can be reduced to one micron or less by polishing, but porosity depends strongly on the powder particle size. Phase composition, Al content, and O content also depend on particle size, so there is a need to tune starting material and process parameters to obtain high-density, quasicrystalline coatings (Dubois et al. 1994a, Besser and Sordelet 1997).

12.3.1.1 Non-Stick Coatings for Cookware. The first proposed and most publicized application of QCs is for cookware coatings. This application depends on a number of properties and behaviors of the Al-Cu-Fe QC. First, as demonstrated by Dubois et al. (1994a), the QC surface is non-stick for steak fat. It is not universally good for all foods (Dubois 1996). Second, Al-Cu-Fe is a higher-temperature material than PTFE, the widely used consumer non-stick coating for cookware. Whereas PTFE decomposes and may release fluorine gas when overheated, the QC is robust against transformation to another structure, as well as against decomposition, at temperatures higher than those encountered in commercial cooking. Third, the QC is hard and has low friction and low wear properties. Spatulas used in cooking will not soon scrape it off the cookware as they do PTFE. Fourth, the elements in the QC are not toxic if ingested in small quantities with food. Fifth, the QC can be made into a thick, firmly attached coating by thermal-spray methods. The absence of any one of these properties would have doomed the proposed application. At the time of the 5th International Conference on Quasicrystals, May 1995, in Avignon, France, a restaurant there was using prototype Al-Cu-Fe-coated cookware for cooking steaks.

12.3.1.2 Other Coating Applications. At room temperature a quasicrystalline Al-Cu-Fe coating is non-stick for water, abrasion resistant, and can be well-bonded. It has been proposed that such coatings might be useful in equipment used to handle and mold concrete (Goldman et al. 1996).

At high temperatures coatings of Al-Cu-Fe and other QCs remain hard and tough, and have increasing ductility which may allow them to remain uncracked even on substrates not well matched in thermal expansion. Besides the aerospace applications, such as coatings for wings, bodies, or turbine blades, there is also a promising use as a thermal-barrier coating on parts that receive heat treatments including quenching from elevated temperatures. Paradoxically, a barrier to thermal conductivity at the surface enhances the cooling rate of a bulk solid that is heated and then plunged into water. Coating a part with a film of quasicrystalline material that has low thermal conductivity can increase the cooling rate of the part. In an experiment, aluminum and nickel cylinders were quenched in room-temperature and nearly boiling water from initial temperatures of 800 °C (Ni) and 500 °C (Al). Uncoated Ni quenched in room-temperature water cooled slowly at first, faster later. The initial slow cooling was due to film boiling; a thin layer of water vapor formed at the Ni surface and separated that surface from the liquid water. Ni cylinders coated with *i* quasicrystalline films of $Al_{70}Cu_9Fe_{10.5}Cr_{10.5}$, with thicknesses ranging from 150 to 1500 μm were also quenched in room-temperature water. The initial cooling rate was highest and there was no film boiling with a 500 μm coating. Al quenched from 500 °C in room-temperature water cooled rapidly, without film boiling, whether uncoated or coated with a film only 145 μm thick. Thicker coatings slowed the cooling. In nearly boiling water uncoated and thinly coated Al cooled from 500 °C with film boiling and a slow initial cooling rate; a 685 μm coating eliminated film boiling and produced a much faster initial cooling rate. QC coatings can be used now to eliminate film boiling and, by adjusting their thicknesses, to control cooling rates during quenching of parts whose properties are being altered by heat treatments (Archambault 1997).

12.3.2 Dispersion Hardening of Crystalline, Quasicrystalline, and Amorphous Alloys

A new steel alloy in the family of maraging steels has high strength in combination with good ductility after the introduction of a high density of *i* QC precipitates. These precipitates resist shearing, presumably by inhibiting dislocation motion. Traditionally, maraging steels are air cooled from the austenitic temperature range, forming an entirely martensitic structure, and then isothermally annealed at 400 to 500 °C to form precipitates of the Laves phases, η−Ni_3Ti, and orthorhombic Ni_3Mo. The new alloy, Sandvik IRK91 produced by AB Sandvik Steel in Sweden, contains Fe, 12 wt% Cr, 9 wt% Ni, 4 wt% Mo, 2 wt% Cu, 0.9 wt% Ti, 0.3 wt% Al, 0.15 wt% Si, and less than 0.05 wt% of either C or N (Liu and Nilsson 1997). It exhibits a maximum strength of 300 MPa. A maximum Vickers hardness of 730 is obtained after 336 h annealing at either 375 or 425 °C. The recommended annealing treatment is 4 h at 475 °C, which produces fine precipitates a few nm in diameter. TEM studies of larger precipitates formed by annealing for 100 h at 475 °C confirmed that they were *i* QCs 10 to 50 nm in diameter containing 50 wt% Mo, 30 wt% Fe, 15 wt% Cr, and traces of Ni or Si. Very resistant to coarsening, they grew no larger, even after annealing for 1000 h. Growing in isothermal conditions, they appear to be thermodynamically stable at 475 °C. This alloy, which is used to make surgical tools, owes its exceptional strength and good ductility to quasicrystalline precipitates (Liu and Nilsson 1997).

Fine-grained quasicrystalline alloys with high fracture strength and ductility sufficient for extrusion formation at 330 to 360 °C have been produced, initially by melt-spinning and, after development, by gas atomization followed by extrusion (Inoue et al. 1997). Lightweight, containing greater than 90 at%. Al, they are commercially available in Japan (Inoue et al. 1998). One extruded bulk alloy has composition $Al_{93}Mn_5Co_2$. In both melt-spun and extruded materials, the microstructure was 20 to 100 nm diameter i QCs, approximately spherical, in a fcc Al matrix. The bulk alloys have fracture strengths in the 500–850 MPa range and elongations before fracture of 7% to 24% at room temperature. They have strengths of about 350 MPa at 200 °C and about 210 MPa at 300 °C, and room-temperature Young's moduli of about 100 GPa (Inoue et al. 1997). High-resolution TEM and nanobeam electron diffraction have shown that the structure of the i particles is disordered at lengths of 1 nm and shorter, but has average i order at lengths of 3 nm. This suggests two-shell Mackay clusters of atoms in the structure that connect to make the i phase, but which are disordered internally, much like the pseudo-Mackay icosahedra proposed by quasicrystallographers (Boudard et al. 1992).

In earlier work with Al-Mn-Ce, Inoue et al. (1992) found tensile fracture strengths exceeding 1000 MPa and good ductility in laboratory samples in which the i grains were equiaxed and 50 to 100 nm in diameter, with the fcc Al grain boundary material only 10 to 15 nm thick (Inoue et al. 1992). Conventional Al alloys have strengths approaching 500 MPa. As the composition was varied, a phase mixture of amorphous and i material was found, but its mechanical properties were not studied. In another study, comparison of microstructures, diffraction, and fracture strengths of Al-Cr-TM, Al-Cr-Ce, and Al-Cr-Ce-TM (TM = transition metal) alloys revealed the conditions required for fracture strengths exceeding 1000 MPa. They are a homogeneous distribution of nanoscale, nearly spherical, i particles surrounded by a layer of fcc Al about 2 nm thick, with a high density of phason defects within the i particles and the absence of internal defects and grain boundaries in the fcc Al layer (Inoue et al. 1994).

12.3.3 Selective Absorbers for Solar-Thermal Converters

A stack of alternating thin films of a dielectric and an Al-Cu-Fe i QC can be optimized for use as a protective and absorbing coating on a solar photothermal converter; a device that efficiently converts solar radiation energy to thermal energy in a working fluid (Eisenhammer 1995). The application requires low reflectivity and strong absorption for wavelengths from 300 to 2000 nm, high reflectivity for longer wavelengths, and stability at temperatures as high as 500 °C (Eisenhammer et al. 1995, Eisenhammer 1997). Randomly inhomogeneous media have performed even better than the stacked thin films. Icosahedrally quasicrystalline Al-Cu-Fe nanoparticles were embedded in an Al_2O_3 matrix by sputtering onto a Cu substrate (high IR reflectivity and good thermal conductivity). The structure deposited on the Cu substrate was first

32 nm alumina, then 14 nm inhomogeneous medium, and finally 37 nm alumina. The inhomogeneous layer was made by sputtering about 4 nm average thickness of Al-Cu-Fe, which produces isolated nanoparticles covering about 30% of the area. Sputtering some alumina before adding more Al-Cu-Fe built up the particle-matrix composite structure (Eisenhammer 1997). The coating had good stability and could be accurately modeled by dynamic effective medium theory (Stroud and Pan 1978). A peak in reflectance of the inhomogeneous layer at a wavelength of 600 nm, reflectance about 0.3, limits absorption by a well-engineered composite to about 80%; the test layer was too thin to achieve this (Eisenhammer 1997).

12.3.4 Thermoelectric Devices

There is always interest in finding more efficient ways of using electrical power to produce and maintain temperature differences, refrigeration and air conditioning for example. The measurement of thermopower can be reversed to use an applied voltage to produce a temperature gradient in a thermoelectric material. Bismuth telluride, for example, has been used in portable beverage coolers that use power from the 12 V dc electrical systems of automobiles (Wu 1997). A good thermoelectric material should conduct electricity well and heat poorly. The electrical conductivity of a QC can be varied by making small changes in its chemical composition (Poon 1993, Wu 1997). Does the thermal conductivity remain usefully low as the electrical conductivity is so varied? There is some evidence that this may be the case (Legault et al. 1997). This possibility is being pursued (Wilson et al. 1997).

12.3.5 Hydrogen Storage and Battery Applications

Key factors for a hydrogen storage material are

- Ability to load a significant amount of hydrogen
- Ability to get the hydrogen into and out of the metal at reasonable values of pressure and temperature and within a reasonable time
- Ability to repeat this cycle many times without degradation of the intermetallic alloy.

As discussed, both the structure and the chemistry of the Ti-Zr-Ni QCs are well-suited to hydrogen storage applications. It absorbs and desorbs significant quantities of hydrogen quickly, and is made of low-cost materials. A brief comparison of the storage characteristics of the $Ti_{45}Zr_{38}Ni_{17}$ *i* phase with those of crystalline materials of current technological interest is given in Table 12.1. Clearly, the total loading capacity of the QC is competitive. It can, for example, absorb hydrogen to a higher weight percent than $LaNi_5$, widely used in hydride batteries, or TiFe, the most promising materials for stationary hydrogen storage applications. While the loading capacity per weight is

Table 12.1. Comparison of hydrogen storage properties of *i*-Ti-Zr-Ni with metal hydrides of technological interest

Material	H/M	Weight % H	Comments
$LaNi_5$	1.1	1.5	Negative electrode in Ni metal-hydride rechargable batteries
TiFe	0.9	1.6	Best material developed for stationary applications; requires high pressure or surface activation
Mg	2.0	7.7	Light, inexpensive, but flammable; unloading temperature higher than typical exhaust gas from internal combustion engine
V	2.0	3.8	Expensive
$Ti_{45}Zr_{38}Ni_{17}$	1.7	2.5	Initial investigations promising

better for Mg or V, Mg is flammable and requires a higher unloading temperature than is typical for the exhaust gas from an internal combustion engine, and the materials costs for V are too high for widespread use. The QC is competitive with materials currently under study.

The desorption of this hydrogen at reasonable temperatures in useful times has been demonstrated. Based on initial desorption data, the binding enthalpy for the least tightly bound hydrogen in the QC is small, of order -0.10 eV/atom. Relevant thermodynamic and kinetic data, including the pressure composition isotherms and the cycling durability are required, however, before a realistic evaluation of their technological usefulness can be made.

12.4 Conclusion

QCs have found commercial applications already as dispersoids in a high-strength surgical steel and as a component of a new kind of composite Al alloy that has both high strength and ductility. Prototype cookware with non-stick, high-temperature, abrasion-resistant coatings of *i*-Al-Cu-Fe have been used in at least one restaurant. QCs show promise for use as photothermal solar absorber coatings, thermal barrier coatings, hydrogen storage materials, and thermoelectric elements for small-scale heating and cooling. These uses are being explored in research laboratories, and their feasibility should be determined in the next few years. As fascinating as QCs were at the time of their discovery in 1985 and still are, because of their unusual structure, they are more than just laboratory curiosities now. The next few years will reveal their true potential as a new class of technological materials.

Acknowledgment

The work of KFK was supported by the US NSF under grant DMR-97-05202.

References

Adamson, A.W. (1990): Physical Chemistry of Surfaces. John Wiley and Sons, New York

Archambault, P. (1997): in New Horizons in Quasicrystals, Goldman, A.I., Sordelet, D.J., Thiel, P.A., Dubois, J.M. (eds). World Scientific, Singapore, p 232

Ashcroft, N.W., Mermin, N.D. (1976): Solid State Physics. Saunders College, Philadelphia

Basov, D.N., Timusk, T., Pierce, F., Volkov, P., Guo, Q., Poon, S.J., Thomas, G.A., Rapkine, D., Kortan, A.R., Barakat, F., Greedan, J., Grushko, B. (1995): in Proceedings of the 5th International Conference on Quasicrystals, Janot, C., Mosseri, R. (eds). World Scientific, Singapore, p 564

Beck L., Mueller W.M. (1968): in Metal Hydrides, Mueller, W.M., Blackledge, J.P., Libowitz, G.G. (eds). Academic Press, New York, p 241

Belin, E, Dankházi, Z., Sadoc, A., Dubois, J.M. (1994): J. Phys. Condens Matter **6**, 8771

Belin, E., Dankhazi, Z., Sadoc, A., Flank, A.M., Poon, S.J., Müller, H., Kirchmayr, H. (1995): in Proceedings of the 5th International Conference on Quasicrystals, Janot, C., Mosseri, R. (eds). World Scientific, Singapore, p 435

Belin-Ferré, E., Fournée, V., Dubois, J.-M. (1997): in New Horizons in Quasicrystals, Goldman, A.I., Sordelet, D.J., Thiel, P.A., Dubois, J.M. (eds). World Scientific, Singapore, p 9

Belin-Ferré, E., Fournée, V., Dubois, J.M., Sadoc, A. (1998): in Proceedings of the 6th International Conference on Quasicrystals, Takeuchi, S., Fujiwara, T. (eds). World Scientific, Singapore, p 603

Besser, M.F., Sordelet, D.J. (1997): in New Horizons in Quasicrystals, Goldman, A.I., Sordelet, D.J., Thiel, P.A., Dubois, J.M. (eds). World Scientific, Singapore, p 288

Boudard, M., de Boissieu, M., Janot, C., Heger, G., Beeli, C., Nissen, H.-U., Vincent, H., Ibberson, R., Audier, M., Dubois J.M. (1992): J. Phys. Condens. Matter **4**, 10 149

Bresson L., (1997): in New Horizons in Quasicrystals, Goldman, A.I., Sordelet, D.J., Thiel, P.A., Dubois, J.M. (eds). World Scientific, Singapore, p 78

Bresson, L., Gratias, D. (1993): J. Non-Cryst. Solids **153-154**, 468

Castle, J.E. (1993): in Concise Encyclopedia of Materials Characterization, Cahn, R.W., Lifshin, E. (eds). Pergamon Press, Oxford, p 74

Chang, S.-L., Zhang, C.-M., Jenks, C.J., Andregg, J.W., Thiel, P.A. (1995): in Proceedings of the 5th International Conference on Quasicrystals, Janot, C., Mosseri, R. (eds). World Scientific, Singapore, p 786

Chernikov, M.A., Bianchi, A., Müller, H., Ott, H.R. (1995a): in Proceedings of the 5th International Conference on Quasicrystals, Janot, C., Mosseri, R. (eds). World Scientific, Singapore, p 569

Chernikov, M.A., Bianchi, A., Ott, H.R. (1995b): Phys. Rev. B **51**, 153

Chernikov, M.A., Edagawa, K., Felder, E., Bianchi, A.D., Gubler, U., Kenzelmann, M., Ott, H.R. (1997): Bull. Am. Phys. Soc. **42**, 499

Chernikov, M.A., Felder E., Bianchi, A.D., Wälti, C., Kenzelman, M., Ott, H.R., Edagawa, K., de Boissieu, M., Janot, C., Feuerbacher, M., Tamura, N., Urban, K. (1998): in Proceedings of the 6th International Conference on Quasicrystals, Takeuchi, S., Fujiwara, T. (eds). World Scientific, Singapore, p 451
Chevrier, J., Cappello, G., Comin, F., Palmari, J.P. (1997): in New Horizons in Quasicrystals, Goldman, A.I., Sordelet, D.J., Thiel, P.A., Dubois, J.M. (eds). World Scientific, Singapore, p 144
Coddens, G., Bellissent, R. (1993): J. Non-Cryst. Solids **153-154**, 557
Coddens, G., Viano, A.M., Gibbons, P.C., Kelton, K.F., Kramer, M.J. (1997): Solid State Commun. **104**, 179
Dubois, J.-M. (1993): Phys. Scr. T **49**, 17
Dubois, J.M. (1996): private communication
Dubois, J.M. (1997): in New Horizons in Quasicrystals, Goldman, A.I., Sordelet, D.J., Thiel, P.A., Dubois, J.M. (eds). World Scientific, Singapore, p 208
Dubois, J.-M., Kang, S.S., Massiani, Y. (1993a): J. Non-Cryst. Solids **153-154**, 443
Dubois, J.M., Kang, S.S., Archambault, P., Colleret, B. (1993b): J. Mater. Res. **8**, 38
Dubois, J.M., Proner, A., Bucaille, B., Cathonnet, P., Dong, C., Richard, V., Pianelli, A., Massiani, Y., Ait Yaazza, S., Belin-Ferre, E. (1994a): Ann. Chim. Fr. **19**, 3
Dubois, J.-M., Kang, S.S., Perrot, A. (1994b): Mater. Sci. Eng. A **179-180**, 122
Dubois, J.M., Plaindoux, P., Belin-Ferre, E., Tamura, N., Sordelet, D.J. (1998): in Proceedings of the 6th International Conference on Quasicrystals, Takeuchi, S., Fujiwara, T. (eds). World Scientific, Singapore, p 733
Eisenhammer, T. (1995): Thin Solid Films **270**, 1
Eisenhammer, T. (1997): in New Horizons in Quasicrystals, Goldman, A.I., Sordelet, D.J., Thiel, P.A., Dubois, J.M. (eds). World Scientific, Singapore, p 304
Eisenhammer, T., Mahr, A., Haugeneder, A., Reichelt, T., Assmann W., (1995): in Proceedings of the 5th International Conference on Quasicrystals, Janot, C., Mosseri, R. (eds). World Scientific, Singapore, p 758
Feuerbacher, M., Rosenfeld, R., Baufeld, B., Bartsch, M., Messerschmidt, U., Wollgarten, M., Urban, K. (1995): in Proceedings of the 5th International Conference on Quasicrystals, Janot, C., Mosseri, R. (eds). World Scientific, Singapore, p 714
Feuerbacher, M., Metzmacher, C., Wollgarten, M., Urban, K., Baufeld, B., Bartsch, M., Messerschmidt, U. (1997): in New Horizons in Quasicrystals, Goldman, A.I., Sordelet, D.J., Thiel, P.A., Dubois, J.M. (eds). World Scientific, Singapore, p 103
Gavatz, M., Rouxel, D., Claudel, D., Pigeat, P., Weber B., Dubois, J.M. (1998): in Proceedings of the 6th International Conference on Quasicrystals, Takeuchi, S., Fujiwara, T. (eds). World Scientific, Singapore, p 765
Gelmann, A. (1996): unpublished
Goldman, A.I., Anderegg, J.W., Besser, M.F., Chang, S.-L., Delaney, D.W., Jenks, C.J., Kramer, M.J., Lograsso, T.A., Lynch, D.W., McCallum, R.W., Shield, J.E., Sordelet, D.J., Thiel, P.A. (1996): Am. Sci. **84**, 230
Gu, T., Goldman, A.I., Pinhero, P.J. (1997a): Bull. Am. Phys. Soc. **42**, 497
Gu T., Goldman, A.I., Pinhero, P., Delaney, D. (1997b): in New Horizons in Quasicrystals, Goldman, A.I., Sordelet, D.J., Thiel, P.A., Dubois, J.M. (eds). World Scientific, Singapore, p 165
Haberkern, R., Fritsch, G. (1995): in Proceedings of the 5th International Conference on Quasicrystals, Janot, C., Mosseri, R. (eds). World Scientific, Singapore, p 460
Hennig, R.G. (1997): private communication

Hennig, R.G., Teichler, H. (1997): Philos. Mag. A **76**, 1053
Inoue, A., Watanabe, M., Kimura, H.M., Takahashi, F., Nagata, A. (1992): Mater. Trans., Jpn. Inst. Met. **33**, 723
Inoue, A., Kimura, H.M., Sasamori, K., Masumoto, T. (1994): Mater. Trans., Jpn. Inst. Met. **35**, 85
Inoue, A., Kimura, H., Kita, K. (1997): in New Horizons in Quasicrystals, Goldman, A.I., Sordelet, D.J., Thiel, P.A., Dubois, J.M. (eds). World Scientific, Singapore, p 256
Inoue, A., Kimura, H.M., Sasamori, K., Kita, K. (1998): in Proceedings of the 6th International Conference on Quasicrystals, Takeuchi, S., Fujiwara, T. (eds). World Scientific, Singapore, p 723
Jenks, C.J., Pinhero, P.J., Chang, S.-L., Andregg, J.W., Besser, M.F., Sordelet, D.J., Thiel, P.A. (1997): in New Horizons in Quasicrystals, Goldman, A.I., Sordelet, D.J., Thiel, P.A., Dubois, J.M. (eds). World Scientific, Singapore, p 157
Kang, S.S., Dubois, J.M. (1992): Philos. Mag. A **66**, 151
Kang, S.S., Dubois, J.M., von Stebut, J. (1993): J. Mater. Res. **8**, 2471
Kelton, K.F., Viano, A.M., Stroud, R.M., Majzoub, E.H., Gibbons, P.C., Misture, S.T., Goldman, A.I., Kramer, M.J. (1997a): in New Horizons in Quasicrystals, Goldman, A.I., Sordelet, D.J., Thiel, P.A., Dubois, J.M. (eds). World Scientific, Singapore, p 272
Kelton, K.F., Kim, W.J., Stroud, R.M. (1997b): Appl. Phys. Lett. **70**, 3230
Kelton, K.F., Kim, J.Y., Majzoub, E.H., Gibbons, P.C., Viano, A.M., Stroud, R.M. (1998): in Proceedings of the 6th International Conference on Quasicrystals, Takeuchi, S., Fujiwara, T. (eds). World Scientific, Singapore, p 261
Kim, J.Y. (1997): private communication
Kim, J.Y., Gibbons, P.C., Kelton, K.F. (1998): J. Alloys Comp. **266**, 311
Kim, W.J. (1997): private communication
Kim, W.J., Gibbons, P.C., Kelton, K.F. (1997): Philos. Mag. Lett. **76**, 199
Kim, W.J., Gibbons, P.C., Kelton, K.F. (1998a): in Proceedings of the 6th International Conference on Quasicrystals, Takeuchi, S., Fujiwara, T. (eds). World Scientific, Singapore, p 47
Kim, W.J., Gibbons, P.C., Kelton, K.F., Yelon, W.B. (1998b): Phys Rev B **58**, 2578
Kim, W.J., Gibbons, P.C., Kelton, K.F. (1998c): Philos. Mag. A **78**, 1111
Kittel, C. (1986): Introduction to Solid State Physics, 6th ed. John Wiley and Sons, New York
Legault, S., Ellman, B., Ström-Olsen, J., Taillefer, L., Kycia, S. (1995): in Proceedings of the 5th International Conference on Quasicrystals, Janot, C., Mosseri, R. (eds). World Scientific, Singapore, p 592
Legault, S., Ellman, B., Ström-Olson, J.O., Taillefer, L., Kycia, S., Lograsso, T., Delaney, D. (1997): in New Horizons in Quasicrystals, Goldman, A.I., Sordelet, D.J., Thiel, P.A., Dubois, J.M. (eds). World Scientific, Singapore, p 224
Libbert, J.L., Kelton, K.F., Goldman, A.I., Yelon, W.B. (1994): Phys. Rev. B **49**, 11 675
Lin, S., Li, G., Zhang, D. (1996): Phys. Rev. Lett. **77**, 1998
Liu, P., Nilsson, J.-O. (1997): in New Horizons in Quasicrystals, Goldman, A.I., Sordelet, D.J., Thiel, P.A., Dubois, J.M. (eds). World Scientific, Singapore, p 264
Lubensky, T.C., Socolar, J.E.S, Steinhardt, P.J., Bancel, P.A., Heiney, P.A. (1986): Phys. Rev. Letters **57**, 1440
Majzoub, E.H. (1996): private communication
Massiani, Y., Ait Yaazza, S., Crousier, J.P., Dubois, J.M. (1993): J. Non-Cryst. Solids **159**, 92

Massiani, Y., Ait Yaazza, S., Dubois, J.-M. (1995): in Proceedings of the 5th International Conference on Quasicrystals, Janot, C., Mosseri, R. (eds). World Scientific, Singapore, p 790
McDonald, J.R. (1987): Impedance Spectroscopy. John Wiley and Sons, New York
Mikulla, R., Krul, F., Gumbsch, P., Trebin, H.-R. (1997): in New Horizons in Quasicrystals, Goldman, A.I., Sordelet, D.J., Thiel, P.A., Dubois, J.M. (eds). World Scientific, Singapore, p 200
Molokanov, V.V., Chebotnikov, V.N. (1990): J. Non-Cryst. Solids **117-118**, 789
Mott, N.F., Davis, E.A. (1979): Electronic Processes in Non-Crystalline Materials. Second Edition. Clarendon Press, Oxford
Mueller, W.M. (1968): in Metal Hydrides, Mueller, W.M., Blackledge, J.P., Libowitz, G.G. (eds). Academic Press, New York, pp 336, 377–378
Perrot, A., Dubois, J.M. (1993): Ann. Chim. Fr. **18**, 501
Perrot, A., Dubois, J.M., Cassart, M., Issi, J.P. (1995): in Proceedings of the 5th International Conference on Quasicrystals, Janot, C., Mosseri, R. (eds). World Scientific, Singapore, p 588
Pierce, F.S., Poon, S.J., Guo, Q. (1993): Science **261**, 737
Pierce, F.S., Guo, Q., Poon, S.J. (1994): Phys. Rev. Lett. **73**, 2220
Poon, S.J. (1992): Adv. Phys. **41**, 303
Poon, S.J. (1993): J. Non-Cryst. Solids **153-154**, 334
Poon, S.J., Pierce, F.S., Guo, Q., Volkov, P. (1995): in Proceedings of the 5th International Conference on Quasicrystals, Janot, C., Mosseri, R. (eds). World Scientific, Singapore, p 408
Rivier, N. (1993): J. Non-Cryst. Solids **153-154**, 458
Rivier, N. (1997): in New Horizons in Quasicrystals, Goldman, A.I., Sordelet, D.J., Thiel, P.A., Dubois, J.M. (eds). World Scientific, Singapore, p 188
Rivier, N., Boose, D. (1995): in Proceedings of the 5th International Conference on Quasicrystals, Janot, C., Mosseri, R. (eds). World Scientific, Singapore, p 802
Rouxel, D., Gavatz, M., Pigeat, P., Weber, B., Plaindoux, P. (1997): in New Horizons in Quasicrystals, Goldman, A.I., Sordelet, D.J., Thiel, P.A., Dubois, J.M. (eds). World Scientific, Singapore, p 173
Shastri, A., Majzoub, E.H, Borsa, F., Gibbons, P.C., Kelton, K.F. (1998): Phys. Rev. B **57**, 5148
Shibuya, T., Hashimoto, T., Takeuchi, S. (1990): Jpn. J. Appl. Phys. **29**, L349
Shield, J.E. (1997): in New Horizons in Quasicrystals, Goldman, A.I., Sordelet, D.J., Thiel, P.A., Dubois, J.M. (eds). World Scientific, Singapore, p 312
Shield, J.E., Kramer, M.J. (1995): in Proceedings of the 5th International Conference on Quasicrystals, Janot, C., Mosseri, R. (eds). World Scientific, Singapore, p 718
Sibirtsev, S.A., Chebotnikov, V.N., Molokanov, V.V., Kovneristyĭ (1988): JETP Lett. **47**, 744
Sordelet, D.J., Kramer, M.J., Anderson, I.E., Besser, M.F. (1995): in Proceedings of the 5th International Conference on Quasicrystals, Janot, C., Mosseri, R. (eds). World Scientific, Singapore, p 778
Sordelet, D.J., Gunderman, L.A., Besser, M.F., Akinc, A.B. (1997): in New Horizons in Quasicrystals, Goldman, A.I., Sordelet, D.J., Thiel, P.A., Dubois, J.M. (eds). World Scientific, Singapore, p 296
Speiser R. (1968): in Metal Hydrides, Mueller, W.M., Blackledge, J.P., Libowitz, G.G. (eds). Academic Press, New York, p 82
Stadnik, Z.M. (1997): Bull. Am. Phys. Soc. **42**, 496
Stadnik, Z.M., Purdie, D., Garnier, M., Baer, Y., Tsai, A.-P., Inoue, A., Edagawa, K., Takeuchi, S. (1996): Phys. Rev. Lett. **77**, 1777

Stadnik, Z.M., Purdie, D., Garnier, M., Baer, Y., Tsai, A.-P., Inoue, A., Edagawa, K., Takeuchi, S. (1998): in Proceedings of the 6th International Conference on Quasicrystals, Takeuchi, S., Fujiwara, T. (eds). World Scientific, Singapore, p 563

Stroud, D., Pan, F.P. (1978): Phys Rev B **17**, 1602

Stroud, R.M., Viano, A.M., Gibbons, P.C., Kelton, K.F., Misture, S.T. (1996): Appl. Phys. Lett. **69**, 2998

Switendick A.C. (1978a): Sandia Laboratories Techn. Rep., NTIS Order No. SAND-78-0250

Switendick A.C. (1978b): in Hydrogen in Metals I, Alefeld, G., Völkl, J. (eds). Springer-Verlag, Berlin, p 106

Takeuchi, T., Yamada, Y., Mizutani, U., Honda, Y., Edagawa, K., Takeuchi, S. (1995): in Proceedings of the 5th International Conference on Quasicrystals, Janot, C., Mosseri, R. (eds). World Scientific, Singapore, p 534

Tamura, R., Kirihara, K., Kimura, K., Ino, H. (1995): in Proceedings of the 5th International Conference on Quasicrystals, Janot, C., Mosseri, R. (eds). World Scientific, Singapore, p 539

Viano, A.M. (1996): Ph.D. Thesis, Washington University, St. Louis

Viano, A.M., Stroud, R.M., Gibbons, P.C., McDowell, A.F., Conradi, M.S., Kelton, K.F. (1995a): Phys. Rev. B **51**, 12 026

Viano, A.M., McDowell, A.F., Conradi, M.S., Gibbons, P.C., Kelton, K.F. (1995b): in Proceedings of the 5th International Conference on Quasicrystals, Janot, C., Mosseri, R. (eds). World Scientific, Singapore, p 798

Viano, A.M., Majzoub, E.H., Stroud, R.M., Kramer, M.J., Misture, S.T., Gibbons P.C., Kelton K.F. (1998): Philos. Mag. A **78**, 131

von Stebut, J., Soro, J.M., Plaindoux, P., Dubois, J.M. (1997): in New Horizons in Quasicrystals, Goldman, A.I., Sordelet, D.J., Thiel, P.A., Dubois, J.M. (eds). World Scientific, Singapore, p 248

Wehner, B.I., Köster, U. (1997): in New Horizons in Quasicrystals, Goldman, A.I., Sordelet, D.J., Thiel, P.A., Dubois, J.M. (eds). World Scientific, Singapore, p 152

Westlake, D.G. (1980): J. Less Common Met. **75**, 177

Willems, J.J.G. (1986): Phillips Tech. Rev. **43**, 22

Wilson, M.L., LeGault, S., Stroud, R.M., Tritt, T.M. (1997): in Thermoelectric Materials – New Directions and Approaches, Tritt, T.M., Kanatzidis, M.G., Lyon, Jr., H.B., Mahan, G.D. (eds). Materials Research Society, Warrendale, Pennsylvania, p 323

Wollgarten, M., Saka, H. (1997): in New Horizons in Quasicrystals, Goldman, A.I., Sordelet, D.J., Thiel, P.A., Dubois, J.M. (eds). World Scientific, Singapore, p 320

Wu, C. (1997): Science News **152**, 152

Zhang, X., Stroud, R.M., Libbert, J.L., Kelton, K.F. (1994): Philos. Mag. B **70**, 927

Subject Index

Springer Series in Solid-State Sciences

Editors: M. Cardona P. Fulde K. von Klitzing H.-J. Queisser

1 **Principles of Magnetic Resonance** 3rd Edition By C. P. Slichter
2 **Introduction to Solid-State Theory** By O. Madelung
3 **Dynamical Scattering of X-Rays in Crystals** By Z. G. Pinsker
4 **Inelastic Electron Tunneling Spectroscopy** Editor: T. Wolfram
5 **Fundamentals of Crystal Growth I** Macroscopic Equilibrium and Transport Concepts By F. E. Rosenberger
6 **Magnetic Flux Structures in Superconductors** By R. P. Huebener
7 **Green's Functions in Quantum Physics** 2nd Edition By E. N. Economou
8 **Solitons and Condensed Matter Physics** Editors: A. R. Bishop and T. Schneider
9 **Photoferroelectrics** By V. M. Fridkin
10 **Phonon Dispersion Relations in Insulators** By H. Bilz and W. Kress
11 **Electron Transport in Compound Semiconductors** By B. R. Nag
12 **The Physics of Elementary Excitations** By S. Nakajima, Y. Toyozawa, and R. Abe
13 **The Physics of Selenium and Tellurium** Editors: E. Gerlach and P. Grosse
14 **Magnetic Bubble Technology** 2nd Edition By A. H. Eschenfelder
15 **Modern Crystallography I** Fundamentals of Crystals Symmetry, and Methods of Structural Crystallography 2nd Edition By B. K. Vainshtein
16 **Organic Molecular Crystals** Their Electronic States By E. A. Silinsh
17 **The Theory of Magnetism I** Statics and Dynamics By D. C. Mattis
18 **Relaxation of Elementary Excitations** Editors: R. Kubo and E. Hanamura
19 **Solitons** Mathematical Methods for Physicists By. G. Eilenberger
20 **Theory of Nonlinear Lattices** 2nd Edition By M. Toda
21 **Modern Crystallography II** Structure of Crystals 2nd Edition By B. K. Vainshtein, V. L. Indenbom, and V. M. Fridkin
22 **Point Defects in Semiconductors I** Theoretical Aspects By M. Lannoo and J. Bourgoin
23 **Physics in One Dimension** Editors: J. Bernasconi and T. Schneider
24 **Physics in High Magnetics Fields** Editors: S. Chikazumi and N. Miura
25 **Fundamental Physics of Amorphous Semiconductors** Editor: F. Yonezawa
26 **Elastic Media with Microstructure I** One-Dimensional Models By I. A. Kunin
27 **Superconductivity of Transition Metals** Their Alloys and Compounds By S. V. Vonsovsky, Yu. A. Izyumov, and E. Z. Kurmaev
28 **The Structure and Properties of Matter** Editor: T. Matsubara
29 **Electron Correlation and Magnetism in Narrow-Band Systems** Editor: T. Moriya
30 **Statistical Physics I** Equilibrium Statistical Mechanics 2nd Edition By M. Toda, R. Kubo, N. Saito
31 **Statistical Physics II** Nonequilibrium Statistical Mechanics 2nd Edition By R. Kubo, M. Toda, N. Hashitsume
32 **Quantum Theory of Magnetism** 2nd Edition By R. M. White
33 **Mixed Crystals** By A. I. Kitaigorodsky
34 **Phonons: Theory and Experiments I** Lattice Dynamics and Models of Interatomic Forces By P. Brüesch
35 **Point Defects in Semiconductors II** Experimental Aspects By J. Bourgoin and M. Lannoo
36 **Modern Crystallography III** Crystal Growth By A. A. Chernov
37 **Modern Chrystallography IV** Physical Properties of Crystals Editor: L. A. Shuvalov
38 **Physics of Intercalation Compounds** Editors: L. Pietronero and E. Tosatti
39 **Anderson Localization** Editors: Y. Nagaoka and H. Fukuyama
40 **Semiconductor Physics** An Introduction 6th Edition By K. Seeger
41 **The LMTO Method** Muffin-Tin Orbitals and Electronic Structure By H. L. Skriver
42 **Crystal Optics with Spatial Dispersion, and Excitons** 2nd Edition By V. M. Agranovich and V. L. Ginzburg
43 **Structure Analysis of Point Defects in Solids** An Introduction to Multiple Magnetic Resonance Spectroscopy By J.-M. Spaeth, J. R. Niklas, and R. H. Bartram
44 **Elastic Media with Microstructure II** Three-Dimensional Models By I. A. Kunin
45 **Electronic Properties of Doped Semiconductors** By B. I. Shklovskii and A. L. Efros
46 **Topological Disorder in Condensed Matter** Editors: F. Yonezawa and T. Ninomiya

Springer Series in Solid-State Sciences

Editors: M. Cardona P. Fulde K. von Klitzing H.-J. Queisser

47 **Statics and Dynamics of Nonlinear Systems**
Editors: G. Benedek, H. Bilz, and R. Zeyher

48 **Magnetic Phase Transitions**
Editors: M. Ausloos and R. J. Elliott

49 **Organic Molecular Aggregates**
Electronic Excitation and Interaction Processes
Editors: P. Reineker, H. Haken, and H. C. Wolf

50 **Multiple Diffraction of X-Rays in Crystals**
By Shih-Lin Chang

51 **Phonon Scattering in Condensed Matter**
Editors: W. Eisenmenger, K. Laßmann, and S. Döttinger

52 **Superconductivity in Magnetic and Exotic Materials** Editors: T. Matsubara and A. Kotani

53 **Two-Dimensional Systems, Heterostructures, and Superlattices**
Editors: G. Bauer, F. Kuchar, and H. Heinrich

54 **Magnetic Excitations and Fluctuations**
Editors: S. W. Lovesey, U. Balucani, F. Borsa, and V. Tognetti

55 **The Theory of Magnetism II** Thermodynamics and Statistical Mechanics By D. C. Mattis

56 **Spin Fluctuations in Itinerant Electron Magnetism** By T. Moriya

57 **Polycrystalline Semiconductors**
Physical Properties and Applications
Editor: G. Harbeke

58 **The Recursion Method and Its Applications**
Editors: D. G. Pettifor and D. L. Weaire

59 **Dynamical Processes and Ordering on Solid Surfaces** Editors: A. Yoshimori and M. Tsukada

60 **Excitonic Processes in Solids**
By M. Ueta, H. Kanzaki, K. Kobayashi, Y. Toyozawa, and E. Hanamura

61 **Localization, Interaction, and Transport Phenomena** Editors: B. Kramer, G. Bergmann, and Y. Bruynseraede

62 **Theory of Heavy Fermions and Valence Fluctuations** Editors: T. Kasuya and T. Saso

63 **Electronic Properties of Polymers and Related Compounds**
Editors: H. Kuzmany, M. Mehring, and S. Roth

64 **Symmetries in Physics** Group Theory Applied to Physical Problems 2nd Edition
By W. Ludwig and C. Falter

65 **Phonons: Theory and Experiments II**
Experiments and Interpretation of Experimental Results By P. Brüesch

66 **Phonons: Theory and Experiments III**
Phenomena Related to Phonons
By P. Brüesch

67 **Two-Dimensional Systems: Physics and New Devices**
Editors: G. Bauer, F. Kuchar, and H. Heinrich

68 **Phonon Scattering in Condensed Matter V**
Editors: A. C. Anderson and J. P. Wolfe

69 **Nonlinearity in Condensed Matter**
Editors: A. R. Bishop, D. K. Campbell, P. Kumar, and S. E. Trullinger

70 **From Hamiltonians to Phase Diagrams**
The Electronic and Statistical-Mechanical Theory of sp-Bonded Metals and Alloys By J. Hafner

71 **High Magnetic Fields in Semiconductor Physics**
Editor: G. Landwehr

72 **One-Dimensional Conductors**
By S. Kagoshima, H. Nagasawa, and T. Sambongi

73 **Quantum Solid-State Physics**
Editors: S. V. Vonsovsky and M. I. Katsnelson

74 **Quantum Monte Carlo Methods in Equilibrium and Nonequilibrium Systems** Editor: M. Suzuki

75 **Electronic Structure and Optical Properties of Semiconductors** 2nd Edition
By M. L. Cohen and J. R. Chelikowsky

76 **Electronic Properties of Conjugated Polymers**
Editors: H. Kuzmany, M. Mehring, and S. Roth

77 **Fermi Surface Effects**
Editors: J. Kondo and A. Yoshimori

78 **Group Theory and Its Applications in Physics**
2nd Edition
By T. Inui, Y. Tanabe, and Y. Onodera

79 **Elementary Excitations in Quantum Fluids**
Editors: K. Ohbayashi and M. Watabe

80 **Monte Carlo Simulation in Statistical Physics**
An Introduction 3rd Edition
By K. Binder and D. W. Heermann

81 **Core-Level Spectroscopy in Condensed Systems**
Editors: J. Kanamori and A. Kotani

82 **Photoelectron Spectroscopy**
Principle and Applications 2nd Edition
By S. Hüfner

83 **Physics and Technology of Submicron Structures**
Editors: H. Heinrich, G. Bauer, and F. Kuchar

84 **Beyond the Crystalline State** An Emerging Perspective By G. Venkataraman, D. Sahoo, and V. Balakrishnan

85 **The Quantum Hall Effects**
Fractional and Integral 2nd Edition
By T. Chakraborty and P. Pietiläinen

86 **The Quantum Statistics of Dynamic Processes**
By E. Fick and G. Sauermann

87 **High Magnetic Fields in Semiconductor Physics II**
Transport and Optics Editor: G. Landwehr

88 **Organic Superconductors** 2nd Edition
By T. Ishiguro, K. Yamaji, and G. Saito

89 **Strong Correlation and Superconductivity**
Editors: H. Fukuyama, S. Maekawa, and A. P. Malozemoff

Springer Series in Solid-State Sciences

Editors: M. Cardona P. Fulde K. von Klitzing H.-J. Queisser

Managing Editor: H. K.V. Lotsch

Springer and the environment

At Springer we firmly believe that an international science publisher has a special obligation to the environment, and our corporate policies consistently reflect this conviction.
We also expect our business partners – paper mills, printers, packaging manufacturers, etc. – to commit themselves to using materials and production processes that do not harm the environment. The paper in this book is made from low- or no-chlorine pulp and is acid free, in conformance with international standards for paper permanency.

Printing: Mercedesdruck, Berlin
Binding: Buchbinderei Lüderitz & Bauer, Berlin